Symbolic Modeling of Multibody Systems

SOLID MECHANICS AND ITS APPLICATIONS
Volume 112

Aims and Scope of the Series

The fundamental questions arising in mechanics are: *Why?, How?,* and *How much?* The aim of this series is to provide lucid accounts written by authoritative researchers giving vision and insight in answering these questions on the subject of mechanics as it relates to solids.

The scope of the series covers the entire spectrum of solid mechanics. Thus it includes the foundation of mechanics; variational formulations; computational mechanics; statics, kinematics and dynamics of rigid and elastic bodies: vibrations of solids and structures; dynamical systems and chaos; the theories of elasticity, plasticity and viscoelasticity; composite materials; rods, beams, shells and membranes; structural control and stability; soils, rocks and geomechanics; fracture; tribology; experimental mechanics; biomechanics and machine design.

The median level of presentation is the first year graduate student. Some texts are monographs defining the current state of the field; others are accessible to final year undergraduates; but essentially the emphasis is on readability and clarity.

For a list of related mechanics titles, see final pages.

Symbolic Modeling of Multibody Systems

by

JEAN-CLAUDE SAMIN

Department of Mechanical Engineering
Center for Research in Mechatronics
Université Catholique de Louvain
Louvain-La-Neuve, Belgium

and

PAUL FISETTE

Department of Mechanical Engineering
Center for Research in Mechatronics
Université Catholique de Louvain
Louvain-La-Neuve, Belgium

KLUWER ACADEMIC PUBLISHERS
DORDRECHT / BOSTON / LONDON

A C.I.P. Catalogue record for this book is available from the Library of Congress.

ISBN 1-4020-1629-8

Published by Kluwer Academic Publishers,
P.O. Box 17, 3300 AA Dordrecht, The Netherlands.

Sold and distributed in North, Central and South America
by Kluwer Academic Publishers,
101 Philip Drive, Norwell, MA 02061, U.S.A.

In all other countries, sold and distributed
by Kluwer Academic Publishers,
P.O. Box 322, 3300 AH Dordrecht, The Netherlands.

Printed on acid-free paper

Printed in the Netherlands.

Contents

Preface

For more than two decades our research team has been deeply involved in modeling and analysis of multibody systems, via a large variety of applications and problems such as, for instance, the stability of satellites (in the seventies), the dynamic behavior of conventional railway systems, the modeling and identification of robots (in the eighties), the lateral stability of non conventional tramways, the study of vehicle suspension performances, the modeling of flexible elements, of cam/follower systems (in the nineties), etc.... All these applications led us to tackle fundamental problems in quite diversified fields: multibody modeling, symbolic generation, numerical methods and computational performance, parallel computation, etc. In working these research topics, it continuously and progressively appeared to us that symbolic generation dedicated to multibody systems was a really attractive approach for students, researchers and engineers wishing to learn, understand or profit from multibody dynamics: this has largely motivated the writing of the present book.

The modeling and analysis of multibody systems require a comprehensive understanding of the kinematics and dynamics of rigid bodies. Hence, in the first part of the book, the relevant fundamental principles are first reviewed in detail and illustrated in conformity with the multibody formalisms that follow. The second part of the book is dedicated to various topics treated by the multibody approach and, finally, the third part consists of a tutorial providing the reader means of tackling potential multibody problems.

Whatever the kind of system (tree-like structures, closed-loop mechanisms, systems containing flexible beams or involving tire/ground contact, wheel/rail contact, etc.), multibody formalisms have a common feature in the proposed approach, viz. the symbolic generation of most of the ingredients needed to set up the model. The symbolic approach chosen, specifically dedicated to multibody systems, affords various advantages: it leads to a simplification of the theoretical formulation of models, a considerable reduction in size of the generated equations and hence in resulting computing time, and also enhanced portability of the multibody models towards other specific environments. Moreover, the generation of a multibody model as symbolic toolboxes proves to be an excellent pedagogical medium in teaching mechanics.

Acknowledgments

Some of the research work performed for this book was sponsored by the Belgian Program on Interuniversity Attraction Poles initiated by the Belgian State - Prime Minister's Office - Science Policy Programming (IAP V/5).

We want to thank the Series Editor, Prof. Graham Gladwell who carefully read the manuscript and corrected many turns of phrase and English mistakes. His relevant comments have helped us in significantly improving the text.

We also wish to thank our families and colleagues who supported us in the writing of this book.

Jean-Claude Samin and Paul Fisette

Part I

Theory

Chapter 1

Fundamental Mechanics

1.1 Mathematical background and notations

The aim of this section is to introduce the notations that will be used throughout this book for representing vectors, tensors as well as their time derivatives. It is assumed that the reader is familiar with elementary vector and matrix calculus. Basic properties of vectors and tensors will thus be recalled without proofs.

1.1.1 Vectors

Orthogonal right-handed coordinate systems are extensively used for representing vectors. Orthogonal right-handed means that the three directions *1, 2* and *3* of the coordinate system are mutually orthogonal and space-oriented according to the right-hand rule shown in figure 1.1. A spatial vector **u** is visualized as

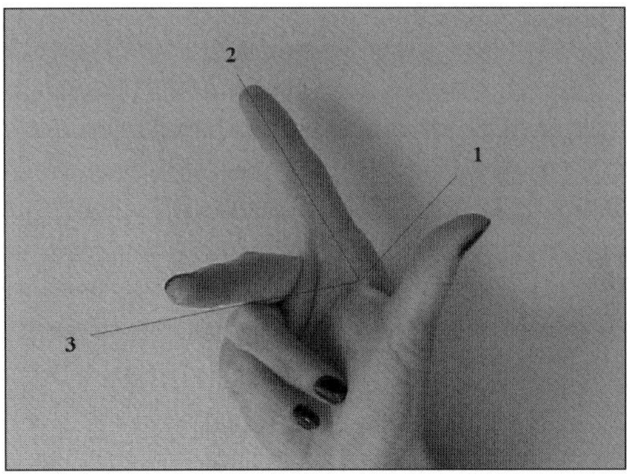

Figure 1.1: Right-hand rule

a *directed line segment* having a specified length or magnitude, as shown in figure 1.2.

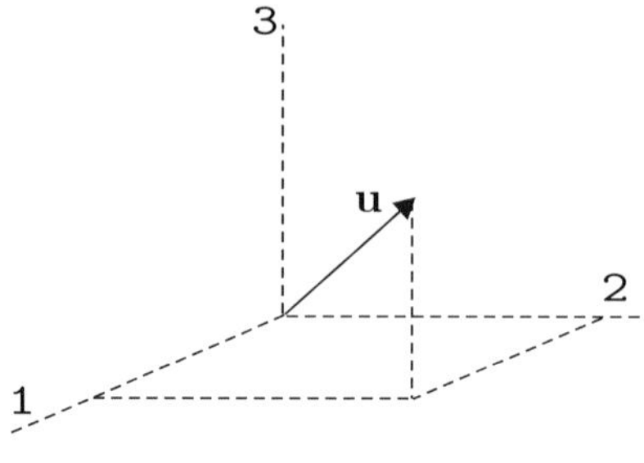

Figure 1.2: Spatial vector in the 3-D space

A spatial vector therefore has an *intrinsic* nature or, in other words, does not depend on a particular choice of coordinate system.

One may distinguish between two kinds of vectors:

- *bound vectors* for which the origin of the directed segment is fixed to a specific point (for instance, as in figure 1.2, the origin of the coordinate system),

- *free vectors* for which the directed segment extends from one arbitrary point to another.

In mechanics, bound vectors are typically used for describing forces. The material point on which a force is acting is of utmost importance for the static and/or dynamical behavior of the system (discussed in section 1.2.7): this point is thus chosen as being the origin of the bound vector, while the amplitude and the direction of the applied force are represented by the directed segment. On the other hand, angular velocity vectors (discussed in section 1.1.9) are typically free vectors.

We consider here that a vector which is simply denoted by a bold character, for instance $\mathbf{u}$, is a free vector. For bound vectors, a superscript will be added to this bold character in order to specify the corresponding origin without ambiguity.

In a right-handed orthogonal coordinate system, a spatial vector $\mathbf{u}$ can be described by its three orthogonal projections u_1, u_2 and u_3 along the three directions $1, 2, 3$ respectively. Its length is given by its norm defined as

$$\| \mathbf{u} \| \triangleq \sqrt{u_1^2 + u_2^2 + u_3^2}$$

A unit vector $\hat{\mathbf{v}}$ (pronounced v-hat) is a vector such that $\| \hat{\mathbf{v}} \| = 1$.

Let us now define the *orthonormal frame* $\{\hat{\mathbf{I}}\}$ by the set which is composed of three *unit base vectors* $\hat{\mathbf{I}}_1, \hat{\mathbf{I}}_2$ and $\hat{\mathbf{I}}_3$ respectively aligned with the principal directions of a right-handed orthogonal coordinate system (see figure 1.3).

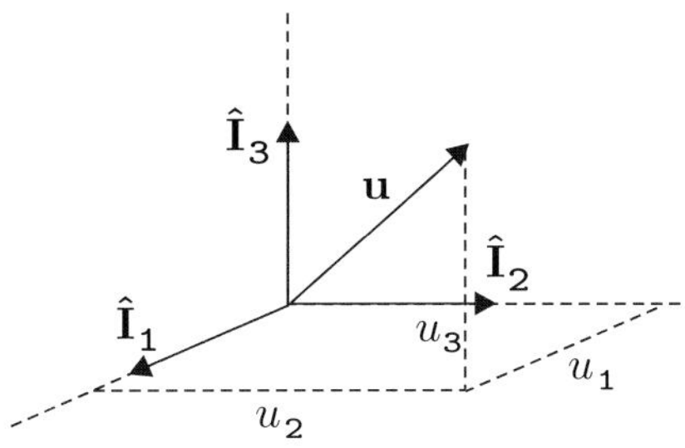

Figure 1.3: Coordinate system

According to elementary vector calculus, any vector $\mathbf{u}$ can be written as

$$\mathbf{u} = u_1\hat{\mathbf{I}}_1 + u_2\hat{\mathbf{I}}_2 + u_3\hat{\mathbf{I}}_3 \tag{1.1}$$

where u_1, u_2, u_3 are the *components* of vector $\mathbf{u}$ in the frame $\{\hat{\mathbf{I}}\}$.

Scalar product

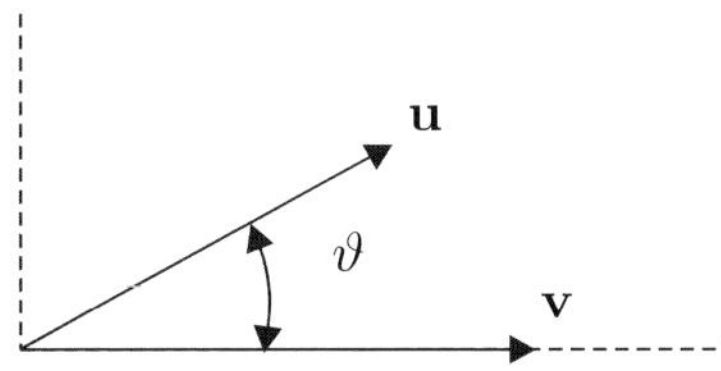

Figure 1.4: Scalar product of two vectors

The *scalar product* - or *dot product* - of two spatial free vectors $\mathbf{u}$ and $\mathbf{v}$ is equal to the real number given by the product

$$\mathbf{u} \,.\, \mathbf{v} \stackrel{\Delta}{=} \| \mathbf{u} \| \, \| \mathbf{v} \| \cos\vartheta \tag{1.2}$$

where the angle ϑ between the two vectors is measured in a plane parallel to the vectors (see figure 1.4). Dot-multiplying a vector with itself thus produces the square of its norm, i.e.

$$\mathbf{u} \,.\, \mathbf{u} = \| \mathbf{u} \|^2 \tag{1.3}$$

One may also verify that the dot product is commutative:

$$\mathbf{u} \,.\, \mathbf{v} = \mathbf{v} \,.\, \mathbf{u}$$

and bi-linear with respect to its arguments $\mathbf{u}$ and $\mathbf{v}$:

$$\begin{aligned}(a_1\mathbf{u}_1 + a_2\mathbf{u}_2)\,.\,\mathbf{v} &= a_1\,(\mathbf{u}_1\,.\,\mathbf{v}) + a_2\,(\mathbf{u}_2\,.\,\mathbf{v})\\ \mathbf{u}\,.\,(a_1\mathbf{v}_1 + a_2\mathbf{v}_2) &= a_1\,(\mathbf{u}\,.\,\mathbf{v}_1) + a_2\,(\mathbf{u}\,.\,\mathbf{v}_2)\end{aligned}$$

Applied to the unit base vectors, the definition of the dot product leads to the property

$$\hat{\mathbf{I}}_\alpha\,.\,\hat{\mathbf{I}}_\beta \stackrel{\Delta}{=} \delta_{\alpha\beta} \qquad \text{with } \alpha,\beta = 1,2,3 \tag{1.4}$$

where $\delta_{\alpha\beta}$ is the "Kronecker delta" whose value is 1 when $\alpha = \beta$ and 0 otherwise. For two vectors expressed in the same frame, namely

$$\mathbf{u} = u_1\hat{\mathbf{I}}_1 + u_2\hat{\mathbf{I}}_2 + u_3\hat{\mathbf{I}}_3 \quad \text{and} \quad \mathbf{v} = v_1\hat{\mathbf{I}}_1 + v_2\hat{\mathbf{I}}_2 + v_3\hat{\mathbf{I}}_3$$

their scalar product is thus given by

$$\mathbf{u}\,.\,\mathbf{v} = u_1v_1 + u_2v_2 + u_3v_3 \tag{1.5}$$

Vector product

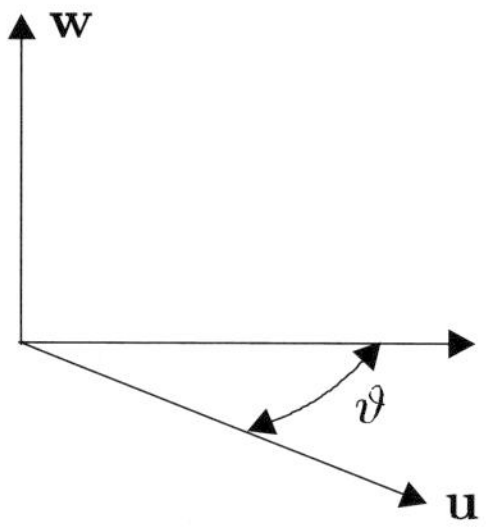

Figure 1.5: Vector product of two vectors **u** and **v**

The *vector product* - or *cross product* - of two spatial vectors **u** and **v** produces a new vector

$$\mathbf{w} \stackrel{\Delta}{=} \mathbf{u} \times \mathbf{v}$$

which is defined as follows (see figure 1.5):

- **w** is orthogonal to both **u** and **v**,
- the triplet $\{\mathbf{u},\,\mathbf{v},\,\mathbf{w}\}$ is space-oriented according to the right-hand rule shown in fig. 1.1,
- $\|\mathbf{w}\| = \|\,\mathbf{u}\,\|\,\|\,\mathbf{v}\,\|\,\sin\vartheta$, where the angle ϑ $(0 \le \vartheta \le \pi)$ between the two vectors **u** and **v** is measured in a plane parallel to the two vectors.

In the same way as for the dot product, one may verify that the vector product is a bi-linear application:

$$\begin{aligned}(a_1\mathbf{u}_1 + a_2\mathbf{u}_2) \times \mathbf{v} &= a_1 (\mathbf{u}_1 \times \mathbf{v}) + a_2 (\mathbf{u}_2 \times \mathbf{v}) \\ \mathbf{u} \times (a_1\mathbf{v}_1 + a_2\mathbf{v}_2) &= a_1 (\mathbf{u} \times \mathbf{v}_1) + a_2 (\mathbf{u} \times \mathbf{v}_2)\end{aligned}$$

but it results from its definition that the cross product is skew-symmetric. Indeed, reversing the order of $\mathbf{u}$ and $\mathbf{v}$ changes the sign of the resulting vector:

$$\mathbf{u} \times \mathbf{v} = -\mathbf{v} \times \mathbf{u} \tag{1.6}$$

The vector product is thus *non-commutative.*
When applied to the base vectors, we obtain from the definition

$$\hat{\mathbf{I}}_\alpha \times \hat{\mathbf{I}}_\beta = \varepsilon_{\alpha\beta\gamma} \hat{\mathbf{I}}_\gamma \tag{1.7}$$

where the Levi-Civita symbol $\varepsilon_{\alpha\beta\gamma}$ is equal to

- $+1$ when α, β, γ is a cyclic permutation of 1, 2, 3
- -1 when α, β, γ is a cyclic permutation of 3, 2, 1
- 0 otherwise (i.e. when two or three indices are equal)

For two arbitrary vectors

$$\mathbf{u} = u_1\hat{\mathbf{I}}_1 + u_2\hat{\mathbf{I}}_2 + u_3\hat{\mathbf{I}}_3 \quad \text{and} \quad \mathbf{v} = v_1\hat{\mathbf{I}}_1 + v_2\hat{\mathbf{I}}_2 + v_3\hat{\mathbf{I}}_3$$

this property of the base vectors leads to

$$\mathbf{u} \times \mathbf{v} = (u_2v_3 - u_3v_2)\,\hat{\mathbf{I}}_1 + (u_3v_1 - u_1v_3)\,\hat{\mathbf{I}}_2 + (u_1v_2 - u_2v_1)\,\hat{\mathbf{I}}_3 \tag{1.8}$$

It is then easy to verify that, when dealing with three different vectors $\mathbf{u}, \mathbf{v}$ and $\mathbf{w}$, the following expressions

$$(\mathbf{u} \times \mathbf{v}) \times \mathbf{w} \quad \text{and} \quad \mathbf{u} \times (\mathbf{v} \times \mathbf{w})$$

lead to different values. Vector product is thus *not associative*.
Moreover, one can also verify in a similar way the following properties

$$(\mathbf{a} \times \mathbf{b}) \,.\, \mathbf{c} = (\mathbf{b} \times \mathbf{c}) \,.\, \mathbf{a} = (\mathbf{c} \times \mathbf{a}) \,.\, \mathbf{b}$$

$$\mathbf{a} \times (\mathbf{b} \times \mathbf{c}) + \mathbf{b} \times (\mathbf{c} \times \mathbf{a}) + \mathbf{c} \times (\mathbf{a} \times \mathbf{b}) = 0$$

The cross product allows us to define the *moment vector* $\mathbf{m}_v$ with respect to a point P of a bound vector $\mathbf{v}$. Figure 1.6 shows a point P and a bound vector $\mathbf{v}$ whose origin coincides with another point Q; if $\mathbf{x} = \overrightarrow{PQ}$, the moment vector of vector $\mathbf{v}$ with respect to P is

$$\mathbf{m}_v(\mathbf{v}) = \mathbf{x} \times \mathbf{v} \tag{1.9}$$

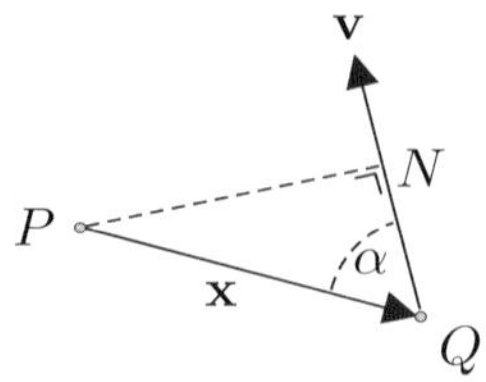

Figure 1.6: Moment of a vector

The geometrical interpretation of this moment vector is the following. If point N is the orthogonal projection of P on vector $\mathbf{v}$, then PN represents the *lever-arm* of vector $\mathbf{v}$ with respect to P. The length of this arm being given by $\|\overrightarrow{PN}\| = \| \mathbf{x} \| \sin\alpha$, the norm of the moment vector can be written as

$$\| \mathbf{m_v}(\mathbf{v}) \| = \|\overrightarrow{PN}\| \; \| \mathbf{v} \|$$

which means that the amplitude of the moment vector is equal to the amplitude of the lever-arm multiplied by the amplitude of the vector itself. The direction of the moment vector $\mathbf{m_v}(\mathbf{v})$ is orthogonal to the arm $\overrightarrow{PN}$ and the vector $\mathbf{v}$, and it is space-oriented according to the right-hand rule.

This notion of moment is useful in mechanics for two kinds of bound vectors, namely the angular momentum and force vectors, as we shall see later on in this chapter.

1.1.2 Tensors

Like vectors, tensors are *intrinsic* quantities which are useful in deriving equations for mechanical systems, although they do not have a direct geometrical visualization. Within the frame of classical mechanics dealing with rigid bodies, only tensors of order 2 are necessary[1]. This is the case for the *inertia tensor* which characterizes a rigid body (see section 1.2.5): this tensor linearly transforms the *angular velocity vector* of the considered body into its *angular momentum vector.*

Generally speaking, a tensor (of second order) can be viewed as an application which transforms any vector $\mathbf{x}$ linearly into another vector $\mathbf{u}$. Since the latter must hold true in any coordinate system, tensors have an *intrinsic* nature like vectors.

A very simple example of a tensor is obtained by forming the *dyadic product* of two vectors $\mathbf{u}$ and $\mathbf{v}$, denoted as *dyad* $\mathbf{uv}$ [33]. A dyad is simply a pair $\mathbf{uv}$ of vectors written in a specific order. The scalar product of a dyad with a vector

[1] In this book, the "order 2" will thus be omitted to denote these tensors.

$\mathbf{x}$, can be performed in two ways, either as

$$\mathbf{uv}\,.\,\mathbf{x} = \mathbf{u}\,(\mathbf{v}\,.\,\mathbf{x}) \tag{1.10}$$

or as

$$\mathbf{x}\,.\,\mathbf{uv} = (\mathbf{x}\,.\,\mathbf{u})\,\mathbf{v}$$

The two products will, in general, not be equal: the scalar product of a dyad with a vector is not commutative. But in both cases, the result of the dot product is a vector (since the term within parenthesis is a scalar), whose value is linearly related to the value of the input vector $\mathbf{x}$. Therefore, using for instance the definition 1.10 of the scalar multiplication between dyads and vectors, one may consider that a dyad is a second order tensor.
It can be proved that tensors - of order 2 - form a vector space whose dimension is equal to 9. A basis of this vector space is provided by the nine dyads one can produce from the base vectors, i.e.: $\hat{\mathbf{I}}_\alpha \hat{\mathbf{I}}_\beta$ with $\alpha, \beta = 1, 2, 3$. Any tensor $\mathbf{T}$ can be written as a linear combination of these nine dyads:

$$\mathbf{T} = \sum_{\alpha,\beta=1,2,3} T_{\alpha\beta}\, \hat{\mathbf{I}}_\alpha \hat{\mathbf{I}}_\beta \tag{1.11}$$

Applied to the vector

$$\mathbf{x} = x_1 \hat{\mathbf{I}}_1 + x_2 \hat{\mathbf{I}}_2 + x_3 \hat{\mathbf{I}}_3 = \sum_{\gamma=1,2,3} x_\gamma \hat{\mathbf{I}}_\gamma$$

and using 1.10 for each term $\hat{\mathbf{I}}_\alpha \hat{\mathbf{I}}_\beta \,.\, \hat{\mathbf{I}}_\gamma$, the tensor $\mathbf{T}$ produces the new vector

$$\mathbf{T}\,.\,\mathbf{x} = \sum_{\alpha,\beta=1,2,3} T_{\alpha\beta}\, x_\beta\, \hat{\mathbf{I}}_\alpha \tag{1.12}$$

1.1.3 Array of vectors

Matrix notations are very convenient and concise tools when dealing with vector and tensor components. For instance, considering for the components of a vector $\mathbf{u}$ the (3×1) column matrix

$$u \triangleq \begin{pmatrix} u_1 \\ u_2 \\ u_3 \end{pmatrix} \tag{1.13}$$

and defining similarly a (3×1) matrix v for the components of a vector $\mathbf{v}$, we can formally write the dot product 1.5 of these two vectors as

$$\mathbf{u}\,.\,\mathbf{v} = u^T v \tag{1.14}$$

where u^T denotes the *transpose* of the column matrix u.

To write the equations of complex multibody systems as concisely as possible, let us introduce the notion of *array of vectors.* We define the (3×1) column array $[\hat{\mathbf{I}}]$ as

$$[\hat{\mathbf{I}}] \triangleq \begin{pmatrix} \hat{\mathbf{I}}_1 \\ \hat{\mathbf{I}}_2 \\ \hat{\mathbf{I}}_3 \end{pmatrix} \tag{1.15}$$

This is not a matrix because its elements are the base vectors $\hat{\mathbf{I}}_\alpha$ instead of real numbers, but standard matrix rules can be used. For instance, we can formally write

$$\mathbf{u} = u^T [\hat{\mathbf{I}}] \tag{1.16}$$

by using the standard matrix product, or equivalently

$$\mathbf{u} = [\hat{\mathbf{I}}]^T u \tag{1.17}$$

by applying to the array $[\hat{\mathbf{I}}]$ the notion of matrix transpose.

Similarly, we can formally define two scalar products between two arrays of vectors $[\hat{\mathbf{I}}]$ and $[\hat{\mathbf{J}}]$ as follows:

$$[\hat{\mathbf{I}}]^T . [\hat{\mathbf{J}}] = \begin{pmatrix} \hat{\mathbf{I}}_1 & \hat{\mathbf{I}}_2 & \hat{\mathbf{I}}_3 \end{pmatrix} . \begin{pmatrix} \hat{\mathbf{J}}_1 \\ \hat{\mathbf{J}}_2 \\ \hat{\mathbf{J}}_3 \end{pmatrix} \triangleq \left(\hat{\mathbf{I}}_1.\hat{\mathbf{J}}_1 + \hat{\mathbf{I}}_2.\hat{\mathbf{J}}_2 + \hat{\mathbf{I}}_3.\hat{\mathbf{J}}_3\right)$$

$$[\hat{\mathbf{I}}] . [\hat{\mathbf{J}}]^T = \begin{pmatrix} \hat{\mathbf{I}}_1 \\ \hat{\mathbf{I}}_2 \\ \hat{\mathbf{I}}_3 \end{pmatrix} . \begin{pmatrix} \hat{\mathbf{J}}_1 & \hat{\mathbf{J}}_2 & \hat{\mathbf{J}}_3 \end{pmatrix} \triangleq \begin{pmatrix} \hat{\mathbf{I}}_1.\hat{\mathbf{J}}_1 & \hat{\mathbf{I}}_1.\hat{\mathbf{J}}_2 & \hat{\mathbf{I}}_1.\hat{\mathbf{J}}_3 \\ \hat{\mathbf{I}}_2.\hat{\mathbf{J}}_1 & \hat{\mathbf{I}}_2.\hat{\mathbf{J}}_2 & \hat{\mathbf{I}}_2.\hat{\mathbf{J}}_3 \\ \hat{\mathbf{I}}_3.\hat{\mathbf{J}}_1 & \hat{\mathbf{I}}_3.\hat{\mathbf{J}}_2 & \hat{\mathbf{I}}_3.\hat{\mathbf{J}}_3 \end{pmatrix} \tag{1.18}$$

Within these definitions, it is to be understood that the arrays are first multiplied as if they were ordinary matrices, and only then are the vector elements combined via dot products. The latter is essential to be consistent with the notations adopted in 1.16 and 1.17 and expression 1.14 of a dot product. This consistency is easily verified:

$$\mathbf{u} . \mathbf{v} = u^T [\hat{\mathbf{I}}] . [\hat{\mathbf{I}}]^T v = u^T v \tag{1.19}$$

since, according to 1.18, $[\hat{\mathbf{I}}] . [\hat{\mathbf{I}}]^T$ is the (3×3) unit matrix.

According to the previous notation, the dyadic product of two vectors can also be written as

$$\mathbf{uv} = [\hat{\mathbf{I}}]^T u v^T [\hat{\mathbf{I}}] \tag{1.20}$$

where, using the usual rule for external multiplication of matrices, uv^T is the following (3×3) matrix:

$$uv^T = \begin{pmatrix} u_1v_1 & u_1v_2 & u_1v_3 \\ u_2v_1 & u_2v_2 & u_2v_3 \\ u_3v_1 & u_3v_2 & u_3v_3 \end{pmatrix}$$

With these matrix notations, and for $\mathbf{x} = [\hat{\mathbf{I}}]^T x$, the product $\mathbf{uv}\,.\,\mathbf{x}$ can be successively rewritten as

$$\mathbf{uv}\,.\,\mathbf{x} = [\hat{\mathbf{I}}]^T uv^T [\hat{\mathbf{I}}]\,.\,[\hat{\mathbf{I}}]^T x = [\hat{\mathbf{I}}]^T uv^T x = [\hat{\mathbf{I}}]^T u\left(v^T x\right) = \mathbf{u}\,(\mathbf{v}\,.\,\mathbf{x})$$

which is consistent with relation 1.10. Similarly, any tensor $\mathbf{T}$ can be formally written

$$\mathbf{T} = [\hat{\mathbf{I}}]^T\, T\, [\hat{\mathbf{I}}] \tag{1.21}$$

where T is the (3×3) matrix containing the nine components $T_{\alpha\beta}$ introduced in 1.11. When applied to an input vector $\mathbf{x} = [\hat{\mathbf{I}}]^T x$, this tensor provides as output

$$\mathbf{T}\,.\,\mathbf{x} = [\hat{\mathbf{I}}]^T\, T\, [\hat{\mathbf{I}}].[\hat{\mathbf{I}}]^T x = [\hat{\mathbf{I}}]^T\, Tx$$

which is also consistent with our previous result 1.12.

If a second tensor - for instance $\mathbf{Q} \triangleq [\hat{\mathbf{I}}]^T\, Q\, [\hat{\mathbf{I}}]$ - is applied to the result $\mathbf{T}\,.\,\mathbf{x}$, one obtains

$$\mathbf{Q}\,.\,(\mathbf{T}\,.\,\mathbf{x}) = [\hat{\mathbf{I}}]^T\, Q\, Tx$$

Therefore, when formulating equations in an intrinsic vector/tensor form, it is useful to introduce and define the dot product of tensors as follows:

$$\mathbf{Q}\,.\,\mathbf{T} \triangleq [\hat{\mathbf{I}}]^T\, Q\, [\hat{\mathbf{I}}]\,.\,[\hat{\mathbf{I}}]^T\, T\, [\hat{\mathbf{I}}] = [\hat{\mathbf{I}}]^T\, Q\, T\, [\hat{\mathbf{I}}] \tag{1.22}$$

While being associative, dot-products of tensors are nevertheless non-commutative (as is the case for matrix products).

1.1.4 Vector pre-product tensor

For two arbitrary vectors $\mathbf{u} = [\hat{\mathbf{I}}]^T u$ and $\mathbf{v} = [\hat{\mathbf{I}}]^T v$, the vector product expression 1.8 leads to

$$\mathbf{w} = \mathbf{u} \times \mathbf{v} = [\hat{\mathbf{I}}]^T \begin{pmatrix} u_2v_3 - u_3v_2 \\ u_3v_1 - u_1v_3 \\ u_1v_2 - u_2v_1 \end{pmatrix} \tag{1.23}$$

or equivalently to

$$\mathbf{w} = \mathbf{u} \times \mathbf{v} = [\hat{\mathbf{I}}]^T\, \tilde{u} v \tag{1.24}$$

by defining the *tilde matrix* $\tilde{u}$ as follows:

$$\tilde{u} \triangleq \begin{pmatrix} 0 & -u_3 & u_2 \\ u_3 & 0 & -u_1 \\ -u_2 & u_1 & 0 \end{pmatrix} \tag{1.25}$$

This matrix is skew-symmetric, implying that $\tilde{u}^T = -\tilde{u}$, and $\tilde{u}u = 0$. Using the skew-symmetric property of the vector product, one also has

$$\tilde{u}v = -\tilde{v}u \tag{1.26}$$

To simplify the vector/dyadic expressions when formulating the kinematic and dynamical equations of multibody systems, we use the tilde matrix $\tilde{u}$ defined in 1.25 to introduce a *vector pre-product tensor*:

$$\tilde{\mathbf{u}} \triangleq [\hat{\mathbf{I}}]^T \, \tilde{u} \, [\hat{\mathbf{I}}] \tag{1.27}$$

This allows us to write the cross product $\mathbf{w} = \mathbf{u} \times \mathbf{v}$ as

$$\mathbf{w} = \mathbf{u} \times \mathbf{v} = \tilde{\mathbf{u}} \,.\, \mathbf{v} \tag{1.28}$$

since

$$\tilde{\mathbf{u}} \,.\, \mathbf{v} = [\hat{\mathbf{I}}]^T \, \tilde{u} \, [\hat{\mathbf{I}}] \,.\, [\hat{\mathbf{I}}]^T \, v = [\hat{\mathbf{I}}]^T \, \tilde{u}v$$

is equivalent to 1.24. As a consequence, we have

$$\tilde{\mathbf{u}} \,.\, \mathbf{v} = -\tilde{\mathbf{v}} \,.\, \mathbf{u}$$

On the other hand, the expression 1.28 clearly illustrates the intrinsic nature of the tensor $\tilde{\mathbf{u}}$ which linearly transforms a vector $\mathbf{v}$ into another vector $\mathbf{w}$.
Finally, using the definition 1.22 of a dot product between tensors, we may write for any vectors $\mathbf{u}, \mathbf{v}$ and $\mathbf{w}$:

$$\mathbf{u} \times (\mathbf{v} \times \mathbf{w}) = \tilde{\mathbf{u}} \,.\, (\tilde{\mathbf{v}} \,.\, \mathbf{w}) = (\tilde{\mathbf{u}} \,.\, \tilde{\mathbf{v}}) \,.\, \mathbf{w}$$

or simply

$$\mathbf{u} \times (\mathbf{v} \times \mathbf{w}) = \tilde{\mathbf{u}} \,.\, \tilde{\mathbf{v}} \,.\, \mathbf{w}$$

which avoids the use of parentheses.

1.1.5 An example: the rotation tensor

Let us consider a vector $\mathbf{u}$ which is rotated around some unit vector $\hat{\mathbf{e}}$ through an angle ϑ. After the rotation, the vector $\mathbf{u}$ becomes a new vector denoted $\mathbf{v}$, as illustrated in figure 1.7_a. One may observe that this operation is *intrinsic* (no reference is made to a particular frame). Moreover, this operation is linear: the image of a vector sum $\mathbf{u}_1+\mathbf{u}_2$ would be equal to the vector sum of the respective

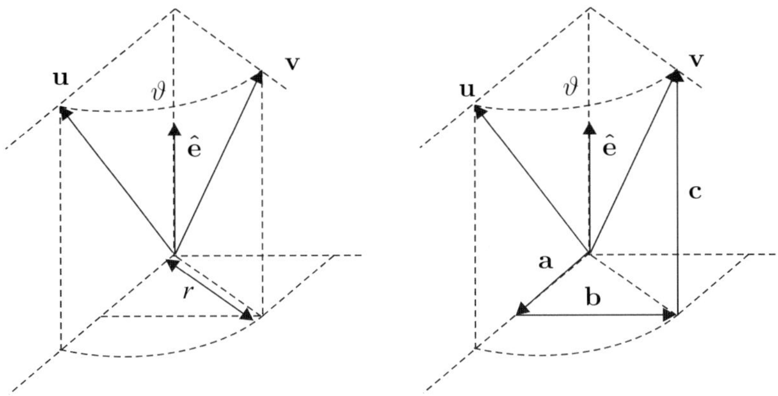

Figure 1.7: Rotation of a vector $\mathbf{u}$

images $\mathbf{v}_1$ and $\mathbf{v}_2$. Therefore, "rotating a vector $\mathbf{u}$ around some unit vector $\hat{\mathbf{e}}$ through an angle ϑ" must be equivalent to applying some tensor $\mathbf{T}$ to $\mathbf{u}$. This is what we will now prove by expressing $\mathbf{T}$ in terms of $\hat{\mathbf{e}}$ and ϑ.
Figure 1.7_b shows that the resulting vector $\mathbf{v}$ can be expressed as the sum of three vectors:

$$\mathbf{v} = \mathbf{a} + \mathbf{b} + \mathbf{c}$$

Being parallel to $\hat{\mathbf{e}}$, vector $\mathbf{c}$ is easily found to be equal to

$$\mathbf{c} = \hat{\mathbf{e}}\ (\hat{\mathbf{e}} \,.\, \mathbf{u})$$

or equivalently to

$$\mathbf{c} = \hat{\mathbf{e}}\,\hat{\mathbf{e}} \,.\, \mathbf{u}$$

The vector difference $\mathbf{u} - \mathbf{c}$ produces the direction of $\mathbf{a}$ with a magnitude equal to the radius r. The magnitude of $\mathbf{a}$ is given by $\| \mathbf{a} \| = r \cos\vartheta$ and we thus obtain

$$\mathbf{a} = (\mathbf{u} - \mathbf{c}) \cos\vartheta$$

or equivalently

$$\mathbf{a} = (\mathbf{E} - \hat{\mathbf{e}}\,\hat{\mathbf{e}}) \cos\vartheta \,.\, \mathbf{u}$$

where $\mathbf{E}$ is the unit tensor ($\mathbf{E} \,.\, \mathbf{u} \triangleq \mathbf{u}$).
Similarly, we observe that vector $\mathbf{b}$ is orthogonal to $\hat{\mathbf{e}}$ and $\mathbf{u}$. Since the magnitude of $\hat{\mathbf{e}} \times \mathbf{u}$ is equal to r, we obtain

$$\mathbf{b} = (\hat{\mathbf{e}} \times \mathbf{u}) \sin\vartheta$$

or equivalently

$$\mathbf{b} = \tilde{\mathbf{e}} \sin\vartheta \,.\, \mathbf{u}$$

Finally, the resulting vector $\mathbf{v}$ can be expressed as

$$\mathbf{v} = \mathbf{a} + \mathbf{b} + \mathbf{c} = (\mathbf{E}\cos\vartheta + (1 - \cos\vartheta)\,\hat{\mathbf{e}}\,\hat{\mathbf{e}} + \tilde{\mathbf{e}}\sin\vartheta)\,.\,\mathbf{u}$$

or as

$$\mathbf{v} = \mathbf{T}\,.\,\mathbf{u} \quad \text{with} \quad \mathbf{T} = \mathbf{E}\cos\vartheta + (1 - \cos\vartheta)\,\hat{\mathbf{e}}\,\hat{\mathbf{e}} + \tilde{\mathbf{e}}\sin\vartheta \tag{1.29}$$

This result is sometimes referred as the *rotation formula* [33] or as the *Rodrigues formula* [53].

1.1.6 Transformation matrices

When dealing with kinematics and dynamics of rigid bodies, it is helpful to define different orthonormal frames. It thus becomes necessary to "switch" from one frame to another. Consider for instance two frames $\{\hat{\mathbf{I}}\}$ and $\{\hat{\mathbf{J}}\}$. The three base vectors of frame $\{\hat{\mathbf{J}}\}$ can be expressed in frame $\{\hat{\mathbf{I}}\}$:

$$\begin{aligned}
\hat{\mathbf{J}}_1 &= R_{11}\hat{\mathbf{I}}_1 + R_{12}\hat{\mathbf{I}}_2 + R_{13}\hat{\mathbf{I}}_3 \\
\hat{\mathbf{J}}_2 &= R_{21}\hat{\mathbf{I}}_1 + R_{22}\hat{\mathbf{I}}_2 + R_{23}\hat{\mathbf{I}}_3 \\
\hat{\mathbf{J}}_3 &= R_{31}\hat{\mathbf{I}}_1 + R_{32}\hat{\mathbf{I}}_2 + R_{33}\hat{\mathbf{I}}_3
\end{aligned}$$

or, in matrix form

$$[\hat{\mathbf{J}}] = R\,[\hat{\mathbf{I}}] \tag{1.30}$$

where

$$R = \begin{pmatrix} R_{11} & R_{12} & R_{13} \\ R_{21} & R_{22} & R_{23} \\ R_{31} & R_{32} & R_{33} \end{pmatrix}$$

is a *transformation matrix*. Using 1.18, one may also observe that

$$R = [\hat{\mathbf{J}}]\,.\,[\hat{\mathbf{I}}]^T$$

Important properties arise from the fact that the elements of matrix R are the orthogonal projections of the base vectors $\hat{\mathbf{J}}_\alpha$ onto the base vectors $\hat{\mathbf{I}}_\beta$. Since the $\hat{\mathbf{J}}_\alpha$ are unit vectors, one has

$$\sum_{\beta=1,2,3} R_{\alpha\beta}^2 = 1 \quad \text{for}\,\alpha = 1,2,3 \tag{1.31}$$

and since they are orthogonal, one also has

$$\sum_{\gamma=1,2,3} R_{\alpha\gamma} R_{\beta\gamma} = 0 \quad \text{for}\,\alpha,\beta = 1,2,3 \quad \text{and}\,\alpha \neq \beta \tag{1.32}$$

Combining 1.31 and 1.32 in matrix form, one then obtains

$$RR^T = E \tag{1.33}$$

where E denotes the unit matrix. This relation indicates that the transformation matrix R is *orthogonal.* As a consequence, we have

$$(\det R)^2 = 1 \tag{1.34}$$

$$\text{and} \quad R^{-1} = R^T \tag{1.35}$$

which implies that a transformation matrix is always regular (invertible). Since its transpose coincides with its inverse, one also has

$$R^T R = E = RR^T \tag{1.36}$$

Relation 1.30 can thus be easily inverted[2]:

$$[\hat{\mathbf{I}}] = R^T [\hat{\mathbf{J}}] \tag{1.37}$$

The property 1.34 is applicable to any orthogonal matrix. However, since it is assumed here that the orthogonal frame $\{\hat{\mathbf{J}}\}$ is right-handed, this property becomes more specifically

$$\det R = 1 \tag{1.38}$$

because when computing the value of the determinant, one may recognize that

$$\det R = \hat{\mathbf{J}}_1 \,.\, \left(\hat{\mathbf{J}}_2 \times \hat{\mathbf{J}}_3\right)$$

which, according to 1.7, must be equal to one.

Knowing the transformation matrix between different frames, we can transform the components of vectors and tensors. Supposing that a vector $\mathbf{x}$ and a tensor $\mathbf{T}$ are described in the frame $\{\hat{\mathbf{J}}\}$:

$$\mathbf{x} = [\hat{\mathbf{J}}]^T x \quad , \quad \mathbf{T} = [\hat{\mathbf{J}}]^T \, T \, [\hat{\mathbf{J}}]$$

one immediately obtains from 1.30 their components in the frame $\{\hat{\mathbf{I}}\}$:

$$\mathbf{x} = [\hat{\mathbf{I}}]^T R^T x \quad , \quad \mathbf{T} = [\hat{\mathbf{I}}]^T \, R^T T R \, [\hat{\mathbf{I}}] \tag{1.39}$$

1.1.7 Euler's theorem on finite rotations

Euler's theorem states the following:

Theorem 1 *Any transformation between two frames $\{\hat{\mathbf{I}}\}$ and $\{\hat{\mathbf{J}}\}$ can be considered as an elementary rotation of amplitude ϑ around a single unit vector $\hat{\mathbf{e}}$.*

---

[2]Some authors introduce the rotation matrix by its inverse. When using the rotation matrix R, we must therefore keep in mind how the relation 1.30 has been defined to avoid any ambiguity.

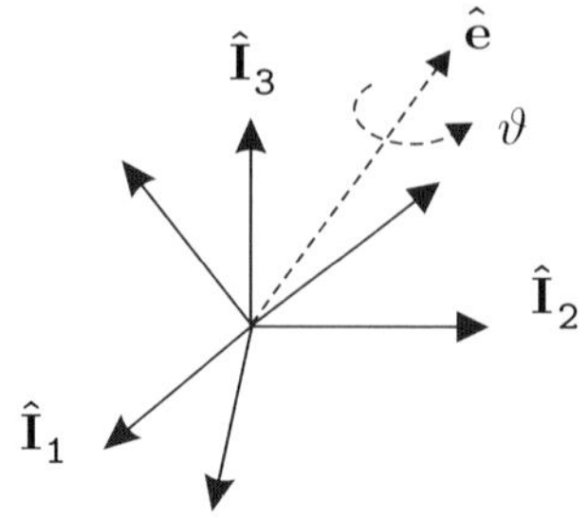

Figure 1.8: General rotation between two frames

In order to prove this theorem, let us consider in figure 1.8 two frames $\{\hat{\mathbf{I}}\}$ and $\{\hat{\mathbf{J}}\}$: if such a rotation axis $\hat{\mathbf{e}}$ exists, it should have the same components in both frames:

$$\hat{\mathbf{e}} = [\hat{\mathbf{I}}]^T \begin{pmatrix} e_1 \\ e_2 \\ e_3 \end{pmatrix} = [\hat{\mathbf{J}}]^T \begin{pmatrix} e_1 \\ e_2 \\ e_3 \end{pmatrix}, \quad \text{or equivalently} \quad e = R\,e.$$

Therefore, we must prove that the transformation matrix R defined by 1.30 has an eigenvalue λ equal to one. The (3×1) column matrix e will then be the associated eigenvector.
The characteristic equation for the matrix R:

$$\det(R - \lambda E) = 0$$

can also be written as

$$(\lambda - \lambda_1)\,(\lambda - \lambda_2)\,(\lambda - \lambda_3) = 0$$

where the λ_α (with $\alpha = 1, 2, 3$) are the eigenvalues of R. From the latter expression, we deduce that the determinant of R, which is equal to one, is equal to the product of the eigenvalues:

$$\det R = \lambda_1 \lambda_2 \lambda_3 = 1$$

On the other hand, the eigenvalues of the inverse A^{-1} of a matrix are the inverses λ_α^{-1} of the eigenvalues λ_α of A. But in the present case, the inverse of R is equal to its transpose R^T which has the same eigenvalues as R. Therefore, if λ_α is an eigenvalue of R, then λ_α^{-1} is also an eigenvalue of R. Supposing for instance that λ' is an eigenvalue of R, we have

$$\det R = \lambda' \left(\lambda'\right)^{-1} \lambda = 1$$

which shows that (at least) one of the eigenvalues or R is equal to one. It can moreover be shown (see [33] for instance) that the other two eigenvalues of R are equal to $\exp\left(\pm i\,\vartheta\right)$.

From Euler's theorem, we may conclude the existence of a rotation axis $\hat{\mathbf{e}}$, and consider that the unit frame vectors $\hat{\mathbf{J}}_\alpha$ are obtained from vectors $\hat{\mathbf{I}}_\alpha$ by means of a rotation tensor $\mathbf{T}$ (see section 1.1.5):

$$\hat{\mathbf{J}}_\alpha = \mathbf{T} \,.\, \hat{\mathbf{I}}_\alpha\,, \quad \text{for } \alpha = 1, 2, 3$$

Let us now consider an arbitrary vector $\mathbf{x}$ which is defined in frame $\{\hat{\mathbf{I}}\}$ as

$$\mathbf{x} = [\hat{\mathbf{I}}]^T x$$

By rotating this vector around axis $\hat{\mathbf{e}}$ with the same amplitude ϑ, we generate a new vector $\mathbf{y} = \mathbf{T} \,.\, \mathbf{x}$ which has in frame $\{\hat{\mathbf{J}}\}$ the same components as $\mathbf{x}$ in $\{\hat{\mathbf{I}}\}$:

$$\mathbf{y} = [\hat{\mathbf{J}}]^T x$$

Expressing tensor $\mathbf{T}$ in frame $\{\hat{\mathbf{I}}\}$, we obtain

$$\mathbf{y} = \mathbf{T} \,.\, \mathbf{x} = [\hat{\mathbf{I}}]^T T [\hat{\mathbf{I}}] \,.\, [\hat{\mathbf{I}}]^T x = [\hat{\mathbf{I}}]^T T x$$

and comparing the last two expressions of $\mathbf{y}$, we find

$$[\hat{\mathbf{J}}]^T = [\hat{\mathbf{I}}]^T T$$

Since the transformation matrix between $\{\hat{\mathbf{J}}\}$ and $\{\hat{\mathbf{I}}\}$ is defined by

$$[\hat{\mathbf{J}}] = R[\hat{\mathbf{I}}]$$

we may finally conclude that

$$R = T^T$$

The value of the transformation matrix R is thus obtained from the components of $\mathbf{T}$ (given in 1.29) in the $\{\hat{\mathbf{I}}\}$ frame:

$$R(\hat{\mathbf{e}}, \vartheta) = E \cos\vartheta + (1 - \cos\vartheta)\, ee^T - \sin\vartheta\, \tilde{e} \tag{1.40}$$

1.1.8 Rotation coordinates

An elementary transformation - or *rotation* - matrix is illustrated in figure 1.9. The frame $\{\hat{\mathbf{J}}\}$ is obtained from $\{\hat{\mathbf{I}}\}$ by a rotation of amplitude ϑ_3 around the third axis, implying the relation

$$[\hat{\mathbf{J}}] = R_3[\hat{\mathbf{I}}] \quad \text{with } R_3 = \begin{pmatrix} \cos\vartheta_3 & \sin\vartheta_3 & 0 \\ -\sin\vartheta_3 & \cos\vartheta_3 & 0 \\ 0 & 0 & 1 \end{pmatrix} \tag{1.41}$$

Similarly, elementary rotations around the first and second axes lead respectively to

$$R_1 = \begin{pmatrix} 1 & 0 & 0 \\ 0 & \cos\vartheta_1 & \sin\vartheta_1 \\ 0 & -\sin\vartheta_1 & \cos\vartheta_1 \end{pmatrix} \quad \text{and } R_2 = \begin{pmatrix} \cos\vartheta_2 & 0 & -\sin\vartheta_2 \\ 0 & 1 & 0 \\ \sin\vartheta_2 & 0 & \cos\vartheta_2 \end{pmatrix}$$

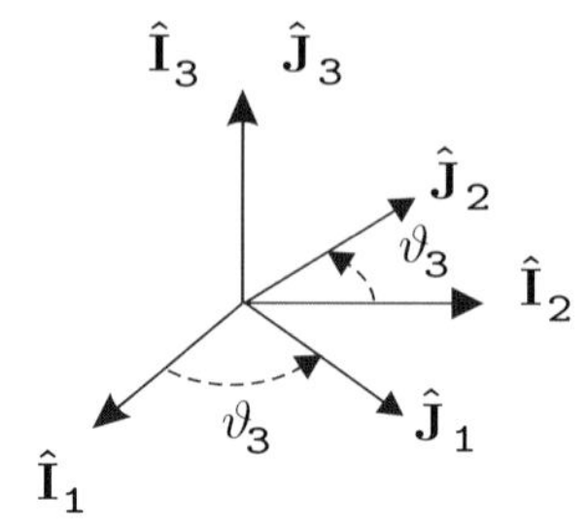

Figure 1.9: Elementary rotation around the 3rd axis

More generally, as shown in figure 1.10, any rotation between two frames can be obtained by a succession of three elementary rotations.
Different sequences of rotations are acceptable. For instance, when the rotations are made successively around axes $1, 2$ and 3 (as in figure 1.10), the corresponding angles ϑ_1, ϑ_2 and ϑ_3 are called the *Tait-Bryan* or *nautical* angles. The resulting rotation matrix R is in this case given by

$$
\begin{aligned}
R &= R_3(\vartheta_3)\, R_2(\vartheta_2)\, R_1(\vartheta_1) \qquad (1.42)\\
&= \begin{pmatrix} c\vartheta_2\, c\vartheta_3 & s\vartheta_1\, s\vartheta_2\, c\vartheta_3 + c\vartheta_1\, s\vartheta_3 & -\, c\vartheta_1\, s\vartheta_2\, c\vartheta_3 + s\vartheta_1\, s\vartheta_3 \\ -\, c\vartheta_2\, s\vartheta_3 & -s\vartheta_1\, s\vartheta_2\, s\vartheta_3 + c\vartheta_1\, c\vartheta_3 & c\vartheta_1\, s\vartheta_2\, s\vartheta_3 + s\vartheta_1\, c\vartheta_3 \\ s\vartheta_2 & -s\vartheta_1\, c\vartheta_2 & c\vartheta_1\, c\vartheta_2 \end{pmatrix}
\end{aligned}
$$

where for compactness, $s\vartheta_a$ and $c\vartheta_a$ respectively denote $\sin\vartheta_a$ and $\cos\vartheta_a$. It should be noted that the elementary rotation matrices R_α appear in the reverse order $3, 2, 1$ in this product.

Knowing the numerical value of the elements of a given rotation matrix R, we may also need to compute the corresponding numerical values of the three Tait-Bryan angles ϑ_α. After some (a little tedious) computations [35], the following formulae are obtained:

$$
\begin{aligned}
\vartheta_1 &= \mathrm{ATAN2}(-\varepsilon R_{32}; \varepsilon R_{33})\\
\vartheta_2 &= \mathrm{ATAN2}\Big(R_{31}; \varepsilon\sqrt{R_{32}^2 + R_{33}^2}\Big)\\
\vartheta_3 &= \mathrm{ATAN2}(-\varepsilon R_{21}; \varepsilon R_{11})
\end{aligned} \qquad (1.43)
$$

where $\varepsilon = \pm 1$ and where ATAN2(arg1; arg2) denotes the four quadrant inverse tangent formula[3] which returns

- an undefined value if the two arguments vanish,
- an angular value ϑ, such that $\tan\vartheta = \frac{\arg 1}{\arg 2}$
    - in the first quadrant (i.e. $0 \le \vartheta \le \frac{\pi}{2}$) if both arguments are positive,

[3] available as FORTRAN®, C and/or MATLAB® instruction

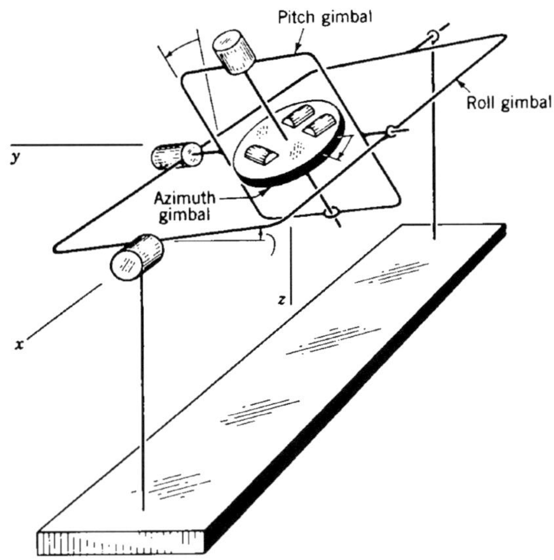

Figure 1.10: Tait-Bryan sequence

- in the second quadrant (i.e. $\frac{\pi}{2} \leq \vartheta \leq \pi$) if the first argument is positive while the second one is negative,
- in the third quadrant (i.e. $-\pi \leq \vartheta \leq -\frac{\pi}{2}$) if both arguments are negative,
- in the fourth quadrant (i.e. $-\frac{\pi}{2} \leq \vartheta \leq 0$) if the first argument is negative while the second is positive.

We must however note that the formulas 1.43 are not always applicable. Indeed, if

$$R_{31} = \pm 1$$

then

$$R_{11} = R_{21} = R_{32} = R_{33} = 0$$

since the rows and columns of R are normalized. In this particular configuration, we can observe that angles ϑ_1 and ϑ_3 are not defined by the formulas 1.43. Moreover $R_{31} = \pm 1$ implies $\vartheta_2 = \pm\frac{\pi}{2}$. This situation corresponds to a mathematical *singularity* problem we may encounter when using a sequence of three rotation angles as orientation coordinates. This singular configuration can also be observed geometrically (see figure 1.11). When $\vartheta_2 = \pm\frac{\pi}{2}$, the 3rd axis of rotation becomes aligned with the 1st one; ϑ_1 and ϑ_3 become undetermined because they are rotation angles around the same direction (only the sum $\vartheta_1 + \vartheta_3$ has a fixed value).

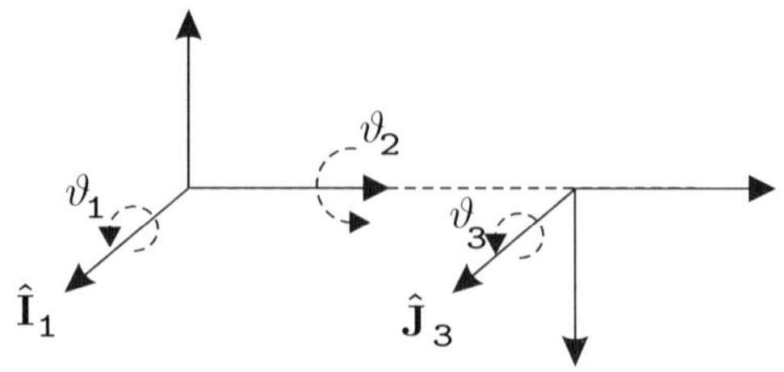

Figure 1.11: Singularity of the Tait-Bryan angles

Euler angles are also proposed in the literature as an alternative choice of angular coordinates (see [33] for instance). The difference consists in the sequence of rotations: the last axis of rotation has the same index as the first one. One example of Euler angles is given by the sequence of rotations $3, 1, 3$ and the corresponding rotation matrix is

$$\begin{aligned} R &= R_3(\psi)\ R_1(\vartheta)\ R_3(\varphi) \\ &= \begin{pmatrix} c\psi\, c\varphi - c\vartheta\, s\varphi\, s\psi & c\psi\, s\varphi + c\vartheta\, c\varphi\, s\psi & s\psi\, s\vartheta \\ -s\psi\, c\varphi - c\vartheta\, s\varphi\, c\psi & -s\psi\, s\varphi + c\vartheta\, c\varphi\, c\psi & c\psi\, s\vartheta \\ s\vartheta\, s\varphi & -s\vartheta\, c\varphi & c\vartheta \end{pmatrix} \end{aligned}$$

With Euler angles, the singularity occurs when the second rotation angle $\vartheta = 0$ or π. Indeed, with $\vartheta = 0$ for instance, the global rotation matrix R reduces to

$$\begin{aligned} R &= R_3(\psi)\ R_3(\varphi) = \begin{pmatrix} c\psi\, c\varphi - s\varphi\, s\psi & c\psi\, s\varphi + c\varphi\, s\psi & 0 \\ -s\psi\, c\varphi - s\varphi\, c\psi & -s\psi\, s\varphi + c\varphi\, c\psi & 0 \\ 0 & 0 & 1 \end{pmatrix} \\ &= \begin{pmatrix} \cos(\psi+\varphi) & \sin(\psi+\varphi) & 0 \\ -\sin(\psi+\varphi) & \cos(\psi+\varphi) & 0 \\ 0 & 0 & 1 \end{pmatrix} = R_3(\psi+\varphi) \end{aligned}$$

and the rotation angles ψ and φ become indistinguishable. The geometrical interpretation is the same as for Tait-Bryan angles: the singularity is due to the fact that the axes of the first and third rotation coincide.

In order to solve this singularity problem which may cause computational difficulties, one may use as rotation coordinates the *Euler parameters* or *quaternions* . The latter are based on Euler's theorem and defined as follows:

$$p \triangleq e_1 \sin\frac{\vartheta}{2}, \quad q \triangleq e_2 \sin\frac{\vartheta}{2}, \quad r \triangleq e_3 \sin\frac{\vartheta}{2} \quad \text{and} \quad s \triangleq \cos\frac{\vartheta}{2} \tag{1.44}$$

Since $\hat{\mathbf{e}}$ is unitary, these parameters are not independent because

$$p^2 + q^2 + r^2 + s^2 = 1 \tag{1.45}$$

Further developments of the relation 1.40 (see for instance [73]) lead to the following expression of the rotation matrix R :

$$R = \begin{pmatrix} 2(p^2+s^2)-1 & 2(pq+rs) & 2(pr-qs) \\ 2(pq-rs) & 2(q^2+s^2)-1 & 2(qr+ps) \\ 2(pr+qs) & 2(qr-ps) & 2(r^2+s^2)-1 \end{pmatrix} \tag{1.46}$$

One may first observe that trigonometric functions are no longer necessary to express the rotation matrix. More important is the fact that, unlike Tait-Bryan or Euler angles, Euler parameters cannot lead to any singularity. On the other hand, they have no physical interpretation. Moreover, one has to deal with four rotational coordinates which are not independent due to relation 1.45.

Euler parameters are used by several authors in multibody dynamics ([39]). This is especially the case when body motions are expressed in absolute coordinates (i.e. the orientations of the various bodies are described with respect to a common fixed frame). Thus the rotation amplitudes can take any arbitrary values and the system can reach a configuration which would be *singular* in terms of Tait-Bryan or Euler angles.

When using *relative* coordinates for modeling multibody systems - as we do in this book - the angular coordinates correspond to the relative orientations between adjacent bodies. The latter are connected via some mechanism (such as telescopic arms, guide rods, hinges, bearing, ...) and, in most cases, we can avoid the singularity problem by choosing an appropriate Tait Bryan sequence[4]. For this reason, and because Euler parameters are not independent and less easy to visualize than Tait-Bryan angles, we will use the latter in this book.

1.1.9 Time derivatives of vectors and tensors

Let us consider a vector $\mathbf{x}$ which is varying with respect to time. This means that, if the frame $\{\hat{\mathbf{I}}\}$ is time-fixed (i.e. has a constant orientation), the components of $\mathbf{x}$ are time-varying functions[5]:

$$\mathbf{x} = [\hat{\mathbf{I}}]^T \, x(t)$$

The first and second time-derivatives of this vector are thus respectively defined as

$$\dot{\mathbf{x}} \triangleq [\hat{\mathbf{I}}]^T \, \frac{d}{dt} x(t) \quad \text{or, more simply as} \quad \dot{\mathbf{x}} = [\hat{\mathbf{I}}]^T \, \dot{x}(t) \tag{1.47}$$

$$\ddot{\mathbf{x}} \triangleq [\hat{\mathbf{I}}]^T \, \frac{d^2}{dt^2} x(t) \quad \text{or} \quad \ddot{\mathbf{x}} = [\hat{\mathbf{I}}]^T \, \ddot{x}(t) \tag{1.48}$$

[4] Practically, a singularity could only occur in a spherical (or ball) joint with unlimited rotational range.

[5] These functions are supposed to be continuously differentiable with respect to time, at least up to the second order.

Similarly, the derivatives of a tensor $\mathbf{T} = [\hat{\mathbf{I}}]^T T(t) [\hat{\mathbf{I}}]$ are defined as

$$\dot{\mathbf{T}} \triangleq [\hat{\mathbf{I}}]^T \dot{T}(t) [\hat{\mathbf{I}}] \quad \text{and} \quad \ddot{\mathbf{T}} \triangleq [\hat{\mathbf{I}}]^T \ddot{T}(t) [\hat{\mathbf{I}}] \tag{1.49}$$

When expressed in a time-fixed frame, time derivatives of vectors and tensors thus reduce to time derivatives of (3×1) and (3×3) matrices:

$$\dot{x}(t) \triangleq \begin{pmatrix} \dot{x}_1(t) \\ \dot{x}_2(t) \\ \dot{x}_3(t) \end{pmatrix} \quad \text{and} \quad \dot{T}(t) \triangleq \begin{pmatrix} \dot{T}_{11}(t) & \dot{T}_{12}(t) & \dot{T}_{13}(t) \\ \dot{T}_{21}(t) & \dot{T}_{22}(t) & \dot{T}_{23}(t) \\ \dot{T}_{31}(t) & \dot{T}_{32}(t) & \dot{T}_{33}(t) \end{pmatrix}$$

Supposing now a vector $\mathbf{u}$ whose components $u(t)$ are given in a frame $\{\hat{\mathbf{J}}\}$ whose orientation with respect to $\{\hat{\mathbf{I}}\}$ is also time varying, i.e.

$$\mathbf{u} = [\hat{\mathbf{J}}]^T u(t) \quad \text{with} \quad [\hat{\mathbf{J}}] = R(t)[\hat{\mathbf{I}}] \tag{1.50}$$

we then have

$$\dot{\mathbf{u}} = [\hat{\mathbf{I}}]^T \frac{d}{dt}\left(R^T(t)\, u(t)\right) = [\hat{\mathbf{I}}]^T \left(\dot{R}^T u + R^T \dot{u}\right) = [\hat{\mathbf{J}}]^T \left(R\dot{R}^T u + \dot{u}\right) \tag{1.51}$$

which involves the time derivative of the rotation matrix R. However, taking the time derivative of the orthogonality property 1.33 of R, namely $R(t)R^T(t) = E$ (which must hold for any time t), we obtain

$$\dot{R}R^T + R\dot{R}^T = 0$$

which shows that the matrix $R\dot{R}^T$ is a skew-symmetric (3×3) matrix. Therefore, $R\dot{R}^T$ can be written as a tilde matrix:

$$R\dot{R}^T \triangleq \begin{pmatrix} 0 & -\omega_3 & \omega_2 \\ \omega_3 & 0 & -\omega_1 \\ -\omega_2 & \omega_1 & 0 \end{pmatrix} = \tilde{\omega} \tag{1.52}$$

and using this, equation 1.51 can be rewritten as

$$\dot{\mathbf{u}} = [\hat{\mathbf{J}}]^T \left(\tilde{\omega}\, u + \dot{u}\right)$$

The time derivative of a vector expressed in a time varying frame thus contains two terms:

- the term $[\hat{\mathbf{J}}]^T \dot{u}$ is due to the fact that vector $\mathbf{u}$ has time varying components $u(t)$ with respect to the frame $\{\hat{\mathbf{J}}\}$ in which it has been defined. When there is no possible confusion as to this frame, this term will be simply noted as

$$\overset{\circ}{\mathbf{u}} \triangleq [\hat{\mathbf{J}}]^T \dot{u} \tag{1.53}$$

and is usually denoted as the *relative* velocity vector (with respect to $\{\hat{\mathbf{J}}\}$).

- the other term arises because frame $\{\hat{\mathbf{J}}\}$ itself does not have a constant orientation with respect to the fixed directions of $\{\hat{\mathbf{I}}\}$. According to 1.24, this term can be rewritten in pure vector form as

$$[\hat{\mathbf{J}}]^T \tilde{\omega}\, u = \boldsymbol{\omega} \times \mathbf{u} \tag{1.54}$$

by defining the *angular velocity vector* of frame $\{\hat{\mathbf{J}}\}$ as being

$$\boldsymbol{\omega} \triangleq [\hat{\mathbf{J}}]^T \omega \tag{1.55}$$

Moreover, since a cross product is involved, it is convenient for later use to introduce the pre-product tensor associated with this vector $\boldsymbol{\omega}$:

$$\tilde{\boldsymbol{\omega}} \triangleq [\hat{\mathbf{J}}]^T \tilde{\omega} [\hat{\mathbf{J}}] \tag{1.56}$$

The time derivative $\dot{\mathbf{u}}$ can thus be written under two equivalent vector/tensor forms

$$\dot{\mathbf{u}} = \overset{\circ}{\mathbf{u}} + \boldsymbol{\omega} \times \mathbf{u} \quad \text{or} \quad \dot{\mathbf{u}} = \overset{\circ}{\mathbf{u}} + \tilde{\boldsymbol{\omega}} \, . \, \mathbf{u} \tag{1.57}$$

In the same way as for a vector, a tensor $\mathbf{Q}$ may also be expressed in a time-varying frame $\{\hat{\mathbf{J}}\}$:

$$\mathbf{Q} = [\hat{\mathbf{J}}]^T Q \, [\hat{\mathbf{J}}] = [\hat{\mathbf{I}}]^T R^T Q \, R \, [\hat{\mathbf{I}}]$$

and its time-derivative is then given by

$$\begin{aligned} \dot{\mathbf{Q}} &= [\hat{\mathbf{I}}]^T \left(\dot{R}^T Q \, R + R^T \dot{Q} \, R + R^T Q \, \dot{R} \right) [\hat{\mathbf{I}}] \\ &= [\hat{\mathbf{J}}]^T \left(R \, \dot{R}^T Q + \dot{Q} + Q \, \dot{R} \, R^T \right) [\hat{\mathbf{J}}] \end{aligned}$$

which, using property 1.22 and definition 1.56 for a tensor $\tilde{\boldsymbol{\omega}}$, becomes

$$\dot{\mathbf{Q}} = \overset{\circ}{\mathbf{Q}} + \tilde{\boldsymbol{\omega}} \, . \, \mathbf{Q} - \mathbf{Q} \, . \, \tilde{\boldsymbol{\omega}} \tag{1.58}$$

in which

$$\overset{\circ}{\mathbf{Q}} \triangleq [\hat{\mathbf{J}}]^T \dot{Q} \, [\hat{\mathbf{J}}]$$

Finally, one may observe that the derivation rules 1.57 and 1.58 can be used recursively to compute time-derivatives of order higher than one. For instance, the second time derivative of a vector $\mathbf{u} = [\hat{\mathbf{J}}]^T u(t)$ is obtained recursively as

$$\begin{aligned} \dot{\mathbf{u}} &= [\hat{\mathbf{J}}]^T \left(\tilde{\omega} \, u + \dot{u} \right) \\ \ddot{\mathbf{u}} &= [\hat{\mathbf{J}}]^T \tilde{\omega} \, \left(\tilde{\omega} \, u + \dot{u} \right) + [\hat{\mathbf{J}}]^T \left(\dot{\tilde{\omega}} \, u + \tilde{\omega} \, \dot{u} + \ddot{u} \right) \\ &= \overset{\circ\circ}{\mathbf{u}} + \dot{\tilde{\boldsymbol{\omega}}} \, . \, \mathbf{u} + 2 \, \tilde{\boldsymbol{\omega}} \, . \, \overset{\circ}{\mathbf{u}} + \tilde{\boldsymbol{\omega}} \, . \, \tilde{\boldsymbol{\omega}} \, . \, \mathbf{u} \end{aligned} \tag{1.59}$$

in which

- $\overset{\circ\circ}{\mathbf{u}} = [\hat{\mathbf{J}}]^T \ddot{u}$ expresses the relative acceleration of vector $\mathbf{u}$ with respect to the moving frame $\{\hat{\mathbf{J}}\}$
- the last three terms are due to the fact that frame $\{\hat{\mathbf{J}}\}$ has a time-varying orientation:
  - $\dot{\tilde{\omega}}\,.\,\mathbf{u}$ results from the angular acceleration of $\{\hat{\mathbf{J}}\}$,
  - $2\,\tilde{\omega}\,.\,\overset{\circ}{\mathbf{u}}$, called the *Coriolis* acceleration (and characterized by the presence of a factor 2), arises from both the relative velocity of $\mathbf{u}$ with respect to $\{\hat{\mathbf{J}}\}$ and the time varying orientation of the latter,
  - $\tilde{\omega}\,.\,\tilde{\omega}\,.\,\mathbf{u}$ is usually called the *centripetal* acceleration.

1.1.10 Angular velocity vector

Although it was introduced algebraically in the previous section, the vector $\boldsymbol{\omega}$ has a physical meaning which is easily understood when considering an elementary rotation matrix. Considering for instance the matrix R_3 we defined in 1.41,

$$R_3 = \begin{pmatrix} c\vartheta_3 & s\vartheta_3 & 0 \\ -s\vartheta_3 & c\vartheta_3 & 0 \\ 0 & 0 & 1 \end{pmatrix},$$

we have

$$\begin{aligned} R_3\dot{R}_3^T &= \begin{pmatrix} c\vartheta_3 & s\vartheta_3 & 0 \\ -s\vartheta_3 & c\vartheta_3 & 0 \\ 0 & 0 & 1 \end{pmatrix} \begin{pmatrix} -s\vartheta_3 & -c\vartheta_3 & 0 \\ c\vartheta_3 & -s\vartheta_3 & 0 \\ 0 & 0 & 0 \end{pmatrix} \dot{\vartheta}_3 \\ &= \begin{pmatrix} 0 & -\dot{\vartheta}_3 & 0 \\ \dot{\vartheta}_3 & 0 & 0 \\ 0 & 0 & 0 \end{pmatrix} \end{aligned}$$

and according to 1.52 and 1.55, the angular velocity vector reads

$$\boldsymbol{\omega} = [\hat{\mathbf{J}}]^T \begin{pmatrix} 0 \\ 0 \\ \dot{\vartheta}_3 \end{pmatrix} = \dot{\vartheta}_3 \hat{\mathbf{J}}_3 \tag{1.60}$$

which indicates that $\boldsymbol{\omega}$ is a vector aligned with the axis of (elementary) rotation and that its amplitude is equal to the rate of change of the rotation angle ϑ_3.

For an arbitrary rotation matrix R, one may also compute the value of the $\boldsymbol{\omega}$ vector by computing $R\dot{R}^T$. For instance for the Tait-Bryan sequence given in 1.42, we obtain

$$\boldsymbol{\omega} = [\hat{\mathbf{J}}]^T \begin{pmatrix} c\vartheta_2\; c\vartheta_3 & s\vartheta_3 & 0 \\ -c\vartheta_2\; s\vartheta_3 & c\vartheta_3 & 0 \\ s\vartheta_2 & 0 & 1 \end{pmatrix} \begin{pmatrix} \dot{\vartheta}_1 \\ \dot{\vartheta}_2 \\ \dot{\vartheta}_3 \end{pmatrix}, \tag{1.61}$$

but the interpretation of this result is not immediately clear.

In order to obtain a physical understanding of this angular velocity vector, let us introduce a new frame $\{\hat{\mathbf{K}}\}$ whose orientation will be described with respect to $\{\hat{\mathbf{J}}\}$. In order to avoid any confusion, superscripts are now used to distinguish the different rotation matrices as well as the different angular velocities:

$$[\hat{\mathbf{J}}] = R^{J,I}[\hat{\mathbf{I}}] \quad \text{and} \quad [\hat{\mathbf{K}}] = R^{K,J}[\hat{\mathbf{J}}]$$

By trivial substitution, one first obtains

$$[\hat{\mathbf{K}}] = R^{K,I}[\hat{\mathbf{I}}] \quad \text{with} \quad R^{K,I} = R^{K,J}R^{J,I} \tag{1.62}$$

which allows us to compute the components of the angular velocity vector related to the new frame $\{\hat{\mathbf{K}}\}$ according to the definitions 1.52 and 1.55:

$$\boldsymbol{\omega}^{KI} = [\hat{\mathbf{K}}]^T\omega^{KI} \quad \text{with } \omega^{KI} \text{ such that } \tilde{\omega}^{KI} = R^{K,I}(\dot{R}^{K,I})^T$$

Then, using 1.62 with the definitions 1.52 and 1.56, one successively obtains

$$\begin{aligned}
\tilde{\boldsymbol{\omega}}^{KI} &= [\hat{\mathbf{K}}]^T R^{K,J}R^{J,I}\left((R^{J,I})^T(\dot{R}^{K,J})^T + (\dot{R}^{J,I})^T(R^{K,J})^T\right)[\hat{\mathbf{K}}] \\
&= [\hat{\mathbf{K}}]^T\tilde{\omega}^{KJ}[\hat{\mathbf{K}}] + [\hat{\mathbf{K}}]^T R^{K,J}\tilde{\omega}^{JI}(R^{K,J})^T[\hat{\mathbf{K}}] \\
&= [\hat{\mathbf{K}}]^T\tilde{\omega}^{KJ}[\hat{\mathbf{K}}] + [\hat{\mathbf{J}}]^T\tilde{\omega}^{JI}[\hat{\mathbf{J}}] \\
&= \tilde{\boldsymbol{\omega}}^{KJ} + \tilde{\boldsymbol{\omega}}^{JI}
\end{aligned}$$

which leads to

$$\boldsymbol{\omega}^{KI} = \boldsymbol{\omega}^{KJ} + \boldsymbol{\omega}^{JI} \tag{1.63}$$

This last relation expresses a very important property of the angular velocity vectors: their *additivity.* In the above case, relation 1.63 states that the absolute angular velocity of frame $\{\hat{\mathbf{K}}\}$ with respect to the time-fixed frame $\{\hat{\mathbf{I}}\}$ is equal to its relative angular velocity with respect to the moving frame $\{\hat{\mathbf{J}}\}$ plus the absolute angular velocity of $\{\hat{\mathbf{J}}\}$ itself.

We can exploit this additivity property recursively. Therefore, for a general rotation matrix expressed as a sequence of Tait-Bryan angles (see 1.42), the expression 1.61 is in fact equivalent to the relation

$$\boldsymbol{\omega} = [\hat{\mathbf{J}}]^T R_3(\vartheta_3) R_2(\vartheta_2)\begin{pmatrix}\dot{\vartheta}_1\\0\\0\end{pmatrix} + [\hat{\mathbf{J}}]^T R_3(\vartheta_3)\begin{pmatrix}0\\\dot{\vartheta}_2\\0\end{pmatrix} + [\hat{\mathbf{J}}]^T\begin{pmatrix}0\\0\\\dot{\vartheta}_3\end{pmatrix} \tag{1.64}$$

whose interpretation is now clear: the angular velocity or frame $\{\hat{\mathbf{J}}\}$ is equal to the vector sum of the relative elementary angular velocity vectors corresponding to the different elementary rotations.

Relation 1.61 again reveals the singularity problem associated with Tait-Bryan angles: when $\vartheta_2 = \pm\frac{\pi}{2}$, the coefficient matrix appearing in 1.61 becomes singular. Since, in this singular configuration the 3rd axis of rotation is aligned with the first one, one cannot distinguish between the angular velocities respectively due to $\dot{\vartheta}_1$ and $\dot{\vartheta}_3$ (only the sum $\dot{\vartheta}_1 + \dot{\vartheta}_3$ has a physical interpretation) and it is impossible to compute their values for given components of $\boldsymbol{\omega}$.

1.2 Rigid body representation

1.2.1 Rigid body definition

We define a *body* as a collection (or set) C of material points which will be denoted by X. This collection can be discrete or continuous. Denoting by dm the mass characterizing these material points X, we obtain the total mass of body C by summing these masses over all the elements:

$$m(C) \triangleq \int_C dm \tag{1.65}$$

This integral notation[6] is used to cover the different types of bodies one may use to model physical systems:

- for solids (3-D bodies), one should write $dm \triangleq \rho_V \, dV$, where $\rho_V \; [kg/m^3]$ is the mass density per unit of volume dV,
- 2-D and 1–D dimensional models are quite often used in practical engineering. The mass measure dm is then related to surface and length units respectively by $dm \triangleq \rho_S \, dS$ where ρ_S has dimensions $[kg/m^2]$, and by $dm \triangleq \rho_L \, dL$ where ρ_L has dimensions $[kg/m]$,
- for a discrete collection of point masses m^i, then $\int_C dm \triangleq \sum_i m^i$.

The universal notation we use in 1.65 allows us to consider bodies which are made up of different parts, each of which belongs to one of the categories described above.

A *rigid* body is characterized by the fact that it cannot be deformed. This means that if there is a *reference point* P which is fixed to the body and a *body-fixed frame* $\{\hat{\mathbf{J}}\}$ which is also rigidly attached to it, the relative position vector $\mathbf{u}$ of any material point X with respect to P has constant components if expressed in frame $\{\hat{\mathbf{J}}\}$ (see figure 1.12):

$$\overrightarrow{PX} \triangleq \mathbf{u}(X) = [\hat{\mathbf{J}}]^T u$$

where $\forall X$, u is constant (time independent). Thus, given some fixed origin O

[6]It is more than a simple notation: mathematically speaking, it is a *Lebesgue* integral over the *measure of mass* represented by dm.

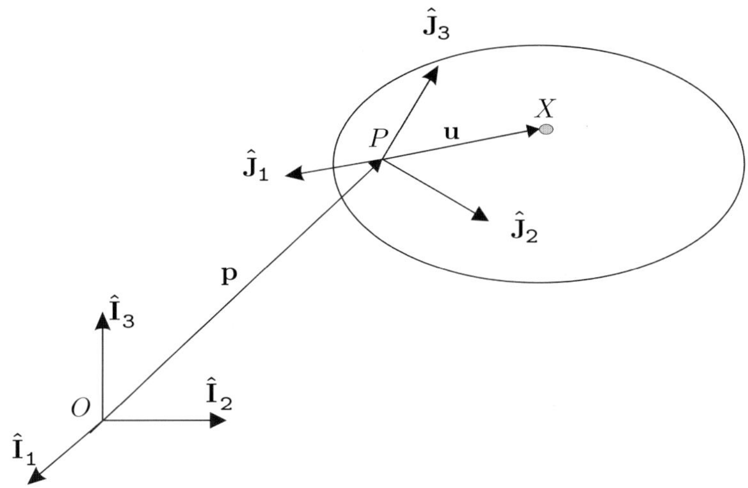

Figure 1.12: Location of a material point in a rigid body

and a fixed frame $\{\hat{\mathbf{I}}\}$, the absolute position $\mathbf{x}$ of any point X on body C can be characterized by

$$\mathbf{x}(X) = \mathbf{p} + \mathbf{u}(X) = [\hat{\mathbf{I}}]^T p + [\hat{\mathbf{J}}]^T u \tag{1.66}$$

where $\mathbf{p}$ is the position vector of P with respect to the origin. Whatever the position and the orientation of body C in space, all its material points are thus localized in space by:

- $\mathbf{p}$, the position vector of the body reference point P with respect to the fixed origin O,

and

- R, the rotation matrix relating the body-fixed frame $\{\hat{\mathbf{J}}\}$ to the fixed frame $\{\hat{\mathbf{I}}\}$, i.e. $[\hat{\mathbf{J}}] = R\,[\hat{\mathbf{I}}]$.

Therefore, the position/orientation description of a rigid body in space requires *6 scalars* (real numbers): three position variables corresponding to the components p of vector $\mathbf{p}$, and three rotation variables (Tait-Bryan or Euler angles) for the rotation matrix R. These 6 scalars are denoted as the *configuration variables* of body C.

1.2.2 Rigid body kinematics

From 1.66, one can easily obtain the (absolute) velocities of all the material points X:

$$\begin{aligned} \dot{\mathbf{x}}(X) &= \dot{\mathbf{p}} + [\hat{\mathbf{J}}]^T \tilde{\omega} u \\ &= \dot{\mathbf{p}} + \tilde{\boldsymbol{\omega}} \,.\, \mathbf{u} \end{aligned} \tag{1.67}$$

Vector $\boldsymbol{\omega} = [\hat{\mathbf{J}}]^T \omega$ is the angular velocity of the body-fixed frame $\{\hat{\mathbf{J}}\}$. Nevertheless, whatever the body-fixed frame whose angular velocity vector is computed, the latter is necessarily unique for a given body C. This is due to the fact that no relative angular (time-varying) motion is possible between any pair of frames fixed to the same rigid body.
Hence, the $\boldsymbol{\omega}$ vector represents the *angular velocity of body* C and the expression 1.67 is intrinsic (i.e. does not depend on a particular frame $\{\hat{\mathbf{J}}\}$).

Relation 1.59 allows us to compute the absolute acceleration of X, by taking into account that $\overset{\circ\circ}{\mathbf{u}} = \mathring{\mathbf{u}} = 0$ for a rigid body:

$$\ddot{\mathbf{x}}(X) = \ddot{\mathbf{p}} + \left(\dot{\tilde{\omega}} + \tilde{\omega} \,.\, \tilde{\omega} \right) .\, \mathbf{u} \tag{1.68}$$

1.2.3 Body center of mass

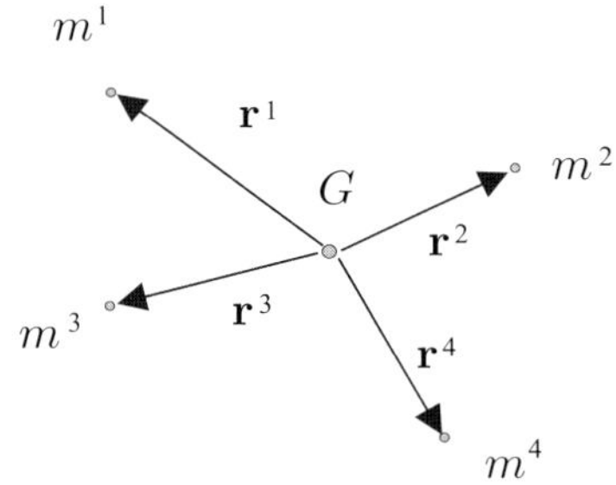

Figure 1.13: Center of mass of a collection of material points

Let us consider, in figure 1.13, material points m^i in three dimensional space. The *center of mass* G of this collection of material points is such that

$$\sum_i m^i \mathbf{r}^i = 0 \tag{1.69}$$

in which $\mathbf{r}^i$ is the position vector of the material point m^i with respect to G. The position of the center of mass is simply a weighted mean of the positions of the various material points.
For a rigid body C, this relation naturally becomes (see figure 1.14)

$$\int_C \mathbf{r} \, dm = 0 \tag{1.70}$$

From equation 1.70, it is easy to express the absolute position of point G. Indeed, on the basis of figure 1.14, the absolute position of a given material point X is given by $\mathbf{x}(X) = \mathbf{x}^G + \mathbf{r}(X)$ and thus

$$\int_C \mathbf{x}(X) \, dm = \int_C \mathbf{x}^G \, dm + \int_C \mathbf{r}(X) \, dm$$

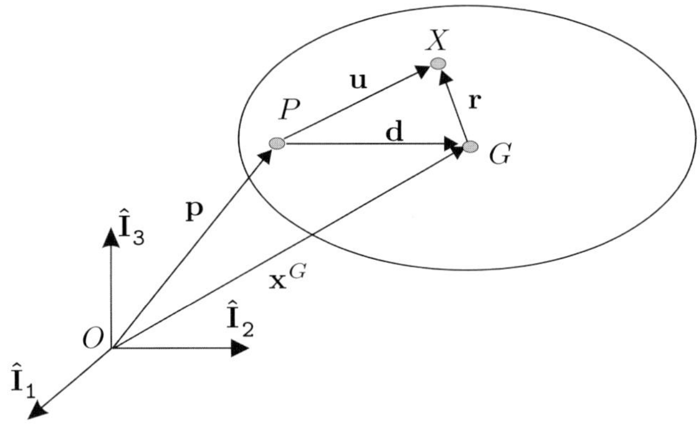

Figure 1.14: Body center of mass G

which implies, using 1.70,

$$m\,(C)\;\mathbf{x}^G = \int_C \mathbf{x}\,dm \tag{1.71}$$

While point G does not necessarily coincide with any material point (ex.: the center of mass of a circular ring), this point is nevertheless fixed with respect to the body. Indeed, by substituting for $\mathbf{x}(X)$ its expression 1.66 under the integral, the relative position of G with respect to the reference point P is found to be

$$\mathbf{x}^G - \mathbf{p} = \frac{1}{m(C)}\int_C \mathbf{u}\,dm = [\hat{\mathbf{J}}]^T\frac{1}{m(C)}\int_C u\,dm$$

which has constant (time invariant) components in the body-fixed frame $\{\hat{\mathbf{J}}\}$. Moreover, the expression 1.71 shows that the center of mass is uniquely defined for a given body C. Thanks to these properties, it is often useful to choose the center of mass to be the body reference point (see figure 1.14). The absolute position of a point X is then

$$\mathbf{x}(X) = \mathbf{x}^G + \mathbf{r}(X) \tag{1.72}$$

1.2.4 Body linear momentum

The *linear momentum vector* of body C is defined as follows:

$$\mathbf{N}(C) \triangleq \int_C \dot{\mathbf{x}}\,dm \tag{1.73}$$

Using G as reference point, one can write

$$\dot{\mathbf{x}}(X) = \dot{\mathbf{x}}^G + \dot{\mathbf{r}}(X)$$

Since relation 1.70 must be satisfied at all times,

$$\int_C \dot{\mathbf{r}}\, dm = 0 \tag{1.74}$$

and thus

$$\mathbf{N}(C) = \int_C \dot{\mathbf{x}}^G\, dm = m\,(C)\, \dot{\mathbf{x}}^G \tag{1.75}$$

The linear momentum vector of a body is thus equal to the product of its mass by the velocity of its center of mass.

1.2.5 Body angular momentum

The moment of the linear momentum of an element of mass dm with respect to the fixed origin O is called its angular momentum with respect to O and is given by the vector

$$\mathbf{x}(X) \times \dot{\mathbf{x}}(X)\, dm$$

The *angular momentum* $\mathbf{H}^O(C)$ of body C with respect to the fixed origin O is then obtained by integrating this vector over the entire mass distribution of body C:

$$\mathbf{H}^O(C) \triangleq \int_C \mathbf{x} \times \dot{\mathbf{x}}\, dm \tag{1.76}$$

Using 1.72, 1.70 and 1.74, we can reduce this angular momentum to

$$\mathbf{H}^O(C) = \mathbf{x}^G \times m(C)\dot{\mathbf{x}}^G + \int_C \mathbf{r} \times \dot{\mathbf{r}}\, dm \tag{1.77}$$

In this equation, the first term represents the moment of the linear momentum of a point mass $m(C)$ located at the center of mass G, while the integral term takes into account the mass distribution around G. The latter is called the *angular momentum* $\mathbf{H}^G(C)$ of body C with respect to its center of mass G:

$$\mathbf{H}^G(C) \triangleq \int_C \mathbf{r} \times \dot{\mathbf{r}}\, dm$$

Recalling that the vectors $\mathbf{r}(X)$ have constant components in the body-fixed frame $\{\hat{\mathbf{J}}\}$ and using tensor notations for the cross-product, we can write this angular momentum as

$$\mathbf{H}^G(C) = \int_C \tilde{\mathbf{r}} \,.\, (\tilde{\boldsymbol{\omega}} \,.\, \mathbf{r})\, dm$$

or equivalently as

$$\mathbf{H}^G(C) = -\int_C \tilde{\mathbf{r}}\,.\,\tilde{\mathbf{r}}\,dm\,.\,\boldsymbol{\omega}$$

since the angular velocity vector is uniquely defined for the whole body. The tensor which is dot-multiplying vector $\boldsymbol{\omega}$ in this relation is defined as the *inertia tensor* of body C with respect to its center of mass:

$$\mathbf{I}^G \triangleq -\int_C \tilde{\mathbf{r}}\,.\,\tilde{\mathbf{r}}\,dm \tag{1.78}$$

and characterizes the mass distribution within the body with respect to its center of mass. The angular momentum with respect to the center of mass G thus finally reads

$$\mathbf{H}^G(C) = \mathbf{I}^G\,.\,\boldsymbol{\omega} \tag{1.79}$$

Introducing this result into 1.77, we may write the angular momentum vector $\mathbf{H}^O$ with respect to a fixed origin O as follows:

$$\mathbf{H}^O(C) = \mathbf{x}^G \times m(C)\dot{\mathbf{x}}^G + \mathbf{I}^G\,.\,\boldsymbol{\omega} \tag{1.80}$$

1.2.6 Inertia tensor

Properties

From equation 1.79, we see that considering the vector $\boldsymbol{\omega}$ as an "input", the inertia tensor $\mathbf{I}^G$ produces as "output" the angular momentum vector $\mathbf{H}^G$. This tensor has several interesting properties:

- its components (units: $[kg.m^2]$) are constant when expressed in the body-fixed frame $\{\hat{\mathbf{J}}\}$:

$$\mathbf{I}^G = [\hat{\mathbf{J}}]^T\, I^G\, [\hat{\mathbf{J}}] \quad \text{where } I^G = -\int_C \tilde{r}\tilde{r}\,dm \text{ is constant} \tag{1.81}$$

- it is symmetric, since $(\tilde{r}\tilde{r})^T = \tilde{r}\tilde{r}$

- its component matrix I^G, or *inertia matrix,* is also symmetric and moreover is positive semi-definite. Indeed, the associated quadratic form $v^T\, I^G v$ where v is an arbitrary (3×1) column matrix can be written, using 1.81, as

$$v^T\, I^G v = -\int_C v^T \tilde{r}\tilde{r} v\,dm = \int_C (\tilde{r}v)^T\, \tilde{r}v\,dm = \int_C \parallel \tilde{r}v \parallel^2\, dm \geq 0$$

Since the inertia matrix is symmetrical and positive semi-definite, it can be diagonalized by means of an orthogonal matrix transformation. Denoting λ_1, λ_2 and λ_3 the eigenvalues of I^G and v_1, v_2 and v_3 the corresponding (normalized) eigenvectors, one obtains

$$\begin{pmatrix} v_1^T \\ v_2^T \\ v_3^T \end{pmatrix} I^G \, (v_1, v_2, v_3) = diag\,(\lambda_1, \lambda_2, \lambda_3) \tag{1.82}$$

$$\text{with} \quad diag\,(\lambda_1, \lambda_2, \lambda_3) \triangleq \begin{pmatrix} \lambda_1 & 0 & 0 \\ 0 & \lambda_2 & 0 \\ 0 & 0 & \lambda_3 \end{pmatrix}$$

By choosing the signs of the normalized eigenvectors such that $\det\,(v_1, v_2, v_3) = 1$, one may consider that the (3×3) matrix $(v_1, v_2, v_3)^T$ is a rotation matrix R. Being constant, this matrix defines a new frame $\{\hat{\mathbf{K}}\}$:

$$[\hat{\mathbf{K}}] = R[\hat{\mathbf{J}}]$$

which, like $\{\hat{\mathbf{J}}\}$, is rigidly attached to the body. Expressing the inertia tensor in this new frame, we obtain

$$\mathbf{I}^G = [\hat{\mathbf{J}}]^T \, I^G \, [\hat{\mathbf{J}}] = [\hat{\mathbf{K}}]^T \, R \, I^G \, R^T \, [\hat{\mathbf{K}}] \tag{1.83}$$

The diagonal matrix $R\,I^G\,R^T = diag\,(\lambda_1, \lambda_2, \lambda_3)$ is called the *principal inertia matrix* of the body, and the three axes passing through the center of mass G and whose directions are parallel to the vectors $\hat{\mathbf{K}}_\alpha$ are called the *principal axes of inertia*. One may observe that when a body is rotating around one of its principal axes of inertia, i.e. when $\boldsymbol{\omega} = \omega_\alpha \hat{\mathbf{K}}_\alpha$ (with $\alpha = 1, 2$ or 3), then the angular momentum

$$\mathbf{H}^G(C) = \mathbf{I}^G \,.\, \boldsymbol{\omega} = \lambda_\alpha \, \omega_\alpha \hat{\mathbf{K}}_\alpha$$

is also aligned with this principal axis of inertia.

Inertia tensor computation

The inertia matrix of a body is evaluated from 1.81:

$$I^G = -\int_C \tilde{r}\tilde{r}\, dm = \begin{pmatrix} \int_C r_2^2 + r_3^2 \, dm & -\int_C r_1 r_2 \, dm & -\int_C r_1 r_3 \, dm \\ -\int_C r_1 r_2 \, dm & \int_C r_1^2 + r_3^2 \, dm & -\int_C r_2 r_3 \, dm \\ -\int_C r_1 r_3 \, dm & -\int_C r_2 r_3 \, dm & \int_C r_1^2 + r_2^2 \, dm \end{pmatrix} \tag{1.84}$$

Each component of I^G requires, for a 3D body, the computation of a multiple integral over the body volume (since $dm = \rho_V \, dV$): several books on fundamental physics give examples of such computations for simple geometries (ex.

rectangular prism, rod, hollow cylinder, ...).
More complex geometries may also be handled by rigidly assembling different sub-bodies which have simpler shapes: if body C results from the union of separate sub-bodies: $C = \cup_i\{C^i\}$, one may use the additivity property of integrals:

$$\mathbf{I}^G = \sum_i \mathbf{I}^G_{C^i} \tag{1.85}$$

which shows how to compute the inertia matrix I^G. However, in equation 1.85, the sub-tensors $\mathbf{I}^G_{C_i}$ of the sub-bodies C^i must be evaluated with respect to the center of mass G of body C. The latter differs from their own centers of mass G^i as illustrated in the small example of figure 1.15. Thus, we need a practical

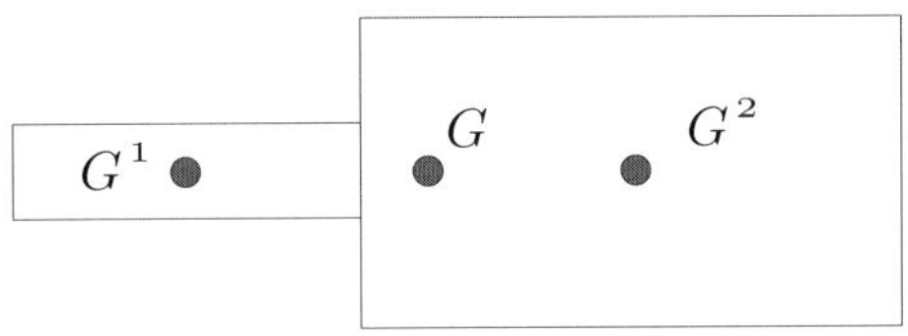

Figure 1.15: Body as a collection of sub-bodies

way to compute an inertia tensor *with respect to some arbitrary reference point* P (see figure 1.14).
Let us denote this tensor $\mathbf{I}^P$:

$$\mathbf{I}^P \triangleq -\int_C \tilde{\mathbf{u}}\,.\,\tilde{\mathbf{u}}\,dm \tag{1.86}$$

in which (see figure 1.14) $\mathbf{u}$ is the relative position vector of X with respect to P. Let us note $\mathbf{d}$ the relative position vector of G also with respect to P. One has thus

$$\mathbf{u}(X) = \mathbf{d} + \mathbf{r}(X), \quad \forall X.$$

By substitution in 1.86 and using 1.70, one obtains

$$\begin{aligned} \mathbf{I}^P &= -\int_C \left(\tilde{\mathbf{d}} + \tilde{\mathbf{r}}\right) . \left(\tilde{\mathbf{d}} + \tilde{\mathbf{r}}\right) dm \\ &= -m(C)\,\tilde{\mathbf{d}}\,.\,\tilde{\mathbf{d}} - \int_C \tilde{\mathbf{r}}\,.\,\tilde{\mathbf{r}}\,dm \end{aligned}$$

and finally

$$\mathbf{I}^P = -m(C)\,\tilde{\mathbf{d}}\,.\,\tilde{\mathbf{d}} + \mathbf{I}^G \tag{1.87}$$

known as the *Steiner formula*.

The computation of the inertia tensor of body $C = \cup_i\{C^i\}$ from equation 1.85 can thus be performed as follows:

1. evaluation of the inertia tensor $\mathbf{I}_{C^i}^{G^i}$ for each sub-body i with respect to its own center of mass G^i

2. for each sub-body, application of the Steiner formula with respect to the global center of mass G: $\mathbf{I}_{C^i}^{G} = -m^i \tilde{\mathbf{d}}^i . \tilde{\mathbf{d}}^i + \mathbf{I}_{C^i}^{G^i}$ with $\mathbf{d}^i = \overrightarrow{GG^i}$.

3. summation of the inertias $\mathbf{I}_{C^i}^{G}$ of the sub-bodies according to 1.85.

1.2.7 Forces and torques acting on a rigid body

Forces and torques are the physical agents which are able to modify the motion of a rigid body. They are generally distributed over surfaces or volumes and can be of various natures:

- aerodynamic forces acting on a wing,
- contact force applied on a tire by the ground,
- gravitational forces acting on any material body,
- ...

For mathematical and modeling convenience, let us start with the idealized notion of a single force.

Force

A single force acting on a body is characterized by *two* elements:

- the material point A of body C at which the force is applied,
- the direction along which the force is acting and its amplitude, giving a vector $\mathbf{F}$.

A force $\mathbf{F}$ is thus a bound vector and it is therefore of utmost importance to be precise as to which material point A this force is acting upon. Figure 1.16 illustrates a force $\mathbf{F}$ applied at two different locations of the same beam: the resulting motion of the beam will of course not be the same.

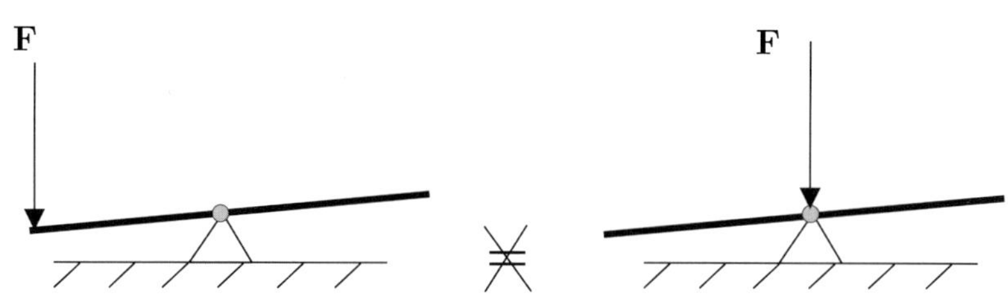

Figure 1.16: Force and application point

If several single forces $\mathbf{F}^i$ are acting at a same point A (see figure 1.17), their effect on the body motion can be represented by a *resultant force* $\mathbf{F} = \sum_i \mathbf{F}^i$ acting at A.

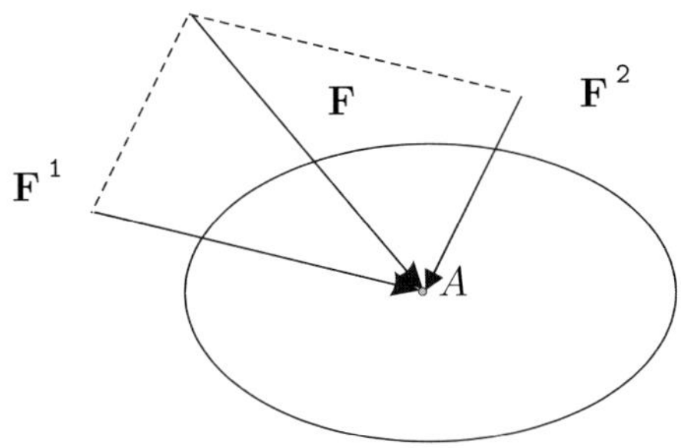

Figure 1.17: Force resultant

Pure torque

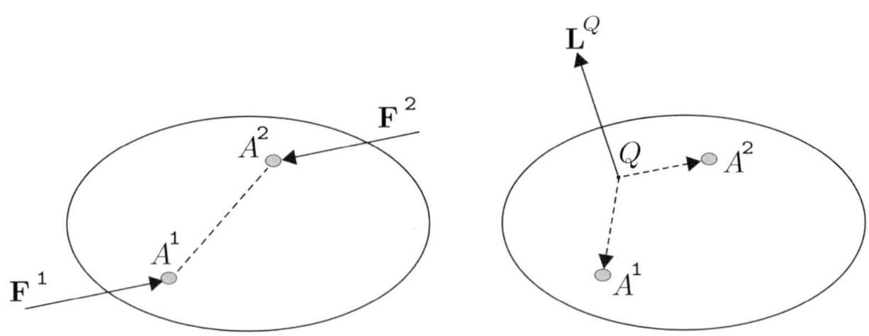

Figure 1.18: Pair of forces replaced by a pure torque

Consider two equal and opposite forces $\mathbf{F}^1$, $\mathbf{F}^2$ acting at two different material points A^1 and A^2 of a body (see fig. 1.18). The resultant force $\mathbf{F} = \mathbf{F}^1 + \mathbf{F}^2 = 0$ but some resulting action is still applied to the body motion (this is for instance the case when unscrewing a wheel with a wheel brace). The latter action is taken into account by summing the vector *moments* of these two forces with respect to some common point Q:

$$\mathbf{L}^Q = \overrightarrow{Q\,A^1} \times \mathbf{F}^1 + \overrightarrow{Q\,A^2} \times \mathbf{F}^2$$

This resulting moment, which is a vector, does not depend on a particular choice of the point Q. Indeed, choosing a different point Q', we find

$$\begin{aligned}
\mathbf{L}^{Q'} &= \overrightarrow{Q' A^1} \times \mathbf{F}^1 + \overrightarrow{Q' A^2} \times \mathbf{F}^2 \\
&= \left(\overrightarrow{Q' Q} + \overrightarrow{Q A^1}\right) \times \mathbf{F}^1 + \left(\overrightarrow{Q' Q} + \overrightarrow{Q A^2}\right) \times \mathbf{F}^2 \\
&= \overrightarrow{Q' Q} \times \left(\mathbf{F}^1 + \mathbf{F}^2\right) + \mathbf{L}^Q = \mathbf{L}^Q \triangleq \mathbf{M}
\end{aligned}$$

This *resultant moment* is thus a free vector $\mathbf{M}$, without any reference to a particular point. Such a vector, which represents the global action of a couple of opposite forces, is called a *pure torque.* By choosing the point Q at A^1, we find the value of this pure torque:

$$\mathbf{M} = \overrightarrow{A^1 A^2} \times \mathbf{F}^2 \tag{1.88}$$

The notion of pure torque is also convenient for "macroscopically" modeling the resultant of local interactions: for instance for a rotor of an electrical motor, the resulting effect of the forces induced by the magnetic field generated by the stator can be modeled as a pure (axial) torque $\mathbf{M}$ applied to the rotor as illustrated in figure 1.19.

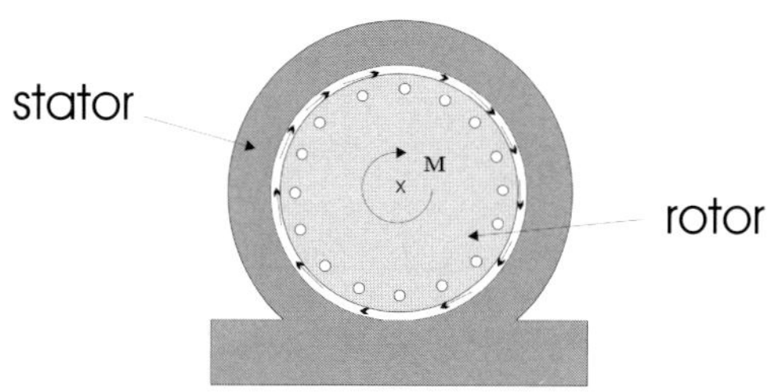

Figure 1.19: Pure torque $\mathbf{M}$ produced by an electrical actuator

Axial torques occur in other situations: the (tangent) friction forces in bearings, clutches, etc.

Force and torque resultant

Consider figure 1.20. A single force $\mathbf{F}$ acts at point A; let us choose a reference point Q attached to the body.

This point could for instance be the center of mass G, but it can also be any point rigidly fixed to the body. Suppose then that we artificially introduce a couple of opposite forces $\mathbf{F}$ and $-\mathbf{F}$ acting at Q. The resultant of the pair of equal and opposite forces respectively acting at A and Q is equal to zero: we can thus replace this pair of forces by an equivalent pure torque $\mathbf{M}^Q = \overrightarrow{QA} \times \mathbf{F}$. The resultant force acting at A now being equal to zero, we may consider that the original force $\mathbf{F}$ acting at A has been replaced by:

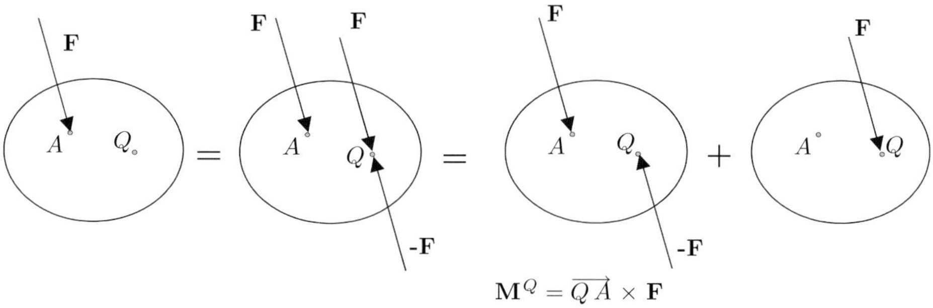

Figure 1.20: Equivalent force and torque resultant

- a force $\mathbf{F}$ acting at Q
- and a pure torque $\mathbf{M}^Q = \overrightarrow{QA} \times \mathbf{F}$

which, as regards the influence on the rigid body motion, are totally equivalent[7] to the original force acting at A.

As a consequence, any system of forces $\{\mathbf{F}^i, A^i\}$ (and pure torques $\mathbf{M}^j$) acting on a given rigid body can be replaced by the equivalent system:

- a *resultant force* $\mathbf{F} \triangleq \sum_i \mathbf{F}^i$ acting at a reference body-fixed point Q
- and a *resultant torque* $\mathbf{L}^Q \triangleq \sum_i \overrightarrow{QA^i} \times \mathbf{F}^i + \sum_j \mathbf{M}^j$

Two particular choices of point Q will be of particular interest later. The first is the center of mass G of the body, for which the resultant torque reads

$$\mathbf{L}^G = \sum_i \overrightarrow{GA^i} \times \mathbf{F}^i + \sum_j \mathbf{M}^j \tag{1.89}$$

and the second one is the fixed origin O (or any other point fixed in space):

$$\mathbf{L}^O = \sum_i \overrightarrow{OA^i} \times \mathbf{F}^i + \sum_j \mathbf{M}^j \tag{1.90}$$

1.2.8 Power considerations

Many mechanical analyses (ex.: machine design) are concerned with the time rate at which energy is transferred. The rate of energy transfer induced by a force $\mathbf{F}$ applied to a material point A of a moving body C is called the

[7] This equivalence is valid only for rigid bodies. For deformable bodies, a force may not be dissociated from its application point.

(instantaneous) mechanical *power* produced by $\mathbf{F}$.
Power is the scalar quantity (units: $[W = N.\frac{m}{\sec} = \frac{J}{\sec}]$) given by the dot product

$$\mathcal{P} = \mathbf{F} \,.\, \dot{\mathbf{x}}^A \tag{1.91}$$

where the vector $\dot{\mathbf{x}}^A$ represents the absolute velocity of the material point A. Considering as before a reference point Q rigidly attached to the body, we can write the absolute position $\mathbf{x}^A$ of point A as

$$\mathbf{x}^A = \mathbf{x}^Q + \overrightarrow{QA}$$

where the vector $\overrightarrow{QA}$ has constant components in a body-fixed frame. The absolute velocity $\dot{\mathbf{x}}^A$ then reads

$$\dot{\mathbf{x}}^A = \dot{\mathbf{x}}^Q + \boldsymbol{\omega} \times \overrightarrow{QA}$$

where $\boldsymbol{\omega}$ is the body angular velocity. By substitution into the expression of power, we obtain

$$\mathcal{P} = \mathbf{F} \,.\, \dot{\mathbf{x}}^Q + \mathbf{F} \,.\, \left(\boldsymbol{\omega} \times \overrightarrow{QA}\right)$$

which can also be written as

$$\begin{aligned} \mathcal{P} &= \mathbf{F} \,.\, \dot{\mathbf{x}}^Q + \boldsymbol{\omega} \,.\, \left(\overrightarrow{QA} \times \mathbf{F}\right) \\ &= \mathbf{F} \,.\, \dot{\mathbf{x}}^Q + \mathbf{L}^Q \,.\, \boldsymbol{\omega} \end{aligned}$$

As a consequence, if a system of forces $\{\mathbf{F}^i, A^i\}$ acting on a given body is replaced by a resultant force $\mathbf{F} = \sum_i \mathbf{F}^i$ acting at a reference body-fixed point Q and a resultant torque $\mathbf{L}^Q = \sum_i \left(\overrightarrow{QA^i} \times \mathbf{F}^i\right)$, the total power produced by these forces acting on the body is given by

$$\mathcal{P} = \sum_i \left(\mathbf{F}^i \,.\, \dot{\mathbf{x}}^{A^i}\right) = \mathbf{F} \,.\, \dot{\mathbf{x}}^Q + \mathbf{L}^Q \,.\, \boldsymbol{\omega} \tag{1.92}$$

This expression also shows that the power produced by a torque $\mathbf{L}$ is given by the dot product of $\mathbf{L}$ with the body angular velocity $\boldsymbol{\omega}$. Therefore, if a system of forces $\{\mathbf{F}^i, A^i\}$ acts on a given body together with a system of pure torques $\{\mathbf{M}^j\}$, the total resulting power is given by 1.92, but in this case with $\mathbf{L}^Q = \sum_i \overrightarrow{QA^i} \times \mathbf{F}^i + \sum_j \mathbf{M}^j$.

1.3 Newton-Euler equations

Newton's three laws of fundamental mechanics can be stated as follows:

1. The linear momentum $\mathbf{N}(C)$ of a body[8] is constant, in the absence of interaction with its environment.

[8] or a collection of bodies C^i

2. The resultant of all the external forces acting on a body is equal to the time derivative of its linear momentum:

$$\dot{\mathbf{N}}(C) = \mathbf{F} \tag{1.93}$$

3. Action and reaction are equal and opposite. Thus for a pair of bodies C^1 and C^2, if a resultant force $\mathbf{F}^{1,2}$ due to body C^1 is applied to body C^2, then a resultant *reaction force* given by $\mathbf{F}^{2,1} = -\mathbf{F}^{1,2}$ is applied to body C^1 by C^2.

The *translational equation* of motion of a body C is given by the second law (equation 1.93) and is known as the *Newton equation.* By means of 1.75, this equation can also be written as

$$m\,(C)\,\ddot{\mathbf{x}}^G = \mathbf{F} \tag{1.94}$$

since $m\,(C)$ is time independent.

The *rotational equation*, known as the *Euler equation,* relates the resultant torque $\mathbf{L}^G$ (given by 1.89) to the time derivative of the angular momentum $\mathbf{H}^G(C)$ of body C *with respect to its center of mass*:

$$\dot{\mathbf{H}}^G(C) = \mathbf{L}^G \tag{1.95}$$

rem.: For rigid bodies, this rotational equation can be deduced from Newton's second law. This is however not possible in general [93] (e.g. for a flexible body, fluids, a system containing an internal angular momentum like a motor, ..., etc.) and it is therefore essential to notice that the Euler equation is independent of the Newton equation.

In the absence of an external resultant torque applied to body C, one can state from equation 1.95 that the angular momentum vector $\mathbf{H}^G(C)$ is constant: this is analogous to Newton's first law about the linear momentum.

There is also an action/reaction principle for torques: if a resultant torque $\mathbf{L}^{1,2}$ (computed according to section 1.2.7 with respect to some reference point Q) due to body C^1 acts on body C^2, then a resultant *reaction torque* applied to body C^1 is given by $\mathbf{L}^{2,1} = -\mathbf{L}^{1,2}$ when computed with respect to the *same* reference point Q.

Using 1.79 and the derivative rule 1.58 for tensors, we find that the time derivative of the angular momentum reduces to

$$\begin{aligned}\dot{\mathbf{H}}^G(C) &= \mathbf{I}^G.\dot{\omega} + \left(\mathring{\mathrm{I}}^G + \tilde{\omega}\,.\,\mathbf{I}^G - \mathbf{I}^G\,.\,\tilde{\omega}\right).\,\omega \\ &= \mathbf{I}^G\,.\,\dot{\omega} + \tilde{\omega}\,.\,\mathbf{I}^G\,.\,\omega\end{aligned}$$

since $\mathring{\mathrm{I}}^G$ vanishes for rigid bodies. Equation 1.95 becomes

$$\mathbf{I}^G\,.\,\dot{\omega} + \tilde{\omega}\,.\,\mathbf{I}^G\,.\,\omega = \mathbf{L}^G \tag{1.96}$$

The Euler equation is also applicable if the resultant torque is computed with respect to the fixed origin (or any other point fixed in space). It then reads

$$\dot{\mathbf{H}}^{O}(C) = \mathbf{L}^{O} \tag{1.97}$$

and, in the particular case for which point O is fixed relative to body C, it becomes

$$\mathbf{I}^{O} \,.\, \dot{\omega} + \tilde{\omega} \,.\, \mathbf{I}^{O} \,.\, \omega = \mathbf{L}^{O} \tag{1.98}$$

where $\mathbf{I}^{O}$ can be computed from the Steiner formula 1.87.

It is important to note that the dynamical equations 1.94, 1.95 and 1.97 involve first (implicitly) the absolute position vectors, with respect to a *fixed origin* O, of the material points X belonging to the rigid body, and secondly the time derivatives of vectors defined in a *fixed frame* $\{\hat{\mathbf{I}}\}$. What can be considered *as being fixed* depends on the application which is under investigation and on the accuracy which is required in predicting the motions. When studying the motion of an industrial robot, we may consider the workshop where it is operating as fixed; but when studying spacecraft applications, one should take into account the daily rotation of the earth around its axis, or even the yearly rotation of the earth around the sun ...

Inertial frames are defined as pairs $\{O, \{\hat{\mathbf{I}}\}\}$ for which the dynamical equations 1.94, 1.95 and 1.97 hold in predicting motion. Unless explicitly mentioned, origins and frames attached to the earth will be assumed to be inertial frames in this book. One should further note that, if $\{O, \{\hat{\mathbf{I}}\}\}$ is an inertial frame, then $\{O', \{\hat{\mathbf{J}}\}\}$ is also inertial if the following two conditions are fulfilled:

- the relative orientation of $\{\hat{\mathbf{J}}\}$ with respect to $\{\hat{\mathbf{I}}\}$ is constant,
- the second derivative of the vector $\overrightarrow{OO'}$ (i.e. the acceleration of O' with respect to O) is equal to zero.

These two conditions mean that the frame $\{O', \{\hat{\mathbf{J}}\}\}$ is either rigidly attached to the frame $\{O, \{\hat{\mathbf{I}}\}\}$ or else undergoes a pure translational motion at constant velocity with respect to $\{O, \{\hat{\mathbf{I}}\}\}$. This would for instance be the case of a frame attached to the cabin of a train moving at constant speed on a straight track. A counter-example is given by the merry-go-round (figure 1.21) turning at constant angular speed: while the origin of frame $\{\hat{\mathbf{J}}\}$ is fixed, this frame has a time-varying orientation with respect to $\{\hat{\mathbf{I}}\}$ and is thus not inertial. Therefore, the Newton/Euler laws are not valid in the frame $\{O, \{\hat{\mathbf{J}}\}\}$.

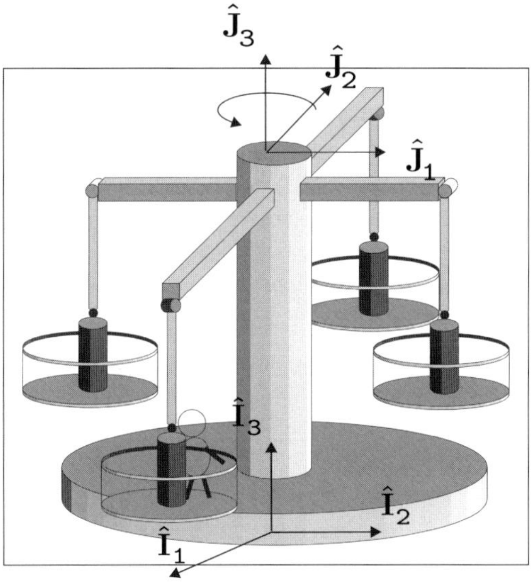

Figure 1.21: A merry-go-round

Chapter 2

Dynamics of rigid bodies

2.1 Introduction

The aim of this chapter is to describe the motion of several bodies. The fundamental notions of generalized coordinates, constraints and degrees of freedom will be introduced and illustrated via simple examples. Then, we shall develop the Newton-Euler procedure which produces the equations of motion relating the motion of the bodies (i.e. the evolution of the generalized coordinates) to the forces/torques applied to the system. Illustrated via an example, this procedure will show its limitations if applied manually, when the number of bodies becomes large. Finally, we shall present a variational approach based on the virtual power principle and use this second approach to treat the same example.

2.2 Generalized coordinates

2.2.1 Generalized coordinates and holonomic constraints

When dealing with several rigid bodies, one needs to describe the *configuration* (see section 1.2.1) of each body using some scalar variables which will play the role of *unknowns* in the set of differential equations of motion. For practical reasons, we wish to use as few unknowns as possible for a given system. The following example – a slider-crank mechanism – illustrates different ways of choosing the unknowns.

Slider-crank mechanism

Slider-crank mechanisms are widely used in machines. As illustrated in figure 2.1, a slider-crank consists of three bodies: a crankshaft rotating at O around the fixed axis $\hat{\mathbf{I}}_3$, a slider B and a connecting rod AB.

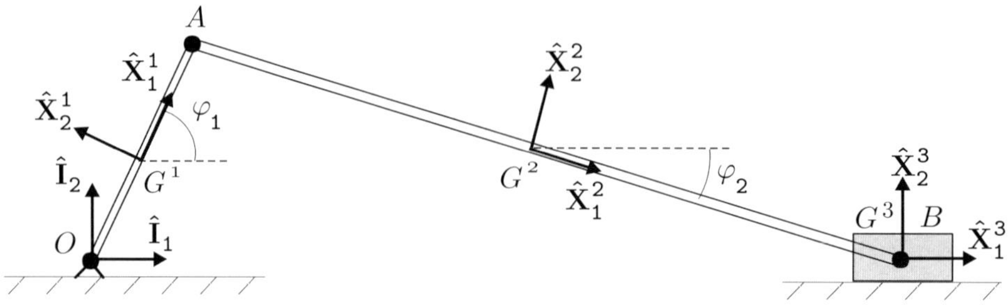

Figure 2.1: A slider-crank mechanism

These three bodies move in the plane $\{\hat{\mathbf{I}}_1, \hat{\mathbf{I}}_2\}$, and each one is located in this plane by 3 configuration variables:

$$\forall \text{ body } i : \quad \{ \ x_i, \quad y_i, \quad \varphi_i \ \} \quad i = 1, 2, 3$$

The *cartesian* variables $\{x_i, y_i\}$ locate the three centers of mass G^i, and the *angular* variables $\{\varphi_i\}$ describe the absolute orientations of the body-fixed frames $\{\hat{\mathbf{X}}^i\}$ with respect to the inertial frame $\{\hat{\mathbf{I}}\}$. Globally, nine configuration variables are thus necessary for this example.

The name *generalized coordinates* is used to denote the variables (of any kind: e.g. cartesian or angular) which may serve to describe a system configuration. These generalized coordinates are denoted by q_i. As illustrated below, we may use different sets of generalized coordinates $\{q_i\}$ for the same mechanism.

Choice 1 We may use, as in [61], [38], the nine absolute variables as generalized coordinates:

$$q = \{x_1, y_1, \varphi_1, x_2, y_2, \varphi_2, x_3, y_3, \varphi_3\} \tag{2.1}$$

However the three moving bodies are assembled in the real system: the length $\| \overrightarrow{OA} \|$ of the crank is r while the length of the connecting rod $\| \overrightarrow{AB} \|$ is l. Our 9 generalized coordinates are thus not independent since, in figure 2.1 (and assuming for simplicity that the mass is uniformly distributed along the crank and the connecting rod), we observe that the following relations must hold:

position of G^1:

$$x_1 = \frac{r}{2} \cos \varphi_1 \tag{2.2}$$

$$y_1 = \frac{r}{2} \sin \varphi_1 \tag{2.3}$$

position of G^2:

$$x_2 = r \cos \varphi_1 + \frac{l}{2} \cos \varphi_2 \tag{2.4}$$

$$y_2 = r \sin \varphi_1 + \frac{l}{2} \sin \varphi_2 \tag{2.5}$$

position of G^3:

$$x_3 = r\cos\varphi_1 + l\cos\varphi_2 \tag{2.6}$$
$$y_3 = r\sin\varphi_1 + l\sin\varphi_2 \tag{2.7}$$

position of B:

$$y_3 = 0 \tag{2.8}$$
$$\varphi_3 = 0 \tag{2.9}$$

Consequently, when *all* the configuration variables of a system are used as generalized coordinates, the latter are often dependent because of the connections between the different bodies.

Choice 2 The simplest choice of generalized coordinate which allows us to describe all the configuration variables is:

$$q = \{\varphi_1\}$$

Indeed, using 2.7 and 2.8, we have

$$\varphi_2 = -\arcsin\left(\frac{r}{l}\sin q\right) \tag{2.10}$$

while $x_1, y_1, x_2, y_2, x_3, y_3$ and φ_3 follows directly from relations (2.2 -2.9).

This choice is optimal as regards the number of variables which will be involved in the equations of motion. However, these equations also require the first and second time-derivatives of the configuration variables in order to compute the velocities and accelerations of the bodies. Looking at the imbedded trigonometric functions which appear in 2.10, we may expect that the computation of these time-derivatives will be cumbersome. Moreover 2.10 may lead to two different solutions for φ_2 : φ_2 and $\pi - \varphi_2$.

Choice 3 In view of these considerations, the "best" choice of generalized coordinates for the slider-crank mechanism is the following trade-off:

$$q = \{\varphi_1, \varphi_2\} \tag{2.11}$$

These two coordinates are not independent, but they allow us to avoid complicated expressions like 2.10. On the other hand, the generalized coordinates are *explicitly* related by the relation:

$$h(q) = r\sin q_1 + l\sin q_2 = 0 \tag{2.12}$$

which physically expresses that the slider is constrained to move along a horizontal line (i.e. $y_3 = 0$). A constraint which is an explicit algebraic relation between generalized coordinates is said to be a *holonomic constraint.* The constraint 2.12 is thus a holonomic constraint.
Here, the assembly of the slider-crank mechanism is taken into account in two different ways:

- on the one hand, the constraint 2.12 will be *explicitly* considered as an additional equation when we shall generate the dynamical model,
- on the other hand, the seven relations (2.2 - 2.6) and (2.8 - 2.9) will be used *implicitly* to compute the configuration variables (and their time derivatives) in terms of the generalized coordinates $\{\varphi_1, \varphi_2\}$.

A constraint which does not involve the time t is said to be *scleronomic*. Thus 2.12 is *scleronomic*. The following example illustrates how time-dependence may appear explicitly in the expressions of the configuration variables and/or the holonomic constraints.

Moving ball in a rotating tube

Consider the system in figure 2.2. A small ball, considered as a material point, slides in a tube. The tube, which is attached to the fixed origin O via a revolute joint, rotates at constant angular velocity ω around axis $\hat{\mathbf{I}}_3$. At initial time ($t = 0$), axis $\hat{\mathbf{X}}_1$ coincides with $\hat{\mathbf{I}}_1$. Since the motion of the tube is imposed, the system

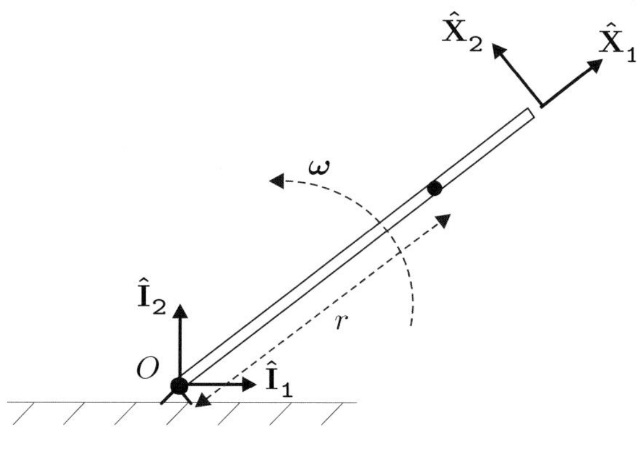

Figure 2.2: Moving ball in a rotating tube

under consideration reduces to the ball. As it is modeled as a (material) point moving in the $\{\hat{\mathbf{I}}_1, \hat{\mathbf{I}}_2\}$ plane, the ball requires only 2 configuration variables $\{x_1, x_2\}$. Two choices of generalized coordinates can be made for this simple 2-D system:

- If we use a single generalized coordinate $q = \{r\}$ (where r is the distance of the ball from the origin), there is no constraint, but the relations between the configuration variables $\{x_1, x_2\}$ and q become time-dependent:

$$x_1 = q\cos(\omega t)$$
$$x_2 = q\sin(\omega t)$$

- If we take all the configuration variables as generalized coordinates:

$$q = \{x_1,x_2\}$$

we must state that the ball remains in the tube. This can be done by requiring that the two vectors $\mathbf{x} = x_1\hat{\mathbf{I}}_1 + x_2\hat{\mathbf{I}}_2$ and $\hat{\mathbf{X}}_2 = -\sin(\omega t)\hat{\mathbf{I}}_1 + \cos(\omega t)\hat{\mathbf{I}}_2$ remain orthogonal:

$$-x_1 \sin(\omega t) + x_2 \cos(\omega t) = 0$$

This equation is a holonomic constraint on the generalized coordinates. A constraint which involves an explicit dependence on time t is said to be *rheonomic*; this last constraint is thus rheonomic.

Dependent and independent constraints

When considering in 2.1 the set $q = \{x_1, y_1, \varphi_1, x_2, y_2, \varphi_2, x_3, y_3, \varphi_3\}$ of generalized coordinates for the slider-crank mechanism, we found eight[1] relations (2.2 - 2.9) constraining these generalized coordinates.
Let us now consider an additional constraint. For instance, we could express that the distance AB between the top of the crank-arm and the slider is constant:

$$(x_3 - 2x_1)^2 + (y_3 - 2y_1)^2 = \| \overrightarrow{AB} \|^2 = l^2 \tag{2.13}$$

This "new" constraint is not independent of the previous ones: if the constraints 2.2 - 2.9 are satisfied, then the new one is also automatically satisfied. Indeed, replacing (x_1, y_1) and (x_3, y_3) in the new constraint by their values (2.2, 2.3) and (2.6, 2.7), we find

$$l^2 \cos^2 \varphi_2 + l^2 \sin^2 \varphi_2 = l^2$$

which is a trivial relation. The additional constraint 2.13 is thus *redundant.*

In this example, only a little algebra is necessary to check that the additional constraint is redundant. However in practical situations, it is sometimes difficult to detect whether a given constraint is independent or not. Therefore, we need a systematic mathematical test. Using 2.1 as generalized coordinates, let us formulate this test by writing the constraints (2.2, 2.3) and (2.6, 2.7) and the additional constraint 2.13 in the matrix form

$$h(q) = \begin{pmatrix} q_1 - \frac{r}{2}\cos q_3 \\ q_2 - \frac{r}{2}\sin q_3 \\ q_7 - r\cos q_3 - l\cos q_6 \\ q_8 - r\sin q_3 - l\sin q_6 \\ (q_7 - 2q_1)^2 + (q_8 - 2q_2)^2 - l^2 \end{pmatrix} = 0$$

[1] intentionally, we confess !

By computing the *Jacobian matrix* defined by

$$\frac{\partial h}{\partial q^T} \triangleq \begin{pmatrix} \frac{\partial h_1}{\partial q_1} & \frac{\partial h_1}{\partial q_2} & \cdots & \frac{\partial h_1}{\partial q_n} \\ \cdots & \cdots & \cdots & \cdots \\ \frac{\partial h_m}{\partial q_1} & \cdots & \cdots & \frac{\partial h_m}{\partial q_n} \end{pmatrix}$$

we obtain[2]

$$\frac{\partial h}{\partial q^T} = \begin{pmatrix} 1 & 0 & \frac{r}{2}\,s_3 & 0 & 0 & 0 & 0 & 0 & 0 \\ 0 & 1 & -\frac{r}{2}\,c_3 & 0 & 0 & 0 & 0 & 0 & 0 \\ 0 & 0 & r\,s_3 & 0 & 0 & l\,s_6 & 1 & 0 & 0 \\ 0 & 0 & -r\,c_3 & 0 & 0 & -l\,c_6 & 0 & 1 & 0 \\ 8q_1-4q_7 & 8q_2-4q_8 & 0 & 0 & 0 & 0 & 2q_7-4q_1 & 2q_8-4q_2 & 0 \end{pmatrix}$$

Because of the constraints $q_7 - 2q_1 = l\cos q_6$ and $q_8 - 2q_2 = l\sin q_6$, the rows of this matrix are linearly dependent. Indeed, one may verify that

$$(2 * Row(1) - Row(3)) * 2lc_6 + (2 * Row(2) - Row(4)) * 2ls_6 + Row(5) = 0$$

Thus the Jacobian matrix of the constraints does not have *full rank*[3].

This is an example of a general result proved in the appendix to this chapter. Therefore, we have at our disposal a practical way to verify that a set $h(q,t)$ of constraint functions are independent:

1. find numerical values of the generalized coordinates q which satisfy the constraints $h(q,t) = 0$,

2. compute the rank of the Jacobian matrix,

3. if this Jacobian matrix does not have full rank, i.e. is rank deficient, then the considered constraints are dependent.

If the Jacobian matrix is rank deficient, there are algebraic techniques that allow us to numerically select a full rank submatrix of the Jacobian matrix: the rows of this submatrix correspond to a subset of constraints which are independent. We will thus assume throughout this book that the *Jacobian matrix* of the considered constraints has full rank, i.e. that the constraints are independent.

---

[2] for brevity, $\sin q_i$ and $\cos q_i$ are written s_i and c_i respectively.

[3] The *full* rank of a $(m \times n)$ matrix A, with $m \leq n$, is equal to m.

2.2.2 Generalized velocities and non-holonomic constraints

The system configuration described by means of n generalized coordinates q_α is given by

$$\begin{aligned} \mathbf{x}^k &= [\hat{\mathbf{I}}]^T x^k(q,t) \\ [\hat{\mathbf{X}}^k] &= R^k(q,t)[\hat{\mathbf{I}}] \end{aligned} \qquad \forall \text{ body } k \tag{2.14}$$

where the elements of $x^k(q,t)$ and $R^k(q,t)$ are algebraic functions of q and t. Taking time derivatives, we first compute the velocity vectors of the centers of mass G^k:

$$\dot{\mathbf{x}}^k = [\hat{\mathbf{I}}]^T \dot{x}^k = [\hat{\mathbf{I}}]^T \left(\sum_{\alpha=1}^{n} \frac{\partial x^k(q,t)}{\partial q_\alpha} \dot{q}_\alpha + \frac{\partial x^k(q,t)}{\partial t} \right) \tag{2.15}$$

Then, recalling the definition of angular velocity (1.55 and 1.52):

$$\boldsymbol{\omega}^k \triangleq [\hat{\mathbf{X}}^k]^T \omega^k \quad \text{with} \quad \tilde{\omega}^k = R^k \left(\dot{R}^k \right)^T$$

and computing the time derivatives of the rotation matrices:

$$\dot{R}^k = \sum_{\alpha=1}^{n} \frac{\partial R^k(q,t)}{\partial q_\alpha} \dot{q}_\alpha + \frac{\partial R^k(q,t)}{\partial t}$$

we conclude that the components ω^k of the angular velocity vectors are function of $(\dot{q}, q, t)$. Moreover, they are linear with respect to the $\dot{q}_\alpha$ and can formally be written in the form

$$\boldsymbol{\omega}^k \triangleq [\hat{\mathbf{X}}^k]^T \left(\sum_{\alpha=1}^{n} a_\alpha^k(q,t) \dot{q}_\alpha + b^k(q,t) \right) \tag{2.16}$$

Both linear and angular velocities of the bodies of the system are thus linear[4] functions of the time-derivatives $\dot{q}_\alpha$ of the generalized coordinates q_α; the $\dot{q}_\alpha$ are called the *generalized velocities*.

Note that if the generalized coordinates are constrained by m' holonomic constraints $h(q,t) = 0$, the generalized velocities are not independent. Since all the constraints must hold during the system motion described by the evolution $q(t)$ of the generalized coordinates with respect to time, the functions $h(q(t),t)$ must vanish at all times. Therefore

$$h(q(t),t) \equiv 0 \quad \Rightarrow \quad \frac{d}{dt}\left(h(q(t),t) \right) = 0$$

which in turn implies

$$\frac{\partial h}{\partial q^T} \dot{q} + \frac{\partial h}{\partial t} = 0 \tag{2.17}$$

[4] linear, but non homogenous if there is an explicit time dependence.

Since we assume that the constraints are independent, the Jacobian matrix $\frac{\partial h}{\partial q^T}$ has full rank m', and only $n - m'$ generalized velocities are independent.

Besides the holonomic constraints which are often of geometrical nature, some additional relations may constrain the generalized velocities, as illustrated by the following example.

Disk rolling on a horizontal plane

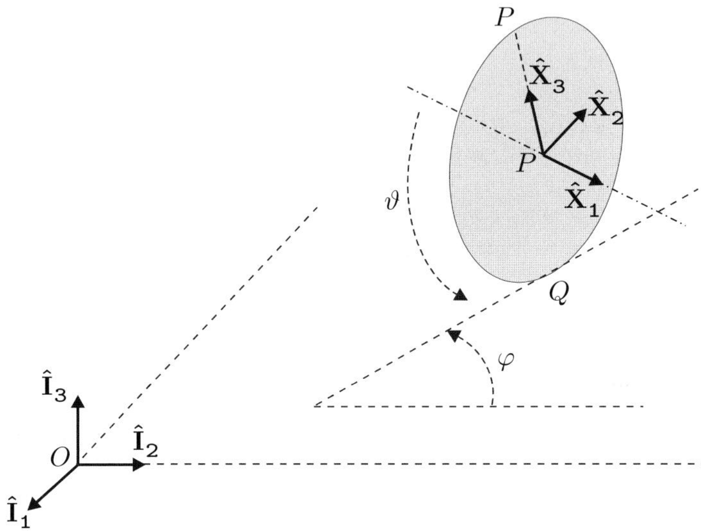

Figure 2.3: Disk rolling on a plane

A thin disk (see figure 2.3) with radius r rolls on the horizontal plane $\{\hat{\mathbf{I}}_1, \hat{\mathbf{I}}_2\}$. The disk is modeled as a two-dimensional body, such that the contact patch between the disk and the plane reduces to a point denoted Q on the figure. The disk is assumed to stay vertical: it is thus rotating around its $\hat{\mathbf{X}}_1$ axis which remains horizontal and forms an angle φ with the inertial axis $\hat{\mathbf{I}}_1$. Under these hypotheses, any configuration of the disk can be described by four generalized coordinates:

$$q = (x, y, \varphi, \vartheta)^T$$

where

- (x, y) locates the contact point Q with respect to the fixed coordinate system $\{O, \{\hat{\mathbf{I}}\}\}$,
- angle φ characterizes the orientation of the vertical plane containing the disk,

- angle ϑ describes the rotation of the disk around its $\hat{\mathbf{X}}_1$ axis.

The system configuration is thus given by

$$\mathbf{x} = [\hat{\mathbf{I}}]^T \begin{pmatrix} x \\ y \\ r \end{pmatrix} \quad \text{and} \quad [\hat{\mathbf{X}}] = R[\hat{\mathbf{I}}] \quad \text{with} \quad R = R_1(\vartheta) R_3(\varphi)$$

where

$$R_1(\vartheta) \triangleq \begin{pmatrix} 1 & 0 & 0 \\ 0 & \cos\vartheta & \sin\vartheta \\ 0 & -\sin\vartheta & \cos\vartheta \end{pmatrix} \quad \text{and} \quad R_3(\varphi) \triangleq \begin{pmatrix} \cos\varphi & \sin\varphi & 0 \\ -\sin\varphi & \cos\varphi & 0 \\ 0 & 0 & 1 \end{pmatrix}$$

If the motion of the disk is a pure rolling motion, no slip may occur between the material point of the disk which occupies the contact position Q and the corresponding point of the fixed horizontal plane. Thus, the absolute velocity of the material point of the disk located at Q must vanish.
We first compute the absolute position of a material point P located at the periphery of the disk (see figure 2.3):

$$\mathbf{x}(P) = \mathbf{x} + r\hat{\mathbf{X}}_3$$

The instantaneous velocity of this point is given by

$$\dot{\mathbf{x}}(P) = \dot{x}\hat{\mathbf{I}}_1 + \dot{y}\hat{\mathbf{I}}_2 + \boldsymbol{\omega} \times r\hat{\mathbf{X}}_3 \quad \text{with} \quad \boldsymbol{\omega} = \dot{\varphi}\hat{\mathbf{I}}_3 + \dot{\vartheta}\hat{\mathbf{X}}_1$$

When $\hat{\mathbf{X}}_3 = -\hat{\mathbf{I}}_3$, this point P occupies the position Q and its velocity

$$\dot{\mathbf{x}}(P) = \dot{x}\hat{\mathbf{I}}_1 + \dot{y}\hat{\mathbf{I}}_2 + r\dot{\vartheta}\hat{\mathbf{X}}_2$$

must vanish. Therefore, by projecting the above expression onto the $\{\hat{\mathbf{X}}\}$ frame in this configuration (i.e. when $\hat{\mathbf{X}}_3 = -\hat{\mathbf{I}}_3$), we obtain

$$\dot{x}\cos\varphi + \dot{y}\sin\varphi = 0 \quad (2.18)$$

$$-\dot{x}\sin\varphi + \dot{y}\cos\varphi + r\dot{\vartheta} = 0 \quad (2.19)$$

These two relations are *non-integrable* constraints on the generalized velocities. By "non-integrable", we mean that it is not possible to find a function $h(q,t)$ so that these constraints have the form 2.17, with integral $h(q,t) = 0$.

Such constraints are called *non-holonomic constraints*. As in the example, they are linear with respect to the generalized velocities for pure mechanical systems[5]. They can thus be written in matrix form as

$$B(q,t)\dot{q} + b(q,t) = 0 \quad (2.20)$$

where B is a $(m'' \times n)$ matrix and b a m'' vector. As for holonomic constraints, non-holonomic constraints are independent if the coefficient matrix B has full rank m''.

[5] This linearity with respect to $\dot{q}$ will be assumed throughout this book.

2.2.3 Degrees of freedom

Consider a system described by:

- n generalized coordinates
- m' holonomic constraints $h(q,t) = 0$
- m'' non-holonomic constraints $B(q,t)\dot{q} + b(q,t) = 0$

Conditions 2.17 and 2.20 globally impose $m \triangleq (m' + m'')$ relations which must be fulfilled by the generalized velocities:

$$\begin{pmatrix} \frac{\partial h}{\partial q^T} \\ B \end{pmatrix} \dot{q} + \begin{pmatrix} \frac{\partial h}{\partial t} \\ b \end{pmatrix} = 0 \tag{2.21}$$

The $(m \times n)$ coefficient matrix of the generalized velocities is the *Jacobian matrix* $J(q,t)$ of the constraints:

$$J(q,t) \triangleq \begin{pmatrix} \frac{\partial h}{\partial q^T} \\ B \end{pmatrix} \tag{2.22}$$

If this Jacobian has full rank, only $(n-m)$ generalized velocities are independent. We thus say that the system has $(n-m)$ motional *degrees of freedom* (also denoted *d.o.f.* for short)[6].

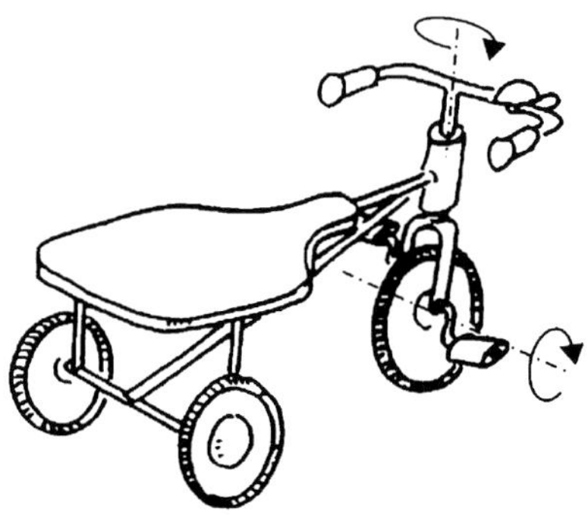

Figure 2.4: A tricycle has two degrees of freedom

It is important to note that the number of degrees of freedom of a given system is *intrinsic:* it does not depend on a particular choice of generalized coordinates. In many cases, we can use our physical intuition to find this number. For instance, the slider-crank mechanism has only one d.o.f.: only a single actuator (acting either on the crankshaft side or the slider side) is necessary

[6] A few authors (see for instance [30], [73]) define $(n - m')$ as the number of *positional* degrees of freedom.

to animate the complete system. Similarly, we can guess that a disk rolling on the plane has two d.o.f. Indeed, comparing this disk with the front wheel of a child's tricycle (see figure 2.4), we know in advance that the child can perform two different actions: cycling (accelerating or decelerating) and turning (left or right). These two actions correspond to the two d.o.f of the disk.
Unfortunately, for more complicated systems, our intuition is sometimes not sufficient: we must first select a set of suitable generalized coordinates, then make a complete inventory of constraints, and finally select a set of independent constraints on the basis of the rank of the Jacobian.

On the other hand, since the Jacobian matrix depends on the configuration q, it may happen that $J(q,t)$ has rank deficiency for some particular value of q. The corresponding configuration is called a *singular* configuration and should be avoided as far as possible in practical applications. For instance, consider the parallelogram of figure 2.5.

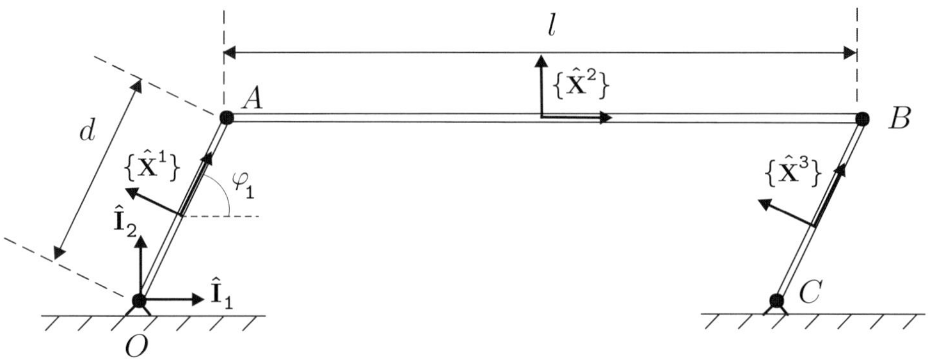

Figure 2.5: A planar parallelogram

Choosing the absolute angles φ_i $(i = 1, 2, 3)$ of frames $\{\hat{\mathbf{X}}^i\}$ as generalized coordinates, we must close the mechanism at point C. This leads to the following two constraints

$$\begin{aligned} h_1(q) &= d\cos\varphi_1 + l\cos\varphi_2 - d\cos\varphi_3 - l = 0 \\ h_2(q) &= d\sin\varphi_1 + l\sin\varphi_2 - d\sin\varphi_3 = 0 \end{aligned}$$

One solution satisfying the constraints, is $\varphi_1 = \varphi_3$ and $\varphi_2 = 0$. The associated Jacobian matrix reads

$$J(q) = \begin{pmatrix} -d\sin\varphi_1 & -l\sin\varphi_2 & d\sin\varphi_3 \\ d\cos\varphi_1 & l\cos\varphi_2 & -d\cos\varphi_3 \end{pmatrix}$$

and has the following form once the constraints are satisfied

$$J(q) = \begin{pmatrix} -d\sin\varphi_1 & 0 & d\sin\varphi_1 \\ d\cos\varphi_1 & l & -d\cos\varphi_1 \end{pmatrix}$$

When $\varphi_1 = 0$, the Jacobian becomes

$$J(q) = \begin{pmatrix} 0 & 0 & 0 \\ d & l & -d \end{pmatrix}$$

and is singular. The rank of the Jacobian being equal to one, the system *locally* has $3 - 1 = 2$ degrees of freedom. The parallelogram lies in the *singular* configuration shown in figure 2.6. The first degree of freedom corresponds to an individual vertical motion of point B (while A stays at rest). Conversely, A is moving while B is at rest for the second d.o.f.

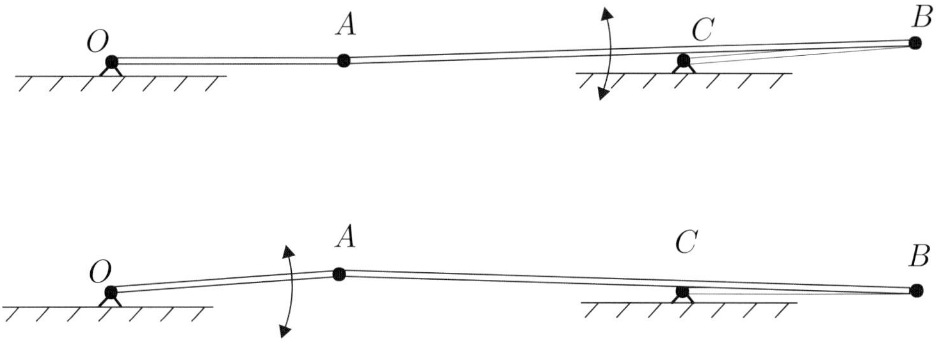

Figure 2.6: Singular configuration of a parallelogram

2.3 Newton-Euler procedure

In this section, we describe the different stages covered in modeling a mechanical system made up of several rigid bodies. From this model, we obtain the equations which allow us to predict the system behavior starting from a given initial state. We also show how to find the values of some unknown forces and/or torques during this motion.

2.3.1 The procedure

Step 1: System analysis

The system is first decomposed into N rigid bodies $C_{i=1,\dots,N}$. These bodies must be clearly identified and, for each one, the center of mass G^i and a body-fixed frame $\{\hat{\mathbf{X}}^i\}$ (aligned if possible with the principal axis of inertia) must be defined. A convenient inertial frame $\{O, \{\hat{\mathbf{I}}\}\}$ is also defined at this step.

Step 2: Choice of generalized coordinates and formulation of constraints

The choice of generalized coordinates governs the resulting form of all the equations required to solve the problem; it will thus influence the ease (or the difficulty) of the final mathematical (and even numerical) treatment. To make this choice, we must deal with two issues which are often contradictory:

- to reduce the number of unknowns, we must try to introduce as few generalized coordinates as possible,
- our set of generalized coordinates must be sufficient to describe the system configuration, i.e. the functions $x^i(q,t)$ and $R^i(q,t)$ with $i = 1, ..., N$. Since time-derivatives of these analytical functions are also required, these functions should preferably have simple forms.

As illustrated in section 2.2.1 for the slider-crank mechanism, accommodation is often needed to balance these two requirements: the number of generalized coordinates is not chosen to be the minimum and the coordinates are not independent.

The maximum number of generalized coordinates we can use is equal to $6N$ ($3N$ for a 2-D problem), i.e. the number of configuration variables. Let us suppose that we choose n generalized coordinates only, because there are $6N - n$ geometrical relations for expressing all the functions $x^i(q,t)$ and $R^i(q,t)$. These $6N - n$ relations are in fact $6N - n$ (holonomic) constraints which are *implicitly satisfied* thanks to our choice of generalized coordinates.

If the system has less than n degrees of freedom, we still have to formulate some additional constraints to model the physics of the actual problem. These constraints *explicitly* involve the generalized coordinates and/or velocities. Suppose that we have:
- m' holonomic constraints on the generalized coordinates: $h(q,t) = 0$,
- m'' non-holonomic constraints on the generalized velocities: $B(q,t)\dot{q} + b(q,t) = 0$,
we must check that the value $n - (m' + m'')$ is equal to the number of degrees of freedom.

Step 3: Kinematic computation

Starting from the expressions of $x^i(q,t)$ and $R^i(q,t)$, we need to compute the following quantities:

- the linear momentum vector $\mathbf{N}^i$ of the various bodies C^i, according to 1.75,
- their angular momentum vector $\mathbf{H}^i$ with respect to their center of mass G^i according to 1.79, or equivalently with respect to some fixed point O by means of 1.80,

- the time-derivatives of these vectors.

Step 4: Inventory of forces and torques, and computation of their resultants

The inventory of forces and torques must be complete. Therefore, we recommend considering each body individually and determining:

- the *external* forces (and torques) applied to the system by the environment, for example, the forces due to gravity, ground contact,...
- the forces (and torques) which are exerted on the body by the other bodies belonging to the system. These forces are *internal* to the system and satisfy Newton's third law (see section 1.3): for any action, there is an equal and opposite reaction. Since the reactions lie within the system, these internal forces (and torques) will thus appear twice, with opposite signs and acting on different bodies, in the complete inventory.

We must also make a distinction between two kinds of forces (and torques):

- some forces (torques) are known functions of the positions (orientations) and/or velocities of some given points of the system. They can thus be expressed analytically in terms of the generalized coordinates and/or velocities by means of *constitutive equations.* This is the case for the gravity force, or for the force exerted by a spring attached to the system.
- other forces (and torques) are *a priori* unknown: they are the *constraint* forces (and torques[7]) which compel the system to satisfy all its constraints. The implicit constraints, as well as the explicit constraints relating the generalized coordinates (or velocities) must be considered here in order to obtain the complete inventory of forces (and torques).
  Examples of such constraint forces can be found in the problem of the disk rolling on a horizontal plane (see figure 2.3). The constraint 2.18 states that the disk does not slip in the transverse direction $\hat{\mathbf{X}}_1$. Physically, no slip occurs in this direction because a force is applied by the ground (the horizontal plane) along this direction. This force is a constraint force: while its *effect* is a priori known, the value necessary to produce this effect is not known[8]. Similarly, the constraint 2.19 states that the disk does not slip longitudinally: another physical force is applied by the ground in the longitudinal direction to prevent this longitudinal slip. More generally, one may consider that every holonomic or non-holonomic (scalar) constraint implies one component of constraint force or constraint torque.

When we have completed the inventory, we must compute for each body i the resultant vector force $\mathbf{F}^i$, as well as the resultant torque $\mathbf{L}^i$ (with respect to its center of mass or some inertially fixed point).

[7] From now on, constraint forces and torques will be simply denoted as "constraint forces" for brevity.

[8] In fact, the value necessary to avoid slippage depends on the system motion.

Step 5: Equations of motion

This step simply consists in writing for each body i the Newton equation 1.94 and one of the Euler equations (1.95 or 1.97). This leads to $2N$ vector equations involving different vectors and tensors. The example we give in the next section shows that choosing appropriate reference frames facilitates the vector/tensor computation within these dynamical equations.

Finally, we must project each of these $2N$ vector equations onto an appropriate frame. This leads to $6N$ scalar equations[9].

Step 6: Solution of the equations

The $6N$ dynamical equations involve the generalized coordinates, their first and second derivatives, as well as the unknown components of the various constraint forces. The total number of equations at our disposal is as follows: $6N$ (or $3N$ for 2-D systems) dynamical equations, plus $m \triangleq (m'+m'')$ constraint equations. Since differential equations are associated with algebraic equations, we are dealing with a set of *differential-algebraic equations* (abbreviated to "DAEs").
On the other hand, the total number of unknowns is given by the n generalized coordinates and the unknown components of constraint forces. There are at least $6N - n$ (or $3N - n$) unknown components which correspond to the implicit constraints, and m components which correspond to the explicit constraints. The total number of unknowns is thus greater than or equal to $6N + m$ (or $3N + m$ for 2-D systems).
Since we have at our disposal $6N$ dynamical equations and m constraint equations, the mathematical model involves a number of unknowns which is at least equal to the number of equations. In general, these two numbers are equal and the system can be mathematically solved.

- In order to predict the system motion resulting from some given initial state, all constraint force and torque components must first be eliminated. Since these components appear linearly in the dynamical equations, this operation can be performed by means of linear combinations. This leads to a reduced system containing ordinary differential equations (ODEs) or differential-algebraic equations (DAEs) in the generalized coordinates q. In most cases, this system must then be integrated numerically. Further details concerning the solution of DAEs will be given in section 4.6.1 and in chapter 10.

- Knowing the system motion, i.e. the functions $q(t)$, $\dot{q}(t)$ and $\ddot{q}(t)$, we can then compute the values of the constraint forces. If the number of unknowns is equal to the number of equations ($6N + m$ or $3N + m$), the system is said to be *isostatic*. The system equations are linear with

[9] For 2-D systems, the number of useful equations reduces to $3N$, i.e. the in-plane components of the translational equations and the orthogonal component of the rotational equations. The remaining $3N$ equations are trivial.

respect to the constraint forces and the solution is unique. If there are more unknowns than equations, the system is *hyperstatic* and some of the constraint forces cannot be uniquely determined[10].

2.3.2 An example

In order to illustrate the Newton-Euler procedure, consider the 2-D example illustrated in figure 2.7. The system consists of three rigid bodies moving in the $\{\hat{\mathbf{I}}_1, \hat{\mathbf{I}}_3\}$ plane:

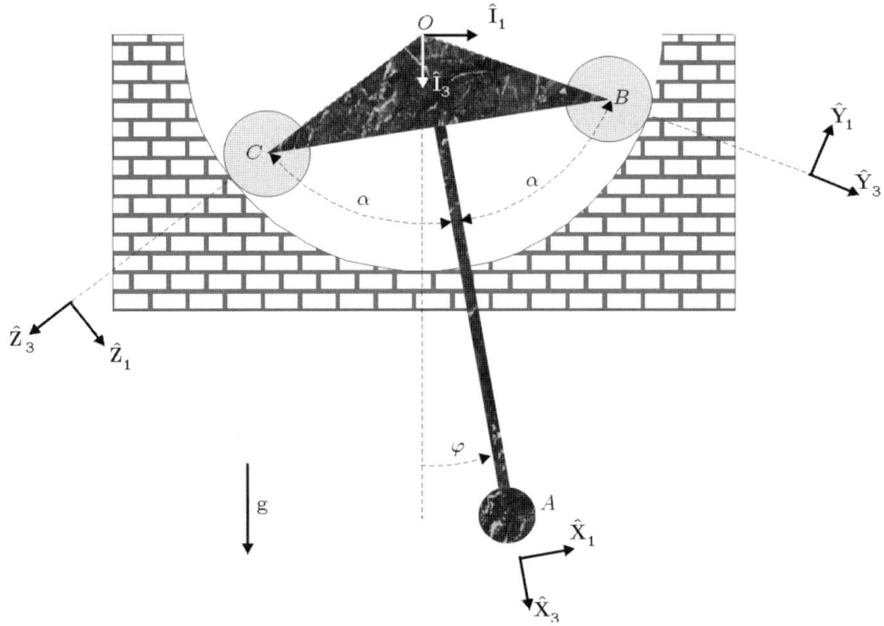

Figure 2.7: Pendulum carried by rollers

- body 1 is a pendulum which consists of a triangular plate OBC, a rod OA and a point mass M located at point A. These three elements are rigidly assembled. Moreover, the plate and the rod are both considered as massless elements (i.e. their masses are considered to be negligible compared with M),
- bodies 2 and 3 are the rollers sustaining the pendulum. They are modeled as homogeneous disks with radius a and mass m.

The rollers are fixed to the pendulum by means of two revolute joints, respectively located at B and C. They roll without slipping on a circular ring.

[10] without dropping the assumption of *rigid* bodies.

The radius of the ring is such that point O, which is fixed to the pendulum, permanently coincides with the center of the ring. Therefore, although it is not mechanically anchored, this point O can be considered as inertially fixed. The whole mechanism lies in a vertical plane, and the gravity field $\mathbf{g}$ is aligned with the vertical direction of $\hat{\mathbf{I}}_3$. Other useful geometrical dimensions are given by:

- b, the length of the segments OB and OC,
- α, the angle $AOB = COA$,
- and l, the length OA of the pendulum.

Suppose that we are interested in knowing the motion of this pendulum for given initial conditions. We follow the different steps described in the previous section.

Step 1: System analysis

This step has already been covered when stating the problem.

Step 2: Choice of generalized coordinates and formulation of constraints

The position and the orientation of body 1 (the pendulum) can both be expressed as functions of the angle φ (see figure 2.7):

$$\mathbf{x}^1 = l\,\hat{\mathbf{X}}_3 \tag{2.23}$$

$$[\hat{\mathbf{X}}] = R_2(\varphi)[\hat{\mathbf{I}}] = \begin{pmatrix} \cos\varphi & 0 & -\sin\varphi \\ 0 & 1 & 0 \\ \sin\varphi & 0 & \cos\varphi \end{pmatrix} [\hat{\mathbf{I}}] \tag{2.24}$$

The positions of the two rollers (bodies 1 and 2) are also described by this angle:

$$\mathbf{x}^2 = b\,\hat{\mathbf{Y}}_3 \quad \text{and} \quad \mathbf{x}^3 = b\,\hat{\mathbf{Z}}_3 \tag{2.25}$$

$$\text{with} \quad [\hat{\mathbf{Y}}] = R_2(\varphi+\alpha)[\hat{\mathbf{I}}] \quad \text{and} \quad [\hat{\mathbf{Z}}] = R_2(\varphi-\alpha)[\hat{\mathbf{I}}] \tag{2.26}$$

Two additional angles are needed to describe the absolute orientations of the rollers: φ_2 for the roller centered at point B, φ_3 for the roller centered at C.

$$[\hat{\mathbf{X}}^2] = R_2(\varphi_2)[\hat{\mathbf{I}}] \quad \text{and} \quad [\hat{\mathbf{X}}^3] = R_2(\varphi_3)[\hat{\mathbf{I}}], \tag{2.27}$$

where the frames $\{\hat{\mathbf{X}}^2\}$ and $\{\hat{\mathbf{X}}^3\}$ (not shown in figure 2.7) are rigidly attached to the right and left rollers respectively.

The set of generalized coordinates $q \triangleq (\varphi, \varphi_2, \varphi_3)$ is sufficient to describe the configuration of the entire system. Note that some constraints have already been taken into account. Since n ($n = 3$) generalized coordinates are used to describe the configurations of N ($N = 3$) bodies moving in a plane, the number of constraints which have been implicitly used is equal to $3N - n = 6$. Indeed, when expressing $\mathbf{x}^1$ we used the fact that point O is inertially fixed in the vertical plane: this corresponds to two constraints. Physically, the constraint forces which compel O to be fixed are the two normal forces (respectively along the $\hat{\mathbf{Y}}_3$ and $\hat{\mathbf{Z}}_3$ directions) exerted by the ring on the rollers in order to support the pendulum. According to our classification, these two forces are "external forces" applied to the system. When expressing the positions $\mathbf{x}^2$ and $\mathbf{x}^3$, we also used the fact that the rollers are fixed on the pendulum by means of revolute joints. The rollers are allowed to rotate freely around their respective centers A and B. However these two centers are rigidly fixed to the pendulum: in the plane, this corresponds to four constraints. The total number of constraints that we have implicitly used is thus 6.

In addition to these implicit constraints, two explicit constraints arise from the non-slipping conditions which are imposed on the rollers. To obtain their analytical expressions, we must state that the velocity of the material point of a roller which is in contact with the ring is zero. For the roller centered at B, the position of a point P located at its periphery is given by

$$\mathbf{x}^P = b\,\hat{\mathbf{Y}}_3 + a\,\hat{\mathbf{X}}_3^2$$

The velocity of this point is thus equal to

$$\begin{aligned} \dot{\mathbf{x}}^P &= \dot{\varphi} b\,\hat{\mathbf{Y}}_2 \times \hat{\mathbf{Y}}_3 + \dot{\varphi}_2 a\,\hat{\mathbf{X}}_2^2 \times \hat{\mathbf{X}}_3^2 \\ &= \dot{\varphi} b\,\hat{\mathbf{Y}}_1 + \dot{\varphi}_2 a\,\hat{\mathbf{X}}_1^2 \end{aligned}$$

This point is in contact with the ring when $\hat{\mathbf{X}}_3^2 = \hat{\mathbf{Y}}_3$ and its velocity then becomes

$$\dot{\mathbf{x}}^P = \dot{\varphi} b\,\hat{\mathbf{Y}}_1 + \dot{\varphi}_2 a\,\hat{\mathbf{Y}}_1$$

For this roller, the non-slipping condition thus implies the constraint

$$\dot{\varphi} b + \dot{\varphi}_2 a = 0 \tag{2.28}$$

and, similarly, we obtain for the second roller

$$\dot{\varphi} b + \dot{\varphi}_3 a = 0 \tag{2.29}$$

Written in matrix form:

$$\begin{pmatrix} b & a & 0 \\ b & 0 & a \end{pmatrix} \begin{pmatrix} \dot{\varphi} \\ \dot{\varphi}_2 \\ \dot{\varphi}_3 \end{pmatrix} = 0 \tag{2.30}$$

these two constraints are clearly independent because the coefficient matrix of the generalized velocities has full row rank. The system under consideration therefore has $3 - 2 = 1$ degree of freedom, as expected from physical reasons.

Step 3: Kinematic computation

For body 1, we obtain

$$\begin{aligned} \mathbf{x}^1 &= l\,\hat{\mathbf{X}}_3 \quad \text{and} \quad [\hat{\mathbf{X}}] = R_2(\varphi)\,[\hat{\mathbf{I}}] \\ \boldsymbol{\omega}^1 &= \dot{\varphi}\hat{\mathbf{X}}_2 = \dot{\varphi}\hat{\mathbf{I}}_2 \\ \dot{\mathbf{x}}^1 &= l\,\dot{\varphi}\,\hat{\mathbf{X}}_1 \\ \ddot{\mathbf{x}}^1 &= l\,\ddot{\varphi}\,\hat{\mathbf{X}}_1 - l\,\dot{\varphi}^2\,\hat{\mathbf{X}}_3 \end{aligned}$$

$\mathbf{H}^{G^1} = 0$, because all the mass is assumed to be concentrated at point A.
Similarly, we obtain for body 2

$$\begin{aligned}\mathbf{x}^2 &= b\,\hat{\mathbf{Y}}_3 \quad \text{and} \quad [\hat{\mathbf{X}}^2] = R_2\,(\varphi_2)\,[\hat{\mathbf{I}}] \\ \boldsymbol{\omega}^2 &= \dot{\varphi}_2 \hat{\mathbf{X}}_2^2 = \dot{\varphi}_2 \hat{\mathbf{I}}_2 \\ \dot{\mathbf{x}}^2 &= b\,\dot{\varphi}\,\hat{\mathbf{Y}}_1 \\ \ddot{\mathbf{x}}^2 &= b\,\ddot{\varphi}\,\hat{\mathbf{Y}}_1 - b\,\dot{\varphi}^2\,\hat{\mathbf{Y}}_3\end{aligned}$$

The inertia tensor of a 2-D homogenous disk lying in a $\{\hat{\mathbf{X}}_3, \hat{\mathbf{X}}_1\}$ plane being given by

$$\mathbf{I}^{G^2} = [\hat{\mathbf{X}}]^T \begin{pmatrix} \frac{ma^2}{4} & 0 & 0 \\ 0 & \frac{ma^2}{2} & 0 \\ 0 & 0 & \frac{ma^2}{4} \end{pmatrix} [\hat{\mathbf{X}}] \tag{2.31}$$

we obtain

$$\mathbf{H}^{G^2} = \frac{ma^2}{2} \dot{\varphi}_2 \hat{\mathbf{X}}_2^2 \quad \text{and} \quad \dot{\mathbf{H}}^{G^2} = \frac{ma^2}{2} \ddot{\varphi}_2 \hat{\mathbf{X}}_2^2$$

The kinematic expressions for the second roller (body 3) are obtained by analogy

$$\begin{aligned}\mathbf{x}^3 &= b\,\hat{\mathbf{Z}}_3 \quad \text{and} \quad [\hat{\mathbf{X}}^3] = R_2\,(\varphi_3)\,[\hat{\mathbf{I}}] \\ \boldsymbol{\omega}^3 &= \dot{\varphi}_3 \hat{\mathbf{X}}_2^3 = \dot{\varphi}_3 \hat{\mathbf{I}}_2 \\ \dot{\mathbf{x}}^3 &= b\,\dot{\varphi}\,\hat{\mathbf{Z}}_1 \\ \ddot{\mathbf{x}}^3 &= b\,\ddot{\varphi}\,\hat{\mathbf{Z}}_1 - b\,\dot{\varphi}^2\,\hat{\mathbf{Z}}_3 \\ \mathbf{H}^{G^3} &= \frac{ma^2}{2} \dot{\varphi}_3 \hat{\mathbf{X}}_2^3 \quad \text{and} \quad \dot{\mathbf{H}}^{G^3} = \frac{ma^2}{2} \ddot{\varphi}_3 \hat{\mathbf{X}}_2^3\end{aligned}$$

Step 4: Inventory of forces and torques, and computation of their resultants

At this point, and for reasons of convenience, each force is expressed in the most appropriate frame.

- Body 1
  - resultant of gravity forces, applied to the center of mass A: $Mg\,\hat{\mathbf{I}}_3$
  - constraint force (action) exerted by body 2 at point B: $\mathbf{S}^2 = S_1^2\,\hat{\mathbf{Y}}_1 + S_3^2\,\hat{\mathbf{Y}}_3$
  - constraint force exerted by body 3 at point C: $\mathbf{S}^3 = S_1^3\,\hat{\mathbf{Z}}_1 + S_3^3\,\hat{\mathbf{Z}}_3$
- Body 2
  - resultant of gravity forces, applied to the center of mass B: $mg\,\hat{\mathbf{I}}_3$

  - constraint force (reaction) exerted by body 1 at point B: $-\mathbf{S}^2$
  - constraint force exerted by the ring at the contact point: $N^2\,\hat{\mathbf{Y}}_3 + T^2\,\hat{\mathbf{Y}}_1$

- Body 3
  - resultant of gravity forces, applied to the center of mass C: $mg\,\hat{\mathbf{I}}_3$
  - constraint force (reaction) exerted by body 1 at point C: $-\mathbf{S}^3$
  - constraint force exerted by the ring at the contact point: $N^3\,\hat{\mathbf{Z}}_3 + T^3\,\hat{\mathbf{Z}}_1$

The moments of these forces around the respective centers of mass lead to the following resultant torques

$$\begin{aligned}
\mathbf{L}^1 &= \overrightarrow{AB} \times \mathbf{S}^2 + \overrightarrow{AC} \times \mathbf{S}^3 \\
\mathbf{L}^2 &= a\,\hat{\mathbf{Y}}_3 \times \left(N^2\,\hat{\mathbf{Y}}_3 + T^2\,\hat{\mathbf{Y}}_1\right) = a\,T^2\,\hat{\mathbf{Y}}_2 \\
\mathbf{L}^3 &= a\,\hat{\mathbf{Z}}_3 \times \left(N^3\,\hat{\mathbf{Z}}_3 + T^3\,\hat{\mathbf{Z}}_1\right) = a\,T^3\,\hat{\mathbf{Z}}_2
\end{aligned}$$

Since only the gravity forces have known constitutive equations, 8 force components are unknown in these expressions: S_1^2, S_3^2, S_1^3, S_3^3, N^2, T^2, N^3 and T^3. Each component corresponds to a constraint which is either explicitly or implicitly applied to the system.

Step 5

For each body, we obtain 2 vector equations of motion by combining the outputs of steps 3 and 4. The vector quantities involved in these equations are still expressed in different frames. Each vector equation will now be projected onto a (convenient) single frame so as to obtain its scalar form.

- Body 1

The translation (Newton) equation reads

$$M\ddot{\mathbf{x}}^1 = Ml\ddot{\varphi}\,\hat{\mathbf{X}}_1 - Ml\dot{\varphi}^2\,\hat{\mathbf{X}}_3 = Mg\,\hat{\mathbf{I}}_3 + S_1^2\,\hat{\mathbf{Y}}_1 + S_3^2\,\hat{\mathbf{Y}}_3 + S_1^3\,\hat{\mathbf{Z}}_1 + S_3^3\,\hat{\mathbf{Z}}_3$$

and will now be expressed in the single frame $\{\hat{\mathbf{X}}\}$:

$$\begin{aligned}
Ml\ddot{\varphi}\,\hat{\mathbf{X}}_1 - Ml\dot{\varphi}^2\,\hat{\mathbf{X}}_3 = {} & \hat{\mathbf{X}}_1\left(-Mg\sin\varphi + \left(S_1^2 + S_1^3\right)\cos\alpha + \left(S_3^2 - S_3^3\right)\sin\alpha\right) \\
& + \hat{\mathbf{X}}_3\left(Mg\cos\varphi + \left(S_1^3 - S_1^2\right)\sin\alpha + \left(S_3^2 + S_3^3\right)\sin\alpha\right)
\end{aligned}$$

Since $\dot{\mathbf{H}}^1 = 0$, the rotation equation (Euler) around the center of mass is given by

$$\begin{aligned}
0 &= \overrightarrow{AB} \times \mathbf{S}^2 + \overrightarrow{AC} \times \mathbf{S}^3 \\
&= \left(-l\,\hat{\mathbf{X}}_3 + b\,\hat{\mathbf{Y}}_3\right) \times \mathbf{S}^2 + \left(-l\,\hat{\mathbf{X}}_3 + b\,\hat{\mathbf{Z}}_3\right) \times \mathbf{S}^3
\end{aligned}$$

which, when expressed in the $\{\hat{\mathbf{X}}\}$ frame, becomes

$$\hat{\mathbf{X}}_2\left((b - l\cos\alpha)\left(S_1^2 + S_1^3\right) + l\sin\alpha\left(S_3^3 - S_3^2\right)\right) = 0$$

- Body 2

The translation equation:

$$m\ddot{\mathbf{x}}^2 = m\,b\,\ddot{\varphi}\,\hat{\mathbf{Y}}_1 - m\,b\,\dot{\varphi}^2\,\hat{\mathbf{Y}}_3 = mg\,\hat{\mathbf{I}}_3 - \mathbf{S}^2 + N^2\,\hat{\mathbf{Y}}_3 + T^2\,\hat{\mathbf{Y}}_1$$

is expressed in the $\{\hat{\mathbf{Y}}\}$ frame, which is the most convenient:

$$\begin{aligned} mb\ddot{\varphi}\,\hat{\mathbf{Y}}_1 - mb\dot{\varphi}^2\,\hat{\mathbf{Y}}_3 &= \hat{\mathbf{Y}}_1\left(-mg\sin(\varphi+\alpha) - S_1^2 + T^2\right) \\ &\quad + \hat{\mathbf{Y}}_3\left(mg\cos(\varphi+\alpha) - S_3^2 + N^2\right) \end{aligned}$$

The rotation equation is given by

$$\dot{\mathbf{H}}^2 = \frac{ma^2}{2}\ddot{\varphi}_2\hat{\mathbf{X}}_2^2 = a\,T^2\,\hat{\mathbf{Y}}_2$$

where the axes $\hat{\mathbf{X}}_2^2$ and $\hat{\mathbf{Y}}_2$ are aligned.

- Body 3

Using symmetry, this is simply a matter of changing the indices, replacing α by $-\alpha$ and replacing the frame $\{\hat{\mathbf{Z}}\}$ by $\{\hat{\mathbf{Y}}\}$ to obtain the translation and rotation equations

$$\begin{aligned} mb\ddot{\varphi}\,\hat{\mathbf{Z}}_1 - mb\dot{\varphi}^2\,\hat{\mathbf{Z}}_3 &= \hat{\mathbf{Z}}_1\left(-mg\sin(\varphi-\alpha) - S_1^3 + T^3\right) \\ &\quad + \hat{\mathbf{Z}}_3\left(mg\cos(\varphi-\alpha) - S_3^3 + N^3\right) \\ \dot{\mathbf{H}}^3 &= \frac{ma^2}{2}\ddot{\varphi}_3\hat{\mathbf{X}}_2^3 = a\,T^3\,\hat{\mathbf{Z}}_2 \end{aligned}$$

where $\hat{\mathbf{X}}_2^3 = \hat{\mathbf{Z}}_2$.

Step 6: Solution of the equations

From these vector equations, we immediately obtain the following 9 scalar equations

$$Ml\ddot{\varphi} = -Mg\sin\varphi + \left(S_1^2 + S_1^3\right)\cos\alpha + \left(S_3^2 - S_3^3\right)\sin\alpha \tag{2.32}$$

$$-Ml\dot{\varphi}^2 = Mg\cos\varphi + \left(S_1^3 - S_1^2\right)\sin\alpha + \left(S_3^2 + S_3^3\right)\sin\alpha \tag{2.33}$$

$$0 = (b - l\cos\alpha)\left(S_1^2 + S_1^3\right) + l\sin\alpha\left(S_3^3 - S_3^2\right) \tag{2.34}$$

$$mb\ddot{\varphi} = -mg\sin(\varphi+\alpha) - S_1^2 + T^2 \tag{2.35}$$

$$-mb\dot{\varphi}^2 = mg\cos(\varphi+\alpha) - S_3^2 + N^2 \tag{2.36}$$

$$\frac{ma^2}{2}\ddot{\varphi}_2 = a\,T^2 \tag{2.37}$$

$$mb\ddot{\varphi} = -mg\sin(\varphi-\alpha) - S_1^3 + T^3 \tag{2.38}$$

$$-mb\dot{\varphi}^2 = mg\cos(\varphi-\alpha) - S_3^3 + N^3 \tag{2.39}$$

$$\frac{ma^2}{2}\ddot{\varphi}_3 = a\,T^3 \tag{2.40}$$

which, with the constraint equations 2.28 and 2.29, lead to a total number of 11 DAE equations for the 3 generalized coordinates and the 8 unknown components of constraint forces.

This set of equations must first be used to determine the system motion (for given initial conditions on the generalized coordinates and velocities). Then, if required, they may also be used to find the values of constraint forces during that motion.

- Equation of motion

Although the constraint equations are due to velocity constraints (non-slipping conditions), they are integrable in the present case[11]:

$$b\dot{\varphi} + a\dot{\varphi}_2 = 0 \quad \Rightarrow \quad \varphi_2(t) - \varphi_2(0) = -\frac{b}{a}\left(\varphi(t) - \varphi(0)\right) \tag{2.41}$$

$$b\dot{\varphi} + a\dot{\varphi}_3 = 0 \quad \Rightarrow \quad \varphi_3(t) - \varphi_3(0) = -\frac{b}{a}\left(\varphi(t) - \varphi(0)\right) \tag{2.42}$$

and one may thus expect that a single equation in the generalized coordinate φ (and its derivatives) will be sufficient to describe the motion. Taking the derivative of the constraint equations 2.28 and 2.29, we obtain the accelerations $\ddot{\varphi}_2$ and $\ddot{\varphi}_3$:

$$\ddot{\varphi}_2 = \ddot{\varphi}_3 = -\frac{b}{a}\,\ddot{\varphi} \tag{2.43}$$

which are easily eliminated from the set of equations 2.32 - 2.40.

The 8 constraint forces also must be eliminated. Since they appear linearly in the equations, the elimination can be performed by means of a linear combination. Indeed, the reader may check that the following combination

$$l * (2.32) + b * (2.35) - \frac{b}{a} * (2.37) + b * (2.38) - \frac{b}{a} * (2.40) \tag{2.44}$$

leads to, using 2.43 and 2.34,

$$\left(Ml^2 + 3mb^2\right)\ddot{\varphi} + g\left(Ml + 2mb\cos\alpha\right)\sin\varphi = 0 \tag{2.45}$$

which is the required equation of motion of the pendulum. Solving this differential equation for a given initial state $(\varphi(0), \dot{\varphi}(0))$ provides the time history of $\varphi(t)$.

- Constraint forces

Knowing the time history of $\varphi(t)$ (and thus of its derivatives $\dot{\varphi}(t)$ and $\ddot{\varphi}(t)$), it becomes possible to compute the value of the constraint forces from the set of dynamical equations 2.32 - 2.40. For instance, the unknown component S_1^2 is obtained by eliminating T^2 from equations 2.35 and 2.37, using 2.43:

$$S_1^2(t) = -\frac{3}{2}mb\ddot{\varphi}(t) - mg\sin\left(\varphi(t) + \alpha\right) \tag{2.46}$$

[11] This holds because we are dealing with a non-slipping condition in a 2-D system.

- Conclusion

This particularly simple example - a small number of rigid bodies moving in a plane - shows that the number of equations which are necessary to solve the problem may grow rapidly and thus become difficult to handle. This example also shows that all these equations are highly coupled. Computer-aided tools are thus necessary to model practical problems which often involve many bodies.

We emphasize that the final equation of motion 2.45 was obtained by eliminating the constraint forces by means of the linear combination 2.44. For more complicated systems, finding such a "good" linear combination is not trivial: this justifies the variational approach we will be presenting in the next section.

2.4 Variational Approach

The method presented below is known as the "principle of virtual power" or "Jourdain's principle" ([64]). This principle is strictly equivalent to d'Alembert's Principle which is generally proposed in the literature. While d'Alembert's Principle uses the notion of virtual infinitesimal displacements, the virtual power principle involves virtual changes in velocities[12]. In the authors' opinion, the latter are easier to visualize and to understand than infinitesimal virtual displacements, particularly when rotational motions are involved.

As in the previous sections, the virtual power principle will be introduced by means of examples in order to underline the advantages of using this variational approach. Rigorous proofs can be found in the literature (see for instance [64]) and will not be repeated here.

2.4.1 Virtual power principle

When dealing with the pendulum example in the previous section, we concluded that even if sufficient attention is paid to reduce the number of generalized coordinates, the total number of equations rapidly becomes large when the number of bodies increases. The total number of unknowns also increases, this mainly being due to the unknown constraint forces which must be eliminated

[12] Generally speaking, this principle represents projections of the (set of local) equations of motion onto the tangent space of the configuration manifold. The various formulations of this principle - for instance as presented by [64] - differ only in the choice of local coordinates in the tangent space; this space being linear, any (regular) linear transformation can be considered, and no restriction to "infinitesimal" coordinates or small virtual motions is necessary.

From a historical point of view, this principle was already known and used in statics during the 18th century. In particular, it appears explicitly in a letter from Jean Bernouilli to P. Vigneron - dated January 26, 1717 and published in a posthumous book by P. Vigneron in 1725 [90]. It can be considered as a basic principle in classical mechanics and leads to the Lagrange equations (in this case the tangent space coordinates are the generalized velocities or equivalently virtual infinitesimal changes in the generalized coordinates). Extensions to generalized pseudo-velocities were made by Bolzmann [3] and Hamel [37] at the beginning of the 20th century. For rigid body dynamics, the equations obtained via this principle - using angular velocity components as pseudo-velocities - are referred to as Kane's equations [46] in the American literature.

during the last solution stage in order to reach the final form of the equations of motion. Up to now, we have concentrated our efforts on the geometrical and kinematic aspects of the problem by making "good choices" of generalized coordinates; the dynamical aspects must now be considered so as to reduce the number of (useless) dynamical equations.

For a mechanical system composed of N rigid bodies, we start from the $2N$ vector equations:
- Newton translation equations:

$$m^i\ddot{\mathbf{x}}^i - \mathbf{F}^i = 0 \quad \text{for } i = 1, ..., N \tag{2.47}$$

where $\mathbf{F}^i$ represents the resultant force (see section 1.3) acting at the body i center of mass G^i
- Euler rotation equations, for instance[13] around the respective bodies centers of mass:

$$\dot{\mathbf{H}}^i - \mathbf{L}^i = 0 \quad \text{for } i = 1, ..., N \tag{2.48}$$

where $\mathbf{L}^i$ is the resultant torque (with respect to the center of mass) applied to body i.

In order to reduce the number of equations, we will start by selecting some linear combinations of these fundamental dynamical equations. Using vector notations, we can write them as

$$\sum_{i=1}^{N}\left(m^i\ddot{\mathbf{x}}^i - \mathbf{F}^i\right).\mathbf{a}^i + \sum_{i=1}^{N}\left(\dot{\mathbf{H}}^i - \mathbf{L}^i\right).\mathbf{b}^i = 0 \tag{2.49}$$

where the (components of the) vectors $\mathbf{a}^i$ and $\mathbf{b}^i$ contain arbitrary coefficients.

When we add quantities in an equation, they must have the same dimensions. To make the two sums in 2.49 have the same dimension, we choose the vectors $\mathbf{a}^i$ to be linear velocities and the vectors $\mathbf{b}^i$ to be angular velocities. These two coefficient vectors will thus be considered as "virtual velocities", or more exactly[14] as *virtual velocity changes*. In view of their dimensions, they will be denoted as

$$\mathbf{a}^i \triangleq \Delta\dot{\mathbf{x}}^i \quad \text{and} \quad \mathbf{b}^i \triangleq \Delta\boldsymbol{\omega}^i \tag{2.50}$$

Special attention must be paid to the fact that these vectors are just arbitrary coefficient vectors: therefore, no confusion should arise between them and the *actual* velocities $\dot{\mathbf{x}}$ and $\boldsymbol{\omega}^i$ of the various bodies. Replacing $\mathbf{a}^i$ and $\mathbf{b}^i$ by their representations 2.50, we state 2.49 as

$$\sum_{i=1}^{N}\left(m^i\ddot{\mathbf{x}}^i - \mathbf{F}^i\right).\Delta\dot{\mathbf{x}}^i + \sum_{i=1}^{N}\left(\dot{\mathbf{H}}^i - \mathbf{L}^i\right).\Delta\boldsymbol{\omega}^i = 0; \tag{2.51}$$

---

[13] The following developments can also be carried out with Euler equations around fixed points O^i. For conciseness, the subscript ^{G} will be omitted here.

[14] for reasons which will become clear later on.

this has the dimension of power. However, since the $\Delta\dot{\mathbf{x}}^i$ and $\Delta\boldsymbol{\omega}^i$ are arbitrary vectors, the quantity

$$\Delta\mathcal{P} \triangleq \sum_{i=1}^{N} \left(\mathbf{F}^i \,.\, \Delta\dot{\mathbf{x}}^i + \mathbf{L}^i \,.\, \Delta\boldsymbol{\omega}^i \right) \tag{2.52}$$

is *not* the actual power produced by the forces and torques acting on the different bodies of the system. According to the definitions 2.50, $\Delta\mathcal{P}$ is a *virtual power change.* The latter explains the name of the method we are presently developing: the *virtual power principle.*

It is clear that 2.51 arises from the dynamical equations of the various bodies. Conversely, if

$$\sum_{i=1}^{N} \left(m^i \ddot{\mathbf{x}}^i - \mathbf{F}^i \right) .\, \Delta\dot{\mathbf{x}}^i + \sum_{i=1}^{N} \left(\dot{\mathbf{H}}^i - \mathbf{L}^i \right) .\, \Delta\boldsymbol{\omega}^i = 0 \tag{2.53}$$

holds for *all* possible arbitrary vectors $\Delta\dot{\mathbf{x}}^i$ and $\Delta\boldsymbol{\omega}^i$, then the Newton-Euler equations 2.47 and 2.48 are automatically satisfied for each body.

If *all possible* vectors $\Delta\dot{\mathbf{x}}^i$ and $\Delta\boldsymbol{\omega}^i$ are used, then 2.53 is strictly equivalent to the system of dynamical equations 2.47 and 2.48. However, since reducing the number of equations (and unknowns) is our goal, we will use only some subsets of the possible vectors $\Delta\dot{\mathbf{x}}^i$ and $\Delta\boldsymbol{\omega}^i$. How to select such subsets efficiently is best explained via a few examples.

A single body: the moving ball in a rotating tube

Consider the system illustrated in figure 2.2 that we used in section 2.2.1. The ball inside the tube is now attached to a linear spring whose other end is connected to the bottom of the tube (see figure 2.8). Besides the kinematics, the dynamical aspects (mass distribution and applied forces and torques) must now be taken into account. For simplicity, the ball is modeled as a point mass: its angular momentum vector, as well as the resultant applied torque, are zero; the Euler rotation equation is trivial. Considering this system as being 2-D and assuming that the tube is rotating horizontally at constant speed $\boldsymbol{\omega}$, we find the forces acting on the ball in the $\{\hat{\mathbf{I}}_1, \hat{\mathbf{I}}_2\}$ plane are[15]:

- the spring force given by $\mathbf{F}_{spring} = -k\, r\, \hat{\mathbf{X}}_1$. For simplicity, we assume that the spring reaches its neutral length when the ball is at O,
- the constraint force which maintains the ball inside the tube: $\mathbf{F}_N = F_N\, \hat{\mathbf{X}}_2$.

[15] The motion of the ball inside the tube is assumed to be frictionless.

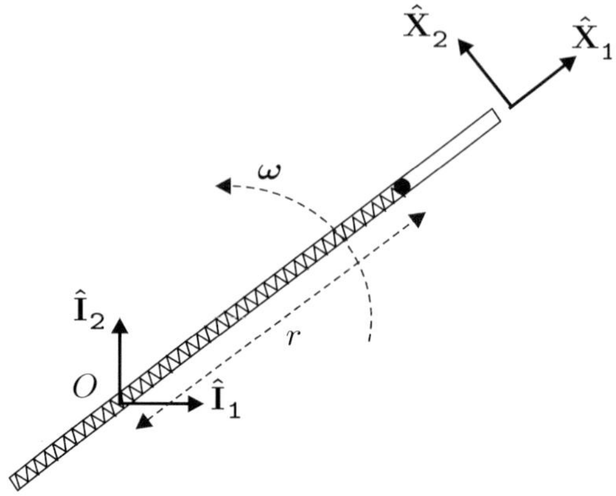

Figure 2.8: Ball in a rotating tube

As it has only one degree of freedom, this system can be characterized by a single generalized coordinate $q = \{r\}$ and its configuration variables are given by

$$\mathbf{x} = q\,\hat{\mathbf{X}}_1 \quad \text{with} \quad [\hat{\mathbf{X}}] = R_3(\omega t)[\hat{\mathbf{I}}] \tag{2.54}$$

The absolute velocity and acceleration of the ball are

$$\dot{\mathbf{x}} = \dot{q}\hat{\mathbf{X}}_1 + \omega q\hat{\mathbf{X}}_2 \quad \text{and} \quad \ddot{\mathbf{x}} = \left(\ddot{q} - \omega^2 q\right)\hat{\mathbf{X}}_1 + 2\omega\dot{q}\hat{\mathbf{X}}_2 \tag{2.55}$$

For this simple example, the principle of virtual power 2.53 reduces to

$$\left(m\ddot{\mathbf{x}} - \mathbf{F}\right) . \Delta\dot{\mathbf{x}} = 0 \quad \forall\ \Delta\dot{\mathbf{x}} \tag{2.56}$$

and it remains to choose the vector coefficient $\Delta\dot{\mathbf{x}}$ in order to obtain as few equations as possible to describe the system motion. As there is only one degree of freedom, a single equation must be sufficient to describe this motion. Therefore, the question is: what virtual velocity change $\Delta\dot{\mathbf{x}}$ should we use?

We can find an answer to this question by examining figure 2.8. Indeed, we may *virtually* consider that this figure is an instantaneous snapshot taken at some arbitrarily given time t_0 when the actual system was moving. At that time, the system was characterized by the values of its generalized coordinate $q(t_0)$ and velocity $\dot{q}(t_0)$. The value $q(t_0)$ is "frozen" on the shot, but when looking at the picture, we can imagine that exactly the same photo *could* have been obtained at the same time t_0 but with the system moving at a *different* velocity. Let us denote this different (or virtual) generalized velocity as $\dot{q}(t_0) + \Delta\dot{q}$. According to the kinematic relations 2.55, the corresponding virtual velocity vector of the ball *would* thus have been at time t_0

$$\dot{\mathbf{x}}(t_0) + \Delta\dot{\mathbf{x}} = (\dot{q}(t_0) + \Delta\dot{q})\,\hat{\mathbf{X}}_1 + \omega q(t_0)\hat{\mathbf{X}}_2 \tag{2.57}$$

Comparing 2.57 and 2.55, we see that a possible virtual velocity change $\Delta\dot{\mathbf{x}}$ that we can imagine from the picture is given by

$$\Delta\dot{\mathbf{x}} = \Delta\dot{q}\hat{\mathbf{X}}_1 \tag{2.58}$$

We emphasize that this particular virtual velocity change is *compatible* with the constraint applied to the system. Indeed, this constraint (i.e.: "the ball is compelled to remain inside the tube") is implicitly taken into account by the generalized coordinate q and the kinematic relations 2.54 and 2.55. As it results from a virtual change of the generalized velocity $\Delta\dot{q}$, the vector $\Delta\dot{\mathbf{x}}$ is thus automatically compatible with this implicit constraint.

Having found this particular value for the virtual velocity change $\Delta\dot{\mathbf{x}}$, we may now determine what equation the virtual power principle will lead to, using this $\Delta\dot{\mathbf{x}}$. Replacing the latter by its value in 2.56, we obtain

$$\left(m\ddot{\mathbf{x}} - \mathbf{F}\right) . \Delta\dot{q}\hat{\mathbf{X}}_1 = 0$$

which reduces to

$$\left\{m\left(\ddot{q} - \omega^2 q\right) + k\,q\right\} \Delta\dot{q} = 0$$

or, equivalently, since this equation must hold for any $\Delta\dot{q}$

$$m\left(\ddot{q} - \omega^2 q\right) + k\,q = 0 \tag{2.59}$$

This is the sought for equation of motion for the single generalized coordinate q. The unknown constraint force $\mathbf{F}_N$ has disappeared because the virtual change of power associated with this force and the chosen virtual change $\Delta\dot{\mathbf{x}}$ vanishes:

$$\Delta\mathcal{P}_{\mathbf{F}_N} = F_N\,\hat{\mathbf{X}}_2 . \Delta\dot{q}\hat{\mathbf{X}}_1 = 0$$

Although it has been illustrated here via a simple example, this is a general property which constitutes the core of the virtual power principle:

The contribution of constraint forces and constraint torques to the virtual power change $\Delta\mathcal{P}$ vanishes
for all virtual velocity changes $\Delta\dot{\mathbf{x}}^i$ and $\Delta\omega^i$ which are compatible with the corresponding constraints.

Choosing the virtual velocity changes $\Delta\dot{\mathbf{x}}^i$ and $\Delta\boldsymbol{\omega}^i$ in the virtual power principle 2.53 *compatible* with the constraints thus leads to those linear combinations of the dynamical equations which eliminate the unknown constraint forces.

In our example, the virtual velocity change is automatically compatible, because it results from a virtual change $\Delta\dot{q}$ of the independent generalized velocity. We can also obtain compatible virtual changes $\Delta\dot{\mathbf{x}}^i$ and $\Delta\boldsymbol{\omega}^i$ even if the system is described by means of non-independent generalized coordinates. Indeed, let us suppose that the moving ball of figure 2.8 is described by two generalized coordinates:

$$q = \{x_1, x_2\}$$

with the following kinematic relations:

$$\begin{aligned} \mathbf{x} &= x_1 \hat{\mathbf{I}}_1 + x_2 \hat{\mathbf{I}}_2 \\ \dot{\mathbf{x}} &= \dot{x}_1 \hat{\mathbf{I}}_1 + \dot{x}_2 \hat{\mathbf{I}}_2 \end{aligned}$$

The "ball in the tube" constraint is no longer implicitly taken into account. The generalized coordinates are not independent, and are related by an explicit constraint which expresses that the ball remains in the tube, i.e. in the present case:

$$\mathbf{x} \, . \, \hat{\mathbf{X}}_2 = -x_1 \sin(\omega t) + x_2 \cos(\omega t) = 0 \tag{2.60}$$

which has the form $h(q, t) = 0$. More generally, when the model of a system involves holonomic and/or non-holonomic constraints, the generalized velocities are related through the velocity constraints 2.21:

$$J(q, t) \, \dot{q} + \begin{pmatrix} \frac{\partial h}{\partial t} \\ b \end{pmatrix} = 0$$

Let us now return to figure 2.8 and its "virtual" interpretation. When the "photograph" of the real system was taken at time t_0, the generalized coordinates and velocities satisfied the velocity constraints:

$$J\left(q(t_0), t_0\right) \, \dot{q}(t_0) + \begin{pmatrix} \frac{\partial h}{\partial t} \\ b \end{pmatrix} = 0$$

Then, if we consider on the basis of this photograph, a set of "virtual" velocity values $\dot{q}(t_0) + \Delta\dot{q}$ which is different from $\dot{q}(t_0)$, the former will be compatible with the physics of the system (and thus compatible with the constraints) so long as

$$J\left(q(t_0), t_0\right) \, \left(\dot{q}(t_0) + \Delta\dot{q}\right) + \begin{pmatrix} \frac{\partial h}{\partial t} \\ b \end{pmatrix} = 0$$

Comparing these last two constraint equations, we obtain the following condition

$$J\left(q(t_0), t_0\right) \, \Delta\dot{q} = 0 \tag{2.61}$$

which must be satisfied by the virtual changes of generalized velocities $\Delta\dot{q}$ in order for them to be *compatible with the constraints.*
For the moving ball, this condition becomes

$$\begin{pmatrix} -\sin(\omega t) & \cos(\omega t) \end{pmatrix} \begin{pmatrix} \Delta\dot{x}_1 \\ \Delta\dot{x}_2 \end{pmatrix} = 0$$

and has as solution

$$\left(\begin{array}{c} \Delta\dot{x}_1 \\ \Delta\dot{x}_2 \end{array} \right) = \left(\begin{array}{c} \cos(\omega t) \\ \sin(\omega t) \end{array} \right) \Delta v$$

where Δv is arbitrary (but has the dimension of a velocity). According to the kinematic equations, the corresponding virtual velocity vector of the ball at time t_0 is

$$\dot{\mathbf{x}} + \Delta\dot{\mathbf{x}} = \left(\dot{x}_1 + \Delta\dot{x}_1\right) \hat{\mathbf{I}}_1 + \left(\dot{x}_2 + \Delta\dot{x}_2\right) \hat{\mathbf{I}}_2$$

and the virtual velocity change $\Delta\dot{\mathbf{x}}$ is thus given by

$$\Delta\dot{\mathbf{x}} = \Delta v \left\{\cos(\omega t)\, \hat{\mathbf{I}}_1 + \sin(\omega t)\, \hat{\mathbf{I}}_2\right\} = \Delta v\, \hat{\mathbf{X}}_1$$

This is clearly equivalent to the expression we found for the virtual velocity change in 2.58. Once introduced into the virtual power principle 2.53, it will therefore also lead to the equation of motion. The latter now reads

$$\left\{m\left(\ddot{x}_1\hat{\mathbf{I}}_1 + \ddot{x}_2\hat{\mathbf{I}}_2\right) + k\sqrt{x_1^2 + x_2^2}\, \hat{\mathbf{X}}_1 - F_N\, \hat{\mathbf{X}}_2\right\} . \Delta v\, \hat{\mathbf{X}}_1 = 0$$

and, after simplifications, becomes

$$m\left(\ddot{x}_1 \cos(\omega t) + \ddot{x}_2 \sin(\omega t)\right) + k\sqrt{x_1^2 + x_2^2} = 0$$

Together with the constraint 2.60, this equation of motion forms a set of two differential/algebraic equations in the two generalized coordinates $q = \{x_1, x_2\}$. The unknown constraint force $\mathbf{F}_N$ has been eliminated.

In this example, the constraint force $\mathbf{F}_N$ is external: it is a force applied to the system (the ball) by its environment (the tube). The following example deals with internal constraints.

Several bodies: the slider-crank mechanism

As mentioned in section 2.2.1, we can model the slider-crank mechanism illustrated in figure 2.9 by means of two generalized coordinates: $q = \{\varphi_1, \varphi_2\}$. These two coordinates are not independent, since they are explicitly related by the constraint 2.12:

$$h(q) = r \sin\varphi_1 + l \sin\varphi_2 = 0 \tag{2.62}$$

which states that the slider B is constrained by its environment to remain on a horizontal line (i.e.: $y_3 = 0$).

On the other hand, together with the set of kinematic relations {2.2 - 2.6, 2.8 and 2.9}, the two generalized coordinates implicitly take into account the assembly of the system. For instance, body 1 (the crank arm) and body 2 (the connecting rod) are connected at A by a revolute joint. Denoting the material

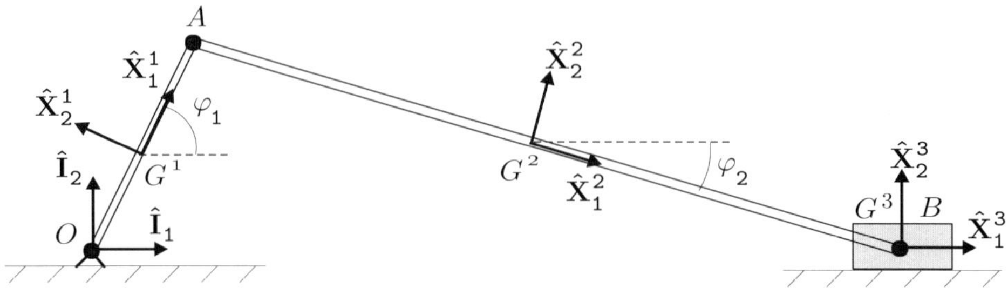

Figure 2.9: The slider-crank mechanism

points of the crank arm and the connecting rod which are located at A by A^1 and A^2 respectively, we may consider that the assembly of the system at A is due to:
- some constraint force $\mathbf{F}^A$ exerted on A^1 by body 2
- and the constraint reaction force $-\mathbf{F}^A$ exerted on A^2 by body 1.

All the forces (and torques) which are applied to the various bodies of a system must be included in the general formulation 2.51 of the virtual power principle

$$\sum_{i=1}^{N}\left(m^i\ddot{\mathbf{x}}^i - \mathbf{F}^i\right).\Delta\dot{\mathbf{x}}^i + \sum_{i=1}^{N}\left(\dot{\mathbf{H}}^i - \mathbf{L}^i\right).\Delta\boldsymbol{\omega}^i = 0$$

Therefore, both the constraint force $\mathbf{F}^A$ and its reaction $-\mathbf{F}^A$ must be taken into account in evaluating the virtual power change we defined as

$$\Delta\mathcal{P} \triangleq \sum_{i=1}^{N}\left(\mathbf{F}^i.\Delta\dot{\mathbf{x}}^i + \mathbf{L}^i.\Delta\boldsymbol{\omega}^i\right)$$

The contributions of $\mathbf{F}^A$ in the resultant force $\mathbf{F}^1$ and torque $\mathbf{L}^1$ applied to the center of mass of body 1 are respectively given by

$$\mathbf{F}^1_{\mathbf{F}_A} = \mathbf{F}^A \quad \text{and} \quad \mathbf{L}^1_{\mathbf{F}_A} = \overrightarrow{G^1A}\times\mathbf{F}^A = \frac{r}{2}\hat{\mathbf{X}}^1_1\times\mathbf{F}^A$$

and its contribution to the virtual power change is thus equal to

$$\Delta\mathcal{P}_{\mathbf{F}_A} = \mathbf{F}^A.\Delta\dot{\mathbf{x}}^1 + \left(\frac{r}{2}\hat{\mathbf{X}}^1_1\times\mathbf{F}^A\right).\Delta\boldsymbol{\omega}^1$$

or equivalently to

$$\Delta\mathcal{P}_{\mathbf{F}_A} = \mathbf{F}^A.\left(\Delta\dot{\mathbf{x}}^1 + \Delta\boldsymbol{\omega}^1\times\frac{r}{2}\hat{\mathbf{X}}^1_1\right)$$

On the other hand, since the absolute velocity of the material point A^1 is given by

$$\dot{\mathbf{x}}^{A^1} = \dot{\mathbf{x}}^1 + \boldsymbol{\omega}^1 \times \frac{r}{2}\hat{\mathbf{X}}_1^1$$

its virtual velocity would be

$$\dot{\mathbf{x}}^{A^1} + \Delta\dot{\mathbf{x}}^{A^1} = \dot{\mathbf{x}}^1 + \Delta\dot{\mathbf{x}}^1 + \left(\boldsymbol{\omega}^1 + \Delta\boldsymbol{\omega}^1\right) \times \frac{r}{2}\hat{\mathbf{X}}_1^1$$

Therefore, the coefficient of $\mathbf{F}^A$ in the expression of $\Delta\mathcal{P}_{\mathbf{F}_A}$ is simply the virtual velocity change of point A^1 which is induced by the virtual changes $\Delta\dot{\mathbf{x}}^1$ and $\Delta\boldsymbol{\omega}^1$. The contribution of $\mathbf{F}^A$ to the resulting virtual power change is thus equal to

$$\Delta\mathcal{P}_{\mathbf{F}_A} = \mathbf{F}^A \,.\, \Delta\dot{\mathbf{x}}^{A^1} \tag{2.63}$$

Although illustrated on a particular example, this result is general:

The contribution of a force $\mathbf{F}$ to the virtual power change due to virtual velocity changes $\Delta\dot{\mathbf{x}}^i$ and $\Delta\boldsymbol{\omega}^i$ is equal to the dot product of this force with the virtual velocity change of its application point resulting from $\Delta\dot{\mathbf{x}}^i$ and $\Delta\boldsymbol{\omega}^i$.

Applied to the reaction force acting on A^2, this result implies

$$\Delta\mathcal{P}_{-\mathbf{F}_A} = -\mathbf{F}^A \,.\, \Delta\dot{\mathbf{x}}^{A^2}$$

The global contribution of the action $\mathbf{F}^A$ and its reaction $-\mathbf{F}^A$ to the global virtual power change $\Delta\mathcal{P}$ thus vanishes if:

$$\Delta\dot{\mathbf{x}}^{A^1} = \Delta\dot{\mathbf{x}}^{A^2}$$

or, in other words, if the virtual velocity changes are compatible with the assembly constraints of the system at A. This shows that internal constraints in a multibody system obey the same rules as external constraints.

Since we modeled the slider crank mechanism using two generalized coordinates $q = \{\varphi_1, \varphi_2\}$ which take the connection conditions into account, the set of forces (and torques) which contribute to $\Delta\mathcal{P}$ and must therefore be considered in the virtual power principle reduces to:
- forces due to gravity,
- a (possible) pure torque which could be externally applied to the crank axis at O,
- possible pure friction torques in the revolute joints O, A and B,
- the normal (constraint) force $\mathbf{F}_N = F_N\,\hat{\mathbf{I}}_2$ applied by the ground to the slider[16],

[16] This force needs to be considered here since the generalized coordinates do not take into account the constraint which compels the slider to remain on the horizontal line $y_3 = 0$.

- and possibly a transverse force $\mathbf{F}_T = F_T\,\hat{\mathbf{I}}_1$ applied by the ground to the slider in the case of friction.

By choosing $\Delta\dot{q} = \{\Delta\dot{\varphi}_1, \Delta\dot{\varphi}_2\}$ as virtual change in the virtual power principle, we obtain a linear combination of the dynamical system equations which has the form

$$\sum_{\alpha=1}^{2} \mathcal{F}_\alpha(t, q, \dot{q}, \ddot{q}, F_N, F_T, g)\Delta\dot{\varphi}_\alpha = 0 \tag{2.64}$$

We can then obtain a reduced number of equations of motion in two different ways:

- If the two variations $\Delta\dot{\varphi}_1$ and $\Delta\dot{\varphi}_2$ are considered as arbitrary (and thus independent), two equations of motion are deduced from this combination

  $$\mathcal{F}_\alpha(t, q, \dot{q}, \ddot{q}, F_N, F_T, g) = 0 \quad \alpha = 1, 2$$

  Together with the constraint 2.62, we thus obtain three differential-algebraic equations for the three unknowns $q = \{\varphi_1, \varphi_2\}$ and F_N. The constraint force F_N is still present in the equations of motion since $\Delta\dot{\varphi}_1$ and $\Delta\dot{\varphi}_2$ have been considered as arbitrary and are thus not (necessarily) compatible with the constraint 2.62.

- If the two virtual changes $\Delta\dot{\varphi}_1$ and $\Delta\dot{\varphi}_2$ are chosen to be compatible with the constraint, i.e. if on the basis of 2.61 they satisfy

  $$\begin{pmatrix} r\cos\varphi_1 & l\cos\varphi_2 \end{pmatrix} \begin{pmatrix} \Delta\dot{\varphi}_1 \\ \Delta\dot{\varphi}_2 \end{pmatrix} = 0$$

  by having for instance the form

  $$\begin{pmatrix} \Delta\dot{\varphi}_1 \\ \Delta\dot{\varphi}_2 \end{pmatrix} = \begin{pmatrix} -l\cos\varphi_2 \\ r\cos\varphi_1 \end{pmatrix} \Delta v \quad ,$$

  then the combination 2.64 leads to the single equation of motion

  $$-l\cos\varphi_2\,\mathcal{F}_1(t, q, \dot{q}, \ddot{q}, F_N, F_T, g) + r\cos\varphi_1\,\mathcal{F}_2(t, q, \dot{q}, \ddot{q}, F_N, F_T, g) = 0$$

  The unknown constraint force F_N now disappears. We thus arrive at an equation of motion which has the form

  $$\mathcal{F}(t, q, \dot{q}, \ddot{q}, F_T, g) = 0$$

  and constitutes, with the constraint 2.62, a set of two differential-algebraic equations (DAEs) in the unknowns $q = \{\varphi_1, \varphi_2\}$.

We must make an additional comment concerning these two possibilities. While in general it is important to eliminate the unknown constraint forces, it may happen that some of these constraint forces are essential in determining the system motion; this happens for instance when the system involves Coulomb friction. According to Coulomb's law, the friction force applied to the slider acts in the opposite direction to the slider velocity with an intensity given by

$$\mid F_T \mid = \mu \, F_N \tag{2.65}$$

Thus even if F_N has been formally eliminated in the second method, it will reappear through the constitutive equation for F_T. Therefore, the computation of F_N cannot be avoided in determining the system's behavior. This means that in such a situation, only the first method can actually be considered.

2.4.2 Generalized forces and Lagrange multipliers

The principle of virtual power 2.51 can be written for N bodies in the equivalent form

$$\sum_{i=1}^{N} m^i \ddot{\mathbf{x}}^i \, . \, \Delta \dot{\mathbf{x}}^i + \sum_{i=1}^{N} \dot{\mathbf{H}}^i \, . \, \Delta \boldsymbol{\omega}^i = \Delta \mathcal{P} \tag{2.66}$$

where

$$\Delta \mathcal{P} \triangleq \sum_{i=1}^{N} \left(\mathbf{F}^i \, . \, \Delta \dot{\mathbf{x}}^i + \mathbf{L}^i \, . \, \Delta \boldsymbol{\omega}^i \right) \tag{2.67}$$

is the *virtual power change* induced by the virtual velocity changes $\Delta \dot{\mathbf{x}}^i$ and $\Delta \boldsymbol{\omega}^i$ of the various bodies. As we know, if these changes are compatible with some constraints, then the associated constraint forces (or torques) do not contribute to the virtual power change $\Delta \mathcal{P}$ and thus disappear from the equations of motion.

Let us now focus on the forces and torques which *contribute* to this virtual power change. Two cases need to be considered, depending on the choice of generalized coordinates.

Independent generalized coordinates

If the system is modeled by n independent coordinates q, all the (physical) constraints are implicitly taken into account. The configuration variables of the system are given by the kinematic equations

$$\begin{array}{l} \mathbf{x}^k = [\hat{\mathbf{I}}]^T x^k(q,t) \\ [\hat{\mathbf{X}}^k] = R^k(q,t)[\hat{\mathbf{I}}] \end{array} \qquad \text{for } \forall \text{ body } k \tag{2.68}$$

Moreover, as mentioned in section 2.2.2, both the linear and angular velocity vectors of the various bodies are linear functions of the generalized velocities:

$$\begin{aligned} \dot{\mathbf{x}}^k(\dot{q}, q, t) &= \sum_{\alpha=1}^{n} \frac{\partial \mathbf{x}^k(q,t)}{\partial q_\alpha} \dot{q}_\alpha + \frac{\partial \mathbf{x}^k(q,t)}{\partial t} \\ \boldsymbol{\omega}^k(\dot{q}, q, t) &= \sum_{\alpha=1}^{n} \mathbf{a}_\alpha^k(q,t)\dot{q}_\alpha + \mathbf{b}^k(q,t) \end{aligned} \tag{2.69}$$

Since $\dot{\mathbf{x}}^k$ and $\boldsymbol{\omega}^k$ are functions of the variables $(\dot{q}, q, t)$, their linear dependency with respect to the $\dot{q}_\alpha$ implies

$$\begin{aligned} \frac{\partial \dot{\mathbf{x}}^k(\dot{q}, q, t)}{\partial \dot{q}_\alpha} &= \frac{\partial \mathbf{x}^k(q,t)}{\partial q_\alpha} \\ \text{and} \quad \frac{\partial \boldsymbol{\omega}^k(\dot{q}, q, t)}{\partial \dot{q}_\alpha} &= \mathbf{a}_\alpha^k(q,t) \end{aligned} \tag{2.70}$$

which allow us to rewrite the kinematic relations 2.69 as follows:

$$\begin{aligned} \dot{\mathbf{x}}^k(\dot{q}, q, t) &= \sum_{\alpha=1}^{n} \frac{\partial \dot{\mathbf{x}}^k}{\partial \dot{q}_\alpha} \dot{q}_\alpha + \frac{\partial \mathbf{x}^k}{\partial t} \\ \boldsymbol{\omega}^k(\dot{q}, q, t) &= \sum_{\alpha=1}^{n} \frac{\partial \boldsymbol{\omega}^k}{\partial \dot{q}_\alpha} \dot{q}_\alpha + \mathbf{b}^k \end{aligned} \tag{2.71}$$

Thus if virtual changes $\Delta\dot{q}$ are introduced on the generalized velocities, these velocity vectors virtually become

$$\begin{aligned} \dot{\mathbf{x}}^k + \Delta\dot{\mathbf{x}}^k &= \sum_{\alpha=1}^{n} \frac{\partial \dot{\mathbf{x}}^k}{\partial \dot{q}_\alpha} (\dot{q}_\alpha + \Delta\dot{q}_\alpha) + \frac{\partial \mathbf{x}^k}{\partial t} \\ \boldsymbol{\omega}^k + \Delta\boldsymbol{\omega}^k &= \sum_{\alpha=1}^{n} \frac{\partial \boldsymbol{\omega}^k}{\partial \dot{q}_\alpha} (\dot{q}_\alpha + \Delta\dot{q}_\alpha) + \mathbf{b}^k \end{aligned}$$

and therefore, the virtual velocity changes of the various bodies can be expressed as

$$\begin{aligned} \Delta\dot{\mathbf{x}}^k &= \sum_{\alpha=1}^{n} \frac{\partial \dot{\mathbf{x}}^k}{\partial \dot{q}_\alpha} \Delta\dot{q}_\alpha \\ \Delta\boldsymbol{\omega}^k &= \sum_{\alpha=1}^{n} \frac{\partial \boldsymbol{\omega}^k}{\partial \dot{q}_\alpha} \Delta\dot{q}_\alpha \end{aligned} \tag{2.72}$$

These two formulae are interesting in practice: since the actual velocities (and accelerations) of the system need to be computed anyway to obtain the linear and angular momenta of the various bodies, the expressions of the virtual velocity changes $\Delta\dot{\mathbf{x}}^k$ and $\Delta\boldsymbol{\omega}^k$ are easily computed concurrently using the partial derivatives $\frac{\partial}{\partial \dot{q}_\alpha}$.

On the other hand, since the virtual velocity changes are linear with respect to the $\Delta\dot{q}_\alpha$, the virtual change of power can be formally written in the form

$$\Delta\mathcal{P} \triangleq \sum_{k=1}^{N} \left(\mathbf{F}^k \,.\, \Delta\dot{\mathbf{x}}^k + \mathbf{L}^k \,.\, \Delta\boldsymbol{\omega}^k \right) = \sum_{\alpha=1}^{n} Q_\alpha \,\Delta\dot{q}_\alpha \tag{2.73}$$

where the coefficients Q_α are called the *generalized forces* associated with the generalized coordinates q_α. We emphasize that these generalized forces depend only on the forces and torques which *contribute* to $\Delta\mathcal{P}$. The constraint forces thus do not influence the generalized forces. With this property in mind, we can compute the generalized force Q_α from

$$Q_\alpha = \sum_{k=1}^{N} \left(\mathbf{F}^k_{contributive} \cdot \frac{\partial \dot{\mathbf{x}}^k}{\partial \dot{q}_\alpha} + \mathbf{L}^k_{contributive} \cdot \frac{\partial \boldsymbol{\omega}^k}{\partial \dot{q}_\alpha} \right) \tag{2.74}$$

In practice however, Q_α is more easily obtained as the coefficient of $\Delta\dot{q}_\alpha$ in the expression for the virtual power change $\Delta\mathcal{P}$.

Finally we observe that, in the case of independent generalized coordinates q_α, the equations of motion provided by the virtual power principle are n ordinary differential equations having the form

$$\Phi_\alpha(t, q, \dot{q}, \ddot{q}) = Q_\alpha \quad \alpha = 1, ..., n \tag{2.75}$$

Dependent generalized coordinates

Consider a system modeled by means of n generalized coordinates q which, being dependent, are related by m holonomic and/or non-holonomic explicit constraints. An arbitrary change $\Delta\dot{q}$ of the generalized velocities is thus not necessarily compatible with these constraints. If this change is not compatible, then

$$J(q,t)\,\Delta\dot{q} \neq 0 \tag{2.76}$$

where J is the Jacobian matrix 2.22 of the constraints. The corresponding constraint forces (and torques) will not be eliminated if such a virtual change $\Delta\dot{q}$ is used in the virtual power principle. They do not contribute to the generalized forces, but will contribute to the virtual power. The virtual power change will thus have the form

$$\Delta\mathcal{P} = \sum_{\alpha=1}^{n} Q_\alpha \,\Delta\dot{q}_\alpha + \sum_{\alpha=1}^{n} Q'_\alpha \,\Delta\dot{q}_\alpha$$

where the Q'_α are the *generalized constraint forces.* Their expressions are similar to 2.74, i.e.:

$$Q'_\alpha = \sum_{k=1}^{N} \left(\mathbf{F}^k_{constraint} \cdot \frac{\partial \dot{\mathbf{x}}^k}{\partial \dot{q}_\alpha} + \mathbf{L}^k_{constraint} \cdot \frac{\partial \boldsymbol{\omega}^k}{\partial \dot{q}_\alpha} \right) \tag{2.77}$$

and the equations of motion now read

$$\Phi_\alpha(t, q, \dot{q}, \ddot{q}) = Q_\alpha + Q'_\alpha \quad \alpha = 1, ..., n$$

or, in matrix form

$$\Phi(t, q, \dot{q}, \ddot{q}) = Q + Q' \quad \text{with: } \Phi, Q \text{ and } Q' \in \Re^n \tag{2.78}$$

While the values of the generalized forces must be computed from 2.74 in order to solve the equations of motion, the following question arises: since the generalized constraint forces Q' involve the unknown constraint forces $\mathbf{F}^k_{constraint}$ and torques $\mathbf{L}^k_{constraint}$ which need not be evaluated to determine the system motion, is it always necessary to compute 2.77 explicitly? Fortunately the response is negative, thanks to the following alternative which is called the *Lagrange multiplier technique.*

We know that the constraint forces (and torques) do not contribute to the virtual power change $\Delta\mathcal{P}$ so long as, instead of being arbitrary, the virtual changes $\Delta\dot{q}$ of the generalized velocities are chosen to be compatible with the constraints. Mathematically speaking, this means that

$$\begin{aligned} \Delta\mathcal{P}_{constraint} \triangleq \sum_{\alpha=1}^{n} Q'_\alpha \, \Delta\dot{q}_\alpha \quad &= \quad (Q')^t \; \Delta\dot{q} = 0 \\ \text{for any } \Delta\dot{q} \text{ which satisfies} \quad &: \quad J(q,t)\, \Delta\dot{q} = 0 \end{aligned}$$

Viewed in the vector space $\Re^n$, this condition implies that any "$\Re^n$-vector" $\Delta\dot{q}$ which is orthogonal to the rows of the Jacobian matrix J, is automatically orthogonal to the "$\Re^n$-vector" Q'. From elementary algebra, this vector is thus necessarily in the vector subspace spanned by the m independent rows of J. The "vector" Q' must therefore be a linear combination of these rows and can be written as follows

$$Q' = J^T(q,t)\,\lambda \tag{2.79}$$

where the components of $\lambda \in \Re^m$ are the coefficients of this linear combination. These coefficients are called the *Lagrange multipliers* associated with the (explicit) constraints. The equations of motion may thus be written in the form

$$\Phi(t, q, \dot{q}, \ddot{q}) = Q + J^T(q,t)\,\lambda \tag{2.80}$$

where the unknown constraint forces are now replaced by the m unknown coefficients λ_β. Together with the m constraint equations, we thus have a system of $n+m$ differential/algebraic equations (DAEs) for the $n+m$ unknowns $q \in \Re^n$ and $\lambda \in \Re^m$.

If necessary, we can eliminate the Lagrange multipliers from this system of equations numerically. Indeed since the Jacobian matrix has full rank m, it can be partitioned, possibly after reordering the coordinates q_α, as

$$J = \begin{pmatrix} J_u & J_v \end{pmatrix} \tag{2.81}$$

where the $(m \times m)$ submatrix J_v is non singular. Reordering the equations of motion in the same way, we can also make the partition

$$\begin{pmatrix} \Phi_u(t,q,\dot{q},\ddot{q}) \\ \Phi_v(t,q,\dot{q},\ddot{q}) \end{pmatrix} = \begin{pmatrix} Q_u \\ Q_v \end{pmatrix} + \begin{pmatrix} (J_u)^T \\ (J_v)^T \end{pmatrix} \lambda \tag{2.82}$$

Since J_v is non singular, we can extract the λ's numerically from the last m equations

$$\lambda = \left((J_v)^T\right)^{-1} \left(\Phi_v(t,q,\dot{q},\ddot{q}) - Q_v\right) \tag{2.83}$$

and substitute the result into the first $n - m$ equations:

$$\Phi_u(t,q,\dot{q},\ddot{q}) - (J_u)^T \left((J_v)^T\right)^{-1} \Phi_v(t,q,\dot{q},\ddot{q}) = Q_u - (J_u)^T \left((J_v)^T\right)^{-1} Q_v \tag{2.84}$$

Together with the m constraints, this last set of $n-m$ equations thus constitutes a system of n differential/algebraic equations for the n unknowns $q \in \Re^n$.

2.4.3 Physical interpretation of the Lagrange multipliers

Although the Lagrange multipliers were introduced via linear algebra considerations, they are related to the physical constraint forces (and torques). This relation comes from the generalized constraint forces that we can now write in two different forms:

$$Q' = \sum_{k=1}^{N} \left(\mathbf{F}^k_{constraint} \cdot \frac{\partial \dot{\mathbf{x}}^k}{\partial \dot{q}} + \mathbf{L}^k_{constraint} \cdot \frac{\partial \boldsymbol{\omega}^k}{\partial \dot{q}} \right) = J^t \, \lambda$$

We have equivalently

$$J J^T \lambda = J \sum_{k=1}^{N} \left(\mathbf{F}^k_{constraint} \cdot \frac{\partial \dot{\mathbf{x}}^k}{\partial \dot{q}} + \mathbf{L}^k_{constraint} \cdot \frac{\partial \boldsymbol{\omega}^k}{\partial \dot{q}} \right)$$

where the $(m \times m)$ matrix JJ^T is non singular[17] because J has full rank m. We can thus obtain the expression of λ in terms of the constraint forces and torques by inverting this matrix:

$$\lambda = \left(JJ^T\right)^{-1} J \sum_{k=1}^{N} \left(\mathbf{F}^k_{constraint} \cdot \frac{\partial \dot{\mathbf{x}}^k}{\partial \dot{q}} + \mathbf{L}^k_{constraint} \cdot \frac{\partial \boldsymbol{\omega}^k}{\partial \dot{q}} \right) \tag{2.85}$$

However, this relation has no practical use. Indeed, when the Lagrange multipliers technique is used, the system motion is first solved in terms of the unknowns $q \in \Re^n$ and $\lambda \in \Re^m$. Then, when the time histories $q(t)$ and $\lambda(t)$ are known

[17] assuming that the constraints are independent !

from the time integration, the knowledge of the time histories of the physical constraint forces and torques might be of practical interest. If this is the case, we must invert the linear system 2.85 to compute the physical constraint force and torque components as functions of $\lambda(t)$ (and $q(t)$). However this inversion is not always possible, in particular for hyperstatic systems[18] (see section 2.3) since the number of unknown constraint components is then greater than m, the number of available equations 2.85.

Nevertheless if attention is paid to the form under which the constraints are expressed, it is often possible to establish the correspondence between the λ and the constraint forces and/or torques. The form in which the constraints are written is of utmost importance because, replacing e.g. two constraint expressions by two equivalent linear combinations completely changes the content of the Jacobian matrix. The meaning of the Lagrange multipliers is then also completely modified.
Practically, the way to find a relation between the λ and a particular constraint force $\mathbf{F}_{constr}$ consists in comparing the available forms of the corresponding virtual power change

$$\Delta\mathcal{P}_{constr} = \lambda^T J_{constr}\,\Delta\dot{q} = \mathbf{F}_{constr}\,.\,\Delta\dot{\mathbf{x}}_{constr} \tag{2.86}$$

where J_{constr} is the Jacobian of the constraint(s) *associated* with $\mathbf{F}_{constr}$, and $\Delta\dot{\mathbf{x}}_{constr}$ is the virtual velocity change of the material point at which $\mathbf{F}_{constr}$ is applied.

For instance, for the slider crank mechanism (see figure 2.9), we may consider that the constraint force $\mathbf{F}_N = F_N\,\hat{\mathbf{I}}_2$ is applied at point B. On the other hand, the associated constraint states that point B cannot move along $\hat{\mathbf{I}}_2$. In terms of the generalized coordinates $\{\varphi_1, \varphi_2\}$, the position of B is given by:

$$\mathbf{x}^B = (r\cos\varphi_1 + l\cos\varphi_2)\,\hat{\mathbf{I}}_1 + (r\sin\varphi_1 + l\sin\varphi_2)\,\hat{\mathbf{I}}_2$$

Its virtual velocity change:

$$\Delta\dot{\mathbf{x}}^B = \left(-r\sin\varphi_1\hat{\mathbf{I}}_1 + r\cos\varphi_1\hat{\mathbf{I}}_2\right)\Delta\dot{\varphi}_1 + \left(-l\sin\varphi_2\hat{\mathbf{I}}_1 + l\cos\varphi_2\hat{\mathbf{I}}_2\right)\Delta\dot{\varphi}_2$$

thus leads, according to property 2.63, to the following power change

$$\Delta\mathcal{P}_{\mathbf{F}_N} = F_N \begin{pmatrix} r\cos\varphi_1 & l\cos\varphi_2 \end{pmatrix} \begin{pmatrix} \Delta\dot{\varphi}_1 \\ \Delta\dot{\varphi}_2 \end{pmatrix}$$

On the other hand, the constraint on the generalized coordinates $\{\varphi_1, \varphi_2\}$ was written as

$$r\sin\varphi_1 + l\sin\varphi_2 = 0$$

[18] Indeed, while the motion of a four-wheeled supermarket trolley on a horizontal plane can be deduced from the dynamic equations, the normal reaction forces due to the ground cannot be determined individually.

such that

$$\Delta\mathcal{P}_{\mathbf{F}_N} = \left(J^T\,\lambda\right)^T\,\Delta\dot{q} = \lambda\,J\,\Delta\dot{q} = \lambda\,\left(\begin{array}{cc} r\cos\varphi_1 & l\cos\varphi_2 \end{array}\right)\left(\begin{array}{c} \Delta\dot{\varphi}_1 \\ \Delta\dot{\varphi}_2 \end{array}\right)$$

We thus obtain by straightforward comparison:

$$F_N = \lambda$$

For this example, it is found that the Lagrange multiplier λ is simply equal to the component F_N of the constraint force. We emphasize that this equality is lost if (equivalently but unfortunately) the constraint is written in the form

$$\left(r\sin\varphi_1 + l\sin\varphi_2\right)^2 = 0$$

2.4.4 An example: a pendulum supported by rollers

We conclude this chapter by illustrating how the virtual power principle may help to find the final form of the equations of motion of a system consisting of several bodies. Let us reconsider the pendulum of section 2.3.2 (see figure 2.10).

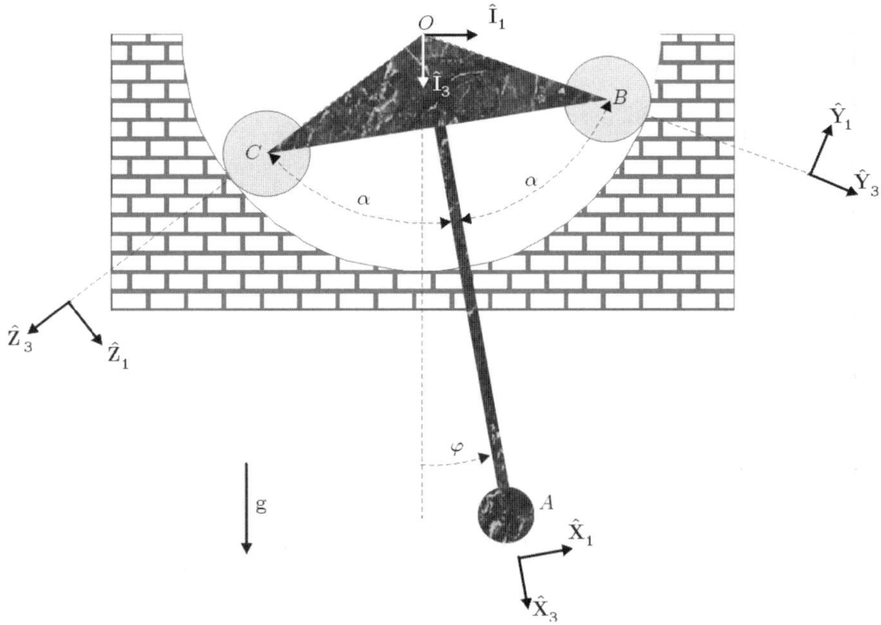

Figure 2.10: The supported pendulum

As in the Newton-Euler approach, the procedure we follow is subdivided into different steps.

Steps 1, 2 and 3

These steps are identical to those of the Newton-Euler method; the three angles $\{\varphi, \varphi_2, \varphi_3\}$ are chosen as generalized coordinates q_i.

Step 4: Expression of the virtual velocity changes

We need to express the virtual velocity changes of the various bodies in terms of virtual changes of the generalized velocities $(\Delta\dot{\varphi}, \Delta\dot{\varphi}_2, \Delta\dot{\varphi}_3)$. These expressions are obtained from the kinematic relations we developed in step 3 (see section 2.3.2):

$$\begin{aligned} \Delta\boldsymbol{\omega}^1 &= \Delta\dot{\varphi}\,\hat{\mathbf{X}}_2 & \qquad \Delta\dot{\mathbf{x}}^1 &= l\,\Delta\dot{\varphi}\,\hat{\mathbf{X}}_1 \\ \Delta\boldsymbol{\omega}^2 &= \Delta\dot{\varphi}_2\,\hat{\mathbf{X}}_2^2 & \qquad \Delta\dot{\mathbf{x}}^2 &= b\,\Delta\dot{\varphi}\,\hat{\mathbf{Y}}_1 \\ \Delta\boldsymbol{\omega}^3 &= \Delta\dot{\varphi}_3\,\hat{\mathbf{X}}_2^3 & \qquad \Delta\dot{\mathbf{x}}^3 &= b\,\Delta\dot{\varphi}\,\hat{\mathbf{Z}}_1 \end{aligned}$$

In order to eliminate all the constraint forces (and torques) from the equations of motion, we choose the virtual changes compatible with the constraints 2.28 and 2.29 acting on the system:

$$\begin{aligned} b\,\Delta\dot{\varphi} + a\,\Delta\dot{\varphi}_2 &= 0 \quad \text{and} \quad b\,\Delta\dot{\varphi} + a\,\Delta\dot{\varphi}_3 = 0 \\ \text{or equivalently} &: \quad \Delta\dot{\varphi}_2 = \Delta\dot{\varphi}_3 = -\frac{b}{a}\,\Delta\dot{\varphi} \end{aligned} \tag{2.87}$$

Step 5: Inventory of forces and torques

This step now reduces to the inventory of forces and torques which will *contribute* to the virtual power change. For the pendulum and the two rollers, we can thus exclude all the constraint forces and torques. Only the gravitational forces contribute:

$$\begin{aligned} &Mg\,\hat{\mathbf{I}}_3 \quad \text{on body 1 (the pendulum)} \\ &mg\,\hat{\mathbf{I}}_3 \quad \text{on bodies 2 and 3 (the rollers)} \end{aligned}$$

Step 6: Virtual power principle

We first compute the left-hand side of the virtual power principle 2.66 by means of the kinematic relations of step 3 and the virtual changes defined in step 4:

$$\begin{aligned} \sum_{i=1}^{3} m^i \ddot{\mathbf{x}}^i . \Delta\dot{\mathbf{x}}^i + \sum_{i=1}^{3} \dot{\mathbf{H}}^i . \Delta\boldsymbol{\omega}^i = \; & Ml^2\,\ddot{\varphi}\,\Delta\dot{\varphi} \quad \text{(for body 1)} \\ & + mb^2\,\ddot{\varphi}\,\Delta\dot{\varphi} + \frac{ma^2}{2}\,\ddot{\varphi}_2\,\Delta\dot{\varphi}_2 \quad \text{(for body 2)} \\ & + mb^2\,\ddot{\varphi}\,\Delta\dot{\varphi} + \frac{ma^2}{2}\,\ddot{\varphi}_3\,\Delta\dot{\varphi}_3 \quad \text{(for body 3)} \end{aligned} \tag{2.88}$$

and, using the constraints, we obtain

$$\sum_{i=1}^{3} m^i \ddot{\mathbf{x}}^i . \Delta\dot{\mathbf{x}}^i + \sum_{i=1}^{3} \dot{\mathbf{H}}^i . \Delta\boldsymbol{\omega}^i = \left(Ml^2 + 3mb^2\right) \ddot{\varphi}\, \Delta\dot{\varphi}$$

On the other hand, the virtual change of power due to the gravity forces is

$$\begin{aligned} \Delta\mathcal{P} &= Mg\, \hat{\mathbf{I}}_3 . l\, \Delta\dot{\varphi}\, \hat{\mathbf{X}}_1 \\ &\quad + mg\, \hat{\mathbf{I}}_3 . b\, \Delta\dot{\varphi}\, \hat{\mathbf{Y}}_1 \\ &\quad + mg\, \hat{\mathbf{I}}_3 . b\, \Delta\dot{\varphi}\, \hat{\mathbf{Z}}_1 \end{aligned}$$

and, after some vector computation, it becomes

$$\begin{aligned} \Delta\mathcal{P} &= \left(-Ml \sin\varphi - mb \sin(\varphi+\alpha) - mb\sin(\varphi-\alpha)\right) g\, \Delta\dot{\varphi} \\ &= -\left(Ml + 2mb\cos\alpha\right) g\, \sin\varphi\, \Delta\dot{\varphi} \end{aligned} \tag{2.89}$$

Equating the two sides of the power principle, we thus obtain, since $\Delta\dot{\varphi}$ is arbitrary,

$$\left(Ml^2 + 3mb^2\right) \ddot{\varphi} = -\left(Ml + 2mb\cos\alpha\right) g\, \sin\varphi \tag{2.90}$$

This is the final equation of motion 2.45 of the system.

Note that the non-trivial "linear combination" 2.44 of the dynamical equations 2.32 - 2.40 has been automatically provided by the virtual power principle.

Step 7 (optional): Constraint forces

If the value of a particular constraint force (or torque) needs to be determined, we can obtain the required equation by choosing virtual velocity changes which "violate" the corresponding constraint in the virtual power principle. For instance, the unknown component S_1^2 (i.e. the component along $\hat{\mathbf{Y}}_1$ of the constraint force applied at B by body 2 on body 1) is obtained using the following velocity changes:

$$\begin{aligned} \Delta\dot{\mathbf{x}}^1 &= 0 & \Delta\boldsymbol{\omega}^1 &= 0 \\ \Delta\dot{\mathbf{x}}^2 &= b\, \Delta\dot{\varphi}\, \hat{\mathbf{Y}}_1 & \Delta\boldsymbol{\omega}^2 &= -\frac{b}{a} \Delta\dot{\varphi}\, \hat{\mathbf{X}}_2^2 \\ \Delta\dot{\mathbf{x}}^3 &= 0 & \Delta\boldsymbol{\omega}^3 &= 0 \end{aligned}$$

These changes correspond to a virtual velocity of the roller: its center is virtually detached from the pendulum (body 1) along the direction $\hat{\mathbf{Y}}_1$, and the roller virtually rolls without slipping on the ring.

For these virtual changes, the power principle becomes

$$\begin{aligned} \sum_{i=1}^{3} m^i \ddot{\mathbf{x}}^i . \Delta\dot{\mathbf{x}}^i + \sum_{i=1}^{3} \dot{\mathbf{H}}^i . \Delta\boldsymbol{\omega}^i &= mb^2\, \ddot{\varphi}\, \Delta\dot{\varphi} + \frac{ma^2}{2} \left(-\frac{b}{a}\right)^2 \ddot{\varphi}\, \Delta\dot{\varphi} \\ \text{and } \Delta\mathcal{P} &= \left(-\mathbf{S}^2 + mg\, \hat{\mathbf{I}}_3\right) . b\, \Delta\dot{\varphi}\, \hat{\mathbf{Y}}_1 \end{aligned}$$

and leads to the equation 2.46:

$$S_1^2 = -\frac{3}{2} mb\,\ddot{\varphi} - mg\sin(\varphi + \alpha) \tag{2.91}$$

2.4.5 Application of the Lagrange multiplier technique

In the previous example, the virtual velocity changes were chosen to be compatible with the constraints. An alternative approach consists in choosing *arbitrary* virtual changes of the generalized velocities. This means for the pendulum that we consider that the three virtual changes

$$\Delta\dot{q} = \{\Delta\dot{\varphi}, \Delta\dot{\varphi}_2, \Delta\dot{\varphi}_3\}$$

are independent.

The left-hand side of the virtual power principle:

$$\begin{aligned}\sum_{i=1}^{3} m^i \ddot{\mathbf{x}}^i \,.\, \Delta\dot{\mathbf{x}}^i + \sum_{i=1}^{3} \dot{\mathbf{H}}^i \,.\, \Delta\boldsymbol{\omega}^i &= \left(Ml^2 + 2mb^2\right) \ddot{\varphi}\,\Delta\dot{\varphi} \\ &\quad + \frac{ma^2}{2}\,\ddot{\varphi}_2\,\Delta\dot{\varphi}_2 \\ &\quad + \frac{ma^2}{2}\,\ddot{\varphi}_3\,\Delta\dot{\varphi}_3\end{aligned}$$

can thus be written in matrix form as

$$\begin{pmatrix}\Delta\dot{\varphi} & \Delta\dot{\varphi}_2 & \Delta\dot{\varphi}_3\end{pmatrix} \begin{pmatrix} \left(Ml^2 + 2mb^2\right)\ddot{\varphi} \\ \frac{ma^2}{2}\ddot{\varphi}_2 \\ \frac{ma^2}{2}\ddot{\varphi}_3 \end{pmatrix} = \Delta\dot{q}^T\,\Phi(t, q, \dot{q}, \ddot{q})$$

On the other hand, the virtual power due to the *contributive* forces (i.e.: the gravity) of the *unconstrained* system is given in step 6 of the previous section by 2.89:

$$\Delta\mathcal{P} = -\left(Ml + 2mb\cos\alpha\right)\, g\,\sin\varphi\,\Delta\dot{\varphi}$$

or equivalently by

$$\Delta\mathcal{P} = \begin{pmatrix}\Delta\dot{\varphi} & \Delta\dot{\varphi}_2 & \Delta\dot{\varphi}_3\end{pmatrix} \begin{pmatrix} -\left(Ml + 2mb\cos\alpha\right)\, g\,\sin\varphi \\ 0 \\ 0 \end{pmatrix} = \Delta\dot{q}^T\, Q$$

However, the *actual* system is constrained (i.e. its generalized coordinates are not independent) and therefore we must also take the generalized constraint forces into account in the virtual power principle. To this end, we use the Lagrange multiplier technique and form 2.80 of the equations of motion:

$$\Phi(t, q, \dot{q}, \ddot{q}) = Q + J^T(q, t)\,\lambda$$

which involves the Jacobian matrix of the constraints $J(q,t)$. This matrix is obtained from 2.87:

$$\begin{pmatrix} b & a & 0 \\ b & 0 & a \end{pmatrix} \begin{pmatrix} \Delta\dot{\varphi} \\ \Delta\dot{\varphi}_2 \\ \Delta\dot{\varphi}_3 \end{pmatrix} = J\,\Delta\dot{q} = 0$$

The form 2.80 of the equations of motion is then

$$\begin{pmatrix} \left(Ml^2 + 2mb^2\right)\ddot{\varphi} \\ \frac{ma^2}{2}\ddot{\varphi}_2 \\ \frac{ma^2}{2}\ddot{\varphi}_3 \end{pmatrix} = \begin{pmatrix} -\left(Ml + 2mb\cos\alpha\right) g \sin\varphi \\ 0 \\ 0 \end{pmatrix} + \begin{pmatrix} b & b \\ a & 0 \\ 0 & a \end{pmatrix} \begin{pmatrix} \lambda_1 \\ \lambda_2 \end{pmatrix} \tag{2.92}$$

where the λ are the (unknown) Lagrange multipliers associated with the two constraints 2.28 and 2.29. Together with these two constraint equations, the three dynamical equations 2.92 form a set of five DAE for the five unknowns q and λ.

As explained in section 2.4.2, we can eliminate the Lagrange multipliers from this set of DAEs. To do this, we look for a non singular submatrix J_v of J, and we partition J as follows

$$J = \begin{pmatrix} J_u & J_v \end{pmatrix} \quad \text{with}\, J_u = \begin{pmatrix} b \\ b \end{pmatrix} \quad \text{and} \quad J_v = \begin{pmatrix} a & 0 \\ 0 & a \end{pmatrix}$$

We then partition the matrices $\Phi(t,q,\dot{q},\ddot{q})$ and Q accordingly:

$$\Phi = \begin{pmatrix} \Phi_u \\ \Phi_v \end{pmatrix} \quad \text{and} \quad Q = \begin{pmatrix} Q_u \\ Q_v \end{pmatrix}$$

where

$$\Phi_u(t,q,\dot{q},\ddot{q}) = \left(Ml^2 + 2mb^2\right)\ddot{\varphi} \quad \text{and} \quad \Phi_v(t,q,\dot{q},\ddot{q}) = \begin{pmatrix} \frac{ma^2}{2}\ddot{\varphi}_2 \\ \frac{ma^2}{2}\ddot{\varphi}_3 \end{pmatrix}$$

$$Q_u = -\left(Ml + 2mb\cos\alpha\right) g \sin\varphi \quad \text{and} \quad Q_v = \begin{pmatrix} 0 \\ 0 \end{pmatrix}$$

Finally, we eliminate the λ's and obtain the reduced set of dynamical equation 2.84:

$$\Phi_u(t,q,\dot{q},\ddot{q}) - (J_u)^T \left((J_v)^T\right)^{-1} \Phi_v(t,q,\dot{q},\ddot{q}) = Q_u - (J_u)^T \left((J_v)^T\right)^{-1} Q_v$$

which reads, in the present case,

$$\left(Ml^2 + 2mb^2\right)\ddot{\varphi} - \frac{b}{a}\left(\frac{ma^2}{2}\ddot{\varphi}_2 + \frac{ma^2}{2}\ddot{\varphi}_3\right) = -\left(Ml + 2mb\cos\alpha\right) g \sin\varphi \tag{2.93}$$

This equation constitutes, with the two constraints, the *reduced* system of DAEs for the three generalized coordinates $\{\varphi, \varphi_2, \varphi_3\}$ of the pendulum.

In general, the constraints are nonlinear with respect to the generalized coordinates. However, in our particular example, we can integrate the constraints 2.28 and 2.29

$$\dot{\varphi}b + \dot{\varphi}_2 a = 0 \quad \text{and} \quad \dot{\varphi}b + \dot{\varphi}_3 a = 0$$

to obtain

$$\begin{aligned} \varphi_2(t) - \varphi_2(0) &= -\frac{b}{a}\left(\varphi(t) - \varphi(0)\right) \\ \varphi_3(t) - \varphi_3(0) &= -\frac{b}{a}\left(\varphi(t) - \varphi(0)\right) \end{aligned}$$

We can thus eliminate the generalized coordinates φ_2 and φ_3 (and their derivatives) analytically from the reduced system 2.93 by means of the constraints. This leads to a single ODE in the independent coordinate φ:

$$\left(Ml^2 + 3mb^2\right)\ddot{\varphi} = -\left(Ml + 2mb\cos\alpha\right) g \sin\varphi$$

which is the equation of motion we found in 2.45.

At this point we should note that, in general, and contrary to the case in the example, it is often not possible to use the constraints to eliminate all the dependent coordinates analytically. In general, we must deal with a set of DAEs containing the dynamical and the constraint equations. A method for numerically solving a system of DAEs will be given in section 4.6.1 and chapter 10.

2.5 Appendix

Consider m' ($m' \leq n$) constraints acting on n generalized coordinates:

$$h(q,t) = 0 \quad \forall t$$

and suppose that the first one, $h_1(q,t)$, is dependent from the others. This means that $\forall t$:

$$h_1(q,t) = 0 \quad \forall q \in \Re^n \text{ satisfying:} \quad h_j(q,t) = 0, \quad j = 2, ..., m' \tag{2.94}$$

Considering a fixed value of time t and a particular value q_0 of q which satisfies

$$h_j(q_0,t) = 0, \quad j = 2, ..., m' \tag{2.95}$$

we may assert that, since 2.94 is satisfied $\forall q \in \Re^n$, it must *a fortiori* be satisfied in a neighborhood of q_0, i.e.:

$$h_1(q_0 + \delta q, t) = 0 \quad \forall \delta q \in \Re^n \text{ satisfying:} \quad h_j(q_0 + \delta q, t) = 0, \quad j = 2, ..., m' \tag{2.96}$$

Since $h_1(q_0, t) = 0$ from 2.95 and 2.94, and taking the infinitesimal character of δq into account, 2.96 also reads:

$$\sum_{\alpha=1}^{n} \frac{\partial h_1}{\partial q_\alpha} \delta q_\alpha = 0 \quad \forall \delta q \in \Re^n \text{ satisfying: } \quad \sum_{\alpha=1}^{n} \frac{\partial h_j}{\partial q_\alpha} \delta q_\alpha = 0, \quad j = 2, ..., m' \tag{2.97}$$

Using the following matrix notation:

$$\frac{\partial h_j}{\partial q^T} \triangleq \left(\frac{\partial h_j}{\partial q_1}, ..., \frac{\partial h_j}{\partial q_n} \right) \triangleq \left(\frac{\partial h_j}{\partial q} \right)^T$$

2.97 can be rewritten as

$$\frac{\partial h_1}{\partial q^T} \delta q = 0 \quad \forall \delta q \in \Re^n \text{ satisfying: } \quad \frac{\partial h_j}{\partial q^T} \delta q = 0, \quad j = 2, ..., m' \tag{2.98}$$

Considering $\Re^n$ as a vector space, this new formulation 2.98 states that *any* vector δq which is orthogonal to vectors $\frac{\partial h_j}{\partial q}$ (for $j = 2, ..., m'$) is automatically orthogonal to vector $\frac{\partial h_1}{\partial q}$. This implies that $\frac{\partial h_1}{\partial q}$ belongs to the vector space spanned by the $\frac{\partial h_j}{\partial q}$, or in other words that $\frac{\partial h_1}{\partial q}$ is a linear combination of the vectors $\frac{\partial h_j}{\partial q}$. Therefore, the rows of the $(m' \times n)$ Jacobian matrix:

$$\frac{\partial h}{\partial q^T} \triangleq \begin{pmatrix} \frac{\partial h_1}{\partial q_1} & \frac{\partial h_1}{\partial q_2} & ... & \frac{\partial h_1}{\partial q_n} \\ ... & ... & ... & ... \\ \frac{\partial h_{m'}}{\partial q_1} & ... & ... & \frac{\partial h_{m'}}{\partial q_n} \end{pmatrix}$$

are linearly dependent.

Conversely, if matrix $\frac{\partial h}{\partial q^T}$ has maximal[19] rank m', neither condition 2.98 nor 2.96 can be satisfied. The linearized forms (in the neighborhood of q_0) of the constraints:

$$\frac{\partial h(q_0, t)}{\partial q^T} \delta q = 0$$

are thus independent. In this case we say that the functions h_i are *functionally independent*[20].

[19] It is assumed here that the number m' of constraints is less than or equal to the number n of coordinates.

[20] The full rank condition on the Jacobian matrix is a necessary condition for the independency of the constraints. It is a sufficient condition for their *functional* independency.

Chapter 3

Tree-like multibody structures

3.1 Definitions, conventions and hypotheses

3.1.1 Fundamental concepts

A multibody system (MBS) is a set of N^{body} *rigid bodies* interconnected by *joints*. As illustrated in figure 3.1, mechanisms, road and railway vehicles, industrial or space manipulators, etc... are typical examples of multibody systems.

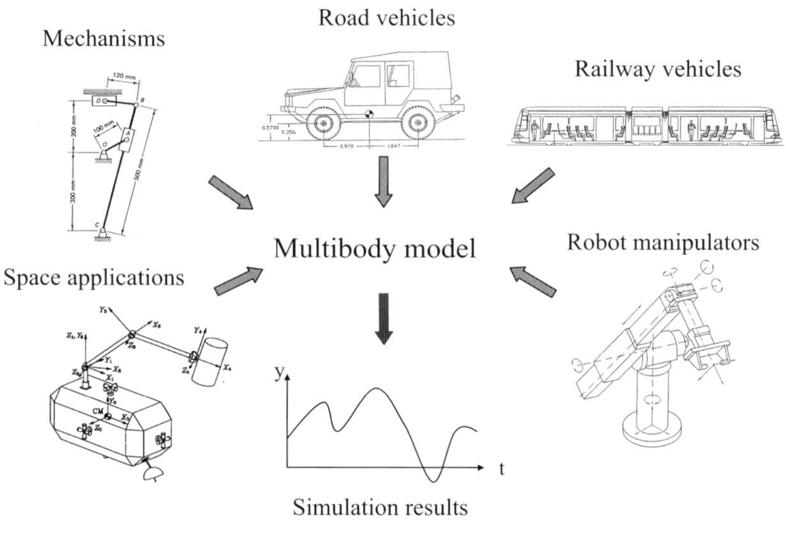

Figure 3.1: Some families of multibody systems

The *joints* (hinges, universal joints, bearings, ...) are the physical devices of the system which allow relative motion between two contiguous bodies. In the multibody model, a joint will be considered as a massless element which connects two bodies at reference *attachment points* (see figure 3.2). At least one joint of the system is directly connected to an inertial body called the *base* (the earth for most practical purposes).

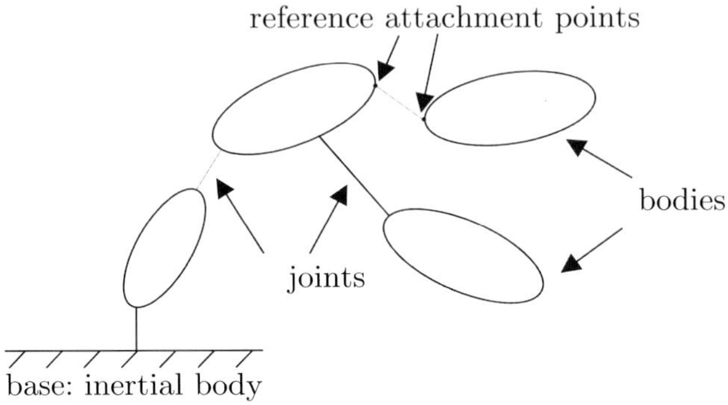

Figure 3.2: Multibody system

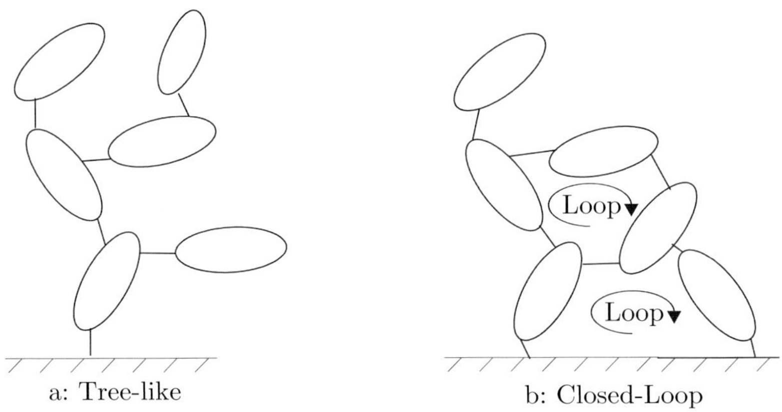

Figure 3.3: Tree-like versus closed-loop structures

The development of theoretical formalisms in multibody dynamics requires the use of specific concepts, such as kinematic chain, body filiation, open structure, which must be clearly defined. For this purpose, *tree-like structures* — i.e. structures without loops (see figure 3.3_a) — will be considered first. The extension of the various formalisms to MBS with kinematic loops (figure 3.3_b) or other types of constraints will be presented in chapter 4.

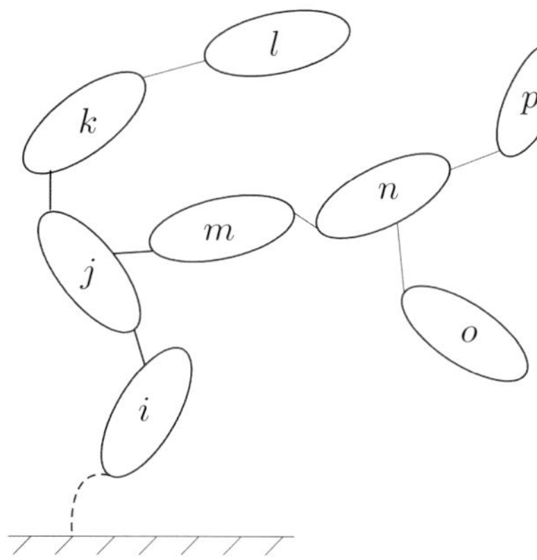

Figure 3.4: Kinematic chain and filiation

A tree-like MBS has the following characteristics:

- the number of rigid bodies (excluding the base) is equal to the number of joints,
- there is one and only one path from one rigid body (including the base) to any other rigid body in the system.

Tree-like structure allows us to use the following *filiation* concepts[1]:

- *Kinematic chain*: an ordered set of several (≥ 1) consecutive bodies. In figure 3.4 for instance, $\{i, j, k\}$, $\{l, k, j, m, n\}$ represent kinematic chains, while $\{j, n, o\}$ and $\{i, k, j\}$ are not.
- *Base body*: the body of the MBS which is assumed to be inertially fixed.
- *Ancestor*: body i is an ancestor of body j if it belongs to the kinematic chain starting from the inertial body (the base) to body j (excluded). The base is an ancestor of all the bodies of the structure.
- *Descendant*: if body i is an ancestor of body j, then body j is said to be a descendant of body i.
- *Parent*: body i is the *parent* of body j if it is the direct ancestor of body j, i.e. the one which is directly connected to it. In figure 3.4, body j is the parent of bodies k and m. In a tree connected to a base, a body can have only one parent; otherwise there is a loop as shown in figure 3.3_b.

[1] Graph matrices, originally introduced in [94], have been abandoned in favour of this more readily usable notion of filiation in order to simplify the introduction of system data [75].

- *Child*: body j is called a *child* of body i, if body i is the parent of body j. By extension, the joint which connects body i to its child j will be qualified as a *child joint* of body i. In a tree, a child has only one parent but a parent may have more than one child. In sections 3.3 and 3.4, the parent of a body i will be often simply denoted by h while the collection of children of body i will be denoted by $\bar{\imath}$.

- *Leaf body*: denotes a terminal body of the tree-like structure. A leaf body has no children.

Let illustrate these definitions on the example of figure 3.4:
- body j is ancestor of bodies k, l, m, n, o, p,
- body n is the parent of bodies o and p,
- body k is a descendant of bodies i, j, and a child of body j,
- the collection $\bar{\jmath}$ of children of body j is $\{k, m\}$,
- bodies l, p, o are leaf bodies.

3.1.2 Topology

As regards the *topology*, i.e. the way the bodies are interconnected in the tree-like structure, the following conventions and definitions are adopted:

Conventions[2]:

1. Bodies are numbered in ascending order starting from the base (index 0) to the leaf bodies, as illustrated in figure 3.5.

2. A joint which precedes a body receives the same index as this body.

Definitions:

- The topological vector *inbody* is defined as the vector whose i^{th} element contains the index of the parent of body i (see the example of figure 3.5).

In the following equations, summations of quantities related to bodies (or joints) of a given kinematic chain will often be used. The following notations are adopted[3]:

- $\sum\limits_{i} a^i$: summation over all bodies ($i = 1, ..., N^{body}$) of the structure,

- $\sum\limits_{i:\, i<j} a^i$: summation over all the ancestors of body j,

- $\sum\limits_{j:\, i<j} a^j$: summation over all the descendants of body i,

[2] These conventions are adopted mainly for convenience in the computer implementation.

[3] Although inequality signs are used in these notations, the reader must take care of the fact that $i < j$ simply means that body i is an ancestor of body j. Thus $i \nless j$ does not imply $i \geq j$. The relationship between bodies is therefore not an ordering in the mathematical sense.

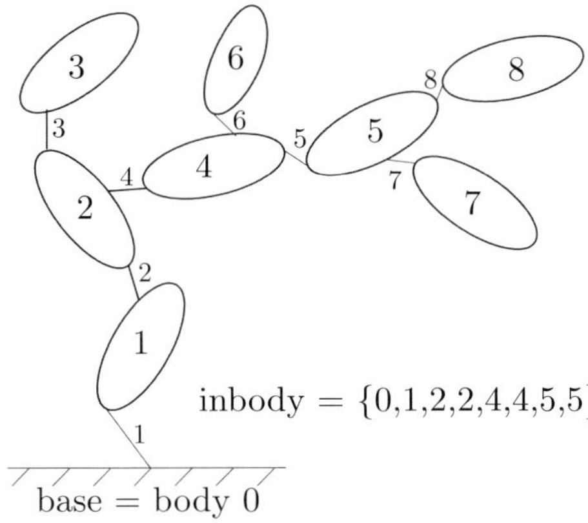

Figure 3.5: Numbering of joints and bodies

- $\sum\limits_{i:\, i \leq j} a^i$, $\sum\limits_{j:\, i \leq j} a^j$: summation over body j and all its ancestors, and summation over body i and all its descendants, respectively,
- $\sum\limits_{i:\, h \leq i < j} a^i$: summation over body h and bodies i which are both descendant of h and ancestors of j,
- $\sum_{j \in \bar{i}}$: summation over the children of body i.

3.1.3 Kinematics: main definitions

Notations

On the basis of figure 3.6, we define the following quantities related to body i, its parent h and its children j, k.

- O^i, O'^i, the reference attachment points of joint i on the parent body h, and its child i respectively,
- $\mathbf{z}^i$, the relative position vector $\overrightarrow{O^i\, O'^i}$, which represents the relative displacement in joint i,
- $\mathbf{d}^{ik}$ (see figure 3.6) is the position vector (with respect to O'^i) of the reference attachment point O^k of joint k, child of body i. This vector, which is fixed on body i, represents the contribution of body i to the kinematic chain which links body k to the base. It also represents the contribution of body i to the kinematic chains which link the descendants of body k (i.e. bodies l and m in figure 3.6) to the base. Thus, by extension, we define $\mathbf{d}^{il} \triangleq \mathbf{d}^{im} \triangleq \mathbf{d}^{ik}$ to denote the same vector $\overrightarrow{O'^i\, O^k}$.
  Since in figure 3.6, body h is not a descendant of body i (and $h \neq i$), then by definition $\mathbf{d}^{ih} \triangleq 0$. Similarly: $\mathbf{d}^{ji} = \mathbf{d}^{jk} = \mathbf{d}^{jl} = \mathbf{d}^{jm} \triangleq 0$,

- $\mathbf{p}^i$, the absolute position vector of the attachment point O^i.

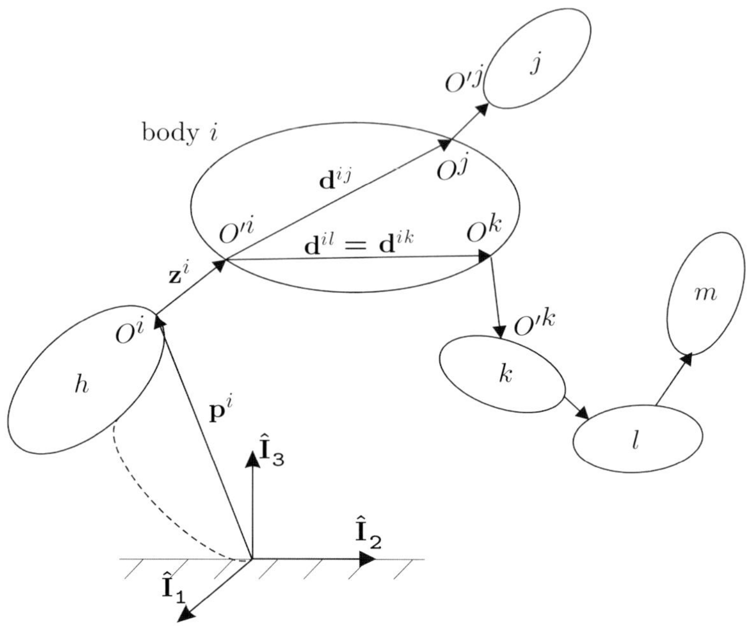

Figure 3.6: Bodies and joints: kinematic notations

Finally, for conciseness, it is convenient to define the following extended geometrical vectors (see figure 3.7):

- $\mathbf{d}_z^{ij} \triangleq \mathbf{z}^i + \mathbf{d}^{ij}$, $\mathbf{d}_z^{ik} \triangleq \mathbf{z}^i + \mathbf{d}^{ik}$, the *extended* position vectors of joints j and k respectively.

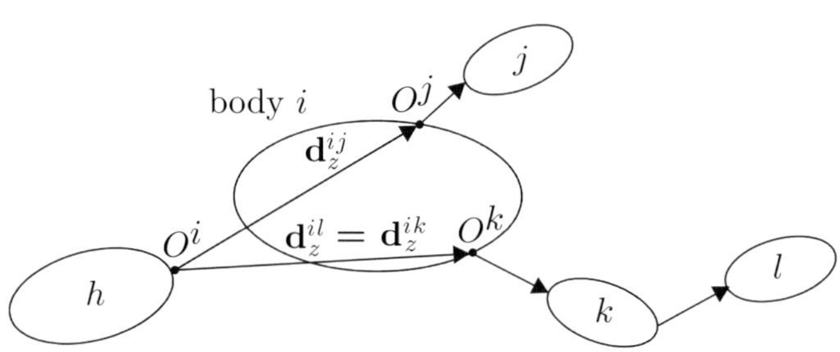

Figure 3.7: Extended geometrical vectors

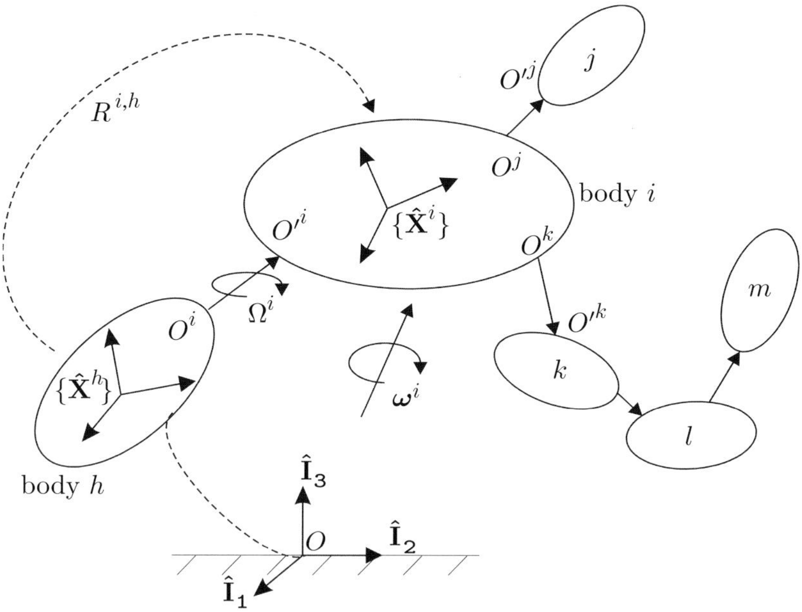

Figure 3.8: Bodies and joints: kinematic notations

The orientation of the various bodies is described by means of the following quantities, illustrated in figure 3.8.

- $\{O, \{\hat{\mathbf{I}}\}\}$, the inertial frame rigidly fixed to the base (body 0),
- $\{\hat{\mathbf{X}}^i\}$, a moving frame rigidly attached to body i. When expressed in this frame, the geometrical vectors $\mathbf{d}^{ij} = [\hat{\mathbf{X}}^i]^T\, d^{ij}$ and $\mathbf{d}^{ik} = [\hat{\mathbf{X}}^i]^T\, d^{ik}$ have constant components d^{ij} and d^{ik} since the bodies are rigid,
- $R^{i,h}$, the rotation matrix[4] such that $[\hat{\mathbf{X}}^i] = R^{i,h}[\hat{\mathbf{X}}^h]$,
- $\boldsymbol{\Omega}^i$, the relative angular velocity vector[5] of body i with respect to its parent h,
- $\boldsymbol{\omega}^i = [\hat{\mathbf{X}}^i]^T\, \omega^i$, the absolute angular velocity vector of the body-fixed frame $\{\hat{\mathbf{X}}^i\}$.

---

[4] As mentioned in section 1.1.6, the notation wich is used for denoting rotation matrices should not be confusing. In our case, $R^{i,h}$ is the rotation matrix defined such that $[\hat{\mathbf{X}}^i] = R^{i,h}[\hat{\mathbf{X}}^h]$, which automatically means that $R^{h,i} = \left(R^{i,h}\right)^T$.

[5] As mentioned in 1.2.2, this vector is unique, whatever is the choice of the body fixed frame.

Joint modeling hypotheses

Joints represent mechanical connection devices such as telescopic arms, guide rods, hinges, bearing, ... The relative motion allowed by a joint arises from its physical nature; this defines its internal number of *relative degrees of freedom* (d.o.f.). Due to the multitude of different kinds of joints one may encounter in practical applications, a general purpose multibody program should normally include an exhaustive library containing all possible kinds of joints. However, the computer implementation of such an exhaustive library of joints often leads to a certain heaviness in the multibody program user interface.

On the other hand, as every physical joint can be modeled, in terms of relative motions, by an appropriate succession of prismatic and revolute joints, we can restrict the library to these two elementary single-degree-of-freedom joints, *prismatic* and *revolute*. This gives us the following modeling hypothesis:

Every connection between bodies is a single degree of freedom joint, revolute or prismatic.

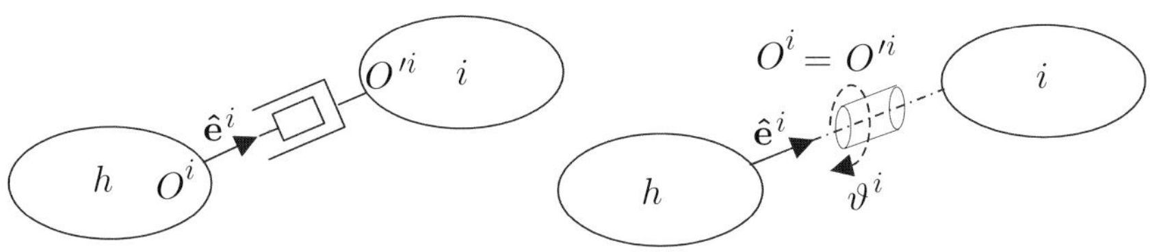

Figure 3.9: Elementary joints

This is illustrated in figure 3.9. The variables describing the relative motion in the joints (linear and angular displacements respectively for prismatic and revolute joints) will be used as *generalized coordinates* q.
For each joint i ($i = 1, ..., N^{body}$), the generalized coordinate q^i thus represents:

- if joint i is prismatic, the amplitude of the relative displacement $\overrightarrow{O^i\,O'^i}$ along the unit joint vector $\hat{\mathbf{e}}^i$,
- if joint i is revolute, the relative rotation angle ϑ^i of body i with respect to its parent around the unit joint vector $\hat{\mathbf{e}}^i$.

The kinematic expressions relating the relative configuration of a child i with respect to its parent are thus as follows:

- for a prismatic joint i :
  - the relative orientation of the body-fixed frame $\{\hat{\mathbf{X}}^i\}$ with respect to the body frame $\{\hat{\mathbf{X}}^h\}$ is given by a constant rotation matrix:

$$[\hat{\mathbf{X}}^i] = R^{i,h}[\hat{\mathbf{X}}^h] \tag{3.1}$$

  - the relative position of the attachment point O'^i, fixed on body i, with respect to the attachment point O^i fixed on the parent h, is given by:

$$\mathbf{z}^i = \overrightarrow{O^i\, O'^i} = q^i \hat{\mathbf{e}}^i$$

    which means that O'^i coincides with O^i when $q^i = 0$.

  - the joint vector $\hat{\mathbf{e}}^i$ has constant components in the $\{\hat{\mathbf{X}}^i\}$ frame, as well as in the $\{\hat{\mathbf{X}}^h\}$ frame since the rotation matrix $R^{i,h}$ is constant.

- for a revolute joint i :

  - the joint vector $\hat{\mathbf{e}}^i$ has constant components in both frames $\{\hat{\mathbf{X}}^i\}$ and $\{\hat{\mathbf{X}}^h\}$,
  - the relative orientation of the body-fixed frame $\{\hat{\mathbf{X}}^i\}$ with respect to the body frame $\{\hat{\mathbf{X}}^h\}$ is given by

$$[\hat{\mathbf{X}}^i] = R^{i,h}(q^i)[\hat{\mathbf{X}}^h] \tag{3.2}$$

    where it should be noticed that the joint angle q^i is the only time varying quantity appearing in matrix $R^{i,h}$.

  - the attachment point O'^i coincides with O^i, at some point arbitrarily chosen but located on the common revolute joint axis and thus (see figure 3.9):

$$\mathbf{z}^i = \overrightarrow{O^i\, O'^i} = 0$$

One can now define the relative velocities of body i with respect to its parent h :

$$\overset{\circ}{\mathbf{z}}{}^i = [\hat{\mathbf{X}}^i]^T \dot{z}^i \quad \text{and} \quad \tilde{\boldsymbol{\Omega}}^i = [\hat{\mathbf{X}}^i]^T R^{i,h} \left(\dot{R}^{i,h}\right)^T [\hat{\mathbf{X}}^i] \tag{3.3}$$

Using the definition of the unit joint vector $\hat{\mathbf{e}}^i$ (see figure 3.9) which has constant components in the $\{\hat{\mathbf{X}}^i\}$ frame for both kinds of joint, we can express the previous expressions simply in vector form:

$$\begin{aligned} \overset{\circ}{\mathbf{z}}{}^i &= \dot{q}^i\, \hat{\mathbf{e}}^i \ , \ \boldsymbol{\Omega}^i = 0 \quad \text{if } i \text{ is prismatic,} \\ \boldsymbol{\Omega}^i &= \dot{q}^i\, \hat{\mathbf{e}}^i \ , \ \overset{\circ}{\mathbf{z}}{}^i = 0 \quad \text{if } i \text{ is revolute.} \end{aligned}$$

For conciseness, two vector symbols, $\boldsymbol{\varphi}^i$ and $\boldsymbol{\psi}^i$, are defined for each joint i as follows:

$$\boldsymbol{\psi}^i \triangleq \hat{\mathbf{e}}^i \text{ and } \boldsymbol{\varphi}^i \triangleq 0 \quad \text{if } i \text{ is prismatic,} \tag{3.4}$$

$$\boldsymbol{\varphi}^i \triangleq \hat{\mathbf{e}}^i \text{ and } \boldsymbol{\psi}^i \triangleq 0 \quad \text{if } i \text{ is revolute.} \tag{3.5}$$

This convention allows us to write all joint relations in one compact form:

$$\overset{\circ}{\mathbf{z}}{}^{i} = \dot{q}^{i}\,\boldsymbol{\psi}^{i} \;\text{ and }\; \boldsymbol{\Omega}^{i} = \dot{q}^{i}\,\boldsymbol{\varphi}^{i}\,. \tag{3.6}$$

and similarly for the accelerations:

$$\overset{\circ\circ}{\mathbf{z}}{}^{i} = \ddot{q}^{i}\,\boldsymbol{\psi}^{i} \;\text{ and }\; \overset{\circ}{\boldsymbol{\Omega}}{}^{i} = \ddot{q}^{i}\,\boldsymbol{\varphi}^{i}\,.$$

3.1.4 Dynamics: main definitions

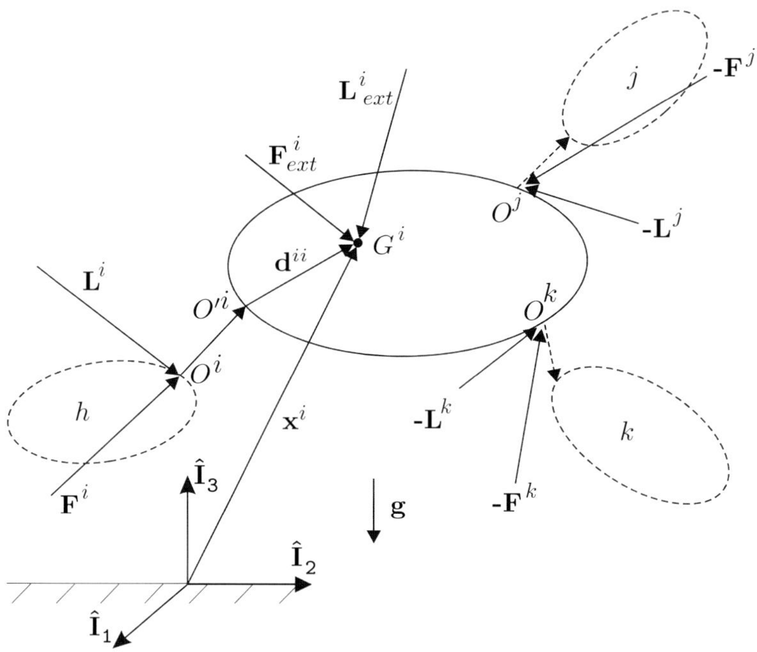

Figure 3.10: Main dynamical notations

Figure 3.10 shows the main dynamical notations which will be used:

- m^i and $\mathbf{I}^i = [\hat{\mathbf{X}}^i]^T\, I^i\, [\hat{\mathbf{X}}^i]$, respectively the mass and the inertia tensor of body i with respect to its center of mass G^i,
- $\mathbf{d}^{ii} = [\hat{\mathbf{X}}^i]^T\, d^{ii}$, the position vector of the center of mass G^i with respect to O'^i. Since the bodies are rigid, the components d^{ii} and I^i are constant in the body-fixed frame $\{\hat{\mathbf{X}}^i\}$,
- $\mathbf{x}^i = [\hat{\mathbf{I}}]^T\, x^i$, the absolute position vector of the center of mass G^i,
- $\mathbf{g} = [\hat{\mathbf{I}}]^T\, g$, the gravity vector[6],

[6] The gravity field is supposed to be uniform; the resultant force due to gravity and applied to the mass center G^i is thus equal to $m^i\mathbf{g}$.

- $\mathbf{F}^i$, $\mathbf{L}^i$, respectively the resultant force and torque[7] applied to body i by its parent h through the joint i, evaluated at point O^i. According to Newton's third law, reactions $-\mathbf{F}^i$ and $-\mathbf{L}^i$ are applied on body h. Similarly in figure 3.10, body i undergoes the reactions $-\mathbf{F}^j$, $-\mathbf{F}^k$, $-\mathbf{L}^j$, and $-\mathbf{L}^k$ due to its children joints,

- $\mathbf{F}^i_{ext}$, $\mathbf{L}^i_{ext}$, the external loads acting on body i (excluding gravity as well as the internal forces and torques transmitted by the joints) under the form of an equivalent resultant force applied to the center of mass G^i and a resultant torque.

Finally, as for the geometrical vectors $\mathbf{d}^{ij}$, it is convenient to define an *extended* vector for the mass center position:

- $\mathbf{d}^{ii}_z \triangleq \mathbf{z}^i + \mathbf{d}^{ii}$, which represents the position vector of the mass center G^i with respect to O^i.

3.1.5 Barycentric quantities

The dynamics of a multibody system depends on the mass distribution of each body. The related parameters, the mass, the location of the center of mass and the inertia tensor could thus be considered individually for each body as in the previous section. However, it will be seen later that these parameters appear in the dynamical equations in constant combinations. The latter, denoted as *barycentric* quantities, are mass characterizations which take the topology of the system into account.

The *barycentric mass* $\bar{m}^i$ of body i is defined as the mass of the body itself, augmented by the total mass of all its descendants:

$$\bar{m}^i \triangleq \sum_{j:\, i \leq j} m^j \tag{3.7}$$

According to this definition, the barycentric mass of a leaf body is equal to the mass of the body itself. Starting from the leaf bodies of the multibody system, we can define the barycentric masses of the other bodies in a (backward) recursive manner as follows:

$$\bar{m}^i \triangleq m^i + \sum_{j \in \bar{i}} \bar{m}^j$$

in which the summation "$j \in \bar{i}$" concerns only the children of body i.

The *augmented body* i is then defined as being the body i augmented by point masses $\bar{m}^j$ which are located at the attachment points O^j of its different children j (see figure 3.11). The mass of this augmented body i is thus equal

[7] As explained in section 1.2.7, any system of forces $\{\mathbf{F}^i, A^i\}$ (and pure torques $\mathbf{M}^j$) acting on a given rigid body can be replaced by a *resultant force* acting on a reference body-fixed point A and a *resultant torque* $\mathbf{L}^A$.

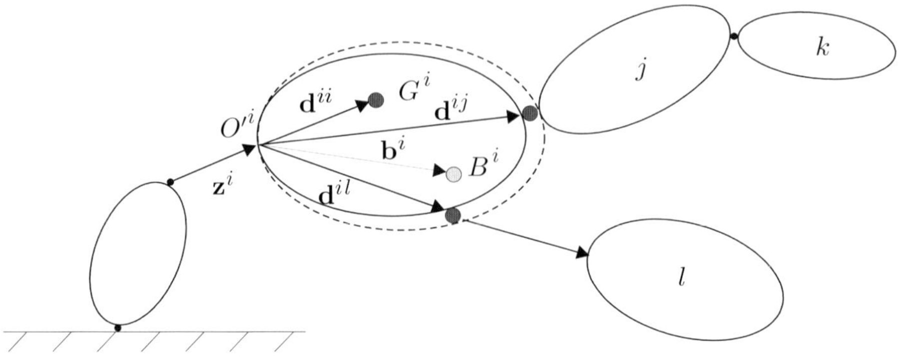

Figure 3.11: Augmented body i

to $\bar{m}^i$. Moreover, this augmented body is rigid: the location $\mathbf{b}^i$ of its center of mass B^i, denoted as the *barycenter*, with respect to the reference point O'^i is given by the *barycentric moment* $\bar{m}^i\mathbf{b}^i$ which is defined as (see equation 1.71)

$$\bar{m}^i\mathbf{b}^i \triangleq m^i\mathbf{d}^{ii} + \sum_{j \in \bar{i}} \bar{m}^j\mathbf{d}^{ij}$$

or, equivalently by

$$\bar{m}^i\mathbf{b}^i \triangleq \sum_{j:\, i \leq j} m^j\mathbf{d}^{ij} \tag{3.8}$$

The inertia tensor of the augmented body with respect to O'^i, called the *barycentric tensor* $\mathbf{K}^i$ is given by[8]

$$\mathbf{K}^i \triangleq \mathbf{I}^i - m^i\tilde{\mathbf{d}}^{ii}.\tilde{\mathbf{d}}^{ii} - \sum_{j \in \bar{i}} \bar{m}^j\tilde{\mathbf{d}}^{ij}.\tilde{\mathbf{d}}^{ij}$$

$$\text{or equivalently by} \quad \mathbf{K}^i \triangleq \mathbf{I}^i - \sum_{j:\, i \leq j} m^j\tilde{\mathbf{d}}^{ij}.\tilde{\mathbf{d}}^{ij} \tag{3.9}$$

Since the augmented body is rigid, it must be emphasized that the barycentric moment vector $\bar{m}^i\mathbf{b}^i$ and the barycentric tensor $\mathbf{K}^i$ have constant components if they are expressed in the body-fixed frame $\{\hat{\mathbf{X}}^i\}$. These constant components can thus be computed at once before implementing the dynamical equations in a computer code.

For conciseness, the following *extended* barycentric quantities (see figure 3.12) will also be used temporarily during the developments of the dynamical equations:

[8] by means of the Steiner formula 1.87

- the barycentric moment with respect to O^i :

$$\bar{m}^i \mathbf{b}_z^i \triangleq \sum_{j:\, i \leq j} m^j \mathbf{d}_z^{ij} = \bar{m}^i \left(\mathbf{b}^i + \mathbf{z}^i \right) \tag{3.10}$$

- the barycentric tensor with respect to O^i :

$$\mathbf{K}_z^i \triangleq \mathbf{I}^i - \sum_{j:\, i \leq j} m^j \tilde{\mathbf{d}}_z^{ij} \,.\, \tilde{\mathbf{d}}_z^{ij} = \mathbf{K}^i - \bar{m}^i \left(\tilde{\mathbf{b}}^i \,.\, \tilde{\mathbf{z}}^i + \tilde{\mathbf{z}}^i \,.\, \tilde{\mathbf{b}}^i + \tilde{\mathbf{z}}^i \,.\, \tilde{\mathbf{z}}^i \right) \tag{3.11}$$

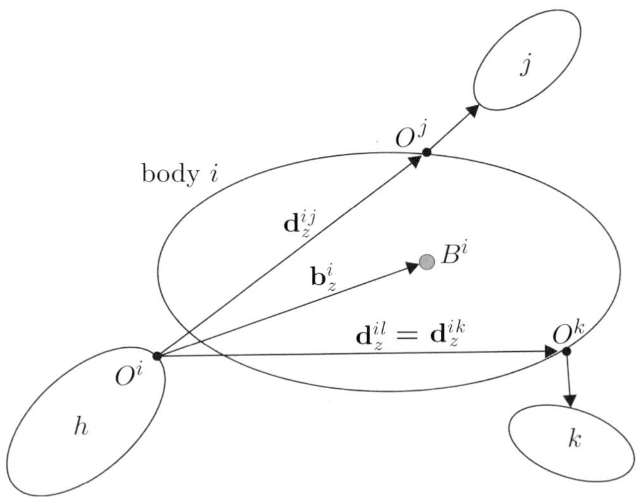

Figure 3.12: Extended barycentric vectors

Like the extended geometrical vectors $\mathbf{d}_z^{ij}$, these quantities do not necessarily have constant components, even if they are expressed in the body-fixed frame $\{\hat{\mathbf{X}}^i\}$.

3.2 Virtual power principle

3.2.1 Introduction

In this section, we use the virtual power principle to generate the dynamical equations of tree-like multibody systems. This approach offers two advantages:

- The generalized coordinates correspond to the relative motions in the joints (linear and angular displacements respectively for prismatic and revolute joints). Therefore, we will expand all the kinematic relations of the MBS in terms of these coordinates (and their derivatives). This will allows us to show explicitly the dependence of the different terms of the dynamical equations on the accelerations, the velocities and generalized coordinates. The structure of these equations will also show which terms correspond to inertia, Coriolis and centrifugal effects; and it will indicate where the forces and torques applied to the MBS influence its dynamical behavior.

- The mathematical structure of the dynamical equations obtained from the virtual power principle will allow us to use additional mathematical tools — such as the Lagrange multiplier technique (see section 2.4.2) — to deal with multibody systems which are more complex than tree-like systems. Such extensions are proposed in chapter 4 for dealing with practical multibody problems.

3.2.2 Kinematics and virtual velocity changes

The absolute position of the center of mass G^j of body j is obtained by running through the kinematic chain which connects this body to the base (see figures 3.6 and 3.7) and is thus given by

$$\mathbf{x}^j = \sum_{i:\, i \leq j} (\mathbf{z}^i + \mathbf{d}^{ij}) = \sum_{i:\, i \leq j} \mathbf{d}_z^{ij} \tag{3.12}$$

Similarly, by additivity of angular (relative) velocities, the absolute angular velocity vector $\boldsymbol{\omega}^j$ of the body-fixed frame $\{\hat{\mathbf{X}}^j\}$ is equal to (see figure 3.8)

$$\boldsymbol{\omega}^j = \sum_{i:\, i \leq j} \boldsymbol{\Omega}^i \tag{3.13}$$

Taking time derivatives, we find the absolute velocity of the center of mass G^j to be

$$\dot{\mathbf{x}}^j = \sum_{i:\, i \leq j} \dot{\mathbf{d}}_z^{ij} = \sum_{i:\, i \leq j} \left(\overset{\circ}{\mathbf{z}}{}^i + \tilde{\omega}^i \,.\, \mathbf{d}_z^{ij} \right) \tag{3.14}$$

where the vector $\overset{\circ}{\mathbf{z}}{}^i = [\hat{\mathbf{X}}^i]^T \dot{z}^i$. Then, the linear and angular accelerations are given by

$$\begin{aligned} \dot{\boldsymbol{\omega}}^j &= \sum_{i:\, i \leq j} \dot{\boldsymbol{\Omega}}^i = \sum_{i:\, i \leq j} \left(\overset{\circ}{\boldsymbol{\Omega}}{}^i + \tilde{\omega}^i \,.\, \boldsymbol{\Omega}^i \right) \\ \ddot{\mathbf{x}}^j &= \sum_{i:\, i \leq j} \ddot{\mathbf{d}}_z^{ij} = \sum_{i:\, i \leq j} \left(\overset{\circ\circ}{\mathbf{z}}{}^i + 2\tilde{\omega}^i \,.\, \overset{\circ}{\mathbf{z}}{}^i + \tilde{\dot{\omega}}^i \,.\, \mathbf{d}_z^{ij} + \tilde{\omega}^i \,.\, \tilde{\omega}^i \,.\, \mathbf{d}_z^{ij} \right) \end{aligned} \tag{3.15}$$

The kinematics developed in this section only takes care of the system topology: no hypothesis is made at this point concerning the nature of the physical joints. Hence, apart from the fact that the relative velocity vector $\overset{\circ}{\mathbf{z}}{}^i$ is supposed to be expressed with respect to the frame $\{\hat{\mathbf{X}}^i\}$ (see equation 3.3), no particular assumption is made on the relations between the vectors $\overset{\circ}{\mathbf{z}}^i$ and $\mathbf{\Omega}^i$ and the generalized joint velocities $\dot{q}^i$. In a preliminary step, we also express all the virtual velocity changes in the virtual power principle in terms of vectors:

$$\Delta\omega^j = \sum_{i:\, i \leq j} \Delta\mathbf{\Omega}^i \tag{3.16}$$

$$\begin{aligned} \Delta\dot{\mathbf{x}}^j &= \sum_{i:\, i \leq j} \left(\Delta \overset{\circ}{\mathbf{z}}^i + \Delta\tilde{\omega}^i \,.\, \mathbf{d}_z^{ij}\right) \\ &= \sum_{i:\, i \leq j} \Delta \overset{\circ}{\mathbf{z}}^i - \sum_{i:\, i \leq j}\ \sum_{k:\, k \leq i} \tilde{\mathbf{d}}_z^{ij} \,.\, \Delta\mathbf{\Omega}^k \end{aligned} \tag{3.17}$$

This deliberate choice will allow us to obtain the vector equations associated with the vector force $\mathbf{F}^i$ and the vector torque $\mathbf{L}^i$ transmitted by the joint i.

3.2.3 Translational vector equation

A vector equation relating the vector force $\mathbf{F}^i$ which is applied through the joint i (at point O^i) on body i by its parent will be obtained by choosing particular virtual velocity changes as follows:

$$\Delta \overset{\circ}{\mathbf{z}}^i \neq 0 \quad \text{and} \quad \begin{cases} \Delta \overset{\circ}{\mathbf{z}}^k = 0 & \forall k \neq i \\ \Delta\mathbf{\Omega}^k = 0 & \forall k \end{cases} \tag{3.18}$$

These changes correspond to the following virtual motion: joint i, body i and all its descendants are "virtually moving" at the same velocity $\Delta\dot{\mathbf{x}}^j = \Delta \overset{\circ}{\mathbf{z}}^i$ ($\forall j : i \leq j$), while the rest of the multibody system is frozen. This virtual motion is illustrated in figure 3.13. In this way, the *internal* joint forces and torques do not produce any virtual power change (thanks to Newton's third law of action and reaction, see section 2.4), except the joint force $\mathbf{F}^i$ which will produce a virtual power change $\Delta\mathcal{P}_{\mathbf{F}_i} = \mathbf{F}^i \,.\, \Delta \overset{\circ}{\mathbf{z}}^i$.
The virtual power principle 2.51:

$$\sum_{j=1}^{N^{body}} m^j \ddot{\mathbf{x}}^j \,.\, \Delta\dot{\mathbf{x}}^j + \sum_{j=1}^{N^{body}} \dot{\mathbf{H}}^j \,.\, \Delta\omega^j = \Delta\mathcal{P}$$

thus reduces with this particular choice of virtual changes to

$$\sum_{j:\, i \leq j} m^j \ddot{\mathbf{x}}^j \,.\, \Delta \overset{\circ}{\mathbf{z}}^i = \mathbf{F}^i \,.\, \Delta \overset{\circ}{\mathbf{z}}^i + \sum_{j:\, i \leq j} \left(\mathbf{F}_{ext}^j + m^j\mathbf{g}\right) \,.\, \Delta \overset{\circ}{\mathbf{z}}^i \tag{3.19}$$

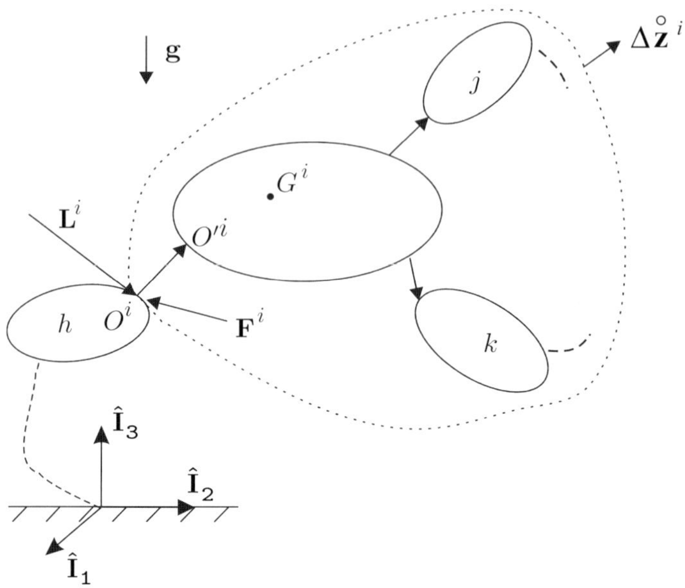

Figure 3.13: Illustration of the virtual velocity field

Since this equation must be satisfied for any vector change $\Delta\overset{\circ}{\mathbf{z}}^{i}$, by using 3.15, we obtain the vector equation

$$\mathbf{F}^i = \sum_{j:\, i \leq j} \left(m^j \sum_{k:\, k \leq j} \ddot{\mathbf{d}}_z^{kj} - \mathbf{F}^j_{ext} - m^j \mathbf{g} \right) \tag{3.20}$$

One may note that this vector equation, which is obtained here by means of the virtual power principle, can also be considered as the global Newton equation for the collection of bodies which is constituted by body i and all its descendants.

The double sum with appears in this equation can be simplified by looking at the respective value of the indices k and i. Indeed

$$\text{if} \quad k < i \leq j, \quad \text{then by definition:} \quad \mathbf{d}_z^{kj} = \mathbf{d}_z^{ki}$$

and the joint vector equation can be rewritten as

$$\mathbf{F}^i = \sum_{j:\, i \leq j} m^j \sum_{k:\, k < i} \ddot{\mathbf{d}}_z^{ki} + \sum_{j:\, i \leq j} \sum_{k:\, i \leq k \leq j} m^j \ddot{\mathbf{d}}_z^{kj} - \sum_{j:\, i \leq j} \mathbf{F}^j_{ext} - \sum_{j:\, i \leq j} m^j \mathbf{g}$$

where in the first term and the coefficient of $\mathbf{g}$, we recognize the barycentric mass $\bar{m}^i$. Moreover, by reordering the indices of the second term:

$$\sum_{j:\, i \leq j} \; \sum_{k:\, i \leq k \leq j} \equiv \sum_{k:\, i \leq k} \; \sum_{j:\, k \leq j} \tag{3.21}$$

we obtain

$$\mathbf{F}^i = \bar{m}^i \sum_{k:\, k<i} \ddot{\mathbf{d}}_z^{ki} + \sum_{k:\, i \le k} \sum_{j:\, k \le j} m^j \ddot{\mathbf{d}}_z^{kj} - \sum_{j:\, i \le j} \mathbf{F}_{ext}^j - \bar{m}^i \mathbf{g}$$

where the (second derivative) of the extended barycentric moment vector $\bar{m}^k \mathbf{b}_z^k$ now appears (see equation 3.10):

$$\mathbf{F}^i = \bar{m}^i \sum_{k:\, k<i} \ddot{\mathbf{d}}_z^{ki} + \sum_{k:\, i \le k} \bar{m}^k \ddot{\mathbf{b}}_z^k - \sum_{j:\, i \le j} \mathbf{F}_{ext}^j - \bar{m}^i \mathbf{g} \tag{3.22}$$

Thanks to the definitions of the barycentric quantities, all double summations have thus been avoided in this final vector form of the translational equation.

3.2.4 Rotational vector equation

In a similar way, a vector rotational equation can be obtained for the joint i. The virtual velocity changes which will isolate the resultant torque $\mathbf{L}^i$ applied through the joint i on body i by its parent are given by

$$\Delta\mathbf{\Omega}^i \neq 0$$

$$\text{and} \quad \left\langle \begin{array}{ll} \Delta\mathbf{\Omega}^k = 0 & \forall k \neq i \\ \Delta \overset{\circ}{\mathbf{z}}^k = 0 & \forall k \end{array} \right. \tag{3.23}$$

which correspond to a virtual motion change of the system during which the body i and all its descendants rotate together around the point O^i, while the remaining bodies of the system stay at rest. In this way, the internal joint forces and torques do not produce any virtual power change, except $\mathbf{L}^i$ which will produce a virtual power change $\Delta\mathcal{P}_{\mathbf{L}_i} = \mathbf{L}^i \,.\, \Delta\mathbf{\Omega}^i$.
During this virtual motion, the absolute velocity changes of body i and its descendants obey the following relations (see equations 3.16 and 3.17):

$$\forall j \ge i: \quad \begin{array}{l} \Delta\boldsymbol{\omega}^j = \Delta\mathbf{\Omega}^i \\ \Delta\dot{\mathbf{x}}^j = \Delta\mathbf{\Omega}^i \times \sum_{k:\, i \le k \le j} \mathbf{d}_z^{kj} \end{array}$$

where $\sum_{k:\, i \le k \le j} \mathbf{d}_z^{kj}$ is the moment vector $\overrightarrow{O^i G^j}$:

$$\sum_{k:\, i \le k \le j} \mathbf{d}_z^{kj} = \mathbf{x}^j - \mathbf{p}^i$$

The virtual power principle 2.51:

$$\sum_{j=1}^{N^{body}} m^j \ddot{\mathbf{x}}^j \,.\, \Delta\dot{\mathbf{x}}^j + \sum_{j=1}^{N^{body}} \dot{\mathbf{H}}^j \,.\, \Delta\boldsymbol{\omega}^j = \Delta\mathcal{P}$$

then becomes

$$\sum_{j:\,i\leq j}\left(m^j\ddot{\mathbf{x}}^j\,.\,\Delta\dot{\mathbf{x}}^j+\dot{\mathbf{H}}^j\,.\,\Delta\boldsymbol{\omega}^j\right)=\mathbf{L}^i\,.\,\Delta\boldsymbol{\Omega}^i$$
$$+\sum_{j:\,i\leq j}\left(\mathbf{L}^j_{ext}\,.\,\Delta\boldsymbol{\omega}^j+\left(\mathbf{F}^j_{ext}+m^j\mathbf{g}\right)\,.\,\Delta\dot{\mathbf{x}}^j\right)$$

Replacing $\Delta\dot{\mathbf{x}}^j$ and $\Delta\boldsymbol{\omega}^j$ by their values, we find that the dot coefficient of $\Delta\boldsymbol{\Omega}^i$ is

$$\begin{aligned}\mathbf{L}^i &= \sum_{j:\,i\leq j}\left(\dot{\mathbf{H}}^j-\mathbf{L}^j_{ext}-\overrightarrow{O^i\,G^j}\times\mathbf{F}^j_{ext}\right)\\ &\quad+\sum_{j:\,i\leq j}\ \sum_{k:\,i\leq k\leq j} m^j\mathbf{d}^{kj}_z\times\left(\ddot{\mathbf{x}}^j-\mathbf{g}\right)\end{aligned}$$

and, by using the property 3.21, we obtain

$$\begin{aligned}\mathbf{L}^i &= \sum_{j:\,i\leq j}\left(\dot{\mathbf{H}}^j-\mathbf{L}^j_{ext}-\overrightarrow{O^i\,G^j}\times\mathbf{F}^j_{ext}\right)\\ &\quad+\sum_{k:\,i\leq k}\ \sum_{j:\,k\leq j} m^j\mathbf{d}^{kj}_z\times\left(\ddot{\mathbf{x}}^j-\mathbf{g}\right)\end{aligned}\tag{3.24}$$

This gives a first expression for the joint rotational equation in its vector form. The various time derivatives have now to be expanded by means of the kinematic relations of section 3.2.2. As in the translational equation, this operation will lead to multiple summations which can be simplified if the barycentric quantities are used.

First of all, we may easily recognize in the coefficient of $\mathbf{g}$ a barycentric quantity which is the extended barycentric moment:

$$\sum_{j:\,k\leq j} m^j\mathbf{d}^{kj}_z=\bar{m}^k\mathbf{b}^k_z$$

and a double summation on these gravity terms can thus be avoided.

On the other hand, when expanding the acceleration term $\ddot{\mathbf{x}}^j$, we obtain a triple summation:

$$\sum_{k:\,i\leq k}\ \sum_{j:\,k\leq j} m^j\mathbf{d}^{kj}_z\times\ddot{\mathbf{x}}^j=\sum_{k:\,i\leq k}\ \sum_{j:\,k\leq j}\ \sum_{l:\,l\leq j} m^j\tilde{\mathbf{d}}^{kj}_z\,.\,\ddot{\mathbf{d}}^{lj}_z$$

Fortunately, we can rearrange these terms by looking at the values of the index l with respect to the value of k (both indices being $\leq j$) :

- if $l<k\,(\leq j)$ then $\mathbf{d}^{lj}_z=\mathbf{d}^{lk}_z$ and the sum over j becomes

$$\sum_{j:\,k\leq j} m^j\tilde{\mathbf{d}}^{kj}_z\,.\,\ddot{\mathbf{d}}^{lj}_z=\sum_{j:\,k\leq j} m^j\tilde{\mathbf{d}}^{kj}_z\,.\,\ddot{\mathbf{d}}^{lk}_z=\bar{m}^k\tilde{\mathbf{b}}^k_z\,.\,\ddot{\mathbf{d}}^{lk}_z$$

- if $k < l\,(\leq j)$ then $\mathbf{d}_z^{kj} = \mathbf{d}_z^{kl}$ and the sum over j becomes

$$\sum_{j:\,l\leq j} m^j \tilde{\mathbf{d}}_z^{kj} \,.\, \ddot{\mathbf{d}}_z^{lj} = \sum_{j:\,l\leq j} m^j \tilde{\mathbf{d}}_z^{kl} \,.\, \ddot{\mathbf{d}}_z^{lj} = \tilde{\mathbf{d}}_z^{kl} \,.\, \bar{m}^l \ddot{\mathbf{b}}_z^{l}$$

The triple sum can thus be split into three double summations (for $l < k$, $k < l$ and $k = l$) :

$$\begin{aligned}\sum_{k:\,i\leq k}\sum_{j:\,k\leq j}\sum_{l:\,l\leq j} m^j \tilde{\mathbf{d}}_z^{kj} \,.\, \ddot{\mathbf{d}}_z^{lj} \;=\; & \sum_{k:\,i\leq k}\left(\sum_{l:\,l<k} \bar{m}^k \tilde{\mathbf{b}}_z^{k} \,.\, \ddot{\mathbf{d}}_z^{lk} + \sum_{l:\,k<l} \tilde{\mathbf{d}}_z^{kl} \,.\, \bar{m}^l \ddot{\mathbf{b}}_z^{l}\right) \\ & + \sum_{k:\,i\leq k}\sum_{j:\,k\leq j} m^j \tilde{\mathbf{d}}_z^{kj} \,.\, \ddot{\mathbf{d}}_z^{kj}\end{aligned}$$

where the last one involves (but with derivatives) the second order moments $\sum_{j:\,k\leq j} m^j \mathbf{d}_z^{kj}\,\mathbf{d}_z^{kj}$. Since these moments are related to the barycentric tensor $\mathbf{K}^k$, we will now fully expand these terms in 3.24, together with the derivatives $\dot{\mathbf{H}}^j$ of the angular momentums.
With the previous results, the complete rotational equation 3.24 becomes

$$\begin{aligned}\mathbf{L}^i \;=\; & \sum_{k:\,i\leq k}\left(\mathbf{I}^k \,.\, \dot{\boldsymbol{\omega}}^k + \tilde{\boldsymbol{\omega}}^k \,.\, \mathbf{I}^k \,.\, \boldsymbol{\omega}^k - \mathbf{L}_{ext}^k - \overrightarrow{O^i\,G^k} \times \mathbf{F}_{ext}^k\right) \\ & - \sum_{k:\,i\leq k} \bar{m}^k \tilde{\mathbf{b}}_z^{k} \,.\, \mathbf{g} \\ & + \sum_{k:\,i\leq k}\left(\sum_{l:\,l<k} \bar{m}^k \tilde{\mathbf{b}}_z^{k} \,.\, \ddot{\mathbf{d}}_z^{lk} + \sum_{l:\,k<l} \tilde{\mathbf{d}}_z^{kl} \,.\, \bar{m}^l \ddot{\mathbf{b}}_z^{l}\right) \\ & + \sum_{k:\,i\leq k}\sum_{j:\,k\leq j} m^j \tilde{\mathbf{d}}_z^{kj} \,.\, \left(\overset{\circ\circ}{\mathbf{z}}{}^k + 2\,\tilde{\boldsymbol{\omega}}^k \,.\, \overset{\circ}{\mathbf{z}}{}^k + \tilde{\dot{\boldsymbol{\omega}}}^k \,.\, \mathbf{d}_z^{kj} + \tilde{\boldsymbol{\omega}}^k \,.\, \tilde{\boldsymbol{\omega}}^k \,.\, \mathbf{d}_z^{kj}\right)\end{aligned}$$

Once again the extended barycentric moments appear, this time in the coefficients of the terms $\overset{\circ\circ}{\mathbf{z}}{}^k + 2\,\tilde{\boldsymbol{\omega}}^k \,.\, \overset{\circ}{\mathbf{z}}{}^k$:

$$\sum_{j:\,k\leq j} m^j \tilde{\mathbf{d}}_z^{kj} \,.\, \left(\overset{\circ\circ}{\mathbf{z}}{}^k + 2\,\tilde{\boldsymbol{\omega}}^k \,.\, \overset{\circ}{\mathbf{z}}{}^k\right) = \bar{m}^k \tilde{\mathbf{b}}_z^{k} \,.\, \left(\overset{\circ\circ}{\mathbf{z}}{}^k + 2\,\tilde{\boldsymbol{\omega}}^k \,.\, \overset{\circ}{\mathbf{z}}{}^k\right)$$

On the other hand, the remaining terms which explicitly involve $\boldsymbol{\omega}^k$ and its derivative in the complete equation are

$$\mathbf{I}^k \,.\, \dot{\boldsymbol{\omega}}^k + \tilde{\boldsymbol{\omega}}^k \,.\, \mathbf{I}^k \,.\, \boldsymbol{\omega}^k + \sum_{j:\,k\leq j} m^j \tilde{\mathbf{d}}_z^{kj} \,.\, \left(\tilde{\dot{\boldsymbol{\omega}}}^k \,.\, \mathbf{d}_z^{kj} + \tilde{\boldsymbol{\omega}}^k \,.\, \tilde{\boldsymbol{\omega}}^k \,.\, \mathbf{d}_z^{kj}\right) \tag{3.25}$$

Using the following tensor-vector property:

$$\tilde{\mathbf{d}} \,.\, \tilde{\boldsymbol{\omega}} \,.\, \tilde{\boldsymbol{\omega}} \,.\, \mathbf{d} = -\tilde{\boldsymbol{\omega}} \,.\, \tilde{\mathbf{d}} \,.\, \tilde{\mathbf{d}} \,.\, \boldsymbol{\omega} \tag{3.26}$$

we can write these terms by means of the extended barycentric tensors 3.11 as

$$\mathbf{K}_z^k \, . \, \dot{\boldsymbol{\omega}}^k + \tilde{\boldsymbol{\omega}}^k \, . \, \mathbf{K}_z^k \, . \, \boldsymbol{\omega}^k$$

Finally, the rotational vector equation can be reorganized in its final form as follows:

$$\begin{aligned}
\mathbf{L}^i \;=\; & \sum_{k:\, i \leq k} \left(\mathbf{K}_z^k \, . \, \dot{\boldsymbol{\omega}}^k + \tilde{\boldsymbol{\omega}}^k \, . \, \mathbf{K}_z^k \, . \, \boldsymbol{\omega}^k \right) \\
& + \sum_{k:\, i \leq k} \left(\sum_{l:\, l < k} \bar{m}^k \tilde{\mathbf{b}}_z^k \, . \, \ddot{\mathbf{d}}_z^{lk} + \sum_{l:\, k < l} \tilde{\mathbf{d}}_z^{kl} \, . \, \bar{m}^l \ddot{\mathbf{b}}_z^l \right) \\
& + \sum_{k:\, i \leq k} \bar{m}^k \tilde{\mathbf{b}}_z^k \, . \, \left(\overset{\circ\circ}{\mathbf{z}}{}^k + 2 \, \tilde{\boldsymbol{\omega}}^k \, . \, \overset{\circ}{\mathbf{z}}{}^k \right) \\
& - \sum_{k:\, i \leq k} \left(\mathbf{L}_{ext}^k + \overrightarrow{O^i \, G^k} \times \mathbf{F}_{ext}^k + \bar{m}^k \tilde{\mathbf{b}}_z^k \, . \, \mathbf{g} \right)
\end{aligned} \tag{3.27}$$

As in the translational vector equation, we see that all the dynamical characteristics of the different bodies (i.e. the masses, the center of mass locations and the inertia tensors) have been replaced by barycentric quantities. This operation allows us to reduce by one the maximum number of summations which are involved in these vector equations.
At this point, it must also be emphasized that contrary to the translational equation, the present rotational equation is *not* a global Euler equation for body i and its descendants. The reason is that point O^i, which is used for computing the equivalent resulting torque due to the applied forces, is neither an inertially fixed point nor the (instantaneous) mass center of this collection of bodies.

3.2.5 Equations of motion

The physical nature (prismatic or revolute) of the different joints, which has not yet been considered to establish the vector form of the translational and rotational equations, should now be taken into account to provide the equations of motion of the multibody system in their scalar form. For this purpose, the virtual velocity changes $\Delta \overset{\circ}{\mathbf{z}}{}^i$ and $\Delta \boldsymbol{\Omega}^i$ will first be expressed in term of virtual changes of the joint generalized velocities $\Delta \dot{q}^i$ according to equation 3.6 in order to be *compatible with the actual joint kinematics*:

$$\Delta \overset{\circ}{\mathbf{z}}{}^i = \Delta \dot{q}^i \, \boldsymbol{\psi}^i \quad \text{and} \quad \Delta \boldsymbol{\Omega}^i = \Delta \dot{q}^i \, \boldsymbol{\varphi}^i \tag{3.28}$$

If such a virtual change is used in the virtual power principle, the virtual power which will be produced by the force $\mathbf{F}^i$ and the torque $\mathbf{L}^i$ acting on body i reduces to

$$\Delta \mathcal{P}_{\mathbf{F}^i} + \Delta \mathcal{P}_{\mathbf{L}^i} = \left(\mathbf{F}^i \, . \, \boldsymbol{\psi}^i + \mathbf{L}^i \, . \, \boldsymbol{\varphi}^i \right) \Delta \dot{q}^i = \mathcal{Q}^i \, \Delta \dot{q}^i \tag{3.29}$$

where $\mathcal{Q}^i$ represents the contribution of the joint force and torque in the generalized force associated with the generalized coordinate q^i. In other words, the generalized force contribution of a one d.o.f. joint is given by:

- either the projection $\mathbf{F}^i \,.\, \hat{\mathbf{e}}^i$ of the vector force $\mathbf{F}^i$ on the joint axis $\hat{\mathbf{e}}^i$ when the joint is prismatic,
- or by the projection $\mathbf{L}^i \,.\, \hat{\mathbf{e}}^i$ of the vector torque $\mathbf{L}^i$ on the joint axis $\hat{\mathbf{e}}^i$ when the joint is revolute.

The other components of $\mathbf{F}^i$ and $\mathbf{L}^i$ will not contribute to the virtual power: indeed, they are (internal) constraint forces and torques applied on body i through the joint i. As explained in section 2.4.1, such constraint forces and torques disappear from the virtual power, and thus from the equations of motion, if the virtual velocities are chosen to be compatible with the constraints. Thus the equations of motion of the multibody system are simply obtained by:

$$\mathcal{Q}^i = \mathbb{F}^i \,.\, \boldsymbol{\psi}^i + \mathbb{L}^i \,.\, \boldsymbol{\varphi}^i \tag{3.30}$$

where $\mathbb{F}^i$ and $\mathbb{L}^i$ are given by the right-hand sides of the vector equations 3.22 and 3.27 respectively.

The second step consists in expanding all the vectors of this equation 3.30 as explicit functions of the generalized coordinates and their derivatives by means of the following kinematic relations:

$$\ddot{\mathbf{d}}_z^{ki} = \overset{\circ\circ}{\mathbf{z}}{}^k + 2\,\tilde{\boldsymbol{\omega}}^k \,.\, \overset{\circ}{\mathbf{z}}{}^k + \dot{\tilde{\boldsymbol{\omega}}}{}^k \,.\, \mathbf{d}_z^{ki} + \tilde{\boldsymbol{\omega}}^k \,.\, \tilde{\boldsymbol{\omega}}^k \,.\, \mathbf{d}_z^{ki} \tag{3.31}$$

$$\ddot{\mathbf{b}}_z^{k} = \overset{\circ\circ}{\mathbf{z}}{}^k + 2\,\tilde{\boldsymbol{\omega}}^k \,.\, \overset{\circ}{\mathbf{z}}{}^k + \dot{\tilde{\boldsymbol{\omega}}}{}^k \,.\, \mathbf{b}_z^{k} + \tilde{\boldsymbol{\omega}}^k \,.\, \tilde{\boldsymbol{\omega}}^k \,.\, \mathbf{b}_z^{k} \tag{3.32}$$

together with

$$\boldsymbol{\omega}^k = \sum_{j:\,j\le k} \boldsymbol{\varphi}^j\,\dot{q}^j \quad , \quad \dot{\boldsymbol{\omega}}^k = \sum_{j:\,j\le k} \left(\boldsymbol{\varphi}^j\,\ddot{q}^j + \tilde{\boldsymbol{\omega}}^j \boldsymbol{\varphi}^j\,\dot{q}^j\right)$$

$$\overset{\circ}{\mathbf{z}}{}^k = \boldsymbol{\psi}^k\,\dot{q}^k \quad , \quad \overset{\circ\circ}{\mathbf{z}}{}^k = \boldsymbol{\psi}^k\,\ddot{q}^k$$

These substitutions lead to relatively complex and tedious computations. However, the manipulations which are involved are systematic and therefore only the final results will be given here below.

- For each prismatic joint, an equation of motion is given by

$$
\begin{aligned}
\mathcal{Q}^i &= \boldsymbol{\psi}^i \,.\, \mathbb{F}^i \\
&= \sum_{k:\,k<i} \bar{m}^i \left(\boldsymbol{\psi}^i \,.\, \boldsymbol{\psi}^k\right) \ddot{q}^k + \sum_{k:\,i\leq k} \bar{m}^k \left(\boldsymbol{\psi}^i \,.\, \boldsymbol{\psi}^k\right) \ddot{q}^k \\
&\quad + \sum_{k:\,k<i} \sum_{j:\,j\leq k} \bar{m}^i \left(\mathbf{d}_z^{ki} \,.\, \tilde{\boldsymbol{\psi}}^i \,.\, \boldsymbol{\varphi}^j\right) \ddot{q}^j \\
&\quad + \sum_{k:\,i\leq k} \sum_{j:\,j\leq k} \bar{m}^k \left(\mathbf{b}_z^k \,.\, \tilde{\boldsymbol{\psi}}^i \,.\, \boldsymbol{\varphi}^j\right) \ddot{q}^j \\
&\quad +2 \sum_{k:\,k<i} \sum_{j:\,j\leq k} \bar{m}^i \left(\boldsymbol{\psi}^i \,.\, \tilde{\boldsymbol{\varphi}}^j \,.\, \boldsymbol{\psi}^k\right) \dot{q}^j \dot{q}^k \\
&\quad +2 \sum_{k:\,i\leq k} \sum_{j:\,j\leq k} \bar{m}^k \left(\boldsymbol{\psi}^i \,.\, \tilde{\boldsymbol{\varphi}}^j \,.\, \boldsymbol{\psi}^k\right) \dot{q}^j \dot{q}^k \\
&\quad + \sum_{k:\,k<i} \sum_{j:\,j\leq k} \sum_{l:\,l\leq j} \bar{m}^i \left(\mathbf{d}_z^{ki} \,.\, \tilde{\boldsymbol{\psi}}^i \,.\, \tilde{\boldsymbol{\varphi}}^l \,.\, \boldsymbol{\varphi}^j\right) \dot{q}^j \dot{q}^l \\
&\quad + \sum_{k:\,k<i} \sum_{l:\,l\leq k} \sum_{j:\,j\leq k} \bar{m}^i \left(\boldsymbol{\psi}^i \,.\, \tilde{\boldsymbol{\varphi}}^l \,.\, \tilde{\boldsymbol{\varphi}}^j \,.\, \mathbf{d}_z^{ki}\right) \dot{q}^j \dot{q}^l \\
&\quad + \sum_{k:\,i\leq k} \sum_{j:\,j\leq k} \sum_{l:\,l\leq j} \bar{m}^k \left(\mathbf{b}_z^k \,.\, \tilde{\boldsymbol{\psi}}^i \,.\, \tilde{\boldsymbol{\varphi}}^l \,.\, \boldsymbol{\varphi}^j\right) \dot{q}^j \dot{q}^l \\
&\quad + \sum_{k:\,i\leq k} \sum_{l:\,l\leq k} \sum_{j:\,j\leq k} \bar{m}^k \left(\boldsymbol{\psi}^i \,.\, \tilde{\boldsymbol{\varphi}}^l \,.\, \tilde{\boldsymbol{\varphi}}^j \,.\, \mathbf{b}_z^k\right) \dot{q}^j \dot{q}^l \\
&\quad - \sum_{j:\,i\leq j} \left(\boldsymbol{\psi}^i \,.\, \mathbf{F}_{ext}^j\right) - \bar{m}^i \left(\boldsymbol{\psi}^i \,.\, \mathbf{g}\right)
\end{aligned} \tag{3.33}
$$

After some additional manipulations based on the following vector-tensor properties:

$$
\begin{aligned}
\mathbf{d}_z^{ki} \,.\, \tilde{\boldsymbol{\psi}}^i \,.\, \tilde{\boldsymbol{\varphi}}^l \,.\, \boldsymbol{\varphi}^j &= \boldsymbol{\psi}^i \,.\, \tilde{\boldsymbol{\varphi}}^l \,.\, \tilde{\boldsymbol{\varphi}}^j \,.\, \mathbf{d}_z^{ki} - \boldsymbol{\psi}^i \,.\, \tilde{\boldsymbol{\varphi}}^j \,.\, \tilde{\boldsymbol{\varphi}}^l \,.\, \mathbf{d}_z^{ki} \\
\mathbf{b}_z^k \,.\, \tilde{\boldsymbol{\psi}}^i \,.\, \tilde{\boldsymbol{\varphi}}^l \,.\, \boldsymbol{\varphi}^j &= \boldsymbol{\psi}^i \,.\, \tilde{\boldsymbol{\varphi}}^l \,.\, \tilde{\boldsymbol{\varphi}}^j \,.\, \mathbf{b}_z^k - \boldsymbol{\psi}^i \,.\, \tilde{\boldsymbol{\varphi}}^j \,.\, \tilde{\boldsymbol{\varphi}}^l \,.\, \mathbf{b}_z^k
\end{aligned}
$$

this equation can be further simplified as follows:

$$
\begin{aligned}
\mathcal{Q}^i &= \sum_{k:\,k<i} \bar{m}^i \left(\boldsymbol{\psi}^i \,.\, \boldsymbol{\psi}^k\right) \ddot{q}^k + \sum_{k:\,i\leq k} \bar{m}^k \left(\boldsymbol{\psi}^i \,.\, \boldsymbol{\psi}^k\right) \ddot{q}^k \\
&\quad + \sum_{k:\,k<i} \sum_{j:\,j\leq k} \bar{m}^i \left(\mathbf{d}_z^{ki} \,.\, \tilde{\boldsymbol{\psi}}^i \,.\, \boldsymbol{\varphi}^j\right) \ddot{q}^j \\
&\quad + \sum_{k:\,i\leq k} \sum_{j:\,j\leq k} \bar{m}^k \left(\mathbf{b}_z^k \,.\, \tilde{\boldsymbol{\psi}}^i \,.\, \boldsymbol{\varphi}^j\right) \ddot{q}^j \\
&\quad +2 \sum_{k:\,k<i} \sum_{j:\,j\leq k} \bar{m}^i \left(\boldsymbol{\psi}^i \,.\, \tilde{\boldsymbol{\varphi}}^j \,.\, \boldsymbol{\psi}^k\right) \dot{q}^j \dot{q}^k
\end{aligned}
$$

$$+2 \sum_{k:\, i \leq k} \sum_{j:\, j \leq k} \bar{m}^k \left(\boldsymbol{\psi}^i \,.\, \tilde{\boldsymbol{\varphi}}^j \,.\, \boldsymbol{\psi}^k \right) \dot{q}^j \dot{q}^k$$
$$+2 \sum_{k:\, k < i} \sum_{j:\, j \leq k} \sum_{l:\, l < j} \bar{m}^i \left(\boldsymbol{\psi}^i \,.\, \tilde{\boldsymbol{\varphi}}^l \,.\, \tilde{\boldsymbol{\varphi}}^j \,.\, \mathbf{d}_z^{ki} \right) \dot{q}^j \dot{q}^l$$
$$+2 \sum_{k:\, i \leq k} \sum_{j:\, j \leq k} \sum_{l:\, l < j} \bar{m}^k \left(\boldsymbol{\psi}^i \,.\, \tilde{\boldsymbol{\varphi}}^l \,.\, \tilde{\boldsymbol{\varphi}}^j \,.\, \mathbf{b}_z^k \right) \dot{q}^j \dot{q}^l$$
$$+ \sum_{k:\, k < i} \sum_{j:\, j \leq k} \bar{m}^i \left(\boldsymbol{\psi}^i \,.\, \tilde{\boldsymbol{\varphi}}^j \,.\, \tilde{\boldsymbol{\varphi}}^j \,.\, \mathbf{d}_z^{ki} \right) \dot{q}^j \dot{q}^j$$
$$+ \sum_{k:\, i \leq k} \sum_{j:\, j \leq k} \bar{m}^k \left(\boldsymbol{\psi}^i \,.\, \tilde{\boldsymbol{\varphi}}^j \,.\, \tilde{\boldsymbol{\varphi}}^j \,.\, \mathbf{b}_z^k \right) \dot{q}^j \dot{q}^j$$
$$- \sum_{j:\, i \leq j} \left(\boldsymbol{\psi}^i \,.\, \mathbf{F}_{ext}^j \right) - \bar{m}^i \left(\boldsymbol{\psi}^i \,.\, \mathbf{g} \right) \tag{3.34}$$

- For each revolute joint, an equation of motion is given by

$$\mathcal{Q}^i = \boldsymbol{\varphi}^i \,.\, \mathbb{L}^i$$
$$= \sum_{k:\, i \leq k} \sum_{j:\, j \leq k} \left(\boldsymbol{\varphi}^i \,.\, \mathbf{K}_z^k \,.\, \boldsymbol{\varphi}^j \right) \ddot{q}^j$$
$$+ \sum_{k:\, i \leq k} \sum_{j:\, j \leq k} \bar{m}^k \left(\boldsymbol{\varphi}^i \,.\, \tilde{\mathbf{b}}_z^k \,.\, \boldsymbol{\psi}^j \right) \ddot{q}^j$$
$$+ \sum_{k:\, i \leq k} \sum_{j:\, k < j} \bar{m}^j \left(\boldsymbol{\varphi}^i \,.\, \tilde{\mathbf{d}}_z^{kj} \,.\, \boldsymbol{\psi}^j \right) \ddot{q}^j$$
$$- \sum_{k:\, i \leq k} \sum_{l:\, l < k} \sum_{j:\, j \leq l} \bar{m}^k \left(\boldsymbol{\varphi}^i \,.\, \tilde{\mathbf{b}}_z^k \,.\, \tilde{\mathbf{d}}_z^{lk} \,.\, \boldsymbol{\varphi}^j \right) \ddot{q}^j$$
$$- \sum_{k:\, i \leq k} \sum_{l:\, k < l} \sum_{j:\, j \leq l} \bar{m}^l \left(\boldsymbol{\varphi}^i \,.\, \tilde{\mathbf{d}}_z^{kl} \,.\, \tilde{\mathbf{b}}_z^l \,.\, \boldsymbol{\varphi}^j \right) \ddot{q}^j$$
$$+ \sum_{k:\, i \leq k} \sum_{j:\, j \leq k} \sum_{l:\, l \leq j} \left(\boldsymbol{\varphi}^i \,.\, \mathbf{K}_z^k \,.\, \tilde{\boldsymbol{\varphi}}^l \,.\, \boldsymbol{\varphi}^j \right) \dot{q}^j \dot{q}^l$$
$$+ \sum_{k:\, i \leq k} \sum_{j:\, j \leq k} \sum_{l:\, l \leq k} \left(\boldsymbol{\varphi}^i \,.\, \tilde{\boldsymbol{\varphi}}^l \,.\, \mathbf{K}_z^k \,.\, \boldsymbol{\varphi}^j \right) \dot{q}^j \dot{q}^l$$
$$+ 2 \sum_{k:\, i \leq k} \sum_{l:\, l \leq k} \sum_{j:\, j \leq l} \bar{m}^k \left(\boldsymbol{\varphi}^i \,.\, \tilde{\mathbf{b}}_z^k \,.\, \tilde{\boldsymbol{\varphi}}^j \,.\, \boldsymbol{\psi}^l \right) \dot{q}^j \dot{q}^l$$
$$+ 2 \sum_{k:\, i \leq k} \sum_{l:\, k < l} \sum_{j:\, j \leq l} \bar{m}^l \left(\boldsymbol{\varphi}^i \,.\, \tilde{\mathbf{d}}_z^{kl} \,.\, \tilde{\boldsymbol{\varphi}}^j \,.\, \boldsymbol{\psi}^l \right) \dot{q}^j \dot{q}^l$$
$$- \sum_{k:\, i \leq k} \sum_{l:\, l < k} \sum_{j:\, j \leq l} \sum_{a:\, a \leq j} \bar{m}^k \left(\boldsymbol{\varphi}^i \,.\, \tilde{\mathbf{b}}_z^k \,.\, \tilde{\boldsymbol{\varphi}}^a \,.\, \tilde{\mathbf{d}}_z^{lk} \,.\, \boldsymbol{\varphi}^j \right) \dot{q}^j \dot{q}^a$$

$$- \sum_{k:\, i \leq k} \sum_{l:\, k<l} \sum_{j:\, j \leq l} \sum_{a:\, a \leq j} \bar{m}^l \left(\varphi^i \,.\, \tilde{\mathbf{d}}_z^{kl} \,.\, \tilde{\varphi}^a \,.\, \tilde{\mathbf{b}}_z^l \,.\, \varphi^j \right) \dot{q}^j \dot{q}^a$$

$$- \sum_{k:\, i \leq k} \varphi^i \,.\, \left(\mathbf{L}_{ext}^k + \overrightarrow{O^i G^k} \times \mathbf{F}_{ext}^k + \bar{m}^k \, \tilde{\mathbf{b}}_z^k \,.\, \mathbf{g} \right) \tag{3.35}$$

The final step one has to perform before implementing these equations in a computer program, consists in evaluating all scalars quantities expressed as dot products of vectors and tensors within the terms under parenthesis. Since the vector/tensor components are given in different body-fixed frames, these computations require the relative rotation matrices $R^{i,j}$ between these frames. The elementary rotation matrices between two contiguous bodies have already be defined in 3.1 for prismatic joints and in 3.2 for revolute joints. The relative rotation matrix between two arbitrary bodies is then computed recursively by following the kinematic chain which connects these two bodies. For the example shown in figure 3.14, the rotation matrix $R^{j,i}$ is given by

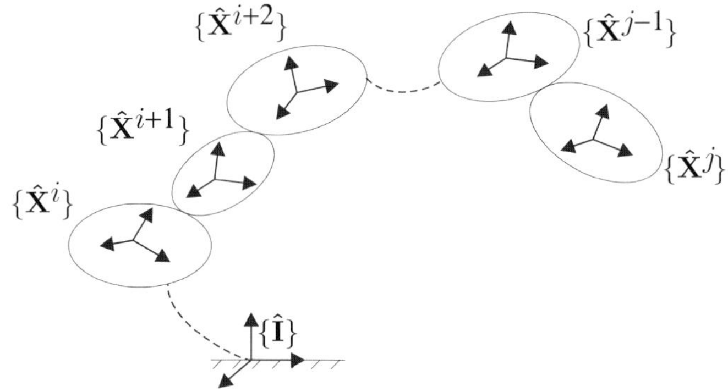

Figure 3.14: Rotation matrix of a kinematic chain

$$[\hat{\mathbf{X}}^j] = R^{j,i}[\hat{\mathbf{X}}^i] \quad \text{with} \quad R^{j,i} = R^{j,j-1} \ \ldots \ R^{i+2,i+1} R^{i+1,i} \tag{3.36}$$

3.2.6 Matrix form of the joint equations

The equations of motion 3.34 and 3.35 are presented in the form equivalent[9] to 2.75:

$$\Phi^i(\ddot{q}, \dot{q}, q, F_{ext}, L_{ext}, g) = \mathcal{Q}^i \quad i = 1, ..., N^{body} \tag{3.37}$$

where the accelerations $\ddot{q}$ are implicitly contained in the expression of Φ^i. However, when fully developed with respect to the $\dot{q}$ and $\ddot{q}$, these equations can also

[9] equivalent but not strictly identical because $\mathcal{Q}^i$ only contains the contribution of the joint forces and torques, the effect of gravity and of external loads being included in Φ^i.

be formulated in matrix form:

$$M(q)\ddot{q} + c(\dot{q}, q, F_{ext}, L_{ext}, g) = \mathcal{Q} \tag{3.38}$$

where the accelerations $\ddot{q}$ appear explicitly. This form is nevertheless denoted as being *semi-explicit* (with respect to the accelerations), since a (linear) system has still to be solved in order to provide the accelerations in terms of the generalized position and velocities $(\dot{q}, q)$ and the applied forces $(\mathcal{Q}, F_{ext}, L_{ext}, g)$ explicitly. The coefficient matrix M of the generalized accelerations is called the *generalized mass* matrix of the system.

From 3.34 and 3.35, we see that if i and j do not belong to a same kinematic chain (i.e. $i \not< j$ and $j \not< i$) then $M_{(i,j)} = M_{(j,i)} = 0$. On the other hand, the (i,j) element of the mass matrix is given by:

if $i < j$:

$$\begin{aligned} M_{(i,j)} \;=\;& \bar{m}^j \left(\boldsymbol{\psi}^i \,.\, \boldsymbol{\psi}^j\right) + \sum_{k:\, j \le k} \left(\boldsymbol{\varphi}^i \,.\, \mathbf{K}_z^k \,.\, \boldsymbol{\varphi}^j\right) \\ & - \sum_{k:\, j \le k} \bar{m}^k \left(\boldsymbol{\psi}^i \,.\, \tilde{\mathbf{b}}_z^k \,.\, \boldsymbol{\varphi}^j\right) + \sum_{k:\, j \le k} \bar{m}^k \left(\boldsymbol{\varphi}^i \,.\, \tilde{\mathbf{b}}_z^k \,.\, \boldsymbol{\psi}^j\right) \\ & + \sum_{k:\, i \le k < j} \bar{m}^j \left(\boldsymbol{\varphi}^i \,.\, \tilde{\mathbf{d}}_z^{kj} \,.\, \boldsymbol{\psi}^j\right) \\ & - \sum_{l:\, j \le l} \sum_{k:\, l < k} \bar{m}^k \left(\boldsymbol{\varphi}^i \,.\, \tilde{\mathbf{b}}_z^k \,.\, \tilde{\mathbf{d}}_z^{lk} \,.\, \boldsymbol{\varphi}^j\right) \\ & - \sum_{l:\, j \le l} \sum_{k:\, i \le k < l} \bar{m}^l \left(\boldsymbol{\varphi}^i \,.\, \tilde{\mathbf{d}}_z^{kl} \,.\, \tilde{\mathbf{b}}_z^l \,.\, \boldsymbol{\varphi}^j\right) \end{aligned}$$

if $j < i$:

$$\begin{aligned} M_{(i,j)} \;=\;& \bar{m}^i \left(\boldsymbol{\psi}^i \,.\, \boldsymbol{\psi}^j\right) + \sum_{k:\, i \le k} \left(\boldsymbol{\varphi}^i \,.\, \mathbf{K}_z^k \,.\, \boldsymbol{\varphi}^j\right) \\ & - \sum_{k:\, j \le k < i} \bar{m}^i \left(\boldsymbol{\psi}^i \,.\, \tilde{\mathbf{d}}_z^{ki} \,.\, \boldsymbol{\varphi}^j\right) \\ & - \sum_{k:\, i \le k} \bar{m}^k \left(\boldsymbol{\psi}^i \,.\, \tilde{\mathbf{b}}_z^k \,.\, \boldsymbol{\varphi}^j\right) + \sum_{k:\, i \le k} \bar{m}^k \left(\boldsymbol{\varphi}^i \,.\, \tilde{\mathbf{b}}_z^k \,.\, \boldsymbol{\psi}^j\right) \\ & - \sum_{k:\, i \le k} \sum_{l:\, j \le l < k} \bar{m}^k \left(\boldsymbol{\varphi}^i \,.\, \tilde{\mathbf{b}}_z^k \,.\, \tilde{\mathbf{d}}_z^{lk} \,.\, \boldsymbol{\varphi}^j\right) \\ & - \sum_{k:\, i \le k} \sum_{l:\, k < l} \bar{m}^l \left(\boldsymbol{\varphi}^i \,.\, \tilde{\mathbf{d}}_z^{kl} \,.\, \tilde{\mathbf{b}}_z^l \,.\, \boldsymbol{\varphi}^j\right) \end{aligned} \tag{3.39}$$

and if $j = i$:

$$\begin{aligned} M_{(i,i)} &= \bar{m}^i + \sum_{k:\, i \leq k} \left(\boldsymbol{\varphi}^i \,.\, \mathbf{K}_z^k \,.\, \boldsymbol{\varphi}^i\right) \\ &- \sum_{k:\, i \leq k} \;\sum_{l:\, i \leq l < k} \bar{m}^k \left(\boldsymbol{\varphi}^i \,.\, \tilde{\mathbf{b}}_z^k \,.\, \tilde{\mathbf{d}}_z^{lk} \,.\, \boldsymbol{\varphi}^i\right) \\ &- \sum_{k:\, i \leq k} \;\sum_{l:\, i \leq k < l} \bar{m}^l \left(\boldsymbol{\varphi}^i \,.\, \tilde{\mathbf{d}}_z^{kl} \,.\, \tilde{\mathbf{b}}_z^l \,.\, \boldsymbol{\varphi}^i\right) \end{aligned} \tag{3.40}$$

One may thus verify that:

$$M_{(i,j)} = M_{(j,i)} \quad \text{in any cases,}$$

which means that the obtained generalized mass matrix is symmetric. It can also be shown that the kinetic energy T of a multibody system can be written in terms of the generalized mass matrix M :

$$T = \frac{1}{2} \dot{q}^T M \dot{q} \tag{3.41}$$

Since the kinetic energy is positive, the quadratic form 3.41 is positive definite. Hence, the matrix M is positive definite, thus non singular.

3.3 Newton-Euler scheme

3.3.1 Introduction

In the equations of motion of the previous section, the kinematic relations were fully expanded: this leads to several multiple summations and a number of operations which grows rapidly with the number of degrees of freedom. This number often becomes excessive for real applications. Hence, to develop an efficient multibody simulation program, we need to formulate the equations with a minimal number of arithmetical operations. That is the purpose of the so-called "recursive Newton-Euler" formalism.

The recursive Newton-Euler formalism was initially developed in robotics for control purposes (see [52], [72]). It allows us to compute, with a minimum of arithmetic operations, the *inverse* dynamical model of serial robot manipulators, i.e. the generalized forces $\mathcal{Q}$ to be applied to the joints as a function of the generalized joint positions q, velocities $\dot{q}$ and accelerations $\ddot{q}$:

$$\mathcal{Q} = \Phi(q, \dot{q}, \ddot{q}, ...) \tag{3.42}$$

These equation are *implicit* with respect to the generalized accelerations $\ddot{q}$, and thus not suitable for simulation purposes. Hence, in this chapter, we will modify the original Newton-Euler scheme in order to obtain recursively the semi-explicit form 3.38 of the dynamical equations of tree-like multibody systems:

$$M(q)\,\ddot{q} + c(q,\, \dot{q},\, F_{ext},\, L_{ext},\, g) = \mathcal{Q} \tag{3.43}$$

Although it will be obtained in a different way, the system 3.43 will have exactly the same mathematical structure as the system 3.38 we obtained by the virtual power principle: M is the generalized mass matrix, c contains the Coriolis, centrifugal and gravity terms as well as the external forces and torques. This equivalence will allow us, in chapter 4, to extend the Newton-Euler recursive approach for complex multibody systems which are not tree-like.

3.3.2 Forward kinematics

Consider in figure 3.15 a rigid body i carried by a parent body h via joint i. For body i, we can write:

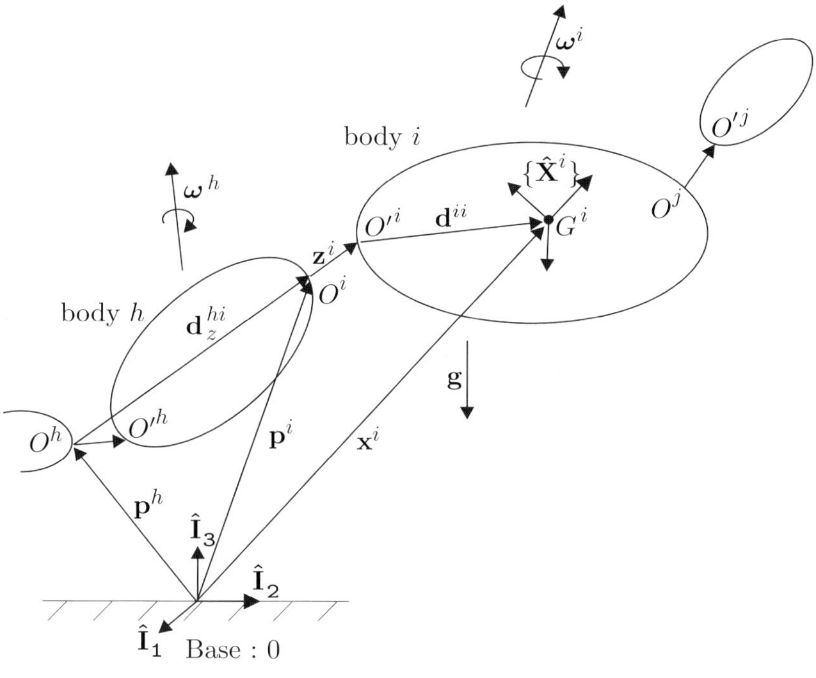

Figure 3.15: Body i — forward kinematics

- Absolute position vectors:

$$\mathbf{p}^i = \mathbf{p}^h + \mathbf{d}_z^{hi}, \text{ for the attachment point } O^i \text{ on body } h, \quad (3.44)$$

$$\mathbf{x}^i = \mathbf{p}^i + \mathbf{d}_z^{ii}, \text{ for the center of mass of body } i. \quad (3.45)$$

- Absolute velocities of body i :

  - angular:

$$\boldsymbol{\omega}^i = \boldsymbol{\omega}^h + \boldsymbol{\Omega}^i = \boldsymbol{\omega}^h + \boldsymbol{\varphi}^i \dot{q}^i \quad (3.46)$$

– linear:

$$\dot{\mathbf{p}}^i = \dot{\mathbf{p}}^h + \tilde{\boldsymbol{\omega}}^h \,.\, \mathbf{d}_z^{hi} + \boldsymbol{\psi}^h \dot{q}^h \tag{3.47}$$
$$\dot{\mathbf{x}}^i = \dot{\mathbf{p}}^i + \tilde{\boldsymbol{\omega}}^i \,.\, \mathbf{d}_z^{ii} + \boldsymbol{\psi}^i \dot{q}^i \tag{3.48}$$

- Absolute accelerations of body i :

– angular:

$$\dot{\boldsymbol{\omega}}^i = \dot{\boldsymbol{\omega}}^h + \tilde{\boldsymbol{\omega}}^i \,.\, \boldsymbol{\varphi}^i \dot{q}^i + \boldsymbol{\varphi}^i \ddot{q}^i \tag{3.49}$$

– linear:

$$\ddot{\mathbf{p}}^i = \ddot{\mathbf{p}}^h + (\tilde{\dot{\boldsymbol{\omega}}}^h + \tilde{\boldsymbol{\omega}}^h \,.\, \tilde{\boldsymbol{\omega}}^h) \,.\, \mathbf{d}_z^{hi} + 2\,\tilde{\boldsymbol{\omega}}^h \,.\, \boldsymbol{\psi}^h \dot{q}^h + \boldsymbol{\psi}^h \ddot{q}^h \tag{3.50}$$
$$\ddot{\mathbf{x}}^i = \ddot{\mathbf{p}}^i + (\tilde{\dot{\boldsymbol{\omega}}}^i + \tilde{\boldsymbol{\omega}}^i \,.\, \tilde{\boldsymbol{\omega}}^i) \,.\, \mathbf{d}_z^{ii} + 2\,\tilde{\boldsymbol{\omega}}^i \,.\, \boldsymbol{\psi}^i \dot{q}^i + \boldsymbol{\psi}^i \ddot{q}^i \tag{3.51}$$

In order to obtain the dynamical equations in a compact form, it is convenient to define the following quantities:

$$\boldsymbol{\beta}^i \triangleq \tilde{\dot{\boldsymbol{\omega}}}^i + \tilde{\boldsymbol{\omega}}^i \,.\, \tilde{\boldsymbol{\omega}}^i \tag{3.52}$$
$$\boldsymbol{\alpha}^i \triangleq \ddot{\mathbf{p}}^i + 2\,\tilde{\boldsymbol{\omega}}^i \,.\, \boldsymbol{\psi}^i \dot{q}^i + \boldsymbol{\psi}^i \ddot{q}^i - \mathbf{g} \tag{3.53}$$

where $\mathbf{g}$ denotes the gravity vector.
Using 3.50, we can write the relation 3.53 in a recursive manner:

$$\boldsymbol{\alpha}^i = \boldsymbol{\alpha}^h + \boldsymbol{\beta}^h \,.\, \mathbf{d}_z^{hi} + 2\,\tilde{\boldsymbol{\omega}}^i \,.\, \boldsymbol{\psi}^i \dot{q}^i + \boldsymbol{\psi}^i \ddot{q}^i \tag{3.54}$$

In order to express the mass matrix in the semi-explicit form 3.43, we need to isolate the generalized accelerations $\ddot{q}$ in the recursive equations 3.49, 3.52 and 3.54 by splitting up the $\dot{\boldsymbol{\omega}}^i, \boldsymbol{\beta}^i$ and $\boldsymbol{\alpha}^i$ quantities as follows:

$$\dot{\boldsymbol{\omega}}^i = \sum_{k:\, k \leq i} \mathbf{O}_M^{ik} \, \ddot{q}^k + \dot{\boldsymbol{\omega}}_c^i \tag{3.55}$$

$$\boldsymbol{\beta}^i = \sum_{k:\, k \leq i} \mathbf{B}_M^{ik} \, \ddot{q}^k + \boldsymbol{\beta}_c^i \tag{3.56}$$

$$\boldsymbol{\alpha}^i = \sum_{k:\, k \leq i} \mathbf{A}_M^{ik} \, \ddot{q}^k + \boldsymbol{\alpha}_c^i \tag{3.57}$$

The recursive computation of equations 3.55 — 3.57 can then be performed as in the original Newton-Euler recursive scheme for the inverse dynamics. The algorithm thus reads in its vector form:

Initialization:

$$\boldsymbol{\alpha}_c^0 = -\mathbf{g} \;\; ; \;\; \boldsymbol{\omega}^0 = 0 \;\; ; \;\; \dot{\boldsymbol{\omega}}_c^0 = 0 \;\; ; \;\; \mathbf{O}_M^{ik} = 0 \;\; ; \;\; \mathbf{A}_M^{ik} = 0$$
$$(\forall\, i = 0 : N^{body}, \forall k = 0 : i)$$

Recursion:
For $i = 1 : N^{body}$

$$h = \text{inbody}(i) \quad \text{(index of the parent body)}$$

$$\boldsymbol{\omega}^i = \boldsymbol{\omega}^h + \dot{q}^i\, \boldsymbol{\varphi}^i \tag{3.58}$$

$$\dot{\boldsymbol{\omega}}_c^i = \dot{\boldsymbol{\omega}}_c^h + \tilde{\boldsymbol{\omega}}^i \,.\, \boldsymbol{\varphi}^i\, \dot{q}^i \tag{3.59}$$

$$\boldsymbol{\beta}_c^i = \tilde{\dot{\boldsymbol{\omega}}}_c^i + \tilde{\boldsymbol{\omega}}^i \,.\, \tilde{\boldsymbol{\omega}}^i \tag{3.60}$$

$$\boldsymbol{\alpha}_c^i = \boldsymbol{\alpha}_c^h + \boldsymbol{\beta}_c^h \,.\, \mathbf{d}_z^{hi} + 2\, \tilde{\boldsymbol{\omega}}^i \,.\, \boldsymbol{\psi}^i\, \dot{q}^i \tag{3.61}$$

For $k = 1 : i$

$$\mathbf{O}_M^{ik} = \mathbf{O}_M^{hk} + \delta^{ki}\, \boldsymbol{\varphi}^i \tag{3.62}$$

$$(\mathbf{B}_M^{ik} = \tilde{\mathbf{O}}_M^{ik}) \tag{3.63}$$

$$\mathbf{A}_M^{ik} = \mathbf{A}_M^{hk} + \tilde{\mathbf{O}}_M^{hk} \,.\, \mathbf{d}_z^{hi} + \delta^{ki} \boldsymbol{\psi}^i \tag{3.64}$$

$$\text{with } \delta^{ki} = 1 \text{ if } k = i, \text{ and } 0 \text{ otherwise} \tag{3.65}$$

end

end.

Finally, for practical computer implementation, all the vector computations of this kinematic algorithm must be transformed into their matrix forms by expressing all vectors and tensors in their appropriate frames[10]. This leads to the following matrix form of the recursive scheme:

Initialization:

$$\alpha_c^0 = -g \;\; ; \;\; \omega^0 = 0 \;\; ; \;\; \dot{\omega}_c^0 = 0 \;\; ; \;\; O_M^{ik} = 0 \;\; ; \;\; A_M^{ik} = 0$$
$$(\forall\, i = 0 : N^{body}, \forall k = 0 : i)$$

Recursion:
For $i = 1 : N^{body}$

$$h = \text{inbody}(i)$$

$$\omega^i = R^{i,h}(q^i)\, \omega^h + \varphi^i\, \dot{q}^i$$

$$\dot{\omega}_c^i = R^{i,h}(q^i)\, \dot{\omega}_c^h + \tilde{\omega}^i \varphi^i\, \dot{q}^i$$

$$\beta_c^i = \tilde{\dot{\omega}}_c^i + \tilde{\omega}^i \tilde{\omega}^i$$

$$\alpha_c^i = R^{i,h}(q^i) \left(\alpha_c^h + \beta_c^h (z^h + d^{hi} \right) + 2\, \tilde{\omega}^i \psi^i\, \dot{q}^i$$

[10] For instance: $\boldsymbol{\psi}^i = [\hat{\mathbf{X}}^i]^T \psi^i$, $\mathbf{z}^h = [\hat{\mathbf{X}}^h]^T z^h, \ldots$

For $k = 1 : i$

$$\begin{aligned} O_M^{ik} &= R^{i,h}(q^i)\, O_M^{hk} + \delta^{ki}\, \varphi^i \\ (B_M^{ik} &= \tilde{O}_M^{ik}) \\ A_M^{ik} &= R^{i,h}(q^i) \left(A_M^{hk} + \tilde{O}_M^{hk}(z^h + d^{hi})\right) + \delta^{ki}\, \psi^i \\ \text{with } \delta^{ki} &= 1 \text{ if } k = i, \text{ and } 0 \text{ otherwise} \end{aligned}$$

end

end.

3.3.3 Backward dynamics

To achieve the dynamical recursion, one can easily obtain the vector equations of motion of body i (see figure 3.16), on the basis of the Newton-Euler equations 1.94 and 1.96.

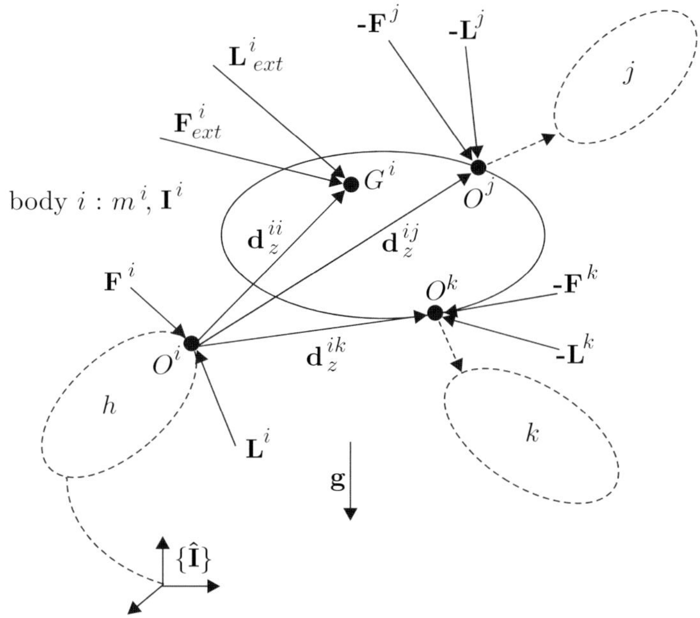

Figure 3.16: Body i — backward dynamics

In accordance with the Newton equation of motion 1.94, the translational vector equation of motion of body i is

$$\mathbf{F}^i - \sum_{j \in \bar{i}} \mathbf{F}^j + \mathbf{F}^i_{ext} + m^i \mathbf{g} = m^i \ddot{\mathbf{x}}^i \tag{3.66}$$

where $\mathbf{F}^i$, defined in section 3.1.4, represents the resultant force acting on body i through the joint i, evaluated at the connection point O^i (see figure.3.16), $\mathbf{F}^i_{ext}$ is the external (except gravity $\mathbf{g}$) resultant force applied to center of mass G^i of body i, and $\sum_{j\,\in\,\bar{\imath}}$ denotes all the children of body i (j and k in figure 3.16).

Using equation 3.51 and the definitions 3.52 and 3.53, we can write equation 3.66 as

$$\mathbf{F}^i = \sum_{j\,\in\,\bar{\imath}} \mathbf{F}^j + m^i\,(\boldsymbol{\alpha}^i + \boldsymbol{\beta}^i\,.\,\mathbf{d}^{ii}_z) - \mathbf{F}^i_{ext}$$

or equivalently

$$\begin{aligned} \mathbf{F}^i &= \sum_{j\,\in\,\bar{\imath}} \mathbf{F}^j + \mathbf{W}^i \\ &\text{with} \quad \mathbf{W}^i \triangleq m^i\,(\boldsymbol{\alpha}^i + \boldsymbol{\beta}^i\,.\,\mathbf{d}^{ii}_z) - \mathbf{F}^i_{ext} \end{aligned} \tag{3.67}$$

Similarly, the rotational equation of motion of body i with respect to its center of mass G^i (Euler equation 1.96) can be written as

$$\mathbf{L}^i - \sum_{j\,\in\,\bar{\imath}} \mathbf{L}^j + \mathbf{L}^i_{ext} - \tilde{\mathbf{d}}^{ii}_z\,.\,\mathbf{F}^i - \sum_{j\,\in\,\bar{\imath}} (\tilde{\mathbf{d}}^{ij}_z - \tilde{\mathbf{d}}^{ii}_z)\,.\,\mathbf{F}^j = \mathbf{I}^i\,.\,\dot{\boldsymbol{\omega}}^i + \tilde{\boldsymbol{\omega}}^i\,.\,\mathbf{I}^i\,.\,\boldsymbol{\omega}^i \tag{3.68}$$

where $\mathbf{I}^i$ is the inertia tensor of body i with respect to its center of mass, $\mathbf{L}^i$ represents the joint resultant torque acting on body i via joint i and $\mathbf{L}^i_{ext}$ is the external resultant torque applied on body i (including the moment of all the external loads with respect to the center of mass G^i).

Using the previous definition of $\mathbf{W}^i$, we can rewrite the rotational equation of motion of body i with respect to its center of mass as

$$\mathbf{L}^i = \sum_{j\,\in\,\bar{\imath}} \{\mathbf{L}^j + \tilde{\mathbf{d}}^{ij}_z\,.\,\mathbf{F}^j\} + \tilde{\mathbf{d}}^{ii}_z\,.\,\mathbf{W}^i - \mathbf{L}^i_{ext} + \mathbf{I}^i\,.\,\dot{\boldsymbol{\omega}}^i + \tilde{\boldsymbol{\omega}}^i\,.\,\mathbf{I}^i\,.\,\boldsymbol{\omega}^i \tag{3.69}$$

Equations 3.67 and 3.69 can be recursively computed from the leaf bodies of the multibody system to the base. This leads to the "classical" backward recursion of the Newton-Euler scheme which provides (after projection on the joint axes) the *inverse dynamics* model of the system under its implicit form 3.42. In order to get the mass matrix M and the vector c (equation 3.43), we need to split up the $\mathbf{F}^i$ and $\mathbf{L}^i$ quantities and isolate the contribution of each generalized acceleration $\ddot{q}^k$:

$$\mathbf{F}^i = \sum_k \mathbf{F}^{ik}_M\,\ddot{q}^k + \mathbf{F}^i_c \tag{3.70}$$

$$\mathbf{W}^i = \sum_k \mathbf{W}^{ik}_M\,\ddot{q}^k + \mathbf{W}^i_c \tag{3.71}$$

$$\mathbf{L}^i = \sum_k \mathbf{L}_M^{ik} \, \ddot{q}^k + \mathbf{L}_c^i \tag{3.72}$$

These new quantities (with a low index "M" and "c") can also be computed in a recursive manner by introducing the relations 3.55, 3.56 and 3.57 into the dynamical equations 3.67 and 3.69. One finally obtains the following scheme:

For $i = N^{body} : 1$

$$\mathbf{W}_c^i = m^i(\boldsymbol{\alpha}_c^i + \boldsymbol{\beta}_c^i \,.\, \mathbf{d}_z^{ii}) - \mathbf{F}_{ext}^i \tag{3.73}$$

$$\mathbf{F}_c^i = \sum_{j \in \bar{i}} \mathbf{F}_c^j + \mathbf{W}_c^i \tag{3.74}$$

$$\mathbf{L}_c^i = \sum_{j \in \bar{i}} (\mathbf{L}_c^j + \tilde{\mathbf{d}}_z^{ij} \,.\, \mathbf{F}_c^j) + \tilde{\mathbf{d}}_z^{ii} \,.\, \mathbf{W}_c^i - \mathbf{L}_{ext}^i + \mathbf{I}^i \,.\, \dot{\boldsymbol{\omega}}_c^i + \tilde{\boldsymbol{\omega}}^i \,.\, \mathbf{I}^i \,.\, \boldsymbol{\omega}^i \tag{3.75}$$

For $k = 1 : i$

$$\mathbf{W}_M^{ik} = m^i \, (\mathbf{A}_M^{ik} + \tilde{\mathbf{O}}_M^{ik} \,.\, \mathbf{d}_z^{ii}) \tag{3.76}$$

$$\mathbf{F}_M^{ik} = \sum_{j \in \bar{i}} \mathbf{F}_M^{jk} + \mathbf{W}_M^{ik} \tag{3.77}$$

$$\mathbf{L}_M^{ik} = \sum_{j \in \bar{i}} (\mathbf{L}_M^{jk} + \tilde{\mathbf{d}}_z^{ij} \,.\, \mathbf{F}_M^{jk}) + \tilde{\mathbf{d}}_z^{ii} \,.\, \mathbf{W}_M^{ik} + \mathbf{I}^i \,.\, \mathbf{O}_M^{ik} \tag{3.78}$$

end

end.

It must be emphasized at this point that the forward kinematic and backward dynamical algorithms allow us to compute, by means of 3.70 and 3.72 the vector force $\mathbf{F}^i$ and torque $\mathbf{L}^i$ applied to the body i through the joint i. Even though they are obtained here in a completely different manner, these equations are fully equivalent to the vector equations of motion 3.22 and 3.27 we obtained by the virtual power principle.

As for the forward kinematic scheme, the backward dynamical algorithm has to be transformed into matrix form for computer implementation. Assuming that the external resultant loads $\mathbf{F}_{ext}^i$ and $\mathbf{L}_{ext}^i$ are given by their components in the $\{\hat{\mathbf{X}}^i\}$ body frame, we obtain the following scheme:

For $i = N^{body} : 1$

$$\begin{aligned}
W_c^i &= m^i(\alpha_c^i + \beta_c^i(z^i + d^{ii}) - F_{ext}^i \\
F_c^i &= \sum_{j \in \bar{i}} R^{i,j}(q^j) \, F_c^j + W_c^i \\
L_c^i &= \sum_{j \in \bar{i}} \left(R^{i,j}(q^j) \, L_c^j + \left(z^i + d^{ij}\right)^{\sim} R^{i,j}(q^j) \, F_c^j\right) \\
&\quad + \left(z^i + d^{ii}\right)^{\sim} W_c^i - L_{ext}^i + I^i \dot{\omega}_c^i + \tilde{\omega}^i I^i \omega^i
\end{aligned}$$

For $k = 1 : i$

$$
\begin{aligned}
W_M^{ik} &= m^i \left(A_M^{ik} + O_M^{ik} (z^i + d^{ii}) \right) \\
F_M^{ik} &= \sum_{j \in \bar{\imath}} (R^{i,j}(q^j)\, F_M^{jk} + W_M^{ik} \\
L_M^{ik} &= \sum_{j \in \bar{\imath}} \left(R^{i,j}(q^j)\, L_M^{jk} + \left(z^i + d^{ij} \right)^{\sim} R^{i,j}(q^j)\, F_M^{jk} \right) \\
&\quad + \left(z^i + d^{ii} \right)^{\sim} W_M^{ik} + I^i O_M^{ik}
\end{aligned}
$$

end

end.

As in section 3.2.5, the ith joint equation is then obtained by projecting the vector equations of motion 3.70 and 3.72 according to 3.30:

$$\mathcal{Q}^i = \mathbf{F}^i \,.\, \psi^i + \mathbf{L}^i \,.\, \varphi^i \tag{3.79}$$

or, equivalently in the corresponding matrix form

$$\mathcal{Q}^i = \left(\psi^i \right)^T F^i + \left(\varphi^i \right)^T L^i$$

Considering separately the acceleration terms during this operation, we obtain the ith equation of motion of the system 3.43 by respectively projecting 3.74, 3.75 and 3.77, 3.78 on the ith joint axis:

$$
\begin{aligned}
c_{(i)} &= \left(\psi^i \right)^T F_c^i + \left(\varphi^i \right)^T L_c^i \quad \forall i, && (3.80) \\
M_{(i,\,j)} &= \left(\psi^i \right)^T F_M^{ij} + \left(\varphi^i \right)^T L_M^{ij} \quad \forall i \text{ and } \forall j : j \leq i. && (3.81)
\end{aligned}
$$

Finally, the ith equation of motion of system 3.43 written in matrix form is

$$\sum_j M_{(i,\,j)}\, \ddot{q}^j + c_{(i)} = \mathcal{Q}_{(i)} \tag{3.82}$$

These scalar equations of motion are fully equivalent to those obtained from the virtual power principle, i.e. equations 3.38. Because of this equivalence, we can assert that the generalized mass matrix M recursively computed by equation 3.81 is symmetric and positive definite, although the recursive nature of the previous scheme makes it difficult to show this.

3.4 Newton-Euler scheme

with barycentric parameters

3.4.1 Introduction

In section 3.2, barycentric parameters have been introduced to generate the dynamical equations in a non-recursive manner[11] from the virtual power prin-

[11] i.e. by writing the equations in "extenso" form without resorting to auxiliary recursive variables as in section 3.3.

ciple. For this formulation, it was shown that the use of barycentric parameters allowed us to simplify the equations by reducing by one the number of summation indexes ($\sum$). Moreover, it is shown in [71], [18] that the barycentric parameters combine in a way which depends on the sequence and type (prismatic, revolute) of the various joints. These parameter combinations obey rules which lead to a *minimal set of parameters* for tree-like systems, and simplify the dynamical model, as shown in [18], [8].

Hence in the present section, we introduce the barycentric parameters in the Newton-Euler recursive formalism of section 3.3. Starting from the recursive formulation with barycentric parameters proposed in [72] to obtain the inverse dynamical model of serial robot manipulators, we generate here the recursive direct model 3.43 for tree-like structures.

The kinematic formulation is similar to the one proposed in the recursive Newton-Euler scheme developed in section 3.3. Concerning the dynamical formulation, we first recall the definition of the barycentric mass, the barycentric moment vector and the barycentric tensor of body i:

$$\begin{aligned} \bar{m}^i &= \sum_{j:\, i \leq j} m^j \\ \bar{m}^i \mathbf{b}^i &= \sum_{j:\, i \leq j} m^j \, \mathbf{d}^{ij} \\ \mathbf{K}^i &= \mathbf{I}^i - \sum_{j:\, i \leq j} m^j \, \tilde{\mathbf{d}}^{ij} . \tilde{\mathbf{d}}^{ij} \end{aligned} \tag{3.83}$$

Since these barycentric quantities involve all the descendants of body i, they cannot appear in a pure (step by step) recursive formulation as in section 3.3. Thus during the backward dynamics, instead of writing the vector equation of one parent as function of the corresponding equations of its children only, we will write this equation as function of the equations of all its descendants. To clarify the developments, we will first establish the dynamical equations in their implicit form, i.e. the inverse dynamics of the multibody structure. Then in the last section, we will establish the extraction of the mass matrix and the final recursive scheme.

3.4.2 Inverse Dynamics

Translational joint equations

Considering the equation 3.66 (section 3.3) for body j:

$$\mathbf{F}^j - \sum_{k \in \bar{j}} \mathbf{F}^k = m^j \ddot{\mathbf{x}}^j - \mathbf{F}^j_{ext} - m^j \mathbf{g} \tag{3.84}$$

we may obtain a translational equation associated with a joint i by summing these equations over i and all its descendants j:

$$\mathbf{F}^i = \sum_{j:\, i \leq j} \left(m^j \ddot{\mathbf{x}}^j - \mathbf{F}^j_{ext} - m^j \mathbf{g} \right)$$

This equation can also be considered as the global Newton equation for the collection of bodies which is constituted by the body i and all its descendants. We obtained such an equation by the virtual power principle in section 3.2 under the form 3.20:

$$\mathbf{F}^i = \sum_{j:\, i \leq j} \left(m^j \sum_{k:\, k \leq j} \ddot{\mathbf{d}}_z^{kj} - \mathbf{F}_{ext}^j - m^j \mathbf{g} \right)$$

and then transformed it into equation 3.22:

$$\mathbf{F}^i = \bar{m}^i \sum_{k:\, k < i} \ddot{\mathbf{d}}_z^{ki} + \sum_{k:\, i \leq k} \bar{m}^k \ddot{\mathbf{b}}_z^{k} - \sum_{j:\, i \leq j} \mathbf{F}_{ext}^j - \bar{m}^i \mathbf{g} \tag{3.85}$$

which is expressed in terms of the barycentric masses $\bar{m}^i$ and (extended) moment vectors $\bar{m}^k \mathbf{b}_z^k$.

To comply with the philosophy of the present Newton-Euler approach, this equation has to be written under a recursive formulation. We may first recognize in the coefficient of the barycentric mass $\bar{m}^i$ the term

$$\sum_{k:\, k < i} \ddot{\mathbf{d}}_z^{ki} - \mathbf{g} = \ddot{\mathbf{p}}^i - \mathbf{g}$$

which according to the expression 3.50 of $\ddot{\mathbf{p}}^i$ and our previous definitions, can be written as

$$\sum_{k:\, k < i} \ddot{\mathbf{d}}_z^{ki} - \mathbf{g} = \boldsymbol{\alpha}^h + \boldsymbol{\beta}^h \,.\, \mathbf{d}_z^{hi}$$

where h, as before, denotes $inbody(i)$, the parent body of i. Extending the second derivatives of the barycentric moments (according to 3.32), we write the ith translational vector equation as

$$\mathbf{F}^i = \bar{m}^i \left(\boldsymbol{\alpha}^h + \boldsymbol{\beta}^h \,.\, \mathbf{d}_z^{hi} \right) + \sum_{k:\, i \leq k} \left(\bar{m}^k \left(\boldsymbol{\beta}^k \,.\, \mathbf{b}_z^k + \boldsymbol{\psi}^k \ddot{q}^k + 2\, \tilde{\boldsymbol{\omega}}^k \,.\, \boldsymbol{\psi}^k \dot{q}^k \right) - \mathbf{F}_{ext}^k \right) \tag{3.86}$$

Finally, defining new symbols for the pure kinematic terms by

$$\boldsymbol{\alpha}^{*i} \triangleq \boldsymbol{\alpha}^h + \boldsymbol{\beta}^h \,.\, \mathbf{d}_z^{hi} \tag{3.87}$$

$$\boldsymbol{\gamma}^k \triangleq 2\, \tilde{\boldsymbol{\omega}}^k \,.\, \boldsymbol{\psi}^k \dot{q}^k + \boldsymbol{\psi}^k \ddot{q}^k \tag{3.88}$$

which (from 3.54) implies that

$$\boldsymbol{\alpha}^i = \boldsymbol{\alpha}^{*i} + \boldsymbol{\gamma}^i \tag{3.89}$$

we find the joint equation

$$\begin{aligned} \mathbf{F}^i &= \bar{m}^i \boldsymbol{\alpha}^{*i} + \mathbf{W}^{*i} \\ &\text{with} \quad \mathbf{W}^{*i} \triangleq \sum_{k:\, i \leq k} \left(\bar{m}^k \left(\boldsymbol{\beta}^k \,.\, \mathbf{b}_z^k + \boldsymbol{\gamma}^k \right) - \mathbf{F}_{ext}^k \right) \end{aligned} \tag{3.90}$$

Once the kinematics is computed, the latter term $\mathbf{W}^{*i}$ can be reformulated in the following backward[12] recursive manner:

$$\mathbf{W}^{*i} = \sum_{j \in \bar{i}} \mathbf{W}^{*j} - \mathbf{F}^i_{ext} + \boldsymbol{\beta}^i \,.\, \bar{m}^i \mathbf{b}^i_z + \bar{m}^i \,\boldsymbol{\gamma}^i$$

Rotational joint equations

To obtain the rotational joint equation of motion, we start from the Euler equation 3.68 for body i:

$$\mathbf{L}^i - \sum_{j \in \bar{i}} \mathbf{L}^j + \mathbf{L}^i_{ext} - \tilde{\mathbf{d}}^{ii}_z \,.\, (\mathbf{F}^i - \sum_{j \in \bar{i}} .\, \mathbf{F}^j) - \sum_{j \in \bar{i}} \tilde{\mathbf{d}}^{ij}_z \,.\, \mathbf{F}^j = \mathbf{I}^i \,.\, \dot{\boldsymbol{\omega}}^i + \tilde{\boldsymbol{\omega}}^i \,.\, \mathbf{I}^i \,.\, \boldsymbol{\omega}^i \tag{3.91}$$

By means of the translational equation 3.67 of the classical Newton-Euler formulation (section 3.3), the fourth term of the left-hand-side of the above equation can be rewritten as

$$-\tilde{\mathbf{d}}^{ii}_z \,.\, (\mathbf{F}^i - \sum_{j \in \bar{i}} .\, \mathbf{F}^j) = -\tilde{\mathbf{d}}^{ii}_z \,.\, \left(m^i(\boldsymbol{\alpha}^i + \boldsymbol{\beta}^i \,.\, \mathbf{d}^{ii}_z) - \mathbf{F}^i_{ext}\right) \tag{3.92}$$

Using the equation 3.90 for $\mathbf{F}^j$, we can rewrite the fifth term of the left-hand-side of 3.91 as

$$-\sum_{j \in \bar{i}} \tilde{\mathbf{d}}^{ij}_z \,.\, \mathbf{F}^j = -\sum_{j \in \bar{i}} \tilde{\mathbf{d}}^{ij}_z \,.\, (\bar{m}^j \,\boldsymbol{\alpha}^{*j} + \mathbf{W}^{*j}) \tag{3.93}$$

where, according to its definition 3.87, $\boldsymbol{\alpha}^{*j}$ involves the parent body of the bodies $j \in \bar{i}$, i.e. body i itself:

$$\boldsymbol{\alpha}^{*j} = \boldsymbol{\alpha}^i + \boldsymbol{\beta}^i \,.\, \mathbf{d}^{ij}_z \quad \forall j \in \bar{i} \tag{3.94}$$

Using the last three relations, we can thus reformulate the rotational equation 3.91 as follows:

$$\mathbf{L}^i - \sum_{j \in \bar{i}} \mathbf{L}^j + \mathbf{L}^i_{ext} + \tilde{\mathbf{d}}^{ii}_z \,.\, \mathbf{F}^i_{ext} - \sum_{j \in \bar{i}} \tilde{\mathbf{d}}^{ij}_z \,.\, \mathbf{W}^{*j} = \mathbf{I}^i \,.\, \dot{\boldsymbol{\omega}}^i + \tilde{\boldsymbol{\omega}}^i \,.\, \mathbf{I}^i \,.\, \boldsymbol{\omega}^i$$
$$+ m^i \,\tilde{\mathbf{d}}^{ii}_z \,.\, \boldsymbol{\alpha}^i + \underbrace{m^i \,\tilde{\mathbf{d}}^{ii}_z \,.\, \boldsymbol{\beta}^i \,.\, \mathbf{d}^{ii}_z} + \sum_{j \in \bar{i}} \bar{m}^j \,\tilde{\mathbf{d}}^{ij}_z \,.\, \boldsymbol{\alpha}^i + \sum_{j \in \bar{i}} \underbrace{\bar{m}^j \,\tilde{\mathbf{d}}^{ij}_z \,.\, \boldsymbol{\beta}^i \,.\, \mathbf{d}^{ij}_z}$$

in which we recognize in the coefficient of $\boldsymbol{\alpha}^i$:

$$m^i \,\mathbf{d}^{ii}_z + \sum_{j \in \bar{i}} \bar{m}^j \,\mathbf{d}^{ij}_z = \bar{m}^i \mathbf{b}^i_z$$

[12] from the leaf bodies to the base

a recursive definition of the (extended) barycentric moment $\bar{m}^i \mathbf{b}_z^i$ equivalent to 3.10.
Similarly the two underlined ("$\underbrace{\quad}$") terms are equivalent to

$$\sum_{j:\, i \leq j} m^j \, \tilde{\mathbf{d}}_z^{ij} \,.\, \boldsymbol{\beta}^i \,.\, \mathbf{d}_z^{ij}$$

because $\bar{m}^i \triangleq m^i + \sum_{j \in \bar{i}} \bar{m}^j = \sum_{j:\, i \leq j} m^j$, and then replacing the tensor $\boldsymbol{\beta}^i$ by its value $\tilde{\dot{\boldsymbol{\omega}}}^i + \tilde{\boldsymbol{\omega}}^i \,.\, \tilde{\boldsymbol{\omega}}^i$, we combine these terms with those involving the inertia tensor $\mathbf{I}^i$, exactly as in 3.25, to provide the terms

$$\mathbf{K}_z^i \,.\, \dot{\boldsymbol{\omega}}^i + \tilde{\boldsymbol{\omega}}^i \,.\, \mathbf{K}_z^i \,.\, \boldsymbol{\omega}^i \tag{3.95}$$

where the barycentric tensor $\mathbf{K}_z^i$ defined in 3.11 now appears.

Using the barycentric moments and tensor thus finally leads to the following recursive form of the rotational equation of motion:

$$\mathbf{L}^i = \sum_{j \in \bar{i}} (\mathbf{L}^j + \tilde{\mathbf{d}}_z^{ij} \,.\, \mathbf{W}^{*j}) - \mathbf{L}_{ext}^i - \tilde{\mathbf{d}}_z^{ii} \,.\, \mathbf{F}_{ext}^i + \mathbf{K}_z^i \,.\, \dot{\boldsymbol{\omega}}^i + \tilde{\boldsymbol{\omega}}^i \,.\, \mathbf{K}_z^i \,.\, \boldsymbol{\omega}^i + \bar{m}^i \tilde{\mathbf{b}}_z^i \,.\, \boldsymbol{\alpha}^i \tag{3.96}$$

Vector equations 3.90 and 3.96 represent a recursive formulation of the inverse dynamics which is equivalent to the classical one (section 3.3), but using barycentric instead of classical dynamical parameters.

3.4.3 Recursive direct dynamics

As we previously did for the classical Newton-Euler recursive scheme (section 3.3), we need to isolate the generalized accelerations in this first recursion by splitting up $\dot{\boldsymbol{\omega}}^i, \boldsymbol{\beta}^i$ and $\boldsymbol{\alpha}^i$ according to 3.55, 3.56 and 3.57 and vectors $\boldsymbol{\alpha}^{*i}$ and $\boldsymbol{\gamma}^i$ as follows:

$$\boldsymbol{\alpha}^{*i} = \sum_{k:\, k \leq i} \mathbf{A}_M^{*ik} \, \ddot{q}^k + \boldsymbol{\alpha}_c^{*i} \tag{3.97}$$

$$\boldsymbol{\gamma}^i = \sum_{k:\, k \leq i} \boldsymbol{\Gamma}_M^{ik} \, \ddot{q}^k + \boldsymbol{\gamma}_c^i \tag{3.98}$$

This leads to the following forward recursive scheme for computing the kinematics:

Initialization:

$$\boldsymbol{\alpha}_c^0 = -g \;\; ; \;\; \boldsymbol{\omega}^0 = 0 \; ; \; \dot{\boldsymbol{\omega}}_c^0 = 0 \;\; ; \;\; \mathbf{O}_M^{ik} = 0 \; ; \; \mathbf{A}_M^{ik} = 0$$
$$(\forall\, i = 0 : N^{body}, \forall k = 0 : i)$$

Recursion:
For $i = 1 : N^{body}$

$$\begin{aligned} h &= \text{index of the parent body} \\ \boldsymbol{\omega}^i &= \boldsymbol{\omega}^h + \dot{q}^i \boldsymbol{\varphi}^i && (3.99) \\ \dot{\boldsymbol{\omega}}_c^i &= \dot{\boldsymbol{\omega}}_c^h + \tilde{\boldsymbol{\omega}}^i \,.\, \boldsymbol{\varphi}^i \dot{q}^i && (3.100) \\ \boldsymbol{\beta}_c^i &= \tilde{\boldsymbol{\omega}}^i \,.\, \tilde{\boldsymbol{\omega}}^i + \tilde{\dot{\boldsymbol{\omega}}}_c^i && (3.101) \\ \boldsymbol{\alpha}_c^{*i} &= \boldsymbol{\alpha}_c^h + \boldsymbol{\beta}_c^h \,.\, \mathbf{d}_z^{hi} && (3.102) \\ \boldsymbol{\gamma}_c^i &= 2\,\tilde{\boldsymbol{\omega}}^i . \boldsymbol{\psi}^i \dot{q}^i && (3.103) \\ \boldsymbol{\alpha}_c^i &= \boldsymbol{\alpha}_c^{*i} + \boldsymbol{\gamma}_c^i && (3.104) \end{aligned}$$

For $k = 1 : i$

$$\begin{aligned} \mathbf{O}_M^{ik} &= \mathbf{O}_M^{hk} + \delta^{ki} \boldsymbol{\varphi}^i && (3.105) \\ (\mathbf{B}_M^{ik} &= \tilde{\mathbf{O}}_M^{ik}) \\ \mathbf{A}_M^{*ik} &= \mathbf{A}_M^{hk} + \tilde{\mathbf{O}}_M^{hk} \,.\, \mathbf{d}_z^{hi} && (3.106) \\ \boldsymbol{\Gamma}_M^{ik} &= \delta^{ki} \boldsymbol{\psi}^i && (3.107) \\ \mathbf{A}_M^{ik} &= \mathbf{A}_M^{*ik} + \Gamma_M^{ik} && (3.108) \\ \text{with } \delta^{ki} &= 1 \text{ if } k = i, \ 0 \ \text{otherwise} \end{aligned}$$

end
end.

To achieve the dynamical recursion, we also need to split up the $\mathbf{F}^i$ and $\mathbf{L}^i$ vectors (like for equations 3.70 — 3.72) as follows:

$$\mathbf{F}^i = \sum_k \mathbf{F}_M^{ik} \ddot{q}^k + \mathbf{F}_c^i \,;\, \mathbf{W}^{*i} = \sum_k \mathbf{W}_M^{*ik} \ddot{q}^k + \mathbf{W}_c^{*i} \,;\, \mathbf{L}^i = \sum_k \mathbf{L}_M^{ik} \ddot{q}^k + \mathbf{L}_c^i \tag{3.109}$$

to isolate the contribution of each generalized accelerations $\ddot{q}^k$.
These new quantities (with a low index "M" and "c") are now computed in a recursive manner by introducing relations 3.55 — 3.57 and 3.97 — 3.98 into the dynamical equations 3.90 and 3.96. We finally obtain the following scheme:

For $i = N^{body} : 1$

$$\begin{aligned} \mathbf{W}_c^{*i} &= \sum_{j \in \bar{i}} \mathbf{W}_c^{*j} + \boldsymbol{\beta}_c^i \,.\, \bar{m}^i \mathbf{b}_z^i + \bar{m}^i \boldsymbol{\gamma}_c^i - \mathbf{F}_{ext}^i && (3.110) \\ \mathbf{F}_c^i &= \bar{m}^i \boldsymbol{\alpha}_c^{*i} + \mathbf{W}_c^{*i} && (3.111) \\ \mathbf{L}_c^i &= \sum_{j \in \bar{i}} (\mathbf{L}_c^j + \tilde{\mathbf{d}}_z^{ij} \,.\, \mathbf{W}_c^{*j}) - \mathbf{L}_{ext}^i - \tilde{\mathbf{d}}_z^{ii} . \mathbf{F}_{ext}^i \\ &\quad + \mathbf{K}_z^i . \dot{\boldsymbol{\omega}}_c^i + \tilde{\boldsymbol{\omega}}^i \,.\, \mathbf{K}_z^i \,.\, \boldsymbol{\omega}^i + \bar{m}^i \tilde{\mathbf{b}}_z^i \,.\, \boldsymbol{\alpha}_c^i && (3.112) \end{aligned}$$

For $k = 1 : i$

$$\begin{aligned}
\mathbf{W}^{*ik}_M &= \sum_{j\,\in\,\bar{\imath}} \mathbf{W}^{*jk}_M + \tilde{\mathbf{O}}^{ik}_M \,.\, \bar{m}^i \mathbf{b}^i_z + \bar{m}^i\, \boldsymbol{\Gamma}^{ik}_M && (3.113)\\
\mathbf{F}^{ik}_M &= \bar{m}^i\, \mathbf{A}^{*ik}_M + \mathbf{W}^{*ik}_M && (3.114)\\
\mathbf{L}^{ik}_M &= \sum_{j\,\in\,\bar{\imath}} (\mathbf{L}^{jk}_M + \tilde{\mathbf{d}}^{ij}_z \,.\, \mathbf{W}^{*jk}_M) + \mathbf{K}^i_z \,.\, \mathbf{O}^{ik}_M + \bar{m}^i \tilde{\mathbf{b}}^i_z \,.\, \mathbf{A}^{ik}_M && (3.115)
\end{aligned}$$

end

end.

Finally, as for the classical scheme (section 3.3), the ith equation of motion of system 3.43 is obtained by projecting the vector equations 3.111, 3.112 and 3.114, 3.115 on the ith joint axis:

$$c_{(i)} = \boldsymbol{\psi}^i \,.\, \mathbf{F}^i_c + \boldsymbol{\varphi}^i \,.\, \mathbf{L}^i_c \qquad \forall i, \tag{3.116}$$

$$M_{(i,\,j)} = \boldsymbol{\psi}^i \,.\, \mathbf{F}^{ij}_M + \boldsymbol{\varphi}^i \,.\, \mathbf{L}^{ij}_M \qquad \forall i \text{ and } \forall j : j \leq i. \tag{3.117}$$

The ith equation of motion of system 3.43 is then:

$$\sum_j M_{(i,\,j)}\, \ddot{q}^j + c_{(i)} = \mathcal{Q}_{(i)} \tag{3.118}$$

and is strictly equivalent to the corresponding equation obtained by the former formalism in 3.82.

Chapter 4

Complex multibody structures

The preceding chapter dealt with multibody systems having a tree-like structure. However, in many practical cases, physical systems are not tree-like. Road vehicle suspensions, anthropomorphic and parallel robots, railway bogies, etc. are multibody systems whose topology is not tree-like.

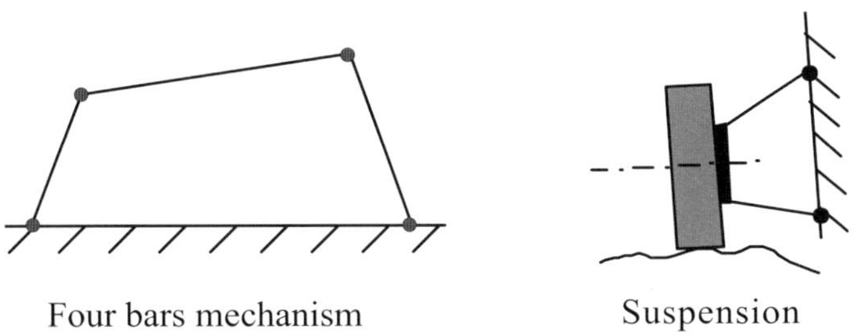

Figure 4.1: Examples of closed-loop structures

The present chapter will first extend the results obtained by the formalism of chapter 3 to deal with these practical applications.

There is a second aim. For convenience, we considered only one degree of freedom joints, i.e. prismatic or revolute joints. Real systems involve more complicated joints. It is almost impossible to foresee all possible joints, so we develop *tools* to help the users to construct models for their particular situations.

Finally, since the present approach for modeling complex multibody systems relies on the use of (nonlinear) constraints and the Lagrange multiplier method (introduced in chapter 2), the model takes the form of differential-algebraic equations. Numerical methods for solving and integrating such equations, as well as for solving particular forms of the rotation constraints, conclude this chapter.

4.1 Closed-loop structures

Figure 4.2$_a$ shows a multibody system which is not a tree. The topology - i.e. the way the bodies and joints are interconnected - contains several *loops* of bodies and joints. During any real motion, these loops have to remain *closed*, and therefore they must be taken into account in both the kinematic relations and the dynamical equations.

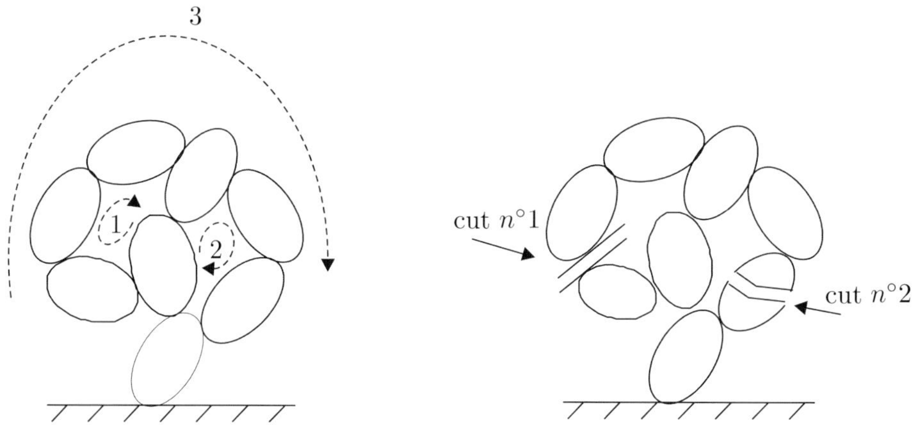

Figure 4.2: a) Closed-loop system b) Open-loop system

Contrary to open-loop structures, i.e. trees, the kinematic chain which connects a body of a closed-loop system and the base is not necessarily unique: the *filiation* concept we used for trees is thus not directly applicable. When using relative joint coordinates – as we do in this book –, we need this filiation concept: it allows us to express unambiguously the absolute position and orientation of the different bodies in terms of a set of generalized (joint) coordinates q.

Facing a closed-loop system, we must first define a *spanning tree*, i.e. a tree which covers all the bodies of the system. Such a spanning tree can be obtained by virtually *cutting* the loops, either by cutting bodies or by disregarding some joints. This procedure, illustrated in figure 4.2$_b$, leads to a tree. Once this tree has been obtained, the bodies and joints can be numbered on the basis of the filiation concept we used in chapter 3.

The first question which arises is: how many cuts are needed to form a spanning tree? One characteristic of closed-loop systems is the fact that the number N^{joint} of joints is greater than the number N^{body} of bodies. The difference $N^{cut} \triangleq N^{joint} - N^{body}$ is called the *cyclomatic* number of the graph topology. Remembering that for a tree the number of joints is equal to the number of bodies and that the spanning tree must cover all bodies, we see that the (minimum) number of cuts which is necessary to form a spanning tree is equal to the cyclomatic number.

A second question concerns the number of loop conditions we should consider in the kinematic and dynamical models in order to ensure that the tree model will behave like the real system. Three different loops can be observed on the example sketched in figure 4.2_a: two internal loops 1 and 2, and an external loop 3. The cyclomatic number, as well as the number of cuts illustrated in figure 4.2_b, is 2, not 3, the number of loops. This is because the loops are not *independent*. Indeed, rejoining the two sides of the cut 1 in figure 4.2_b ensures the closure of the loop 1, and rejoining the cut 2 closes the loop 2. But rejoining both cuts automatically ensures the closure of the external loop 3 : the latter is thus not independent and does not have to be considered in establishing the closure conditions on the system. On the other hand, closing loop 1 does not ensure the closure of loop 2 : loops 1 and 2 are independent. Hence, we obtain a full set of independent loop conditions by looking at the cuts and the N^{cut} associated loops.

The loop closure conditions we have to impose on the spanning tree will lead to (holonomic) *algebraic constraints* on the generalized coordinates q. Generally, these constraints are nonlinear and they cannot be used to express analytically some dependent coordinates in terms of independent ones. We are thus obliged to resort to the Lagrange multiplier technique explained in section 2.4.2. The dynamical equations of the (unconstrained) tree having been obtained by one of the virtual or recursive approaches (see chapter 3, equation 3.38) under the form[1]:

$$M(q,t)\ddot{q} + c(\dot{q},q,t,F_{ext},L_{ext},g) = \mathcal{Q}$$

the equations of the closed-loop system will read

$$\begin{aligned} M(q,t)\ddot{q} + c(\dot{q},q,t,F_{ext},L_{ext},g) &= \mathcal{Q} + J^T(q,t)\,\lambda \\ h(q,t) &= 0 \end{aligned} \tag{4.1}$$

where the algebraic constraints $h(q,t) = 0$ arise from the loop closure conditions. Since the Jacobian of the constraints is also needed for the dynamical equation (as the coefficients of the Lagrange multipliers), the closure conditions have also to be expressed at the velocity level, i.e.:

$$\dot{h}\,(\dot{q},q,t) = J(q,t)\dot{q} + \frac{\partial h\,(q,t)}{\partial t} \tag{4.2}$$

Moreover, as will be seen in section 4.6.1 which deals with the numerical aspects for solving these differential-algebraic motion equations, the second derivatives of the constraints are also needed. The closure conditions must therefore also be generated at the acceleration level:

$$\ddot{h}\,(\ddot{q},\dot{q},q,t) = J(q,t)\ddot{q} + \left(\dot{J}(\dot{q},q,t) + \frac{\partial^2 h\,(q,t)}{\partial t\,\partial q}\right)\dot{q} + \frac{\partial^2 h\,(q,t)}{\partial t^2} \tag{4.3}$$

[1] Time may appear if some joint of the multibody system is kinematically driven and given as a function of time (see section 4.2.2).

For each independent loop, the kinematic equations characterizing the relative position and orientation of the two sides of the cut will serve to provide the constraint equations. As for the dynamical equations, suitable formalisms will be established in section 4.5 to generate these closure conditions. Although many different ways of cutting the loops could be envisaged (as done for instance in [73]), only three of them will be considered here: they have been chosen for their simplicity, while being suitable for covering all possible practical cases.

4.1.1 Cut of a body

This procedure is the most general one and is suitable for any kind of kinematic loop. It consists in cutting a body which belongs to the loop into two parts (see figure 4.3). The two parts are respectively denoted as the *original* body and its

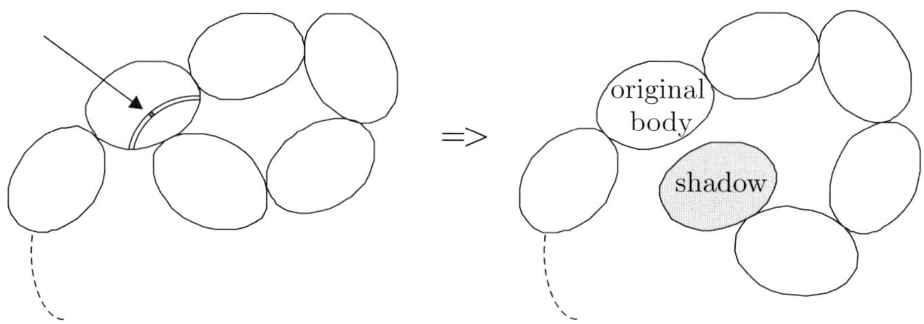

Figure 4.3: Cut in a body

shadow, the latter being added to the original structure. The mass and inertia characteristics of the cut body must be distributed over the original and the shadow bodies. One may for instance decide that the shadow has neither mass nor inertia, while the original body keeps all the mass distribution. Closing such a cut loop involves six constraints (see figure 4.4):

- Three translational constraints expressing that the positions of the two bodies coincide in some reference point $P^o \equiv P^s$ (which can for instance be the center of mass G^o). Defining the position vectors of P^o and P^s respectively by $\mathbf{x}^o = [\hat{\mathbf{I}}]^T x^o$ and $\mathbf{x}^s = [\hat{\mathbf{I}}]^T x^s$, we write these three translational closing conditions in matrix form:

$$h_t(q,t) \triangleq x^s(q,t) - x^o(q,t) = 0 \tag{4.4}$$

- Three rotational constraints expressing that the two frames $\{\hat{\mathbf{X}}^o\}$ and $\{\hat{\mathbf{X}}^s\}$ respectively associated with the original body and its shadow remain aligned. In other words, if one defines the rotation matrix $R(q,t)$ by $[\hat{\mathbf{X}}^s] \triangleq R(q,t)[\hat{\mathbf{X}}^o]$, these closing conditions can be written as

$$R(q,t) - E = 0 \tag{4.5}$$

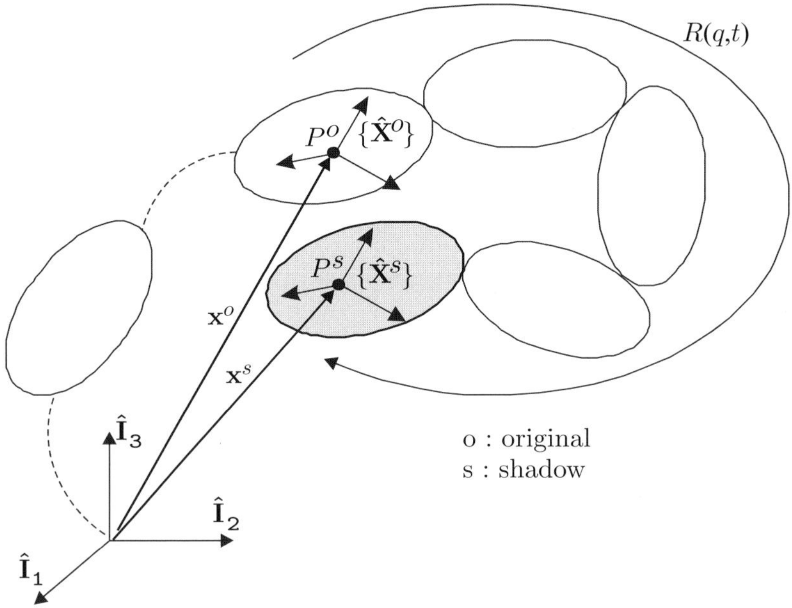

Figure 4.4: Loop closure conditions

where E is the unit matrix. Formally, this matrix equation corresponds to nine scalar constraints, but since the matrix $R(q,t)$ is a rotation matrix which can be expressed in terms of three independent rotation variables (see section 1.1.8), only three of these scalar constraints have to be considered as being independent. As will be justified in section 4.6.2, these three independent rotation constraints can be selected by choosing three off-diagonal elements of the matrix relation 4.5 as follows:

$$h_r(q,t) \triangleq \begin{pmatrix} -R_{(3,2)}(q,t) \\ R_{(3,1)}(q,t) \\ -R_{(2,1)}(q,t) \end{pmatrix} = 0 \tag{4.6}$$

The first and second derivatives of the translational constraints are simply obtained by computing the velocities and accelerations of the two points P^o and P^s :

$$\begin{aligned} \dot{h}_t(\dot{q},q,t) \triangleq \dot{x}^s(\dot{q},q,t) - \dot{x}^o(\dot{q},q,t) &= 0 \\ \ddot{h}_t(\ddot{q},\dot{q},q,t) \triangleq \ddot{x}^s(\ddot{q},\dot{q},q,t) - \ddot{x}^o(\ddot{q},\dot{q},q,t) &= 0 \end{aligned} \tag{4.7}$$

and since the absolute velocities $\dot{x}^s$ and $\dot{x}^o$ are linear functions of the (joint) generalized velocities, one will obtain $\dot{h}_t(\dot{q},q,t)$ under the form 4.2 which allows us to extract the corresponding Jacobian matrix $J_t(q,t)$. This Jacobian will be the coefficient of the Lagrange multipliers λ_t associated with the constraints 4.4.

These Lagrange multipliers have a physical meaning that we may deduce from the virtual power principle. Indeed, the physical effect which compels the point P^s to be located at P^o is a (internal) constraint force $\mathbf{F} = [\hat{\mathbf{I}}]^T F$ applied to the shadow body at P^s by the original body. The reaction $-\mathbf{F}$ is applied by the shadow to the original body. For a virtual change $\Delta\dot{q}$ of the generalized velocities which is *not compatible* with the considered constraints – i.e. a virtual velocity change for which P^s does not remain collocated with P^o – the constraint force $\mathbf{F}$ and its reaction would produce a virtual power change given by

$$\Delta\mathcal{P}_{constraint} = \mathbf{F} \,.\, \Delta\dot{\mathbf{x}}^s - \mathbf{F} \,.\, \Delta\dot{\mathbf{x}}^o = F^T \, J_t(q,t) \, \Delta\dot{q}$$

On the other hand, from section 2.4.3, we know that the virtual power change due to the generalized constraint forces Q' expressed in terms of Lagrange multipliers is given by

$$\Delta\mathcal{P}_{constraint} = \Delta\dot{q}^T Q' = \Delta\dot{q}^T \, J_t{}^T(q,t) \, \lambda_t \tag{4.8}$$

Thus comparing these two expressions for $\Delta\mathcal{P}_{constraint}$, we obtain the equality

$$\lambda_t = F \tag{4.9}$$

which means that the Lagrange multipliers associated with the translational loop closure constraints are equal to the components[2] of the constraint force $\mathbf{F}$ which maintains the point P^s fixed at P^o.

Concerning the rotational loop closure conditions, it will be shown in section 4.6.2 that equating to zero the relative angular velocity of frame $\{\hat{\mathbf{X}}^s\}$ with respect to $\{\hat{\mathbf{X}}^o\}$:

$$\boldsymbol{\omega}^s - \boldsymbol{\omega}^o = 0$$

plays the role of a non-holonomic constraint which is functionally equivalent to the original rotation constraint. The angular velocities being also linear functions of the generalized velocities, a *pseudo-gradient* matrix $J_r(q,t)$ will be obtained by computing $\boldsymbol{\omega}^s$ and $\boldsymbol{\omega}^o$:

$$\boldsymbol{\omega}^s(\dot{q},q,t) - \boldsymbol{\omega}^o(\dot{q},q,t) = [\hat{\mathbf{X}}^s]^T \left(J_r(q,t)\dot{q} + b(q,t)\right)$$

(where time t and the column matrix b only appear if some revolute joint is kinematically driven in the loop). In turn, the angular accelerations will provide equivalent second derivatives of the rotation constraints 4.6. It will also be shown in section 4.6.2 that the Lagrange multipliers λ_r associated with these pseudo-constraints are the components (in frame $\{\hat{\mathbf{X}}^o\}$) of the constraint torque $\mathbf{L}$ which maintains the frame $\{\hat{\mathbf{X}}^s\}$ aligned with the frame $\{\hat{\mathbf{X}}^o\}$.

[2] The reader should observe that if the constraints are expressed in the body fixed frame $\{\hat{\mathbf{X}}^o\}$, then the Lagrange multipliers would be the components of the constraint force $\mathbf{F}$ expressed in this body fixed frame.

4.1.2 Cut in a ball joint

This second procedure is applicable only if the loop contains a ball joint which can be considered as *ideal*, meaning that this ball joint does not transmit any torque. Such cases often occur in mechanisms, for instance vehicle suspensions, so long as the friction in a ball joint of the loop can be reasonably neglected. In this case, even if the previous procedure (cut of a body) remains applicable, the present one is preferable since only three constraints (instead of six) will be generated. Figure 4.5 illustrates the procedure: first the closed-loop is cut by

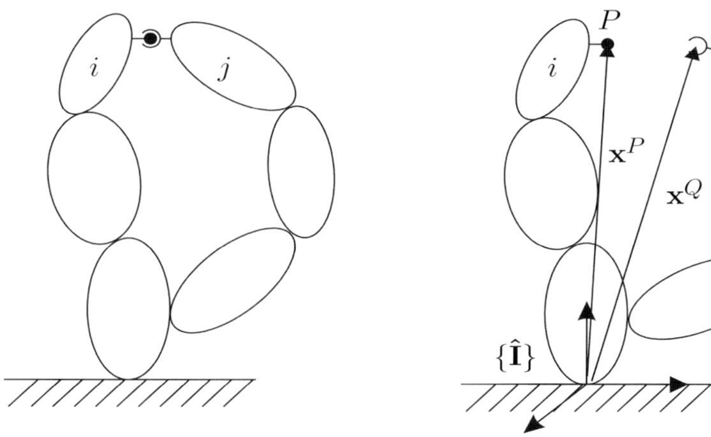

Figure 4.5: Cut in a ball joint

simply disregarding the ball joint, then a translation loop-closure constraint is introduced in order to ensure that the two points P and Q, located at the ball joint center and respectively attached to bodies i and j, coincide. Denoting by $\mathbf{x}^P = [\hat{\mathbf{I}}]^T x^P(q,t)$ and $\mathbf{x}^Q = [\hat{\mathbf{I}}]^T x^Q(q,t)$ their respective position vectors, we find the three closure constraints:

$$h(q,t) \triangleq x^Q(q,t) - x^P(q,t) = 0 \tag{4.10}$$

They (as well as their derivatives) are thus strictly similar to the three translational loop-closure conditions 4.4 we used in the previous section when opening the loop by cutting a body.

Compared with the previous procedure which is more general, opening the loop by cutting a ball joint presents two advantages:

- since the ball joint does not belong to the open-loop structure, the three rotation variables which characterize the relative motion occurring in this joint are *not* included in the set of generalized coordinates, and the dimension of the system of dynamical equations decreases.

- the closing conditions involve only three constraints, allowing us to avoid unnecessary computations. Moreover, numerical convergence is generally better for translational constraints than for rotational ones.

4.1.3 Cut of a connecting rod

This procedure is particularly efficient when a loop is closed by a connecting rod whose mass and inertia can be considered as being negligible. The rod must be attached to two bodies of the structure by means of two *ideal* ball joints, or equivalently by one ideal ball joint on one side and an ideal universal joint on the other side. In this situation, the rod has essentially a kinematic effect: the distance between the two points P and Q of the structure (see figure 4.6) remains constant. Physically, this kinematic effect is produced by the

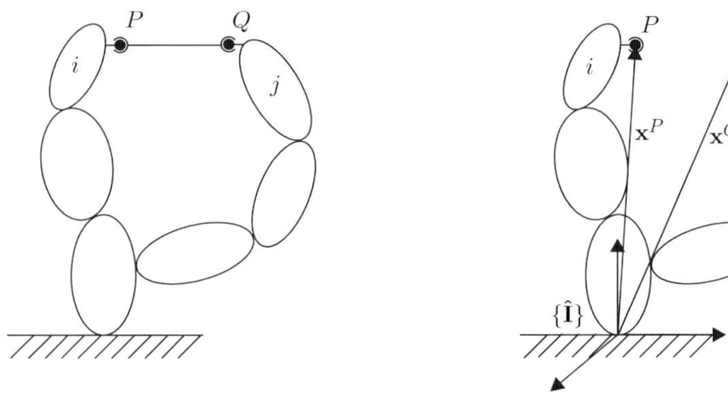

Figure 4.6: Opening a loop by disregarding a connecting rod

force acting in traction or compression within the rod. This internal constraint force, transmitted to the two neighboring bodies i and j, is aligned with the rod itself. Except for this particular component, there is no other component of force or torque applied to the neighboring bodies since the connecting joints are assumed to be ideal. In practical applications, connecting rods are frequently used to transmit motions: examples can be found in vehicle suspensions, steering mechanisms, etc...

The closing condition must guarantee that the distance between the attachment points of the rod remains equal to its length l. Thus, denoting by $\mathbf{x}^P(q,t)$ and $\mathbf{x}^Q(q,t)$ the respective position vectors of these attachment points and defining the vector difference by $\mathbf{x}(q,t) \triangleq \mathbf{x}^Q(q,t) - \mathbf{x}^P(q,t)$, we have the closure constraint:

$$\| \mathbf{x}(q,t) \| - l = 0$$

However, for computation efficiency this closure condition is better written as

$$h(q,t) \triangleq \frac{1}{2} \| \mathbf{x}(q,t) \|^2 - \frac{l^2}{2} = \frac{1}{2} \mathbf{x}(q,t) \,.\, \mathbf{x}(q,t) - \frac{l^2}{2} = 0 \qquad (4.11)$$

This allows us to avoid any square root evaluation and the time derivative of the constraint is then obtained in a straightforward manner as follows:

$$\dot{h}(\dot{q}, q, t) = \mathbf{x}(q, t) \,.\, \dot{\mathbf{x}}(\dot{q}, q, t) = 0 \tag{4.12}$$

Since the absolute velocities $\dot{\mathbf{x}}^P$ and $\dot{\mathbf{x}}^Q$ are linearly dependent on the (joint) generalized velocities (see for instance equation 2.69), this time derivative will have the form

$$\dot{h}(\dot{q}, q, t) = \mathbf{x}(q, t) \,.\, \left(\frac{\partial \dot{\mathbf{x}}(q, t)}{\partial \dot{q}^T} \dot{q} + \frac{\partial \mathbf{x}(q, t)}{\partial t} \right) \tag{4.13}$$

which shows that the Jacobian matrix of this constraint is easily computed by isolating the coefficients of the $\dot{q}_\alpha$ in the expression of the relative velocity vector $\dot{\mathbf{x}} = \dot{\mathbf{x}}^Q - \dot{\mathbf{x}}^P$:

$$J(q, t) = \mathbf{x}(q, t) \,.\, \frac{\partial \dot{\mathbf{x}}(q, t)}{\partial \dot{q}^T} \tag{4.14}$$

The second derivative of the closure constraint is obtained in a similar manner:

$$\ddot{h}(\ddot{q}, \dot{q}, q, t) = \dot{\mathbf{x}}(\dot{q}, q, t) \,.\, \dot{\mathbf{x}}(\dot{q}, q, t) + \mathbf{x}(q, t) \,.\, \ddot{\mathbf{x}}(\ddot{q}, \dot{q}, q, t) \tag{4.15}$$

and thus simply requires the computation of the relative acceleration vector $\ddot{\mathbf{x}}$.

Using the square of the distance l in the formulation 4.11 of the closing constraint influences the physical meaning of the associated Lagrange multiplier. Indeed, supposing that the constraint force applied to body j along the rod direction is given by

$$\mathbf{F} = F \frac{\mathbf{x}}{\| \mathbf{x} \|} \tag{4.16}$$

the virtual power change which will occur during a virtual velocity change which is *not compatible* with the loop closure will be equal to

$$\Delta \mathcal{P}_{constraint} = \mathbf{F} \,.\, \Delta \dot{\mathbf{x}} = \frac{F}{\| \mathbf{x} \|} \mathbf{x} \,.\, \Delta \dot{\mathbf{x}}$$

Looking at the value of the Jacobian matrix 4.14, we see that this power change can also be written as

$$\Delta \mathcal{P}_{constraint} = \frac{F}{\| \mathbf{x} \|} J(q, t) \Delta \dot{q} \tag{4.17}$$

and by comparison with 4.8, we thus obtain

$$\lambda = \frac{F}{\| \mathbf{x} \|} = \frac{1}{l} F \tag{4.18}$$

In this particular case, λ represents the force per unit length $[Newton/m]$ transmitted by the rod.

To conclude this section, it appears that cutting a connecting rod to open a closed-loop presents several computational advantages:

- the number of dynamical equations decreases, since the angular variables between the rod and the neighboring bodies are not considered in the set of generalized coordinates,

- the closure of the loop involves a single (scalar) constraint only.

On the other hand, care must be taken by using this procedure: the mass and the inertia of the rod are neglected and the connections to the neighboring bodies are supposed to be ideal.

4.2 User joints/constraints

Constraints in a multibody system arise in various ways, not only from its closed-loop structure. Instead of cataloging all possible situations, we present a few examples to show how to model a realistic structure by using a standard multibody program.
First of all, whatever the joint types included in the multibody program, the "library" of joints is never exhaustive. A typical example is the wheel-rail contact phenomenon in railway dynamics: with respect to the rail, a wheel has five degrees of freedom. A wheel-rail contact could thus be modeled (as is done in [74]) by a five degrees of freedom joint, if both the wheel and the rail are included in the multibody model. The kinematics as well as the (generalized) joint forces and torques of this contact joint strongly depend on the wheel and rail profiles. One may thus understand that a general purpose multibody program would not necessarily include a *black-box* model for this type of phenomenon. In chapter 7 we will show how, by using a multibody program as a *tool-box*, a user may develop a model for this particular situation.
In the following examples, we suppose — as we did in chapter 3 — that the multibody formalism deals with two types of joint only: revolute and prismatic, both having only one degree of freedom.

4.2.1 Helicoidal joint

Suppose one has to model the system in figure 4.7. This mechanism is driven by a pure torque M acting on the horizontal screw: the rotational motion is transformed into a translational motion of the nut which is connected to the upper left attachment point of the pantograph (whose upper right attachment point is fixed). All joints are revolute, except the helicoidal joint which connects the screw and the nut.

Such a helicoidal joint could easily have been considered in the multibody formalisms we developed in chapter 3, but it can also be modeled by a revolute and a prismatic joint whose coordinates obey a constraint relation. This model is sketched in figure 4.8. The constraint relating the two joint coordinates is simply given by:

$$h(q,t) \triangleq z^j - \rho\,(\vartheta^i - \vartheta^i_{\text{o}}) = 0 \tag{4.19}$$

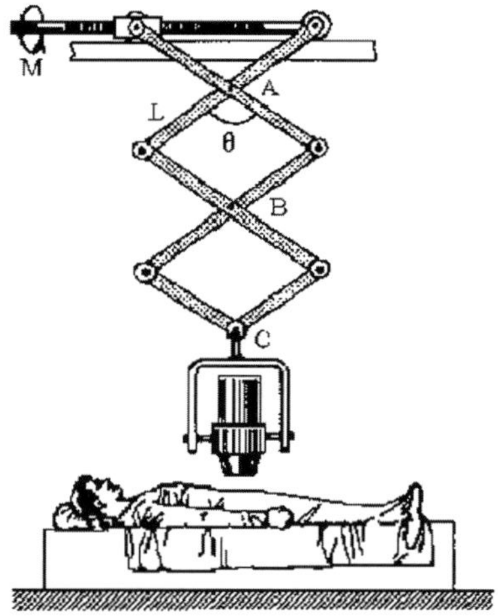

Figure 4.7: Mechanism driven by a helicoidal joint

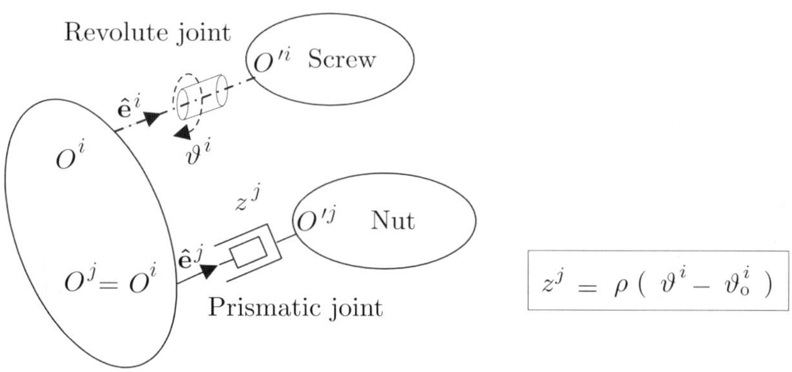

Figure 4.8: Model of a helicoidal joint

where ρ is the screw thread. The derivatives and the Jacobian of this constraint are evident, and the reader can verify that, when the constraint is written in this form, the associated Lagrange multiplier is equal to the force which is required to actuate the nut.

4.2.2 Kinematically driven joint

Sometimes, the relative motion within a joint is prescribed to follow a given time function.

For example in vehicle dynamics, ride comfort in a vehicle on a damaged road is studied by assuming a constant (longitudinal) speed. An easy way to model such a vehicle situation with a multibody approach consists in introducing fictitious joints linking the chassis of the vehicle to the (fixed) ground. The chassis normally has six (absolute) degrees of freedom: the longitudinal, lateral and vertical displacements of its center of mass and three absolute rotations: yaw, pitch and roll. The multibody kinematic chain linking the ground to the chassis is thus made of a succession of six (fictitious) one degree of freedom joints, either prismatic or revolute. Note that the order of these elementary joints is of utmost importance for their physical interpretation. To study the ride at constant speed on a straight road, we must chose a prismatic joint (in the longitudinal direction) for the first joint in the kinematic chain and then require that the longitudinal displacement q_1 in this joint is given as a (linear) function of time:

$$q_1 = \alpha t + \beta$$

More generally, when some joint coordinate q_α is *kinematically driven* and given as time function:

$$q_\alpha = f(t)$$

one may equivalently consider that the system is constrained by

$$h(q,t) \triangleq q_\alpha - f(t) = 0 \tag{4.20}$$

But since the generalized coordinate linearly appears in the constraint function, the Jacobian matrix related to this constraint has the particular form

$$J = \begin{pmatrix} 0 & 0 & \cdots & 1 & \cdots & 0 & 0 \end{pmatrix} \tag{4.21}$$

where the value 1 appears in the αth position. In the matrix dynamical equations 4.1:

$$M(q,t)\ddot{q} + c(\dot{q}, q, t, F_{ext}, L_{ext}, g) = \mathcal{Q} + J^T(q,t)\,\lambda$$

the related Lagrange multiplier is thus only present in the αth row. This means that the αth equation of our system only serves to find the value of this Lagrange multiplier. By disregarding this α^{th} equation (which corresponds to the joint equation related to q_α), we can thus eliminate this Lagrange multiplier from our system. Then by replacing $q_\alpha, \dot{q}_\alpha$ and $\ddot{q}_\alpha$ by their given values of time $f(t), \dot{f}(t)$ and $\ddot{f}(t)$, we reduce by one the number of generalized coordinates in the equations and we obtain the reduced system

$$M_{red}(q_{red}, t)\ddot{q}_{red} + c_{red}(\dot{q}_{red}, q_{red}, t, F_{ext}, L_{ext}, g) = Q_{red}$$

which is purely differential. This reduced system has the same number of equations and unknowns q_{red} and can thus be solved by a standard integration technique. On the other hand, the dynamical equation which has been suppressed allows us to compute afterwards, if necessary, the value of the Lagrange multiplier. Due to the particular form of the Jacobian, it is easily shown by means of the virtual power principle that this Lagrange multiplier is equal to the joint force (resp. torque) which is necessary for driving the prismatic (resp. revolute) joint accordingly to 4.20 during the system motion.

4.2.3 Transmission by pulley

The system illustrated in figure 4.9 has one degree of freedom. Its multibody

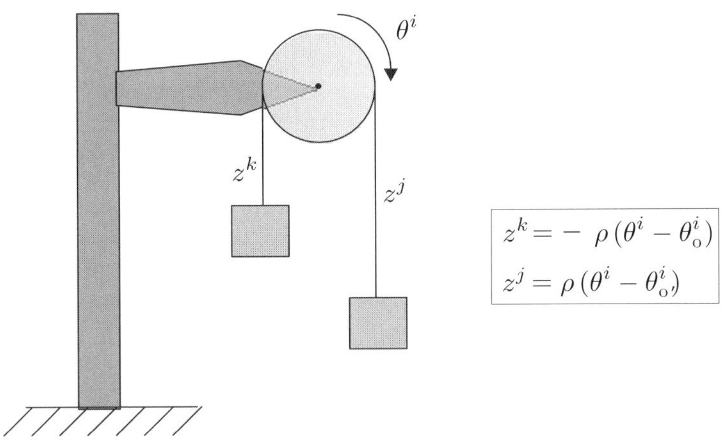

Figure 4.9: Transmission by cable and pulley

model could be made of three bodies respectively connected to the fixed base by a rotational and two translational joints. The three joint variables are constrained by the two relations

$$h_1(q,t) \triangleq z^j - \rho(\vartheta^i - \vartheta^i_o) = 0 \quad \text{and} \quad h_2(q,t) \triangleq z^k + \rho(\vartheta^i - \vartheta_o) = 0 \quad (4.22)$$

whose derivatives are again obtained straightforwardly.

4.2.4 Gear transmission

A fourth example is a (simple) robot model needed for control purposes. Figure 4.10 illustrates a serial link manipulator in which the different arms are

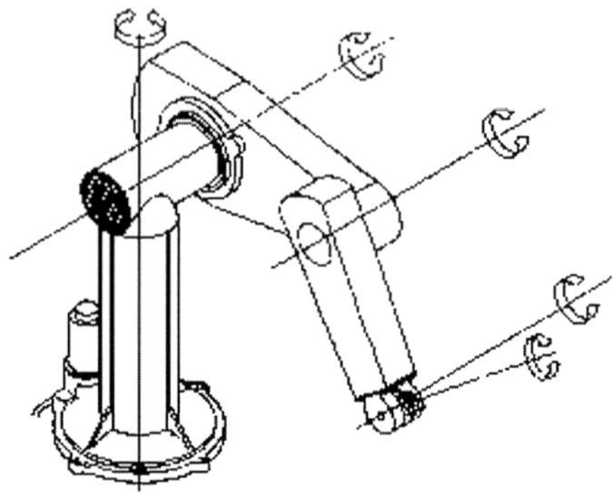

Figure 4.10: Serial-link manipulator

connected by means of revolute joints. These arms are actuated by electric DC motors. A multibody model could be restricted to the arms only, but due to the large transmission gains between the joints and their actuators, the inertia of the motors must also be taken into account in the model. A first approximation consists in adding to the inertia of each arm the rotary inertia of its actuator multiplied by the square of the transmission ratio. However, one must notice that the motor which is driving the link i is generally mounted on the arm $i-1$: the mass and the inertia of the motor are thus involved as soon as the latter arm moves. A better approximation is thus given by the multibody model sketched in figure 4.11; in this model each motor is mounted by a revolute joint on the

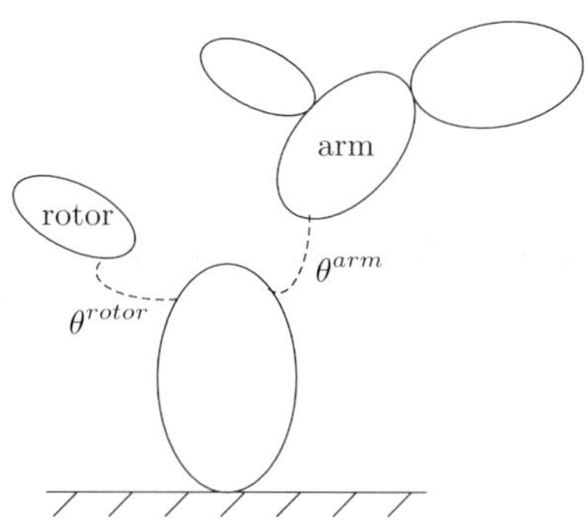

Figure 4.11: Model of a robot manipulator

arm where it is physically located. Then, a constraint relation

$$h(q,t) \triangleq \vartheta^{arm} - \alpha\,(\vartheta^{rotor} - \vartheta^{rotor}_{\mathrm{o}}) = 0$$

which takes the gear transmission α into account, is written between the angular motion of the actuator rotor (within its stator) and the relative angular motion of the driven arm.

4.3 Point-to-point links

The feature of point-to-point links, which is inspired by vehicle suspension problems, is introduced here as a practical means for reducing the dimension of a multibody model in order to save computational effort, when possible. A typical example is illustrated in figure 4.12 (from [41]) which represents a five-bar left rear suspension system of a road vehicle. Considering only one fourth of the

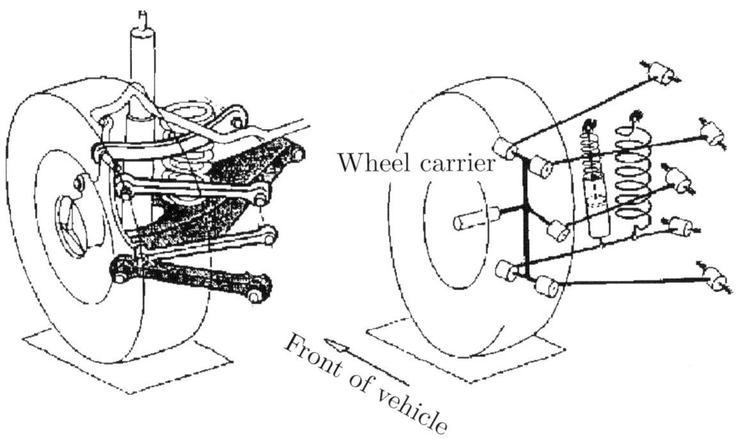

Figure 4.12: Five-bar suspension system

complete vehicle, we see that the main parts of the system are the chassis, the wheel-carrier and the wheel. With respect to the chassis, the wheel-carrier has only one degree of freedom since it is connected to the chassis by means of five connecting rods. These produce four closed loops in the system, and these loops have to be cut in the model according to the procedure explained in section 4.1. In addition, the main rod is connected to the chassis by the main suspension elements: a spring and a damper in parallel.
For modeling the suspension mechanism, one could consider that the spring and damper elements are internal joints, each having six degrees of freedom (since they are mounted on both sides by means of ball joints). In this way, the model would include additional joints and two additional closed-loops.

A simpler model, which avoids these additional closed-loops, can be used if the mass and inertia of the suspension elements are negligible (with respect to the remainder of the vehicle). Indeed, the spring and damper are force elements which do not constrain the kinematic motion between the main rod and the chassis. Under the previous assumption, their dynamical effects can be modeled by means of *external*[3] *forces (and/or torques)* respectively applied to the chassis and the rod. This multibody representation is sketched in figure 4.13. The chassis is connected to the fixed ground by a fictitious six degrees of free-

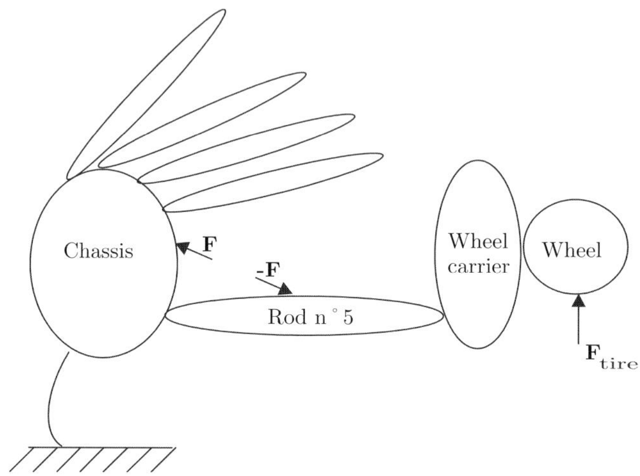

Figure 4.13: Multibody model of suspension

dom joint (a succession of three prismatic and three revolute joints). The five rods are mounted on the chassis by universal joints (a succession of two revolute joints). The ball-joints at the ends of the rods 1 to 4 have been disregarded for opening the loops. The wheel-carrier is connected to main rod 5 by a ball joint on one side, and to the wheel by a revolute joint on the other side.
In this model, the spring and damper elements are introduced by taking into account the force **F** they apply to the chassis, and the reaction $-\mathbf{F}$ acting on the rod 5. Assuming that these suspension elements are mounted by means of *ideal* joints, we may consider that the vector force **F** is aligned with the direction of the spring (the damper). Both this force and its reaction will be modeled as *external forces/torques* $(\mathbf{F}^i_{ext}, \mathbf{L}^i_{ext})$ when generating the equations of motion. The user is thus asked to provide the values of these vector forces and torques in terms of the generalized coordinates and velocities.
Since these coordinates and velocities are those occurring in the joints of the multibody structure, the task facing the user is far from obvious: it is the development of the kinematics relating the deformations of the spring and the damper to the joint deformations. On the contrary, this task is greatly sim-

[3] "external", as opposed to *joint forces*

plified if the required kinematics is obtainable from *tool-boxes* in the computer multibody code, as will be shown in section 4.5.

Real springs can be characterized by numerous types of mathematical laws. For simplicity, suppose that the suspension has a linear characteristic given by its stiffness k and a neutral length l_o. The value F_{spr} of the vector force $\mathbf{F}_{spr}$ applied to the chassis and due to the spring is thus

$$F_{spr} = -k(l - l_o) \tag{4.23}$$

where l is the actual length of the spring. This vector force is applied at the material point of the chassis which serves as attachment point of the spring, and is aligned with the spring direction. Based on this constitutive equation, which is supposed to be provided by the user, one may now compute the spring contribution to the resultant vector force $\mathbf{F}_{ext}^{chas}$ and torque $\mathbf{L}_{ext}^{chas}$ applied to the chassis. For this purpose, we denote by $\mathbf{x}_{spr}^{chas}(q,t)$ the absolute vector position of the attachment point of the spring on the chassis, and by $\mathbf{x}_{spr}^{rod}(q,t)$ the vector position of its attachment point on the rod 5. The vector force is given by

$$\begin{aligned} \mathbf{F}_{spr} &= -k\frac{(l - l_o)}{l}\left(\mathbf{x}_{spr}^{chas} - \mathbf{x}_{spr}^{rod}\right) \\ \text{where} \quad l &= \parallel \mathbf{x}_{spr}^{chas} - \mathbf{x}_{spr}^{rod} \parallel \end{aligned} \tag{4.24}$$

and the contributions to $\mathbf{F}_{ext}^{chas}$ and $\mathbf{L}_{ext}^{chas}$ are given by

$$\begin{aligned} \mathbf{F}_{ext}^{chas} &= \mathbf{F}_{spr} \\ \mathbf{L}_{ext}^{chas} &= \left(\mathbf{x}_{spr}^{chas} - \mathbf{x}^{chas}\right) \times \mathbf{F}_{spr} \end{aligned} \tag{4.25}$$

where $\mathbf{x}^{chas}$ denotes (as usual) the position vector of the center of mass G^{chas} of the body *chassis*. Similarly, the contributions on the external resultant force/torque on the body *rod* read:

$$\begin{aligned} \mathbf{F}_{ext}^{rod} &= -\mathbf{F}_{spr} \\ \mathbf{L}_{ext}^{rod} &= -\left(\mathbf{x}_{spr}^{rod} - \mathbf{x}^{rod}\right) \times \mathbf{F}_{spr} \end{aligned} \tag{4.26}$$

Thus in order to transform any constitutive equation of type 4.23 characterizing an internal *point-to-point link* of the multibody system into *external forces/torques* ($\mathbf{F}_{ext}^{i}, \mathbf{L}_{ext}^{i}$) applied to the involved bodies, the user needs the expressions for the position vectors of the two attachment points (in addition to the G^i position vectors) as function of the joint generalized coordinates.

With slight modifications, the method we used for the spring can be extended for modeling the suspension damper. Assuming a linear characteristic of the damping device:

$$F_{damp} = -c\,\frac{dl}{dt}$$

we find that the force applied by the damper on the chassis is given by

$$F_{damp} = -\,c\left(\dot{\mathbf{x}}_{damp}^{chas} - \dot{\mathbf{x}}_{damp}^{rod}\right) \cdot \frac{\left(\mathbf{x}_{damp}^{chas} - \mathbf{x}_{damp}^{rod}\right)}{\parallel \mathbf{x}_{damp}^{chas} - \mathbf{x}_{damp}^{rod} \parallel} \tag{4.27}$$

which is easily transformed into external resultant forces/torques $(\mathbf{F}^i_{ext}, \mathbf{L}^i_{ext})$ with formulae similar to 4.24 and 4.26. In addition to the positions, the user needs the expressions for the velocities $\dot{\mathbf{x}}^{chas}_{damp}$ and $\dot{\mathbf{x}}^{rod}_{damp}$ in terms of the joint coordinates q and derivatives $\dot{q}$.

4.4 Sub-system segmentation

Relative (joint) coordinates have been deliberately used in chapter 3 for modeling multibody systems and generating their equations of motion. This *a priori* choice has been made in view of the advantages of this type of coordinates:

- low number of generalized coordinates and thus of dynamical equations,
- generally closer to the natural description of an articulated mechanical system than absolute coordinates,
- no kinematic constraints for tree-like structures,
- the ease with which relative coordinates may be changed into absolute ones when these are more appropriate (as will be shown below).

Relative coordinates also present some drawbacks:

- high nonlinearity of the dynamical and (loop) constraint equations,
- the generalized mass matrix M and the constraint Jacobian J are full matrices, so that sparse-matrix methods (see [62]) are not applicable during the numerical process,
- the related formalisms are relatively complex and thus difficult to implement in a computer code.

An alternative way, which has not been yet considered in this book, consists in choosing *absolute coordinates* (as in [61], [38]). The absolute position and orientation variables of the different bodies are used as generalized coordinates of the multibody systems. For a 3-D problem with N^{body} bodies, one has $6N^{body}$ generalized coordinates. Then each joint connecting two bodies introduces several constraints on these generalized coordinates: for instance a one degree of freedom joint (revolute or prismatic) implies five independent constraints. The joint force and/or torque components which depend on the relative motions occurring in the joints, are further introduced by computing these relative motions from the absolute motions of the concerned bodies. This approach has drawbacks:

- a large number of coordinates, as well as a large number of dynamical, kinematic and constraint equations; leading to DAE[4] in most situations,

[4] Differential-Algebraic Equations

- a large amount of (coherent) data for describing the system in a coherent reference configuration,
- the need of an extensive library of joints, since complex joints cannot be modeled as combinations of elementary ones,
- some possible numerical round-off error when computing small relative displacements from (possibly large) absolute quantities,

and some advantages:

- tree-like and closed-loop systems are modeled in the same way,
- easy or even trivial generation of the dynamical equations, the latter being easy to implement in a computer code,
- low degree of nonlinearity of the kinematic and dynamical equations,
- quasi-diagonal mass matrix and quasi sparse Jacobian of the constraints ([62]).

In view of the wide range of multibody applications, it is impossible to assert which approach, relative or absolute, is the most appropriate. For instance, serial-link robots, manipulator arms, cranes, etc... should preferably be modeled by means of relative coordinates. But absolute coordinates should be preferred for mechanisms like windshield wipers, engines, etc... which involve many loops which are closed by the chassis-frame and for which most of the bodies have few – even sometimes only one– absolute degrees of freedom with respect to the chassis.

To combine the advantages of both approaches, we can use a mixed-variables approach. Relative (joint) coordinates are easily transformed into absolute coordinates by means of *fictitious joints and bodies.* Moreover, the drawbacks of the pure relative approach - the complexity of the equations - grow with the length of the kinematic chains which are present in the system. Cutting long chains into sub-chains and introducing fictitious 6 d.o.f. joints to reconnect the resulting *sub-systems* with the base, as illustrated in figure 4.14, gives a simple means to reduce the complexity of these kinematic equations, and thus of the dynamical model. Once these judicious cuts have been performed (see fig. 4.14), each resulting sub-system, internally described in terms of relative coordinates, can be considered as one entity whose global motion is governed by absolute variables (for the 6 degrees of freedom). In practical applications, it often appears that the topology of large mechanical systems reveals distinct entities, such as a set of bodies and joints forming an isolated branch or loop, or even a real "multibody sub-system" (as for instance the trailer for a complete tractor-trailer truck).

In conclusion, sub-system segmentation amounts to choosing a good compromise between a fully absolute or a fully relative coordinates approach. One can make the most of this segmentation approach in parallel computation, as will be briefly discussed in this section and illustrated in sections 6.5.2 and 7.4.4.

4.4.1 Equations of motion (without constraints between sub-systems)

To set up the equations of motion, assume first that the various sub-systems (see fig. 4.14) interact with each other only via point-to-point links which in-

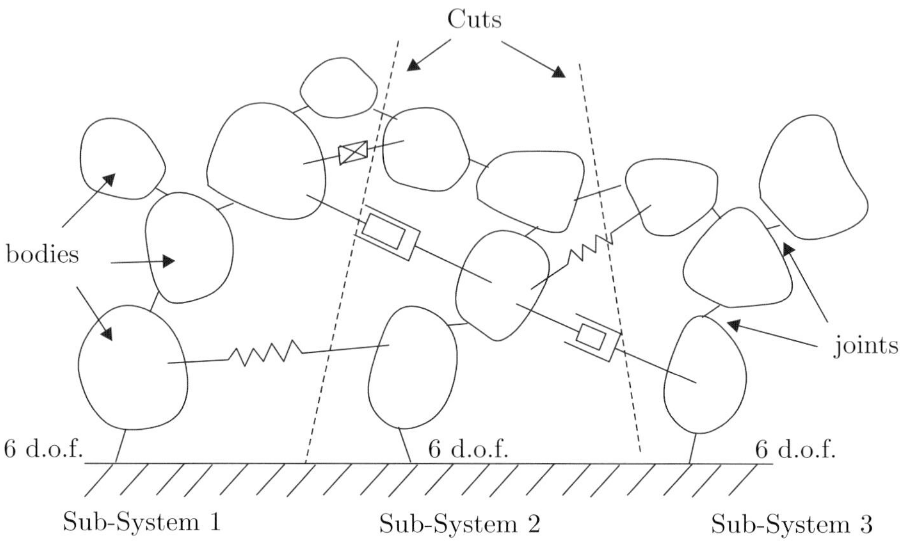

Figure 4.14: Sub-system segmentation

volve contributive forces (e.g.: spring, damper or actuator forces) with known constitutive equations. In other words, any kinematic constraints are assumed to arise only inside the n sub-systems. If necessary, fictitious joints having six degrees of freedom are added to the structure so that each sub-system remains connected to the common base. These sub-systems being considered individually as multibody systems, the equations of motion of the whole system can be written in the following matrix form:

$$\begin{pmatrix} M^1 & 0 & 0 \\ 0 & \dots & 0 \\ 0 & 0 & M^n \end{pmatrix} \begin{pmatrix} \ddot{q}^1 \\ \dots \\ \ddot{q}^n \end{pmatrix} + \begin{pmatrix} c^1(\dot{q}^1, q^1, t, F_{ext}, L_{ext}, g) \\ \dots \\ c^n(\dot{q}^n, q^n, t, F_{ext}, L_{ext}, g) \end{pmatrix} = \begin{pmatrix} \mathcal{Q}^1 \\ \dots \\ \mathcal{Q}^n \end{pmatrix} + \begin{pmatrix} (J^1)^T \lambda^1 \\ \dots \\ (J^n)^T \lambda^n \end{pmatrix} \tag{4.28}$$

$$\begin{pmatrix} h^1(q^1, t) \\ \dots \\ h^n(q^n, t) \end{pmatrix} = 0 \tag{4.29}$$

In this first case, the coupling between the sub-systems equations is ensured by the presence of the interactive forces and torques (which are part of the vectors $\mathbf{F}_{ext}$ and $\mathbf{L}_{ext}$ in equation 4.28) and satisfy the action/reaction principle. While simple in concept, this approach allows us to reduce, when necessary, the length of the different kinematic chains. It is also really attractive in parallel computation. Indeed, consider, as in fig. 4.15, a time simulation scheme in which the DAE[5] system 4.28, 4.29 is reduced to an ODE[6] form by using coordinate partitioning. Without entering into any details — this will be the purpose of section 4.6.1— we see that the method applied to sub-system i consists in:

1. partitioning the generalized coordinates q^i into dependent coordinates v^i and independent coordinates u^i, on the basis of the constraints 4.29_i,
2. solving the constraints 4.29_i at position[7], velocity and acceleration levels: $v^i = v^i(u^i, t)$, $\dot{v}^i = \dot{v}^i(\dot{u}^i, q^i, t)$, $\ddot{v}^i = \ddot{v}^i(\ddot{u}^i, \dot{q}^i, q^i, t)$,
3. reducing the systems $4.28_i, 4.29_i$ on the basis of point 2, by eliminating both the Lagrange multipliers λ^i and the dependent coordinates v^i.

This finally leads to a pure ODE system in terms of the independent coordinates u^i:

$$M^i_{red}(u^i, t)\ddot{u}^i + c^i_{red}(\dot{u}^i, u^i, t, F_{ext}, L_{ext}, g) = Q^i_{red} \qquad \forall i = 1, ...n \tag{4.30}$$

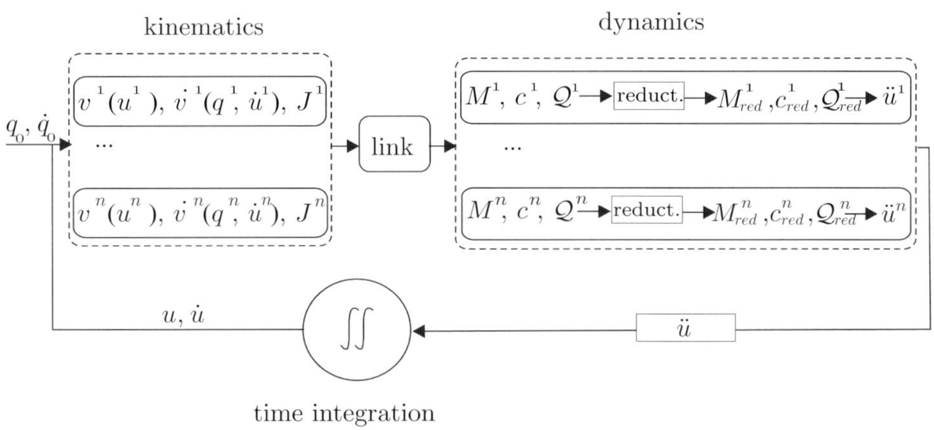

Figure 4.15: Sub-system parallel computation

Applying this reduction to each sub-system and introducing it into a time integration scheme, we obtain the flow-sheet of fig. 4.15 which reveals two large

[5] Differential-Algebraic Equations
[6] Ordinary Differential Equations
[7] using a Newton/Raphson method

blocks, candidates for a fully parallel computation : a *kinematic* and a *dynamical* block. Between them, a connection point ("link") is necessary to compute the interactive linking forces.

Despite the restriction related to the nature of the point-to-point links (between sub-systems), this mixed approach suits practical problems. For instance, it was successfully applied in railway dynamics (see [24]) by splitting railway vehicles into one "carbody" and three "bogie" sub-systems (see fig. 4.16).

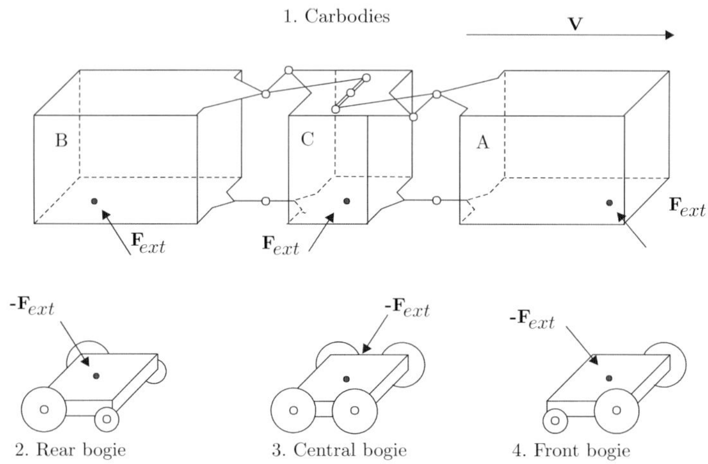

Figure 4.16: System segmentation applied in railway dynamics

In these vehicles, the connection between a carbody and a bogie is ensured by suspension elements which are sufficiently "soft" in the three directions to allow the use of spring/damper force laws, without resorting to any kinematic constraint.

4.4.2 Equations of motion (with constraints between sub-systems)

If it is not possible to apply the sub-system segmentation approach without introducing kinematic constraints between the sub-systems, the degree of parallelism depicted in fig. 4.15 can nevertheless be safeguarded if one accepts the following restriction :

Constraints between sub-systems involve only generalized coordinates belonging to the set of coordinates $\{\{u^1\}, ..., \{u^n\}\}$ *which are, or have been selected as independent with respect to the sub-system internal constraints.*

As we shall see, this hypothesis allows us to separate both the constraints resolution and the equations reduction into two distinct steps: one at the level of the sub-systems and the other at the level of the global system.

If $h(u) = 0$ denotes the set of (independent) constraints *between* the sub-systems variables $u \triangleq \{\{u^1\}, ..., \{u^n\}\}$, the reduced equations of motion 4.30 of sub-system i become

$$M^i_{red}\,\ddot{u}^i + c^i_{red} = \mathcal{Q}^i_{red} + \left(H^i\right)^T \mu \tag{4.31}$$

where μ are the Lagrange multipliers associated with the constraints $h(u) = 0$ and $H^i \triangleq \frac{\partial h}{\partial (u^i)^T}$ denotes the columns-block of the Jacobian matrix $H = \frac{\partial h}{\partial u^T}$ related to sub-system i.
Considering the whole system, we can write

$$M_{red}\,\ddot{u} + c_{red} = \mathcal{Q}_{red} + H^T \mu \tag{4.32}$$

with

$$M_{red} = \begin{pmatrix} M^1_{red} & 0 & 0 \\ 0 & \dots & 0 \\ 0 & 0 & M^n_{red} \end{pmatrix}$$

The system 4.32 is the result of the first reduction step; it can be parallelized sub-system by sub-system. Since constraints $h(u)$ induce a new interdependence

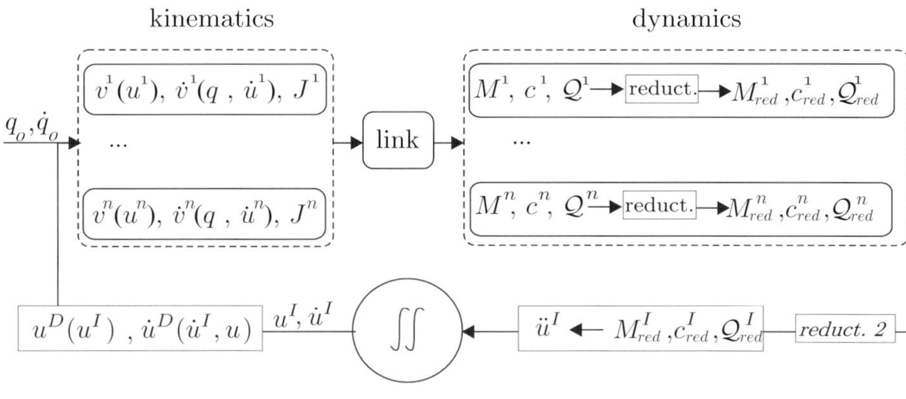

Figure 4.17: Sub-system segmentation

between the variables u, we use a second reduction step by partitioning u into dependent variables u^D and independent variables u^I, and write

$$u^D = u^D(u^I, t),\ \dot{u}^D = \dot{u}^D(\dot{u}^I, u, t),\ \ddot{u}^D = \ddot{u}^D(\ddot{u}^I, \dot{u}, u, t) \tag{4.33}$$

This second reduction based on 4.33 is applied to 4.32 and finally leads to the following ODE system in terms of the minimal set of variables u^I :

$$M^I_{red}\,\ddot{u}^I + c^I_{red} = \mathcal{Q}^I_{red} \tag{4.34}$$

Fig. 4.17 describes the general flow-chart for this case, leading to 4.34 and its time integration scheme. Note that the parallelized blocks of fig. 4.15 are preserved. The treatment of the constraints between sub-systems — i.e. reduction (eq. 4.34) and solution (eq. 4.33) is performed *downstream* and *upstream* from these blocks respectively.

As for systems without constraints between sub-systems, the present procedure has practical applications despite the restriction which states that the constraints between the sub-systems involve the variables u^i only. The following example, which was selected as an international benchmark for multibody dynamics software ([49]), illustrates this potentiality. It relates to the ILTIS mil-

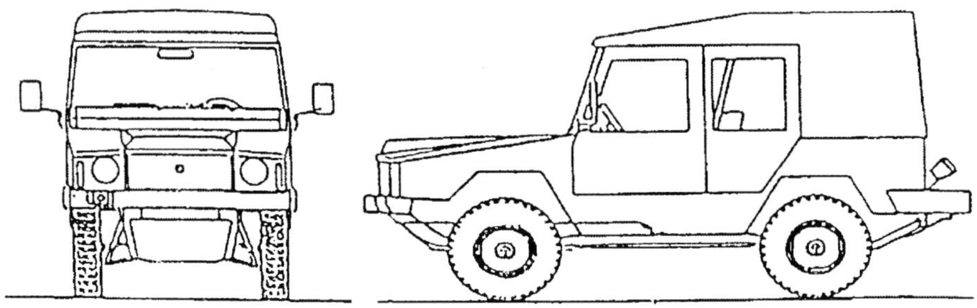

Figure 4.18: The ILTIS Jeep

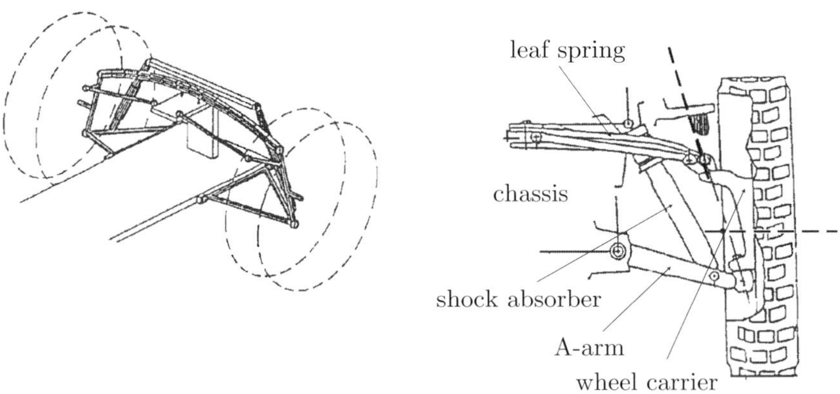

Figure 4.19: Details of the Iltis suspensions

itary vehicle[8], represented in figs. 4.18 and 4.19, which has a similar suspension arrangement at the front and rear : a leaf spring, a wheel-carrier and a A−arm forming a four bar mechanism with the chassis. A shock absorber connects the A−arm to the chassis and an "equivalent" vertical spring acts between the chassis and the leaf spring, the latter being modeled as a rigid element. The wheel/ground contact is modeled via external forces and torques whose constitutive equations are based on the models described in chapter 6, in the normal and tangent directions. They require the on-line computation of the kinematics of each wheel and of its contact point on the ground.

Seen as a whole, the ILTIS jeep can be considered as a global multibody system for which a relative coordinates approach seems to be appropriate as depicted in fig. 4.20_a. In the present context however, we propose to cut the

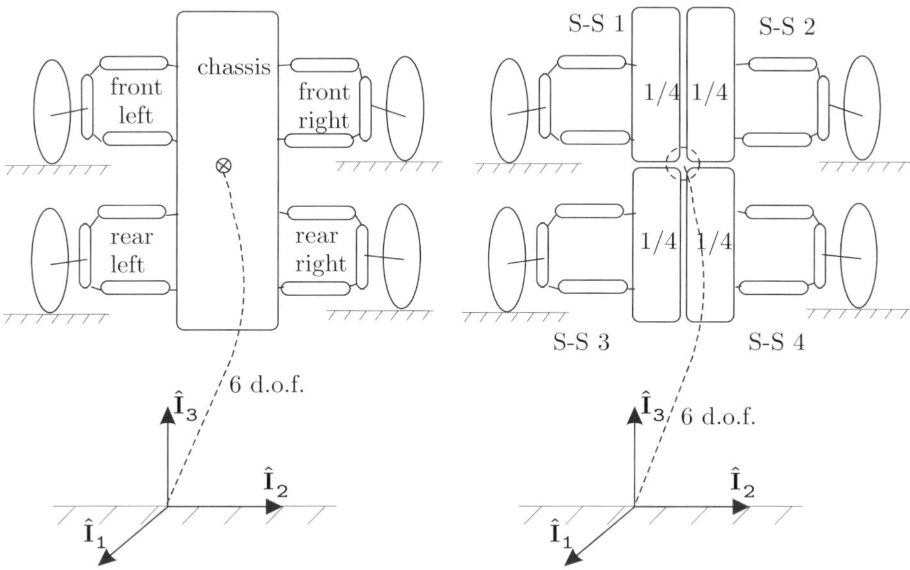

Figure 4.20: a) Global multibody model b) Segmented multibody model

chassis into 4 parts leading to 4 sub-systems which consist of $\frac{1}{4}$ chassis + 1 suspension (see fig. 4.20_b). The real model is then obtained by applying the following elementary constraints between the 4 sub-systems, involving the 4×6

[8]from the Bombardier Company (CANADA)

d.o.f. which connect the four $\frac{1}{4}$ chassis to the ground:

$$\begin{pmatrix} u_1^1 \\ u_2^1 \\ u_3^1 \\ u_4^1 \\ u_5^1 \\ u_6^1 \end{pmatrix} = \begin{pmatrix} u_1^2 \\ u_2^2 \\ u_3^2 \\ u_4^2 \\ u_5^2 \\ u_6^2 \end{pmatrix} = \begin{pmatrix} u_1^3 \\ u_2^3 \\ u_3^3 \\ u_4^3 \\ u_5^3 \\ u_6^3 \end{pmatrix} = \begin{pmatrix} u_1^4 \\ u_2^4 \\ u_3^4 \\ u_4^4 \\ u_5^4 \\ u_6^4 \end{pmatrix} \tag{4.35}$$

and represent a set of 18 independent constraints.
In this way, the model sticks to the flow-chart of fig. 4.17 apart from the fact that there is no need for a contributive link in the present case, and thus no synchronization point between the kinematic and dynamical blocks. This makes integration by parallel processing easier (see [15]).

4.5 Complementary multibody kinematics

The modeling of complex multibody systems which involve closed-loop structures, user-joint constraints, point-to-point links and possible segmentations requires additional tools for generating the equations of motion 4.1. These tools, however, concern only the kinematic part of the generation. They can be the same, whatever is the formalism used for obtaining the dynamics of the open-loop structure, whether the virtual power principle or the Newton-Euler recursive formulation.

As indicated in the previous sections, to establish loop closure conditions, user-joint constraints (at position, velocity and acceleration levels) or point-to-point links, we need expressions for the positions of some particular points P and for the orientations of some particular frames $\{\hat{\mathbf{X}}^i\}$ in terms of the generalized joint coordinates.

4.5.1 Loop closure kinematics

In term of kinematic needs, the most general case of the loop closure methods presented in section 4.1 corresponds to the situation for which a body has been cut. In this case, there are translational as well as rotational closure conditions.

For simplicity, the base and the inertial frame $\{\hat{\mathbf{I}}\}$ are used as references in figure 4.4 for defining the vector positions of the two points $P's$ and the orientation of the two body-fixed frames $\{\hat{\mathbf{X}}\}'s$. However for computer efficiency, a better choice is illustrated in figure 4.21. It consists in using as references the attachment point O'^i and the frame $\{\hat{\mathbf{X}}^i\}$ of the body i which is the closest common ancestor of the two bodies. The two sub-chains, between this ancestor and the two bodies, are then used separately for generating the positions and orientations quantities which are needed to establish the constraints as well as the corresponding Jacobian matrices. For point P on body j for instance, one has to generate:

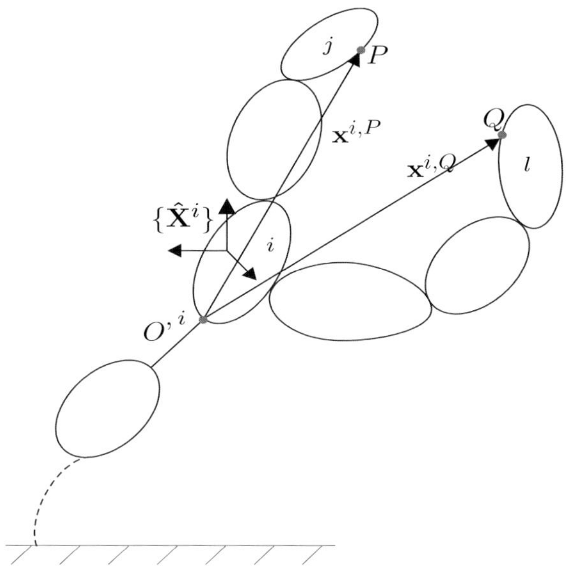

Figure 4.21: Closure kinematics

- the (relative) position vector $\overrightarrow{O'^i P}$ of P with respect to O'^i, expressed in the $\{\hat{\mathbf{X}}^i\}$ frame: $\mathbf{x}^{i,P} \triangleq \overrightarrow{O'^i P} = [\hat{\mathbf{X}}^i]^T x^{i,P}(q,t)$,

- the (relative) orientation matrix $R^{j,i}$ defined by: $[\hat{\mathbf{X}}^j] = R^{j,i}(q,t)[\hat{\mathbf{X}}^i]$,

- the *relative* velocity vector $\overset{\circ}{\mathbf{x}}^{i,P}$ defined by: $\overset{\circ}{\mathbf{x}}^{i,P} = [\hat{\mathbf{X}}^i]^T \dot{x}^{i,P}(\dot{q},q,t)$,

- the *relative* angular velocity vector associated with the rotation matrix $R^{j,i}$: $\boldsymbol{\omega}^{j,i} = [\hat{\mathbf{X}}^i]^T \omega^{j,i}(\dot{q},q,t)$,

- and the corresponding acceleration vectors: $\overset{\circ\circ}{\mathbf{x}}^{i,P} = [\hat{\mathbf{X}}^i]^T \ddot{x}^{i,P}(\ddot{q},\dot{q},q,t)$, $\overset{\circ}{\boldsymbol{\omega}}^{j,i} = [\hat{\mathbf{X}}^i]^T \dot{\omega}^{j,i}(\ddot{q},\dot{q},q,t)$,

A recursive formalism, similar to that for the forward kinematics we used in section 3.3, is used to obtain these expressions. The difference lies in two facts:

- the pair $\{O'^i, \{\hat{\mathbf{X}}^i\}\}$ now plays the role of the inertial frame $\{O, \{\hat{\mathbf{I}}\}\}$ for computing the time derivatives. In other words, in the present algorithm $\overset{\circ}{\mathbf{u}} \triangleq [\hat{\mathbf{X}}^i]^T \dot{u}(\dot{q},q,t)$ and $\overset{\circ\circ}{\mathbf{u}} \triangleq [\hat{\mathbf{X}}^i]^T \ddot{u}(\ddot{q},\dot{q},q,t)$ whatever is the vector $\mathbf{u} = [\hat{\mathbf{X}}^i]^T u(q,t)$,

- for saving computation effort in the evaluation of the Jacobian, the translational part of the kinematics is conducted backwards.

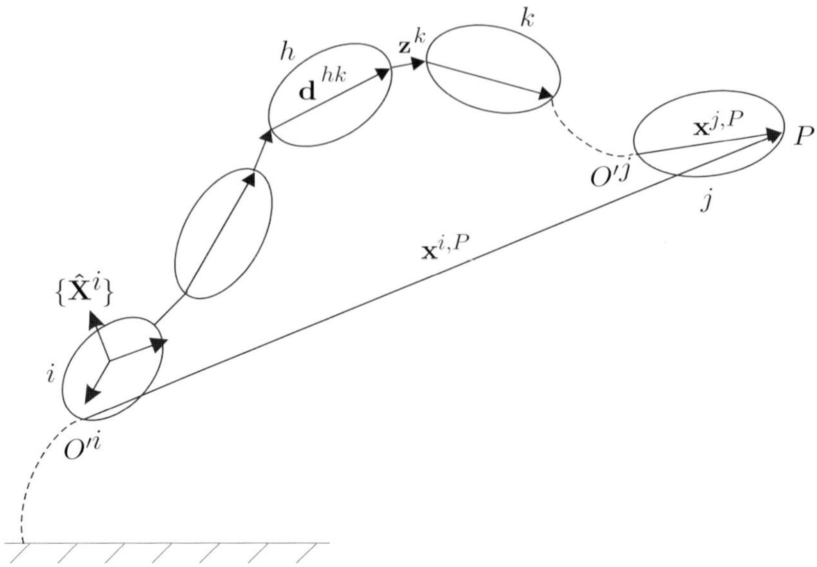

Figure 4.22: Sub-chain kinematics

To develop these recursive algorithms, consider the kinematic chain which links the point P to body i. As illustrated in figure 4.22, the index of the parent body of a body k is denoted by h. All notations are consistent with the definitions we made in figure 3.6 of section 3.1.3 and in figure 3.15 of section 3.3.2. The different algorithms are then as follows.

1. By (backward) recursive use of the *inbody* vector (see section 3.1.2), determine the set of indices of the joints which belong to the kinematic chain linking bodies i and j. Denote this set by $\{i+1 : j\}$.

2. Computation of the rotation matrix $R^{i,k}$ from the elementary joint rotation matrices $R^{h,k}$:

   Initialization: $R^{i,i} = E$ (unit matrix)

   Forward recursion:

   For $k = i+1 : j$

$$h = \text{inbody}(k)$$
$$R^{i,k} = R^{i,h}\, R^{h,k}$$

   end

3. Forward computation of angular velocity and acceleration vectors:

   Initialization:

$$\boldsymbol{\omega}^{i,i} = \overset{\circ}{\boldsymbol{\omega}}{}^{i,i} = 0 \tag{4.36}$$

Forward recursion:

For $k = i+1 : j$

$$\begin{aligned} h &= \text{inbody}(k) \\ \boldsymbol{\omega}^{k,i} &= \boldsymbol{\omega}^{h,i} + \boldsymbol{\varphi}^k \dot{q}^k \\ \overset{\circ}{\boldsymbol{\omega}}{}^{k,i} &= \overset{\circ}{\boldsymbol{\omega}}{}^{h,i} + \tilde{\boldsymbol{\omega}}^{k,i} \,.\, \boldsymbol{\varphi}^k \dot{q}^k + \boldsymbol{\varphi}^k \ddot{q}^k \end{aligned}$$

end

4. Backward computation of position, velocity and acceleration vectors:

Initialization:

$$k = j \quad , \quad \mathbf{x}^{k,P} = \mathbf{x}^{j,P}$$

$$\overset{\circ}{\mathbf{x}}{}^{k,P} = \tilde{\boldsymbol{\omega}}^{j,i} \,.\, \mathbf{x}^{j,P} \quad , \quad \overset{\circ\circ}{\mathbf{x}}{}^{k,P} = \tilde{\overset{\circ}{\boldsymbol{\omega}}}{}^{j,i} \,.\, \mathbf{x}^{j,P} + \tilde{\boldsymbol{\omega}}^{j,i} \,.\, \tilde{\boldsymbol{\omega}}^{j,i} \,.\, \mathbf{x}^{j,P}$$

Recursion[9]:

while $k \neq i$:

$$\begin{aligned} h &= \text{inbody}(k) \\ \mathbf{x}^{h,P} &= \mathbf{x}^{k,P} + \mathbf{d}^{hk} + \mathbf{z}^k \\ \overset{\circ}{\mathbf{x}}{}^{h,P} &= \overset{\circ}{\mathbf{x}}{}^{k,P} + \tilde{\boldsymbol{\omega}}^{h,i} . (\mathbf{d}^{hk} + \mathbf{z}^k) + \boldsymbol{\psi}^k \dot{q}^k \\ \overset{\circ\circ}{\mathbf{x}}{}^{h,P} &= \overset{\circ\circ}{\mathbf{x}}{}^{k,P} + \left(\overset{\circ}{\boldsymbol{\omega}}{}^{h,i} \right)^{\sim} . (\mathbf{d}^{hk} + \mathbf{z}^k) + \tilde{\boldsymbol{\omega}}^{h,i} \,.\, \tilde{\boldsymbol{\omega}}^{h,i} \,.\, (\mathbf{d}^{hk} + \mathbf{z}^k) \\ &\quad + 2\, \tilde{\boldsymbol{\omega}}^{h,i} \,.\, \boldsymbol{\psi}^k \dot{q}^k + \boldsymbol{\psi}^k \ddot{q}^k \\ k &= h \end{aligned}$$

end

5. Computation of the Jacobian matrices

As mentioned in section 4.1, the Jacobian matrix J_t of the translational constraints is obtained by isolating the coefficients of the generalized velocities $\dot{q}$'s in the expressions for the translational velocities. It will be shown in section 4.6.2 that the coefficients of the $\dot{q}$'s in the expressions of the angular velocities play the role of *pseudo-gradients* for the rotation constraints. A vector form, which is called the *vector Jacobian* [35], of the contribution of the sub-chain to the (pseudo) Jacobian is thus obtained by writing

$$\begin{pmatrix} \overset{\circ}{\mathbf{x}}{}^{i,P} \\ \boldsymbol{\omega}^{j,i} \end{pmatrix} = \begin{pmatrix} \mathbf{J}_t^{i,P} \\ \mathbf{J}_r^{i,P} \end{pmatrix} \dot{q} = \mathbf{J}^{i,P} \dot{q} \tag{4.37}$$

[9] when computing the derivatives, we use the fact that $\boldsymbol{\omega}^{h,i} = \boldsymbol{\omega}^{k,i}$ when joint k is prismatic ($\mathbf{z}^k \neq 0$). For revolute joints, $\boldsymbol{\omega}^{h,i} \neq \boldsymbol{\omega}^{k,i}$ but $\mathbf{z}^k = 0$.

After some few manipulations, the kth column $\mathbf{J}^{i,P}_{(k)}$ of this vector matrix, i.e. the coefficient of $\dot{q}^k$, is found to be

$$\mathbf{J}^{i,P}_{(k)} = \begin{pmatrix} \boldsymbol{\psi}^k - \tilde{\mathbf{x}}^{k,P} \cdot \boldsymbol{\varphi}^k \\ \boldsymbol{\varphi}^k \end{pmatrix} \tag{4.38}$$

if k belongs to the kinematic chain $\{i+1 : j\}$, and equal to zero otherwise. This latter expression shows that $\mathbf{J}^{i,P}$ can be computed within the backward algorithm of point 4.

Once the previous algorithms have been run for the two subchains of the closed-loop (see figure 4.21), the closure conditions (equivalent to 4.4 and 4.6) are given by

$$\begin{aligned} h_t(q,t) \triangleq x^{i,P}(q,t) - x^{i,Q}(q,t) &= 0 \\ h_r(q,t) \triangleq \begin{pmatrix} -R_{(3,2)}(q,t) \\ R_{(3,1)}(q,t) \\ -R_{(2,1)}(q,t) \end{pmatrix} &= 0 \quad \text{with: } R = R^{j,i}(q,t)R^{i,l}(q,t) \end{aligned}$$

and the associated Jacobian is given by the components[10] of the vector matrix

$$\mathbf{J} \triangleq \mathbf{J}^P - \mathbf{J}^Q \tag{4.39}$$

4.5.2 User joint/constraints and point-to-point links

These recursive algorithms can also be used to provide the expressions which are necessary for user-defined constraints or joints, as well as point-to-point links. An example of the use of such algorithms for setting user-defined joints is given in chapter 7, applied to the wheel-rail contact problem. On the other hand, their use for point-to-point links is straightforward: for instance for the five bar suspension system considered in section 4.3, the roles of $\mathbf{x}^{chas}$ and $\mathbf{x}^{rod}$ in the constitutive equations 4.24 and 4.27 of the spring and the damper are played by $\mathbf{x}^P$ and $\mathbf{x}^Q$ respectively in the algorithms.

4.6 Numerical aspects

4.6.1 Coordinate partitioning

When the joint coordinates of a multibody system are not independent, the system is *constrained* and its equations of motion must be written according to the form 2.78:

$$\Phi(\ddot{q}, \dot{q}, q, t) = Q + Q' \tag{4.40}$$

[10] For consistency, the components of $\mathbf{J}_t$ must be computed in the $\{\hat{\mathbf{X}}^i\}$ frame of the common ancestor, because the translational constraints are expressed in this frame. On the other hand, the components of $\mathbf{J}_r$ must be computed in the cut-body fixed frame $\{\hat{\mathbf{X}}^j\}$ (or $\{\hat{\mathbf{X}}^l\}$).

where Q' represents the additional (unknown) generalized constraint forces which compel the system to fulfill the constraints. The first part of these equations, i.e.:

$$\Phi(\ddot{q}, \dot{q}, q, t) = Q$$

are the motion equations of the *unconstrained* system, i.e. the system which is obtained if the constraints are not applied to the joint coordinates. Vector Q represents the generalized forces (i.e. the joint forces and torques) associated with the unconstrained joint coordinates. These unconstrained equations of motion can be obtained either by the virtual power principle or by a recursive formalism as explained in chapter 2.
When the m constraints are due to loop closures, to joint or to user constraints, they generally[11] can be written in the form:

$$h(q, t) = 0 \quad \text{with } h \in \Re^m \tag{4.41}$$

and, according to equation 2.80 of section 2.4.2, the generalized constraint forces can be written as:

$$Q' = J^T(q, t)\,\lambda \tag{4.42}$$

where the m by n matrix $J(q, t)$ is the Jacobian of the constraints:

$$J(q, t) = \frac{\partial h(q, t)}{\partial q^T}$$

and where the m unknown coefficients λ_β are the Lagrange multipliers.
Together with the m constraint equations 4.41, the n equations of motion 4.40 form a system of $n+m$ differential/algebraic equations for the $n+m$ unknowns $q \in \Re^n$ and $\lambda \in \Re^m$.

In its semi-explicit form in the multibody context, this system of equations has the form:

$$\begin{aligned} M(q, t)\ddot{q} + c(\dot{q}, q, t, F_{ext}, L_{ext}, g) &= \mathcal{Q} + J^T(q, t)\,\lambda \\ h(q, t) &= 0 \end{aligned} \tag{4.43}$$

When the constraint equations on the generalized coordinates are nonlinear, it is generally not possible to express analytically the dependent coordinates in terms of independent ones – despite the fact that the implicit function theorem is applicable – and one is led to a set of mixed differential-algebraic equations (DAE).

For independent constraints, the reduction to a purely differential form can be obtained by means of the *coordinate partitioning method* [92].
Indeed, since the constraints are independent, the Jacobian matrix J has full row rank m. According to the implicit function theorem, this implies that,

[11] For simplicity, we ignore non-holonomic constraints.

using the constraints, m coordinates can be expressed *locally* as functions of the $(n-m)$ others. In this way, it becomes possible to reduce the original DAE system to a set of $(n-m)$ differential equations in these $(n-m)$ independent coordinates. This reduced set will represent the equations of motion of the constrained multibody system.

Possibly after reordering the variables, the vector q of generalized coordinates can be partitioned[12] as

$$q = \begin{pmatrix} u \\ v \end{pmatrix} \tag{4.44}$$

where u represents the subset of $(n-m)$ independent coordinates. The matrix J is also partitioned into

$$J = \begin{pmatrix} J_u & J_v \end{pmatrix} \tag{4.45}$$

where J_v is a m by m matrix. This matrix is non singular if v is correctly chosen as being the subset of dependent coordinates.

As shown in section 2.4.2, the m Lagrange multipliers λ can then be eliminated from the relation 4.43 – leaving a system of $n-m$ differential equations in the n unknown q and their derivatives. For this purpose, first the generalized mass matrix M and the vector c are also partitioned:

$$\begin{pmatrix} M_{uu} & M_{uv} \\ M_{vu} & M_{vv} \end{pmatrix} \begin{pmatrix} \ddot{u} \\ \ddot{v} \end{pmatrix} + \begin{pmatrix} c_u \\ c_v \end{pmatrix} = \begin{pmatrix} \mathcal{Q}_u \\ \mathcal{Q}_v \end{pmatrix} + \begin{pmatrix} J_u{}^T \\ J_v{}^T \end{pmatrix} \lambda \tag{4.46}$$

Since the matrix J_v is non singular, one obtains by eliminating the unknowns λ

$$\begin{aligned} \begin{pmatrix} M_{uu} & M_{uv} \end{pmatrix} \begin{pmatrix} \ddot{u} \\ \ddot{v} \end{pmatrix} + B_{vu}{}^T \begin{pmatrix} M_{vu} & M_{vv} \end{pmatrix} \begin{pmatrix} \ddot{u} \\ \ddot{v} \end{pmatrix} \\ + c_u + B_{vu}{}^T c_v = \mathcal{Q}_u + B_{vu}{}^T \mathcal{Q}_v \end{aligned}$$

where

$$B_{vu} \triangleq -(J_v)^{-1} J_u \tag{4.47}$$

Then partitioning the first and second derivatives of the constraints 4.41:

$$\begin{pmatrix} J_u & J_v \end{pmatrix} \begin{pmatrix} \dot{u} \\ \dot{v} \end{pmatrix} + \frac{\partial h}{\partial t} = 0$$

$$\begin{pmatrix} J_u & J_v \end{pmatrix} \begin{pmatrix} \ddot{u} \\ \ddot{v} \end{pmatrix} + \left(\dot{J} + \frac{\partial^2 h}{\partial t\, \partial q} \right) \dot{q} + \frac{\partial^2 h}{\partial t^2} = 0$$

we obtain the generalized velocities and accelerations $\dot{v}$ and $\ddot{v}$ respectively:

$$\begin{aligned} \dot{v} &= B_{vu}\dot{u} - (J_v)^{-1} \frac{\partial h}{\partial t} \\ \ddot{v} &= B_{vu}\ddot{u} - (J_v)^{-1}\, b \end{aligned}$$

$$\text{with } b \triangleq \left(\dot{J} + \frac{\partial^2 h}{\partial t\, \partial q} \right) \dot{q} + \frac{\partial^2 h}{\partial t^2}$$

---

[12] The latter can be performed by means of a L-U factorization of the Jacobian matrix $J(q)$ with full row and column pivoting.

They can then be eliminated from the differential equations of motion. This produces the *reduced* system

$$\left(M_{uu} + M_{uv}B_{vu} + {B_{vu}}^T M_{vu} + {B_{vu}}^T M_{vv}B_{vu}\right)\ddot{u}$$
$$- \left(M_{uv} + {B_{vu}}^T M_{vv}\right)(J_v)^{-1}\, b + c_u + {B_{vu}}^T c_v = \mathcal{Q}_u + {B_{vu}}^T \mathcal{Q}_v$$

which can formally be written as

$$M_{red}(u,v,t)\ddot{u} + c_{red}(\dot{u},u,v,t) = \mathcal{Q}_{red} \tag{4.48}$$

The algebraic constraints still have to be solved in order to eliminate the variables v themselves from 4.48. In many practical situations[13], this can be done numerically only by a (Newton-Raphson type) algorithm, because the constraint equations are nonlinear in these variables. The following iterative algorithm can be used for successive estimations of v:

$$v^{k+1} = v^k - (J_v)^{-1}\, h \tag{4.49}$$

where the right hand side is evaluated for $v = v^k$ and the given values of u and time.
Because of this final numerical elimination, the set of purely differential equations 4.48 constitutes the equations of motion of the constrained system described in terms of the $n - m$ independent generalized coordinates u. The flowchart of the numerical integration scheme is shown in figure 4.23.

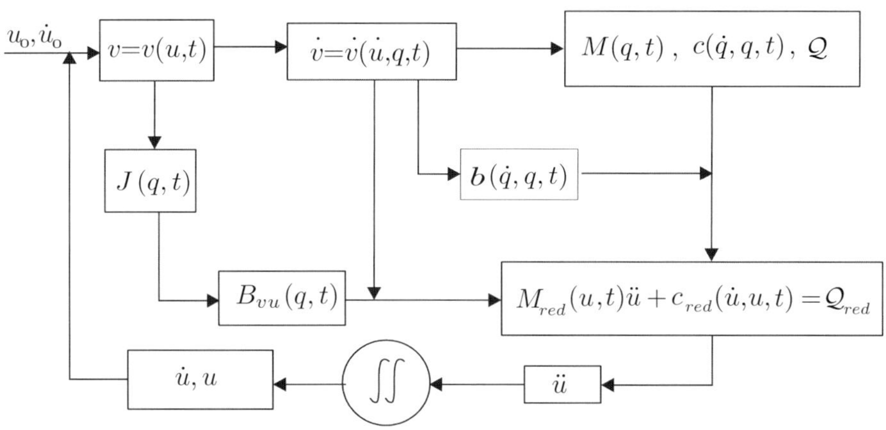

Figure 4.23: Coordinate partitioning and integration scheme

The system motion being computed by this numerical integration, we may be interested in finding the values of some constraint forces during this motion. For this, we must resort to the values taken by the Lagrange multipliers since

[13] see for instance the five-bar suspension system of section 4.3

they replace the constraint forces in our dynamical model (see section 2.4.3). By isolating the λ in the second row of system 4.46:

$$\lambda = \left((J_v)^T\right)^{-1}\left(\begin{pmatrix} M_{vu} & M_{vv} \end{pmatrix}\begin{pmatrix} \ddot{u} \\ \ddot{v} \end{pmatrix} + c_v - \mathcal{Q}_v\right)$$

and using the relation $\ddot{v} = B_{vu}\ddot{u} - (J_v)^{-1}\, b$, we obtain the expression

$$\lambda = \left((J_v)^T\right)^{-1}\left\{(M_{vu} + M_{vv}\, B_{vu})\, \ddot{u} + c_v - \mathcal{Q}_v - M_{vv}\, (J_v)^{-1}\, b\right\} \qquad (4.50)$$

which allows us to compute the constraint forces via the procedure explained in section 2.4.3.

4.6.2 Pseudo rotation constraints

Dependent constraint arise generally[14] when expressing the rotational closure of loops in a multibody system (see section 4.1). Considering the set $\{q_\alpha;\ \alpha = 1, N^{loop}\}$ of revolute joint coordinates which belong to a particular loop, we can write the rotational closure condition in the form

$$R(q,t) - E = 0 \qquad (4.51)$$

where R is a (3×3) rotation matrix depending on the joint coordinates q_α (and possibly on time if a revolute joint in the loop is kinematically driven, i.e. its motion is given as time function) and where E represents the unit matrix. The matrix R relates the angular position of the terminal (shadow) body of the loop - let say body j - with respect to the initial one we denote by i, i.e.:

$$[\hat{\mathbf{X}}^j] = R(q,t)[\hat{\mathbf{X}}^i] \qquad (4.52)$$

The closure condition thus formally implies 9 constraints but since any rotation matrix can be expressed in terms of 3 independent rotation variables (see section 1.1.8 of chapter 1), only 3 of these constraints are independent. Therefore, we shall focus on the problem of selecting a subset of independent constraints and on a method for obtaining the corresponding (full rank) Jacobian matrix.

The 9 rotation constraints 4.51 can be rewritten in matrix form as

$$H(q,t) \triangleq E - R(q,t) = 0$$

and their time derivatives are then given by (see equation 1.52)

$$\dot{H} = \tilde{\omega}\, R$$

where ω represents the components in frame $\{\hat{\mathbf{X}}^j\}$ of the angular velocity vector associated with R. By denoting the columns of H and R as follows

$$H = \begin{pmatrix} H_{(1)} & H_{(2)} & H_{(3)} \end{pmatrix} \quad \text{and} \quad R = \begin{pmatrix} R_{(1)} & R_{(2)} & R_{(3)} \end{pmatrix} \qquad (4.53)$$

[14] User constraints could also be dependent if they are not suitably formulated (for instance redundant constraints must be disregarded). A general discussion on dependent constraints can be found in [76].

we can write these derivatives as:

$$\begin{aligned} \dot{H}_{(i)} &= \tilde{\omega}\, R_{(i)} \\ \text{or equivalently as} \quad \dot{H}_{(i)} &= -\,\tilde{R}_{(i)}\, \omega \quad \text{for } i = 1, 2, 3 \end{aligned} \tag{4.54}$$

In this way, the 9 constraints and their derivatives can be written column-wise in the usual form

$$h(q,t) \triangleq \begin{pmatrix} H_{(1)}(q,t) \\ H_{(2)}(q,t) \\ H_{(3)}(q,t) \end{pmatrix} = 0 \tag{4.55}$$

and:

$$\dot{h}(\dot{q},q,t) = D(q,t)\,\omega = 0 \tag{4.56}$$

$$\text{with} \quad D(q,t) \triangleq - \begin{pmatrix} \tilde{R}_{(1)}(q,t) \\ \tilde{R}_{(2)}(q,t) \\ \tilde{R}_{(3)}(q,t) \end{pmatrix} = \begin{pmatrix} 0 & R_{(3,1)} & -R_{(2,1)} \\ -R_{(3,1)} & 0 & R_{(1,1)} \\ R_{(2,1)} & -R_{(1,1)} & 0 \\ 0 & R_{(3,2)} & -R_{(2,2)} \\ -R_{(3,2)} & 0 & R_{(1,2)} \\ R_{(2,2)} & -R_{(1,2)} & 0 \\ 0 & R_{(3,3)} & -R_{(2,3)} \\ -R_{(3,3)} & 0 & R_{(1,3)} \\ R_{(2,3)} & -R_{(1,3)} & 0 \end{pmatrix} \tag{4.57}$$

On the other hand, the angular velocity components ω_i can be computed from the kinematics of the multibody system: vector $\boldsymbol{\omega}$ is indeed the relative angular velocity of the frame $\{\hat{\mathbf{X}}^j\}$ with respect to the frame $\{\hat{\mathbf{X}}^i\}$:

$$\boldsymbol{\omega} = \boldsymbol{\omega}^j - \boldsymbol{\omega}^i = [\hat{\mathbf{X}}^j]^T \omega \tag{4.58}$$

but it can also be obtained by running through the kinematic chain which forms the loop from body i to body j. This implies that the vector $\boldsymbol{\omega}$ is the vector sum of the relative angular velocities which occur in the different revolute joints belonging to the loop:

$$\boldsymbol{\omega} = \sum_{\alpha \in loop} \boldsymbol{\Omega}^\alpha \tag{4.59}$$

Since these relative angular vectors are linear functions of the joint velocities $\dot{q}_\alpha$, the components of $\boldsymbol{\omega}$ itself has the linear form

$$\omega = B(q,t)\dot{q} + b(q,t) \tag{4.60}$$

(where time t and the column matrix b only appear if some revolute joint is kinematically driven in the loop). Combining the relations 4.56 and 4.60, we find the Jacobian matrix of the rotational constraints $h(q,t) = 0$ to be the product

$$J(q,t) = D(q,t)B(q,t) \tag{4.61}$$

If the loop is really 3-dimensional[15], i.e. if the terminal body j has globally three independent rotational degrees of freedom with respect to the initial body i when the closure conditions are not imposed, then the 3 by N^{loop} matrix B has full rank 3[16]. When the constraints are satisfied, then $R_{(i,j)} = \delta_{ij}$[17] and the rank of matrix D is also equal to 3. A subset of independent constraints is thus obtained by selecting three independent rows of matrix D and the corresponding rows of $h(q,t)$. One may chose for instance as subset of independent constraint functions

$$h_r(q,t) \triangleq \begin{pmatrix} h_{(6)} \\ -h_{(3)} \\ h_{(2)} \end{pmatrix} \equiv \begin{pmatrix} -R_{(3,2)} \\ R_{(3,1)} \\ -R_{(2,1)} \end{pmatrix}$$

with the corresponding sub-matrix D_r:

$$D_r(q,t) = \begin{pmatrix} R_{(2,2)} & -R_{(1,2)} & 0 \\ -R_{(2,1)} & R_{(1,1)} & 0 \\ -R_{(3,1)} & 0 & R_{(1,1)} \end{pmatrix} \tag{4.62}$$

The Jacobian matrix for this subset of constraints is given by

$$J_r(q,t) = D_r(q,t)B(q,t) \tag{4.63}$$

The selection of independent constraints has two consequences: the first concerns the formulation of the constraint generalized forces in terms of Lagrange multipliers, and the second influences the behavior of the Newton-Raphson algorithm for solving numerically the dependent coordinates in terms of the independent ones.

- Equations of motion[18]

When using the constraints $h_r(q,t) = 0$, we find the unreduced system of motion equations 4.43 to be

$$M(q,t)\ddot{q} + c(\dot{q},q,t) = \mathcal{Q} + J_r{}^T(q,t)\,\lambda_r \tag{4.64}$$

where the Jacobian matrix is given by 4.63. However, as soon as the *full* set of constraints $h(q,t) = 0$ is satisfied, then the matrix $D_r(q,t)$ is equal to the

[15] For instance in two-dimensional problems involving in-plane rotations, a loop closure condition would involve a single independent constraint and the rank of matrix B would be equal to one.

[16] except in *singular* configurations due to the particular choice of the rotation coordinates (see section 1.1.8).

[17] where δ is the Kronecker delta

[18] For simplicity, only rotational constraints are considered here.

unit matrix and this Jacobian $J_r(q,t)$ reduces to $B(q,t)$. Therefore, *under the condition that the loop is closed*, the motion and constraint equations of the constrained system are given by

$$\begin{aligned} M(q,t)\ddot{q} + c(\dot{q},q,t) &= \mathcal{Q} + B^T(q,t)\,\lambda_r & (4.65)\\ h_r(q,t) &= 0 & (4.66) \end{aligned}$$

for the unknowns q and λ_r. Since the selected constraints are independent and matrix B has full rank, the coordinate partitioning method can be used to provide the reduced system 4.48.

- Elimination of the dependent coordinates

The Newton-Raphson method for solving the dependent coordinates v in terms of the independent ones u should normally be based on the constraint functions $h_r(q,t)$ and their Jacobian, according to the iterative algorithm given in 4.49. However, as for the equations of motion, the matrix B can be used instead of the exact Jacobian matrix 4.63. This leads to the iterative equation

$$v^{k+1} = v^k - (B_v)^{-1}\,h_r \qquad (4.67)$$

and the related procedure is a *pseudo-gradient* algorithm since B_v is not the analytical Jacobian of h_r. Using the pseudo-gradient instead of the exact Jacobian allows us to avoid the computation of the matrix $D_r(u,v^k,t)$ at each step. However, the convergence of this modified Newton-Raphson algorithm to the desired solution, i.e. the solution which closes the loop, now has to be verified.

For this convergence analysis, let v^0 be the values of the dependent coordinates which ensure the loop closure for given values of u and t. In other words, we have

$$h(u,v^0,t) = 0 \quad \text{and consequently} \quad h_r(u,v^0,t) = 0 \qquad (4.68)$$

In the neighborhood of the exact solution v^0, the column matrix h_r can be approximated by

$$\begin{aligned} h_r(u,v,t) &= D_r(u,v^0,t)B_v(u,v^0,t)\,\left(v^k - v^0\right)\\ &= B_v(u,v^0,t)\,\left(v^k - v^0\right) & (4.69) \end{aligned}$$

since $D_r(u,v^0,t) = E$.
The pseudo-gradient algorithm 4.67 will then provide

$$v^{k+1} = v^k - \left(B_v(u,v^k,t)\right)^{-1} B_v(u,v^0,t)\,\left(v^k - v^0\right)$$

Expanding $B_v(u,v^0,t)$ around v^k:

$$B_v(u,v^0,t) = B_v(u,v^k,t) + O\left(\parallel v^k - v^0 \parallel\right)$$

we obtain

$$\begin{aligned} v^{k+1} &= v^k - \left(B_v(u, v^k, t)\right)^{-1} \left(B_v(u, v^k, t) \left(v^k - v^0\right) + O\left(\| v^k - v^0 \|^2\right)\right) \\ &= v^0 + O\left(\| v^k - v^0 \|^2\right) \end{aligned}$$

which shows that after one step the algorithm give

$$v^{k+1} = v^0$$

at least to first order, i.e. within an error of the order $\| v^k - v^0 \|^2$.

A priori, the algorithm 4.67 may also converge towards values of v for which the complete set of constraints, $h(q,t) = 0$, is not satisfied, because it is only based on the subset h_r of constraint functions. Such undesirable convergence is avoided by the use of the pseudo-gradient. Indeed for any solution of $h_r(q,t) = 0$, we have

$$R_{(3,1)} = R_{(3,2)} = 0$$

which implies $R_{(3,3)} = \pm 1$ because matrix R is orthonormal. In turn, and for the same reason,

$$R_{(3,1)} = R_{(2,1)} = 0 \tag{4.70}$$

implies $R_{(1,1)} = \pm 1$. Moreover, since $\det(R) = 1$, the undesirable solutions of $h_r(u,v,t) = 0$ reduce to values of v for which

$$R = \begin{pmatrix} 1 & 0 & 0 \\ 0 & -1 & 0 \\ 0 & 0 & -1 \end{pmatrix} \quad \text{or} \quad \begin{pmatrix} -1 & 0 & 0 \\ 0 & 1 & 0 \\ 0 & 0 & -1 \end{pmatrix} \quad \text{or} \quad \begin{pmatrix} -1 & 0 & 0 \\ 0 & -1 & 0 \\ 0 & 0 & 1 \end{pmatrix} \tag{4.71}$$

Let us denote by v^1 the value of v corresponding to the first of these undesirable solutions. The corresponding value of the matrix $D_r(u, v^1, t)$ is given by

$$D_r(u, v^1, t) = \begin{pmatrix} R_{(2,2)} & -R_{(1,2)} & 0 \\ -R_{(2,1)} & R_{(1,1)} & 0 \\ -R_{(3,1)} & 0 & R_{(1,1)} \end{pmatrix} = \begin{pmatrix} -1 & 0 & 0 \\ 0 & 1 & 0 \\ 0 & 0 & 1 \end{pmatrix} \tag{4.72}$$

In the neighborhood of this undesirable solution v^1, the matrix h_r can be approximated by

$$h_r(u, v, t) = D_r(u, v^1, t) B_v(u, v^1, t) \left(v^k - v^1\right)$$

and the iterative equation of the pseudo-gradient algorithm then takes the form of the difference equation

$$v^{k+1} - v^1 = v^k - v^1 - \left(B_v(u, v^k, t)\right)^{-1} D_r(u, v^1, t) B_v(u, v^1, t) \left(v^k - v^1\right) \tag{4.73}$$

or equivalently

$$\left(v^{k+1} - v^1\right) = \left(E - \left(B_v(u, v^k, t)\right)^{-1} D_r(u, v^1, t) B_v(u, v^1, t)\right) \left(v^k - v^1\right)$$

$$\text{which has the form} \quad z^{k+1} = \Gamma(u, v^1, t)\, z^k$$

and is divergent because expanding $B_v(u, v^1, t)$ around v^k shows that the iteration matrix:

$$\Gamma(u, v^1, t) = \begin{pmatrix} 2 & 0 & 0 \\ 0 & 0 & 0 \\ 0 & 0 & 0 \end{pmatrix} + O\left(\| v^k - v^1 \|^2\right)$$

has an eigenvalue larger than one in the neighborhood of the undesirable solution v^1.

For the two other undesirable solutions, the pseudo-gradient algorithm will also diverge because the corresponding values of the matrix D_r:

$$\begin{pmatrix} 1 & 0 & 0 \\ 0 & -1 & 0 \\ 0 & 0 & -1 \end{pmatrix} \quad \text{and} \quad \begin{pmatrix} -1 & 0 & 0 \\ 0 & -1 & 0 \\ 0 & 0 & -1 \end{pmatrix}$$

lead to eigenvalues larger than one.

To conclude, the subset h_r of independent constraint functions we selected ensure the convergence of the pseudo-gradient algorithm towards the desired solution for which $h(q,t) = 0$ and *avoids undesirable convergence* towards solutions for which the complete set of constraints would not be satisfied. Moreover, the algorithm does not require additional computation for obtaining the gradient of the constraints: the latter is indeed replaced by the matrix $B(q,t)$ which, by 4.60, relates angular and generalized (joint) velocities. This relation is obtained in a straightforward manner from the loop kinematics. In this way, the pseudo-gradient algorithm improves the generality of the coordinate partitioning method which has been shown ([60]) to be one of the safest methods for solving differential-algebraic equations.

For the following subset of equations 4.56:

$$\dot{h}_r(\dot{q}, q, t) = D_r(q,t)\, \omega = 0$$

it also appears that the components ω of the angular velocity are functionally equivalent to the original set of orientation constraints (at the velocity level): this property allows us to confer a physical meaning on the Lagrange multipliers λ_r. Indeed for a virtual change of the generalized velocities which is not compatible with the constraints, the virtual power change due to the generalized constraint forces (see section 2.4.3):

$$\Delta \mathcal{P}_{constraint} = \sum_{\alpha=1}^{n} Q'_\alpha\, \Delta \dot{q}_\alpha = \Delta \dot{q}^T Q' \tag{4.74}$$

is, in the present case, given by

$$\Delta\mathcal{P}_{constraint} = \Delta\dot{q}^T \, J_r{}^T(q,t)\, \lambda_r \tag{4.75}$$

and, when the loop is closed, reduces to

$$\Delta\mathcal{P}_{constraint} = \Delta\dot{q}^T \, B^T(q,t)\, \lambda_r \tag{4.76}$$

Looking at 4.60 which relates the generalized velocities and the components ω, we then observe that

$$\Delta\mathcal{P}_{constraint} = \Delta\omega^T \, \lambda_r \tag{4.77}$$

On the other hand, the physical effect which compels the terminal body frame $\{\hat{\mathbf{X}}^j\}$ to be aligned with the initial frame $\{\hat{\mathbf{X}}^i\}$ is a torque $\mathbf{L} = [\hat{\mathbf{X}}^i]^T L$ applied to body j by body i. During a virtual velocity change which implies a relative angular motion of body j with respect to body i (i.e. not compatible with the rotation closure constraints), this constraint torque would produce a virtual power change

$$\Delta\mathcal{P}_{constraint} = \mathbf{L} \, . \, \Delta\left(\boldsymbol{\omega}^j - \boldsymbol{\omega}^i\right) = \Delta\omega^T \, L \tag{4.78}$$

Compared with the expression 4.77, one immediately may conclude from 4.78 that

$$\mathbf{L} = [\hat{\mathbf{X}}^i]^T \lambda_r \tag{4.79}$$

The Lagrange multipliers are thus simply the components in the frame $\{\hat{\mathbf{X}}^i\}$ of the constraint torque, and the relation 4.77 shows that, instead of being associated with the original orientation constraint functions $h(q,t)$ (or even $h_r(q,t)$), λ_r is associated with the components of ω. In other words, at the velocity level, one may consider that $\omega = 0$ represent the (non-holonomic) independent constraints which are functionally equivalent to the original orientation constraints $h(q,t) = 0$.

Chapter 5

Symbolic generation

5.1 Introduction

Before the appearance of efficient computer architectures for scientific numerical computations, only analytical methods were available for modeling systems. The analyses often used rather restrictive hypotheses (truncated and/or linearized models, for instance). The emergence of powerful processors and reliable and user-friendly languages and software led the scientific community to develop numerical programs able to cover a wide range of applications in a given field (ex.: structural dynamics, multibody dynamics, fluid mechanics, electronic circuits, ...). In particular, numerous *multibody* programs were developed all over the world as from the seventies [80], each of them being described as a general purpose multibody code, although, in reality, they all imposed some restrictions on both *modeling* and *analysis*.

In this book, we emphasize this distinction between *modeling* and *analysis*. The *modeling* phase is the analytical, numerical or symbolical process which, once and for all, sets up the equations of motion describing a multibody application for a given set of system parameters and generalized coordinates. For instance, the direct dynamical model is illustrated in figure 5.1 as a block whose inputs are the system parameters (m^i, $\mathbf{d}^{ii}$, $\mathbf{d}^{ij}$, ...) and the generalized positions q^i and velocities $\dot{q}^i$ and whose outputs are the generalized accelerations $\ddot{q}^i$ and the Lagrange multipliers λ^i.

Figure 5.1: Direct dynamics : main inputs/outputs

Among the equations constituting the model, we find (for the formalisms developed in the previous chapters):

- the joint dynamical equations (ex.: eq. 3.38),
- the constraint equations (and derivatives) for closed loop systems (ex.: equations 4.4 and 4.6) and their solutions at position, velocity and acceleration levels,
- the kinematic and dynamical equations relating to specific models (ex.: wheel/rail contact, cam/follower devices, ...: see Part II "Special topics"),
- the constitutive equations for specific forces (ex.: equation 4.24),
- the full reduction of the system on the basis of the coordinate partitioning method (see chapter 4),
- the computation of any desired output: absolute kinematics, constraint or point-to-point forces, ...

All these mathematical ingredients are contained in the block of figure 5.1. Once the latter is finalized, it can be used (i.e. called) thousands of time during the subsequent analysis phase.

The *analysis* phase denotes any numerical process which uses the previous *model* (or part of it) to generate the expected results, for instance:

- equilibrium solution,
- modal analysis,
- time simulation,

and in an enlarged context:

- control,
- sensitivity analysis and optimization,
- dimensioning,
- etc. ...

All of these numerical processes require a repeated evaluation of the model, often more than a million times in a time integration or optimization process. This means that, whatever the chosen formalism and the available computer resources, the computation of the model must be as efficient as possible.

The classical way of implementing these two phases, modeling and analysis, in a computer program is to do so in a purely numerical manner. This is mandatory for the analysis phase. Indeed, modal analysis or time integration of complex multibody systems unavoidably relies on numerical solvers. But this is not the only possibility for the modeling phase.

Before thinking about generating it automatically using a computer, the most natural way to model a system such as that of figure 5.1, is to do it *manually* — using a pen or some general purpose program — on the basis of a formalism described in chapters 3 and 4. While appearing to be archaic in the computer era, and not really reliable, this method is still used for simple systems or simplified models. Although it is subject to human error, manual generation yields optimized mathematical equations, in terms of arithmetical and trigonometrical expressions: this remark is fundamental for our later discussions.

Numerical generation means that, each time a computation is required by some analysis process, the model is rebuilt by numerous calls of subroutines, each involved in some specific parts of the equations or specific vector/tensor operations (library subroutines). This multitude of subroutines is the consequence of the universal nature of multibody programs, and the multiple calls are partially responsible for the heaviness of the numerical generation, compared to manual generation. Experiments in multibody dynamics show that for most applications, many data values, e.g. components of geometrical vectors, inertia parameters, joint velocities, etc., are zero. All these zero quantities are treated in a numerical process in the same way as the non-zero ones: a manual generation of the model would eliminate all the unnecessary arithmetical operations beforehand, yielding an optimal form of the equations of motion.

Symbolic generation of multibody models tries to take advantage of both the numerical and manual techniques. So-called symbolic multibody programs manipulate only arithmetical operators $(+, *, -, /)$ and strings of alphanumeric characters (mi, qdi, dij, ...) to generate — in a set of files — the *analytical* form of the equations using a desired syntax (ex.: C, Fortran, MATLAB, SIMULINK). For a given multibody application, this symbolic generation is performed only once, as in manual generation. From the multibody modeling point of view, these symbolic generators ([80], [50], [79], [67], [55], [84], ...) exhibit the same level of generality as their numerical competitors in the sense that they can handle large systems with any topology and containing up to 100 degrees of freedom. However, they allow drastic simplifications, from the most trivial (addition/multiplication by zero) to the most complex ones (simplification of long trigonometric expressions). Comparison tests [12] showed that, all other things being equal, a symbolic multibody model performs a time simulation between five and ten times faster than a purely numerical one. This is not at all negligible.

This speed enhancement motivated the development of symbolic multibody codes in the eighties [80] since computers were at that time quite inefficient in simulating even medium-sized systems (around 10 d.o.f.). The computational superiority of the symbolic approach still remains attractive since the exponential increase in modern computer performances naturally provides the opportunity to investigate larger systems or more sophisticated models [70]. More recently, another interesting aspect of the symbolic approach has emerged in the framework of the analysis of hybrid mechatronic systems involving *en-*

larged multibody models (i.e. including electrical actuators, hydraulic devices, ...) and requiring *enlarged* analysis (control, optimization, ...). Indeed, since symbolic multibody programs clearly separate the *modeling* part from the *analysis* part, it is quite straightforward to extract from them the symbolic model in order to export it towards another environment (ex.: for control or optimization purposes). In other words, symbolic multibody models are portable, and this is very attractive for the analysis of *mechatronic* applications [17].

There do exist commercial general purpose symbolic computation packages like Maple® and Mathematica®. Why we do not use these? There are two reasons for this.
The first and most important one relates to the amount of computer memory they require for generating medium-sized and large models symbolically. For instance, for multibody systems with flexible bodies (see chapter 9), the number of symbolic expressions needed to extract the dynamical equations and the tangent matrices exceeds the present capabilities of these packages.
The second reason concerns the possible simplifications of the symbolic expressions. Although the simplification capabilities of commercial packages are powerful, the condensation of multibody equations relies on specific rules which can be applied more easily by developing a dedicated symbolic program; this was the main motivation to develop the multibody program ROBOTRAN [19]. Its capabilities in terms of symbolic manipulation will be the subject of the present chapter:

- Structure, manipulation and simplification of symbolic expressions (section 5.2)
- Computer Memory: allocation and freeing (section 5.3)
- Trigonometric simplifications (section 5.4)
- Symbolic implementation of recursive formalisms (section 5.5)
- Recursive symbolic differentiation (section 5.6)
- Comparison of multibody formalisms performances (section 5.7)
- Computer implementation of multibody formalisms (section 5.8)
- A short example (section 5.9)

5.2 Symbolic mathematical expressions

5.2.1 Tree representation

A mathematical expression in multibody equations uses arithmetical operators $+, -, *, /, (), =$ and functions: mainly sin() and cos(), occasionally: $sqrt()$, $atan()$, ... When combined in a given expression, these elements obey fundamental priority rules summarized in table 5.1.

Operator	Symbol	Priority	Type
parentheses	()	1	-
sine / cosine / $\sqrt{}$ ...	sin, cos, *sqrt*, ...	2	unary
multiplication/division	$*, /$	3	binary
addition/subtraction	$+, -$	4	binary
assignment	$=$	5	binary

Table 5.1: Priority of operators

These priority rules allow us to consider an expression as a recursive tree whose nodes represent the *nature* of the (sub-)expression, and *points* towards its argument(s). Let us for example consider the following expression

$$2 * a - g * \cos(q1 + q2) \tag{5.1}$$

which, when applying the priority rules of table 5.1, is equivalent to

$$(2 * a) - (g * (\cos(q1 + q2))) \tag{5.2}$$

This expression can be represented by the tree shown in figure 5.2. For instance, the second multiplication is an expression whose nature is $*$ and points towards two arguments: the expressions g and cos(). In a tree representation, the leaves represent alphanumeric symbols, e.g. 2, a, g, ... in example 5.1. They are also considered as expressions (or "leaf expressions") but they have no sub-expression to which they point: they simply contain the string they represent.

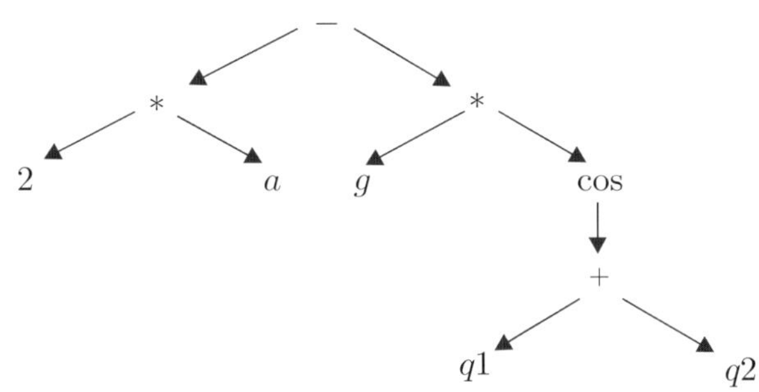

Figure 5.2: Mathematical expression: tree representation

In multibody dynamics, equations are assignments of the type $a = expression$ whose left hand side is the assigned variable, and whose right hand side is a mathematical expression. According to the previous priority rules, the latter is composed of one or more *terms* linked by $-$ or $+$, each term being the product $*$ or the ratio / of several factors. A factor can be a numerical coefficient, an alphanumeric parameter, a trigonometrical function or ... a full mathematical expression between brackets. This clearly reveals the recursive nature of an expression and naturally leads to the use of recursive programming to generate, manipulate and simplify complex equations. The tree representation and the dynamical increase in size of the expressions during the elaboration of the equations leads us to represent an expression by a C-structure, *dynamically* generated and handled via *pointers*. At the root of the program, there is the C-structure *expression*[1] which contains at least the following fields:

1. the *nature* of the expression,
2. the address(es) (i.e. pointers) of the *argument*(s) to which the expression points: these arguments are either strings — for a leaf expression — or other expressions which have been created via a previous dynamical memory allocation.

5.2.2 Expression organization

At this level, it is already possible to build a complex symbolic expression in the memory by allocating the appropriate structures and linking them according to the tree representation: this is the starting point of the symbolic generation.

Once an expression is symbolically evaluated, one could print it in a file according to the operator priorities mentioned in table 5.1: this would lead to a correct ... but quite rudimentary and non optimized symbolic generation. Indeed, before printing any expression, we should try everything possible to simplify it by symbolic manipulation.

For this purpose, the first and most important step is to *organize* factors and terms inside any expression on the basis of well-defined rules. In multibody dynamics, thousands of symbolic expressions may appear, but they are derived from just a few quantities, i.e. masses, components of vectors, tensors, etc. We thus organize the basic symbols as follows:

- numerical coefficients: $0, 1, 2, 4, \frac{1}{2}$. These are the primary ones; others may sometimes appear.
- body masses m^i, components of barycentric vectors $\bar{m}^i \mathbf{b}^i$ and of inertia/barycentric tensors $\mathbf{I}^i, \mathbf{K}^i$
- components of position vectors $\mathbf{d}^{ii}, \mathbf{d}^{ij}$
- external forces and torques: components of vectors $\mathbf{F}^i_{ext}, \mathbf{L}^i_{ext}$
- generalized coordinates: $q^i, \dot{q}^i, \ddot{q}^i$

Each kind of expression, e.g. a parameter, a sin() function, a multiplication, is assigned a different *nature* which is used to organize them inside the expressions in which they will appear. These organization rules yield symbolic equations which are mathematically simplified, while maintaining a measure of legibility with respect to the nature of their operands.

[1] From now on, the term *structure* will refer to this C-structure, i.e. the computer representation of any symbolic expression.

Expression	Nature value
numerical coefficients $(0, 1, 2, ...)$	1
dynamical parameters $(m^i, \bar{m}^i b^i, I^i, K^i)$	2
position vectors (d^{ii}, d^{ij})	3
force/torque parameters $(F^i_{ext}, L^i_{ext}, g)$	4
generalized coordinates $(q^i, \dot{q}^i, \ddot{q}^i)$	5
cosine $(\cos(x))$	6
sine $(\sin(x))$	7
sqrt, asin, acos, atan (x)	8
multiplication $(*)$	9
division $(/)$	10
addition $(+)$	11
subtraction$(-)$	12

Table 5.2: Nature of multibody expressions

Table 5.2 summarizes the full set of elements one can encounter in multibody equations, together with their associated *nature* value. The ordering differs from that of Table 5.1 which governs the sequence for *evaluating* an expression, rather than *ordering and writing* one.

The precedence relations based on the nature value given in Table 5.2 are illustrated as follows:

$$4 < m^i < d^{ij} < F^j_{ext} < \dot{q}^i < b * c < (a - d) \tag{5.3}$$

This pointer organization will be used during the final printing of the results to order the expressions from left to right. During this final printing process, the priority of the operators defined in Table 5.1 is taken into account by placing appropriate brackets.

When two items a and b have the same nature, then the order obeys the *lexicographical* sequence of the ASCII table of characters, by covering the alphanumeric characters of both elements from left to right. For instance:

$$K^{ii} < m^i \quad ; \quad 2 < 3 \quad ; \quad d^{ii} + d^{ij} < d^{ii} + d^{jk} \quad ; \quad etc. \tag{5.4}$$

Superscripts and subscripts of variables are considered similarly to other strings; for instance, symbols mi, dij and $Fjext$ respectively stand for m^i, d^{ij} and F^j_{ext}.

To apply these priority rules correctly on a very long and complex expression, the latter must first be re-organized on the basis of the distributivity properties of the involved operators, as illustrated with the example of figure 5.3. In this example, in order to obtain expression C whose final simplification (expression D) is rather obvious, the original expression $A = (a+b) - (a+c)$ which involves sub-expressions in parentheses is transformed (into expression B) in such a way that each right-hand-side of all the operators $+$ and $-$ becomes an *elementary* symbol[2]. In other words, all parentheses have disappeared in expression B

[2]More generally, during this process, each right-hand-side becomes an expression whose nature is less than the addition/subtraction (in the sense of Table 5.2)

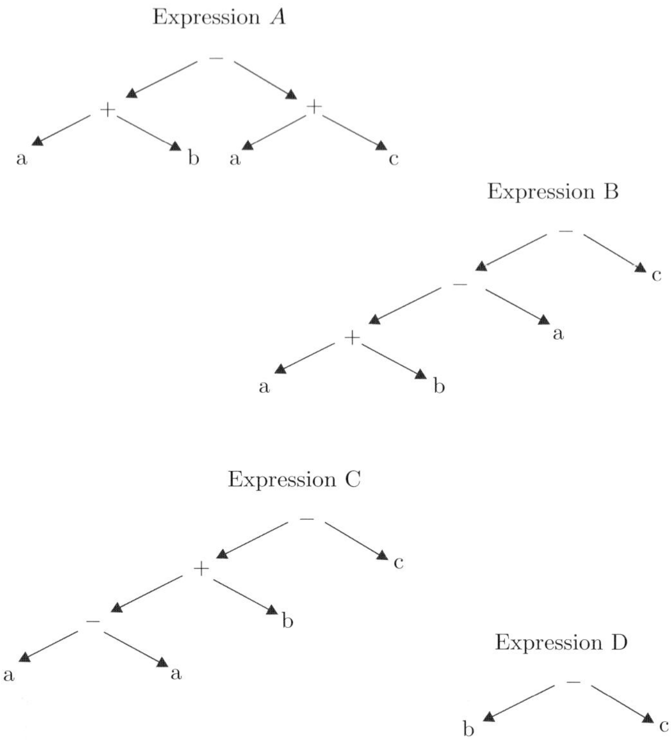

Figure 5.3: Expression reduction by using the lexicographical order

which now has a suitable form to apply – recursively on the right-hand-sides – the lexicographical priority rules mentioned above.

This procedure ensures that the final form (expression D) will be purged of consecutive equal terms with opposite signs.

After presenting this basic simplification process, we now explain its program implementation which is composed of four distinct layers.

Layer 1: symbolic comparison of expressions

This is the lowest level and contains two boolean library functions: $equal(a, b)$ and $less(a, b)$.

- The first one $equal(a, b)$ returns TRUE if expressions a and b are symbolically equal (i.e. either they have the same memory address, or their natures and contents are strictly identical). It returns FALSE otherwise.
- The second one $less(a, b)$ returns TRUE if expression a is "less" than expression b in accordance with the lexicographical rules stated above (see examples 5.3, 5.4), and returns FALSE otherwise.

Layer 2: symbolic scalar operations: *, / +, -, sine, cosine, $\surd$, ...

This second level in the symbolic library contains the following elementary functions:

- $c = mult(a, b)$ and $c = div(a, b)$ respectively evaluate the symbolic multiplication $a * b$ and division a/b, and store the result in a new expression c. For instance, depending on the symbolic contents of a and b, the $mult()$ function acts so that:
  - automatic simplifications are detected: for instance if $a = 0$ and $b = g$, the returned expression is $c = 0$, and no structure needs to be created,
  - re-ordering of the terms is performed: if $a = g$ and $b = m$, the returned expression will be $c = m * g$ and not $g * m$, according to the ordering rules applied by the first layer,
  - a recursive call of $mult()$ is often necessary: for instance, if $a = q^j$ and $b = m^i * q^i$, $mult()$ calls itself by evaluating and returning $mult(m^i, mult(q^i, q^j))$.

  These three elementary processes are only part of those included in the $mult()$ function which also systematically recursively re-organizes the ordering of parentheses that can take place when a and/or b represent long and complex expressions.
  All these internal symbolic manipulations rely strongly on the first layer ($equal()$ and $less()$).

- $c = add(a, b)$ and $c = sub(a, b)$ respectively evaluate the sum $a + b$ and the difference $a - b$. Sub-processes similar to those of the $mult()$ procedure are parts of these functions, their goal being to simplify and/or re-organize the resulting expression c on the basis of the priority rules: extensive use is made of the first layer.
  Recursive calls are also used and can involve $add()$, $sub()$ and $mult()$ at the same time. For instance, to simplify $(x - 3 * y) + (2 * y)$ into $x - y$, the $add()$ procedure detects this specific form, temporarily returns $add(x, mult(sub(2, 3), y))$ which is easily recognized as being $sub(x, y)$.

- $c = \cos(q^i)$, $c = \sin(q^j)$ return in c the auxiliary variables Ci and Si respectively, which will be assigned at the beginning of the resulting file in the following way:

$$\text{Ci} = \cos(q^i) \quad ; \quad \text{Si} = \sin(q^j), \tag{5.5}$$

  to avoid multiple evaluation of the same trigonometric functions which generally appear at various places in multibody equations.

Layer 3: Vector and tensor operations

This level, mainly based on the previous layers, represents the practical multibody library. Among the functions involved, we find these:

- $v = gvadd(v1, v2)$, the symbolic addition of two (3×1) vectors $v1$ and $v2$
- $s = gvvprd(v1, v2)$, the symbolic scalar product of two (3×1) vectors $v1, v2$
- ...
- $m = gmmprd(a1, a2)$, the symbolic multiplication of two (3×3) matrices $a1, a2$
- $m = gmtil(v1, v2)$, the symbolic multiplication of the (3×3) tilde matrices respectively associated with two (3×1) vectors $v1$ and $v2 : m = \tilde{v}1\,\tilde{v}2$
- ...

From the symbolic programing point of view, vectors and matrices correspond to new types of C-variables respectively defined by (3×1) and (3×3) arrays of expressions.

Layer 4: Generation of multibody equations

This is the "top" layer in which we find the following functions, built on the basis of the previous layers, each of which is dedicated to a specific multibody formalism, for instance:

- *kine_rec*(): evaluates a sub-chain recursive kinematics (see section 4.5)
- *dyna_ppp*(): evaluates the dynamics of the multibody system on the basis of the virtual power principle (see section 3.2)
- *dyna_flex*(): evaluates the dynamics of multibody systems containing flexible beams, according to the formalism given in chapter 9.
- etc...

Note that all the previous processes manipulate, re-organize and simplify expressions *within the computer memory*, without printing the symbolic equations. In fact, the printing operation is somewhat trivial for ROBOTRAN since, once the full symbolic evaluation is finished, the entire tree of all the symbolic equations is fixed (i.e. "frozen") in the computer memory and thus ready to be printed according to the operator priorities (see table 5.1) and using a desired computer syntax (e.g.: MATLAB, C).

5.3 Computer memory: allocation and freeing

The symbolic generation of multibody equations requires dynamical memory allocation to create and store new expressions (C-structures). The symbolic processes associated with layers 1 and 2, which recursively re-organize any new expressions in accordance with the ordering rules, can generate in the memory thousands of auxiliary expressions which are not necessary in the final tree representation of a given (complex) expression. Figure 5.3 illustrates this phenomenon and practical experiments showed that, whatever the multibody formalism used, these short-lived expressions induced by symbolic manipulations represent more than 90 % of the whole set of expressions generated. This explains why symbolic programs can lead to an explosive use of computer memory. To solve this delicate problem, one must develop a specific procedure which removes each superfluous expression, or more precisely, each expression which is useless for the final results.

We could try to cut long equations into segments and evaluate and print them separately. Then, after dealing with each segment, we would clear all the generated expressions from the memory. This solution is not only mediocre but also erroneous, as illustrated in the counterexample of figure 5.4: we must evaluate two expressions A and B consecutively, which have common nodes (expressions $a1 * a2$ and $a4$ in figure 5.4).

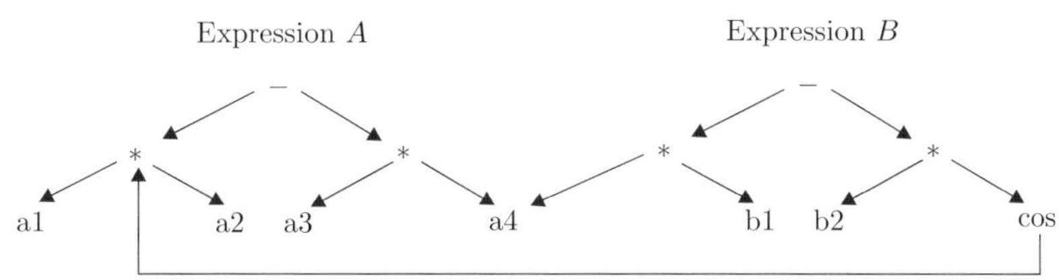

Figure 5.4: Symbolic expression inter-dependency

1. Processing and printing A and B separately is *erroneous* because the elimination from the memory of expression A just after printing it, is relevant with respect to A but not with respect to B which relies on expressions $a1 * a2$ and $a4$.

2. In terms of memory space, the process is *mediocre* because it only clears from the memory the expressions of the *final form* of the tree associated with an expression A. All the short-lived intermediate expressions created by layers 1 and 2 during the elaboration of A are no longer accessible, once the final tree is established. Since these short-lived expressions represent more than 90 % of the generated expressions used to build A, this technique is inefficient to minimize the memory space requirement.

To solve this problem, the following three elements were introduced into ROBOTRAN.

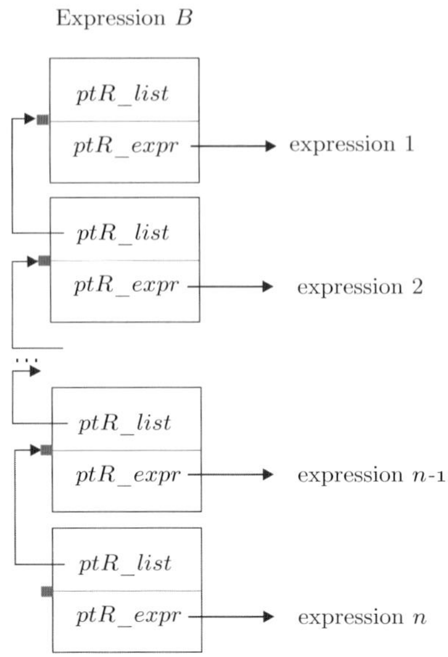

Figure 5.5: Dynamic linked list of expressions

1. During the computation of a symbolic equation (i.e. in fact a complex expression), in order to keep track of each new symbolic expression (short-lived or not) created by some function ($add(), mult(), sub()$, ...), the address of this expression is systematically stored in a *dynamical linked-list,* which grows as and when it is needed, as shown in figure 5.5.

2. In order to protect any expression (e.g. A, $a2$, cos() in the above example), we add to the corresponding C-structure, a boolean field *lock*. The value $lock$ = TRUE tells ROBOTRAN that the corresponding expression cannot be removed from memory, whereas $lock$ = FALSE states that it can; this is the default value.

3. We introduce the procedure $expr\_lock(E)$ to lock a given expression E, i.e. by setting to TRUE the *lock* field of each node belonging to the *final* tree of expression E. The procedure is thus intrinsically recursive as illustrated in figure 5.6.
   Companion procedures $vec\_lock(vE)$ and $mat\_lock(mE)$ respectively lock vectors and matrices of expressions.

By thus locking a given expression E in its *final* form, we do not lock the intermediate "short-lived" expressions created during the evaluation of E, but which do not contribute to its final tree: for these expressions, *lock* keeps its default value. This can be illustrated via the example of figure 5.3: the locking procedure will be applied only to the final result (expression D) and will lock D, b and c ($lock$ = TRUE), while all the intermediate expressions (A, B, C and

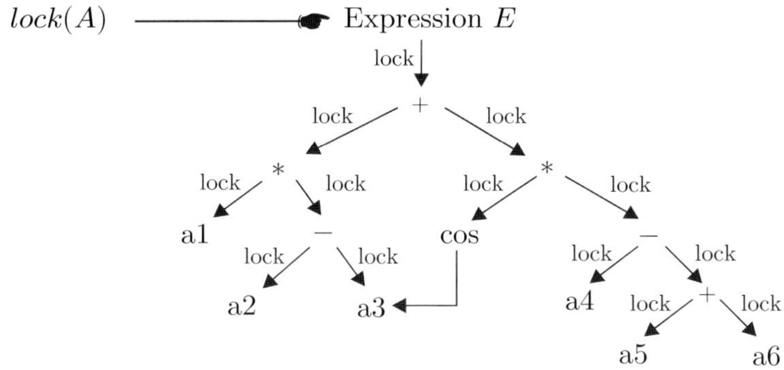

Figure 5.6: Locking an expression in memory

all their nodes except b and c) will be removed from the memory (on the basis of the fact that $lock$ = FALSE for those expressions).

To free the memory in an optimal way, i.e. by removing all the superfluous expressions ... but only those, the program operates as follows: the symbolic evaluation of a multibody formalism is performed "part by part", in a *sequential* way. Each part represents an equation (i.e. a complex expression) whose tree, once symbolically optimized, is stored in memory (or possibly printed). Practically, the memory freeing process is the following (let us remember the example of figure 5.4):

1. For each expression x internally created during the evaluation of a given equation B, we add to the linked list a new element which points to this expression x. At this level, it is not known *a priori* whether x is useful or not for the final tree of B.

2. Once the evaluation of B is finished, and its tree has been purged of superfluous operations, the process carries out the following operations:

    - recursively protects the final form of expression B ($lock(B)$),
    - frees the memory by removing every expression x which has not been locked during the previous step. For that purpose, the list is covered from tail to head, thanks to the linkage pointer ptR_list, and any structure whose field $lock$ is FALSE is entirely removed from the memory. Each element of the list is also removed to end up with an empty list ... ready for the evaluation of a subsequent equation.

With this methodology, all the short-lived expressions are removed from memory without any danger to subsequent evaluations. Moreover, once an expression A is evaluated, its immediate printing is not mandatory because its tree representation is frozen for ever in the memory.

The memory space required by and during a ROBOTRAN process is illustrated in figure 5.7 and has thus a "toothed" shape, rather than a monotonic growing one as in most commercial symbolic packages. The maximum memory space ($SP_{\max}$) is reached by the greediest equation.

This freeing process is of the utmost importance in multibody dynamics since it allows us to eliminate the most critical bottleneck (i.e. the memory space) of the generation of large multibody models.

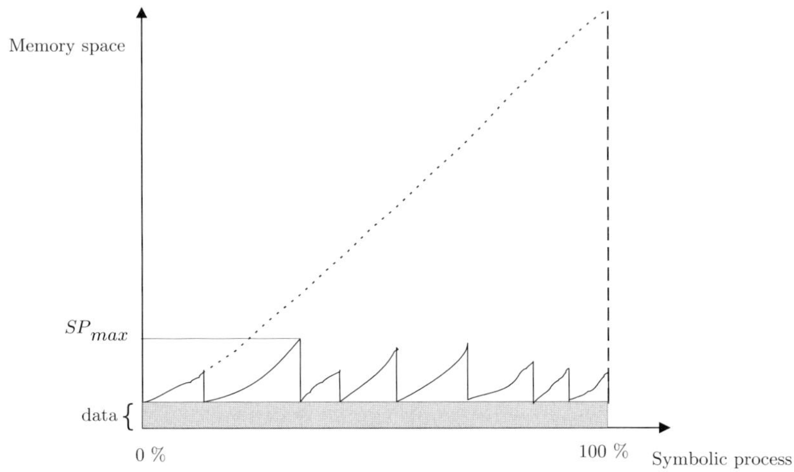

Figure 5.7: Memory allocation for a ROBOTRAN process

5.4 Trigonometric expressions

5.4.1 Introduction

A revolute joint i, with generalized joint coordinate q^i, of a multibody system induces an elementary rotation matrix which contains the trigonometric functions $\cos(q^i)$ and $\sin(q^i)$ (see equation 1.41). The evaluation of rotation matrices between any pair of bodies i and j, such that j is a descendant of i, is obtained by multiplying the elementary rotation matrices associated with each revolute joint k such that $i < k \leq j$. For a kinematic chain $\{i, i+1, i+2, ..., j-1, j\}$ for instance:

$$R^{j,i} = R^{j,j-1}\, R^{j-1,j-2}\, ...\, R^{i+1,i} \tag{5.6}$$

These matrices are indispensable for some of the formalisms presented above (ex.: the virtual power principle based formalism of section 3.2) which must evaluate scalar and/or vector products between vectors and/or tensors whose components are provided in body-frames which are quite "far" from each other.

Thus, we need an optimized trigonometric library for this specific problem: the condensation of complex trigonometric expressions whose terms contain factors $\cos(q^i)$ and/or $\sin(q^j)$. In the following expression, for instance,

$$\text{K212*qpp8*(C2*C2*C3*S4+S2*S24-S2*S34*S5+S2*S2*C3*S4)} \tag{5.7}$$

Cj, Sk and Sjk represent $\cos(q^j)$, $\sin(q^k)$ and $\sin(q^j + q^k)$ respectively.

As stated above (cfr. equation 5.5), the variables Cj, Sk, ... are systematically assigned and printed in the output file — before the multibody equations — in order to avoid multiple appearances of the same trigonometric function. For instance, the following set of preliminary instructions is typical:

$$\begin{aligned}
\text{S3} &= \sin(q3)\\
\text{C3} &= \cos(q3)\\
q3_4 &= q3 - q4\\
\text{S3_4} &= \sin(q3 - q4)\\
q456 &= q4 + q5 + q6\\
\text{C456} &= \cos(q456)
\end{aligned} \tag{5.8}$$

To simplify the trigonometric calculations, ROBOTRAN systematically organizes a given expression according to the symbolic manipulations detailed in section 5.2, so that all the trigonometric terms are isolated and gathered together as illustrated in the expression 5.7.

5.4.2 Symbolic process

The trigonometric simplification process has two main levels.

Level 1: elementary trigonometric formulae

The first one (the "lowest" level) is quite easy to implement, thanks to the symbolic ordering tools described in section 5.2. It consists in a large procedure called *trigo*() which is systematically called at each occurrence of a given expression, whatever its contents. The goal of the procedure is simply to perform the fundamental trigonometric simplifications according to well-known formulae. For instance, let us consider a symbolic expression a:

- if $a =$ C1*C1+S1*S1, *trigo*(a) returns 1,
- if $a =$ C1*C1−S1*S1, *trigo*(a) creates and returns the auxiliary variable C11,
- if $a =$ 2*C1*S1, *trigo*(a) creates and returns the auxiliary variable S11,
- if $a =$ C1*C2−S1*S2, *trigo*(a) creates and returns the auxiliary variable C12, which stands for $\cos(q^1 + q^2)$,

...

- if $a = -$C1*C2$-$S1*S2, *trigo*(a) creates the auxiliary variable C1_2, which stands for $\cos(q^1 - q^2)$ and returns $-$C1_2,
- etc.

By "create", we mean that a new expression (for instance S11) is created in memory but also that the definition of the corresponding variable is printed, if not yet present, in the header of the output file (ex.: S11 = 2*C1*S1, see examples among equations 5.8).

Level 2: simplification of complex trigonometric expressions

While level 1 is able to detect and simplify elementary trigonometric formulae, it is not able to deal with expressions like

$$
\begin{array}{c}
\text{C2*C4*C56*C56*S8+C2*C4*S56*S56*S8} \\
\text{+C2*S4*S56*C8+S2*C4*S56*C8} \\
\text{-S2*S4*C56*C56*S8-S2*S4*S56*S56*S8}
\end{array} \tag{5.9}
$$

for which judicious groupings and factorings must be performed in order to make a maximum of trigonometric formulae appear, which can then be simplified by the *trigo*() procedure. This is the purpose of the second level of the process which consecutively involves the following procedures, detailed later on:

- *simp_trig*(A) recursively simplifies pairs of trigonometric terms of expressions A like 5.9 and finally replaces A by a simplified expression A^*.
- *angle_double*(A^*) spots and recursively simplifies terms in A^* which obey formulae involving *sine* and *cosine* of double angles. We denote the resulting simplified expression by A^{**}.
- *evid_trig*(A^{**}) recursively factors out a maximum of factors in expression A^{**} and replaces them by the final factorized form A^{***}.

a. Trigonometric simplifications: procedure *simp_trig*(A)
For a candidate expression, the following process takes place:

1. Storage in a temporary list, of all the terms (from 1 to N) appearing in the trigonometric expression A. Thus for the symbolic expression in brackets of equation 5.7 (C2*C2*C3*S4+S2*S24-S2*S34*S5+S2*S2*C3*S4), N equals 4 and thus:

$$
\begin{aligned}
list[1] &= \text{C2*C2*C3*S4} \\
list[2] &= \text{S2*S24} \\
list[3] &= -\text{S2*S34*S5} \\
list[4] &= \text{S2*S2*C3*S4}
\end{aligned} \tag{5.10}
$$

2. Processing of each *pair* of terms stored in the list ($list[i] \leftrightarrow list[j], i = 1, N-1\ ;\ j = i+1, N$), endeavouring to simplify them using the *trigo*() procedure.
   For instance, for the pair $list[1], list[4]$: (C2*C2*C3*S4)+(S2*S2*C3*S4):
   - endeavour to factor out a maximum of factors: C3*S4*(C2*C2+S2*S2),
   - call the *trigo*() procedure which returns 1 for the expression in brackets,
   - call the *mult*() procedure which returns the simplified expression C3*S4.

Since this method of factoring out elements in a given expression A is not unique and therefore leads to different final expressions (regarding the degree of simplification), the procedure has been refined by associating with each formula processed by the *trigo*() procedure, a so-called *priority level* such that:

- formulae of type: Cx*Cx+Sx*Sx = 1, −Cx*Cx−Sx*Sx = −1 have the highest priority, i.e. $pr = 2$,
- other ones such as Cx*Cy−Sx*Sy = Cxy receive a lower priority, $pr = 1$.

Indeed, the first ones ($pr = 2$) finally reduce to expressions 1, −1 or 0 which will be symbolically eliminated later. Hence, they are stronger than the others ($pr = 1$) which lead to a new symbol which will be the factor inside a new expression. According to this, the simplification of pairs $list[i], list[j]$ is now performed in two steps: the first identifies and simplifies pairs with priority 2, and the second simplifies the remaining pairs with priority 1.

Once each pair of elements of the list has been processed in this way, the resulting simplified expression ... is a new trigonometric expression in which some new combinations of terms can lead to additional simplifications: this justifies that the *simp_trig*() procedure is called *recursively*[3] as long as at least one trigonometric simplification is detected by the process.

b. Sine and cosine of double angle: procedure $angle\_double(A^*)$

The following two trigonometric simplifications are performed as soon as the $simp\_trig(A)$ procedure is finished.

1. Detection and recursive simplification of all terms in A^* which are *sine* functions of a double angle:
   Formula: Sxx = 2*Cx*Sx
   Example (recursive process):

$$\begin{aligned} & 4{*}C2{*}S2{*}C3{*}S3{*}C4 \\ \rightarrow\ & 2{*}S22{*}C3{*}S3{*}C4 \\ \rightarrow\ & S22{*}S33{*}C4 \end{aligned}$$

2. Detection of the Carnot formulae and recursive simplification
   Formulae: 1+Cxx = 2*Cx*Cx and 1−Cxx = 2*Sx*Sx

[3] i.e. called by itself

Example (recursive process):

$$\begin{aligned} & 4*C2*C2*S3*S3*C5 \\ \rightarrow\ & 2*S3*S3*C5*(1+C22) \\ \rightarrow\ & C5*(1+C22)*(1-C33) \end{aligned}$$

At this level, we can claim that the original expression A has been drastically simplified into an equivalent expression A^{**}. This can now undergo the last step of simplification, more common in symbolic programming: the factoring out of terms.

c. Factoring out expression A^{**}: procedure *evid_trig*(A^{**})

The process of the *evid_trig*() function is somewhat similar to that performed by *simp_trig*() and can be summarized as follows:

1. Storage of all the terms of A^{**} in a linked list
2. Analysis of each pair of terms, endeavouring to factor out some elements ex.: *list*[*i*] = C2*S4*C5 and *list*[*j*] = −C3*S4*C5 can be replaced by S4*C5*(C2−C3).
   As for the trigonometric simplification, a priority level is attached to each possible factoring, the highest priority being associated with pairs in which a maximum number of common factors are detected. In other words, this second step:
   (a) evaluates $N^{\max}$, the maximum number of common factors between pairs of terms of expression A^{**},
   (b) factors out the $N^{\max}$ factors of each concerned pairs of terms,
   (c) eliminates these terms from the list,
   (d) returns to step (a) as long as $N^{\max} > 0$.
3. Recursive call of *evid_trig*(): indeed, as for *simp_trig*(), the resulting expression still contains pairs of terms with common factors. This justifies the recursive nature of the process, to end up with the final simplified expression A^{***}.

The sequence 5.11 summarizes the global simplification process *systematically performed* by ROBOTRAN for each complex symbolic expression A manipulated by the *mult*(), *add*() and *sub*() functions during the generation of the equations. This can lead to an enormous amount of symbolic manipulations (mainly due to the recursive nature of the procedures). However, the corresponding increase in time[4] needed to generate the model via ROBOTRAN is quite reasonable for all practical multibody applications. Furthermore, as

[4] with respect to a symbolic generation which completely disregards trigonometric simplifications

mentioned in the introduction, the symbolic process runs only once for a given application.

$$
\begin{array}{lll}
A & \rightarrow & simp\_trig(A) \\
 & \rightarrow & ...(recursion)... \\
 & \rightarrow & simp\_trig(A) \qquad \rightarrow A^{*} \\
 & \rightarrow & angle\_double(A^{*}) \\
 & \rightarrow & ...(recursion)... \\
 & \rightarrow & angle\_double(A^{*}) \quad \rightarrow A^{**} \\
 & \rightarrow & evid\_trig(A^{**}) \\
 & \rightarrow & ...(recursion)... \\
 & \rightarrow & evid\_trig(A^{**}) \\
 & \rightarrow & A^{***} = A \text{ simplifed}
\end{array}
\tag{5.11}
$$

5.4.3 Illustrative examples

The following examples illustrate the power of the method. The trigonometric expressions result from a dynamical formalism applied to a railway bogie (the "BA2000" bogie - see chapter 7)

- C2*C4*C56*C56*S8+C2*C4*S56*S56*S8+C2*S4*S56*C8
  +S2*C4*S56*C8−S2*S4*C56*C56*S8−S2*S4*S56*S56*S8
  becomes: C24*S8+S24*S56*C8
- C2*S2*C4*C56*C56*S8+C2*S2*C4*S56*S56*S8+C2*S2*S4*S56*C8
  +S2*S2*C4*S56*C8−S2*S2*S4*C56*C56*S8−S2*S2*S4*S56*S56*S8
  becomes: S2*(C24*S8+S24*S56*C8)
- −C24*C24*C56*C56*S8+C24*C24*S56*S56*C8*C8*S8+
  C24*C24*S56*S56*S8*S8*S8+C24*S24*S56*C8*C8*C8
  +C24*S24*S56*C8*S8*S8
  becomes: C24*(−C24*C5656*S8+S24*S56*C8)

5.5 Recursive scheme condensation

5.5.1 Introduction

In chapter 3, three multibody formalisms were proposed in order to obtain the joint dynamical equation in matrix form (equation 3.38), for tree-like structures. In the first formalism, based on the virtual power principle (section 3.2) all the

kinematic quantities are written out in full, i.e. *in extenso* without using intermediate variables. The other two formalisms extensively use auxiliary variables on account of their recursive nature.
For instance, the angular velocity vector of body i in the multibody structure is written as:

$$\boldsymbol{\omega}^i = \sum_{h:\, h \leq i} \Omega^h \quad \text{or} \quad \boldsymbol{\omega}^i = \boldsymbol{\omega}^{inbody(i)} + \Omega^i \tag{5.12}$$

for a full, i.e. *in extenso*, and a recursive formalism, respectively.

From a purely symbolic generation point of view, another important feature distinguishes recursive formalisms of sections 3.3 and 3.4, from those based on the virtual power principle (section 3.2). For the latter, the velocity fields 3.18 and 3.23 chosen to obtain the dynamical equation associated with joint i, are such that the joint forces $\mathbf{F}^j$ and torques $\mathbf{L}^j$ (for $j : i < j$) do not appear explicitly in the vector equations of motion 3.22 and 3.27. Indeed, physically, these velocity fields "freeze" all joints j which are descendants of body i: the vector equations 3.22 and 3.27 are the dynamical equations of the "multi-body" sub-system composed of body i and its descendants j for which $\mathbf{F}^j$ and $\mathbf{L}^j$ are perceived as internal forces. Hence, to obtain the ith scalar dynamical equation, only the projection onto the joint axis $\hat{\mathbf{e}}^i$ (equation 3.30) needs to be evaluated, since the other projections (five for a one d.o.f. joint) are never required explicitly by the formalism.

On the other hand, for recursive schemes, the scalar equation of motion 3.79 associated with joint i is obtained by projection onto the joint axis of the vector equations of motion 3.67 and 3.69 of body i which explicitly involve the joint forces $\mathbf{F}^j$ and torques $\mathbf{L}^j$ of each of its children j. Body i being the child of another body ($inbody(i)$), the three components of $\mathbf{F}^i$ and $\mathbf{L}^i$ must be evaluated during the backward recursion, because they are liable to be used in other joint equations (related to the kinematic chain $\{0...i\}$).

In other words, although recursive formulations intrinsically have a far more compact form than *in extenso* formulations, they paradoxically perform superfluous evaluations: depending on the type and succession of joints of the application, some components of the vector equations are superfluous for the final scalar form.

Whereas a general purpose multibody program which generates the model *numerically* is not able to detect these superfluous equations, a symbolic multibody program can do so. Thus, in addition to removing useless terms in equations (see the previous sections), *entire* equations can be detected symbolically as being superfluous: they can be removed before printing the results.

5.5.2 Recursive symbolic computation

To solve this problem systematically (i.e. such that a recursive scheme of any nature is purged of all superfluous equations), the following procedure has been adopted.
First of all, to clarify the following explanations, we note that a *recursive*

scheme denotes any formalism (kinematic, dynamical, using classical or barycentric parameters ...) written as one, or more algorithmic loops covering the N^{body} bodies. For instance, the Newton/Euler formalism of section 3.3 represents a recursive scheme consisting of two algorithmic loops: one for the forward kinematics ($for\ i = 1 : N^{body}$), the second for the backward dynamics ($for\ i = N^{body} : 1$). The so-called Order-N formalisms ([81] for instance) perform three recursions: forward kinematics, backward dynamics and forward kinetics.
Recall that the computer implementation of a given formalism deals with the matrix form of the vector equations evaluated in appropriate body-fixed frames. For instance, according to 3.46, the relation which expresses the angular velocity of a given body 3 with respect to its parent, body 2, is written in vector form

$$\boldsymbol{\omega}^3 = \boldsymbol{\omega}^2 + \boldsymbol{\varphi}^3 \dot{q}^3 \tag{5.13}$$

If joint 3 is revolute and aligned with the X direction of the fixed frames in both bodies 2 and 3, the matrix form of this equation reads in the $\{\hat{\mathbf{X}}^3\}$ frame

$$\begin{aligned} \text{OM13} &= \text{qd(3)+OM12} \\ \text{OM23} &= \text{OM22*C3+OM32*S3} \\ \text{OM33} &= -\text{OM22*S3+OM32*C3} \end{aligned} \tag{5.14}$$

where OMij denotes the ith component of the jth body angular velocity in frame $\{\hat{\mathbf{X}}^i\}$, qd(j) denotes $\dot{q}^j$ and Cj, Sj represent $\cos(q^j), \sin(q^j)$ respectively.

Returning to the purpose of this section, we note that each recursive equation of type 5.14, once symbolically evaluated, is stored in a dynamical list linked by C-pointers as indicated in figure 5.8. By "symbolically evaluated", we mean that the right hand side of each equation is a tree of sub-expressions linked by pointers in the memory. Moreover, the expression is already well-organized and fully simplified (see sections 5.2, 5.4) and all the superfluous temporarily generated sub-expressions have been removed from the memory (see section 5.3).

The process can be summarized as follows:

1. A recursive equation LHS=RHS [5] (ex.: OM33= −OM22*S3+OM32*C3) is symbolically evaluated. The LHS (OM33) becomes a new *auxiliary* variable (leaf expression) which is generated as soon as the RHS is evaluated.

2. The list is dynamically increased by one unit in which:

   - pointer *ptr_previous* links the unit to the previous one in the recursive algorithm
     (*ptr_previous* =NULL for the first element),

[5] LHS : Left Hand Side ; RHS : Right Hand Side

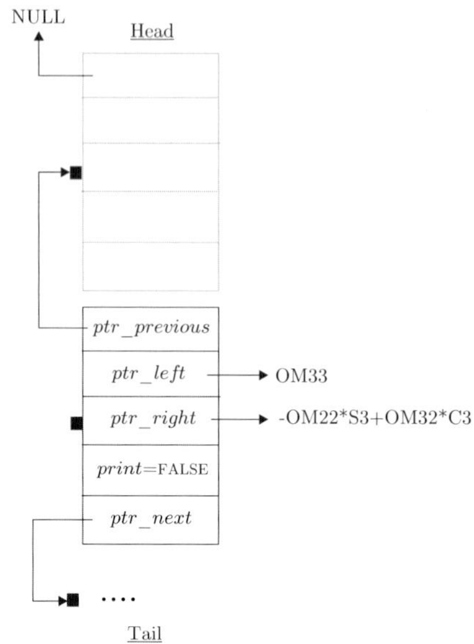

Figure 5.8: C linked-list of recursive equations

- pointer *ptr_next* will link the unit to the next one (*ptr_next* =NULL for the last element),
- pointer *ptr_left* points to the LHS leaf expression,
- pointer *ptr_right* points to the RHS expression,
- The boolean flag *print* is a priori set to FALSE because the corresponding expression do not a priori contribute to the expected results.

rem.: The creation of an auxiliary LHS variable and its storage in the list concern only equations whose RHS contains at least two operands. For instance, $a = b$, $a = -b$ or $a = \cos(b)$ will never lead to a new auxiliary variable a.

5.5.3 Elimination process

The linked list is complete as soon as all the final results have been symbolically evaluated (ex.: all the elements of the mass matrix $M(q)$ and of vector $c(q, \dot{q}, ...)$ for the Newton/Euler recursive scheme, ...)
Each result (ex.: $M_{(i,j)}$, $i = 1 : N^{body}$, $j = i : N^{body}$) having the same structure as the previous recursive equations (i.e. LHS = RHS) is also stored in the list, but of course with the *print* flag set to TRUE. Once they have been evaluated, the selection process can start by scanning the list from tail to head, and can be summarized as follows:

1. Start from the tail element of the list (i.e. the last symbolically evaluated equation (ex.: $M_{(1,1)}$),

2. Analyze the contents of the equation:

   - if the corresponding *print* flag is TRUE, then for each auxiliary variable encountered in the RHS of the equation, find and mark, by setting *print* to TRUE, the recursive equation in the list which symbolically evaluates this variable,

   - else, go directly to step 3, because the corresponding equation is superfluous,

3. Go to the previous element of the list whose address is given by the pointer *ptr_previous*:

   - if the address is the NULL pointer, the head has been reached and the process is finished: go to step 4,

   - else, return to step 2,

4. Print in the output file, by covering the list from head to tail, the equations whose *print* flag is TRUE.

Figure 5.9 summarizes this process using an academic example.

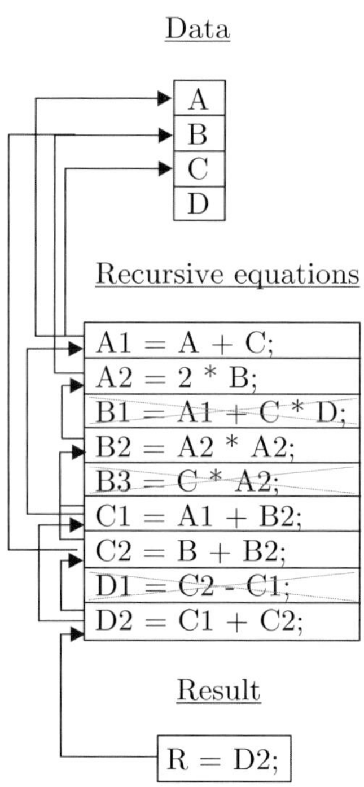

Figure 5.9: Recusive elimination of superfluous equations

This elimination process is systematically performed by ROBOTRAN for any recursive scheme it deals with (kinematics, dynamics, for rigid or flexible bodies,...). We note that more than 30% of the equations are superfluous and this percentage increases for formalisms containing flexible elements (see chapter 9).
In terms of symbolic computation performances, the multiple scanning of the list (required by step 2 in the process) can be quite time consuming for large systems because string comparison is performed thousands of times to find and mark interdependencies. The use of supplementary flags in the list elements, not detailed here, allows us to find and mark the useful equations in an immediate manner. This allows us to reduce the elimination process to a few seconds, even for 100 d.o.f. systems.

5.5.4 Scheme vectorization

Once the list of equations has been purged of all the superfluous ones, a very high level of condensation has been obtained for the generated formalism (each equation also being simplified by the internal symbolic library described in sections 5.2 and 5.4). Of course, some deep investigations of the recursive equations revealed that by *atomizing* them into binary elemental equations (i.e. $a = b + c$, $a = b - c$ or $a = b * c$), we still retain some twins, obviously in distinct equations (ex.: $a = b + c$ and, later on, $r = b + c$), in such a way that the symbolic library cannot detect and simplify them. However, this concerns very few equations and we decided not to track these possible simplifications: indeed, the atomizing process creates thousands of local auxiliary variables (one per binary equation) and the stack of the generated function could reach a size incompatible with standard compilers.

When no further arithmetical simplifications can be made by the symbolic library, we can apply another interesting process to the recursive scheme: the goal is not to reduce the number of arithmetical operations, but to re-organize the equations in a so-called *vector* form which could be evaluated in parallel by computer vector units.

Indeed, as shown in [48], [22] and [15], it is quite apparent that in any multibody recursive scheme, the result of a given recursive step (i.e. the LHS of the corresponding equation) does not necessarily depend on the *closest* previous recursive steps. We can thus imagine a sequence of evaluations in which, starting from the data, each recursive equation is evaluated *as soon as possible* ("asap" re-organization), i.e. as soon as all the terms of on its RHS are known. Such a process leads to a new recursive scheme which is made up of *vector* steps, each of these being composed of several equations resulting from the original serial scheme. For instance, in the academic example depicted in figure 5.10, the evaluations of C1 and C2 are independent of each other in the serial scheme (on the left). They can thus be performed simultaneously, leading to a vector instruction C1 = ...; C2 = ... (on the right).

This technique is implemented in ROBOTRAN so that the process is optimal in the sense that the number of vectorized steps is minimal for a given multibody

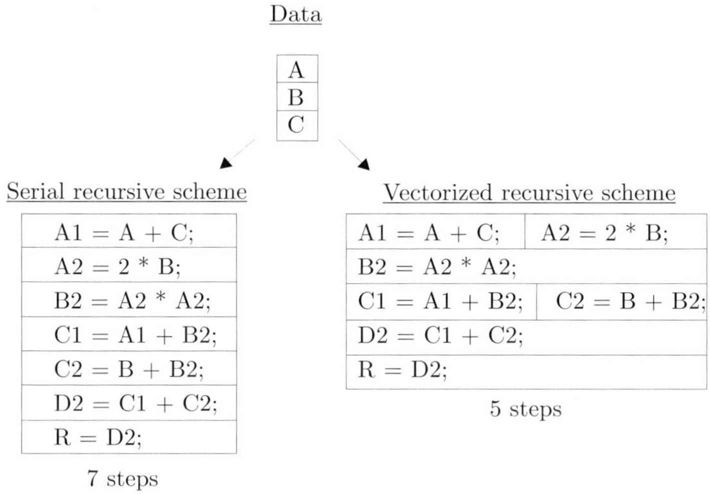

Figure 5.10: Vectorization of multibody equations

application.

The procedure is the following: for each equation of the original serial scheme (on the left in the example of figure 5.10), the program assigns to the LHS a so-called vector step index denoted vs, defined as the index of the step in the vector form (on the right) to which this equation will belong. This computation is based on the following formula:

$$vs(\text{LHS}) = 1 + \max\left\{vs(j^{th} \text{ term of the RHS})\right\} \tag{5.15}$$

For instance, for equation C2 = B + B2 in the serial scheme, the vs of its RHS terms B and B2 are:

- $vs(\text{B}) = 0$,
- $vs(\text{B2}) = 2$,
  being evaluated previously.

Thus, equation 5.15 gives $vs(\text{C2}) = 3$, which means that C2 can be part of the third vector step, i.e. when all terms of the RHS are available. In this way, we ensure that the number of vector steps is minimal. Nevertheless, at this stage, we are quite aware that the process ideally assumes that the available degree of parallelism has no limitation.

Consider the symbolic model of the articulated railway bogie (BA2000) studied in chapter 7, which features a 23 d.o.f. multibody structure. For this system, the Newton/Euler recursive scheme yields 1695 serial equations which are re-organized by the vector procedure into only 18 vector steps! This example illustrates that the reduction of the number of steps, from the serial to the vector scheme, is amazing. However, to avoid any confusion, recall that this technique does not reduce the number of operations, but only re-organizes the recursive equations into a vector form, suitable for parallel computation.

Although leading to a minimum number of steps, the *asap* organization procedure described above generates an ill-balanced scheme in terms of vector step width. Fortunately, as shown in [69], the minimal number of vector steps can also be obtained via an *as late as possible* ("alap") re-organization, starting from the results (tail of the list) rather than from the data. Therefore, between these two extreme solutions (*asap* and *alap* organizations of the equations), there are numerous possible organizations which lead to the – same – minimal number of steps. In fact, each equation can *slide* within the vector formulation between an *asap* and an *alap* step. Among all possible organizations, the one which leads to the minimum step width for the whole scheme should be selected (see [69]). This process could be exploited in the numerical phase, on vector computers or transputer network architectures, as in [48] for instance.

5.6 Recursive symbolic differentiation

5.6.1 Introduction

Although, as claimed in section 5.2, the arithmetical operators $+, -$ and $*$ are the only mathematical ingredients which are necessary to generate multibody models, various numerical processes require the derivatives of multibody (kinematic or dynamical) equations with respect to a given parameter or variable or a set of these. This is for instance the case for (see [17]):

- *Numerical integration* of stiff systems

We write the equations of motion to be time integrated in the implicit form 3.37,

$$\mathcal{Q} = \Phi(q, \dot{q}, \ddot{q}, F_{ext}, L_{ext}, g) \tag{5.16}$$

which corresponds to the pure inverse dynamics of the system. Implicit integrators (see chapter 10 for more details) require the so-called tangent matrices of the equations with respect to the variables q, $\dot{q}$ and $\ddot{q}$ to ensure the convergence of the iterative process. For the equations of motion in their implicit form 5.16, these tangent matrices are

$$M \triangleq \frac{\partial \Phi}{\partial \ddot{q}^T}\text{, the mass matrix} \quad ; \quad G \triangleq \frac{\partial \Phi}{\partial \dot{q}^T} \quad \text{and} \quad K \triangleq \frac{\partial \Phi}{\partial q^T} \tag{5.17}$$

- *Control* of multibody systems

In this case, it is often interesting to have the *linearized* state space model at one's disposal, namely:

$$\dot{x} = Ax + Bu \tag{5.18}$$

$$y = Cx + Du \tag{5.19}$$

where x denotes the state space vector, u and y are the control and output vectors respectively. Thus, the equations of motion 3.38 of unconstrained multibody systems, i.e.:

$$M(q)\,\ddot{q} + c(q, \dot{q}, F_{ext}, L_{ext}, g) = \mathcal{Q}(q, \dot{q})$$

must be linearized around a nominal motion $q^*, \dot{q}^*, \ddot{q}^*$, leading to

$$M\,\Delta\ddot{q} + G\,\Delta\dot{q} + K\;\Delta q = U\,u \tag{5.20}$$

where $\Delta\ddot{q} \stackrel{\Delta}{=} \ddot{q} - \ddot{q}^*$, $\Delta\dot{q} \stackrel{\Delta}{=} \dot{q} - \dot{q}^*$, $\Delta q \stackrel{\Delta}{=} q - q^*$ and u represents the input vector.
Defining the state vector $x \stackrel{\Delta}{=} \begin{pmatrix} \Delta q \\ \Delta\dot{q} \end{pmatrix}$, we can write equations 5.18 as follows:

$$\dot{x} = \begin{pmatrix} O & E \\ -M^{-1}K & -M^{-1}G \end{pmatrix} x + \begin{pmatrix} 0 \\ M^{-1}U \end{pmatrix} u \tag{5.21}$$

in which the tangent matrices $G \stackrel{\Delta}{=} \frac{\partial(c-\mathcal{Q})}{\partial\dot{q}^T}$ and $K \stackrel{\Delta}{=} \frac{\partial(c-\mathcal{Q})}{\partial q^T}$ are required.

- *Optimization* of multibody systems

Without going into details, we note that most of the optimization methods, namely the deterministic ones, require the evaluation of the tangent matrices of the system with respect to the so-called *design parameters* p^i. For the dynamical equations 5.16 for instance, the matrix to be evaluated will have the form $\frac{\partial\Phi}{\partial p^T}$, and will largely participate in the evaluation of the Jacobian of the objective function.

- *Identification* of multibody systems [18]

In [18], systematic rules are established to find the minimal set of dynamical parameters δ of multibody systems, starting from the barycentric formulation of chapter 3. Once they have been identified, we write the inverse dynamical model 5.16 in the following form:

$$\mathcal{Q} = \Psi(q, \dot{q}, \ddot{q}, ...)\,\delta \tag{5.22}$$

so that, if the joint generalized forces $\mathcal{Q}$ and the configuration $\{q, \dot{q}, \ddot{q}\}$ are measured during some exciting trajectories, the minimal set of dynamical parameters can be identified by linear regression techniques (see [71], [8]). It is thus necessary to have an explicit form of the tangent matrix $\Psi = \frac{\partial\Phi}{\partial\delta^T}$ at one's disposal.

For all these applications, the availability of the corresponding tangent matrices *in a purely symbolic* form is important for various reasons:

1. *computer performance*
   For numerical solvers and optimization methods, these tangent matrices are evaluated thousands of times, and provide the temporal bottleneck in these processes. Their symbolic form is all the more interesting since the CPU time reduction factor between a numerical and a symbolical generation is much higher for tangent matrices than for the equations of motion (for which the observed gain is already between 5 and 10 as mentioned in section 5.1).

2. *computer accuracy*
   Symbolic differentiation represents the generalization to large systems of a purely analytical differentiation. For various criteria (convergence, linearization, etc.), an exact computation of tangent matrices of nonlinear systems will be preferable to an approximate value resulting from a numerical method.

3. *computer portability*
   This advantage is non negligible since, for control and optimization purposes for instance, the corresponding processes are generally activated in a different software or even computer environment than that of the original multibody code. The portability from one environment to the other is greatly facilitated if the tangent matrices are available in symbolic form.

5.6.2 Recursive scheme differentiation

All these observations led us to develop a specific procedure to differentiate a given recursive scheme with respect to a given (set of) variable(s) or parameter(s). In the context of differentiation, the equations produced by recursive multibody formalisms can be considered as interwoven functions $(f(g(h(...(x)))))$. However, if the corresponding differentiation rules are applied blindly to a recursive scheme, they produce a very large non-optimized symbolic output, even for medium-sized multibody models: this will be illustrated later on. Therefore, we use the elimination procedure described in section 5.5.3 to solve this problem.

When evaluating a given recursive scheme, we assume that each equation depends, explicitly or not, on the set of system parameters or variables (ex.: $\dot{q}^1 ... \dot{q}^{N_{body}}$) with respect to which the differentiation must be performed. For instance in the following equation:

$$\text{AUXJ} = \text{AUXI} + 2 * \text{VAR} \tag{5.23}$$

AUXJ explicitly depends on the variable VAR via the second term. A priori, it may also depend implicitly on VAR via the first term AUXI.
We thus systematically create and evaluate a new recursive variable, for instance AUXJ_V, for the partial derivative of the current equation with respect to VAR

$$\text{AUXJ\_V} = \text{AUXI\_V} + 2$$

even if, in the end, it appears that this new auxiliary variable is 0 or simply useless. If it is useless, the symbolic library and the elimination process described in section 5.5.3 will detect it and remove the corresponding equation from the list, before printing. Such a technique gives rise to a *compact recursive computation* of the required derivatives.

To illustrate this, let us consider in 5.24 and 5.25, the symbolic evaluation of one element ($J_{(3,1)}$) of the Jacobian matrix of a position vector $p(q)$ associated with a kinematic chain composed of 9 joints.
In 5.24, the classical differentiation rule applied to $p(q)$, i.e. the resulting equation ($J_{(3,1)} = ...$), is far more consuming in terms of operators (220 $\{*, +, -\}$) than when obtained via a recursive differentiation as in 5.25 (62 $\{*, +, -\}$).

```
J(3,1) = q8*(C1*C7*(-C3*S4*C5+S3*S5)-S1*(S2*(S3*S4*(-C5*C7+S5*C6*S7)
         +S7*(-C3*C5*C6+S3*C4*S6))+C7*(C2*C4*C5-S2*C3*S5))
         +S7*(C1*(C3*C4*S6+C6*(C3*S4*S5+S3*C5))-S1*C2*(-C4*S5*C6+S4*S6)))
         +D13*S1*S2+D14*(C1*S3+S1*S2*C3)+D15*(C1*S3+S1*S2*C3)
         +D16*(C1*(C3*S4*S5+S3*C5)+S1*(C2*C4*S5-S2*(-C3*C5+S3*S4*S5)))
         +D17*(C1*C6*(C3*S4*S5+S3*C5)+S1*(-S2*S3*(C4*S6+S4*S5*C6)
         +C6*(C2*C4*S5+S2* C3*C5))+S6*(C1*C3*C4-S1*C2*S4))
         +D18*(C1*S7*(C3*S4*C5-S3*S5)-S1*(S2*(S3*S4*(C5*S7+S5*C6*C7)
         +C7*(-C3*C5*C6+S3*C4*S6))+S7*(-C2*C4*C5+S2*C3*S5))
         +C7*(C1*(C3*C4*S6+C6*(C3*S4*S5+S3*C5))-S1*C2*(-C4*S5*C6+S4*S6)))
         +D19*(C1*S7*(C3*S4*C5-S3*S5)-S1*(S2*(S3*S4*(C5*S7+S5*C6*C7)
         +C7*(-C3*C5*C6+S3*C4*S6))+S7*(-C2*C4*C5+S2*C3*S5))
         +C7*(C1*(C3*C4*S6+C6*(C3*S4*S5+S3*C5))
         -S1*C2*(-C4*S5*C6+ S4*S6)));
```
(5.24)

```
RO22 = S1*S2; RO32 = -C1*S2; RO82 = -S1*C2; RO92 = C1*C2;
RO23 = RO22*C3+C1*S3; RO33 = RO32*C3+S1*S3; RO53 = -RO22*S3+C1*C3;
RO63 = -RO32*S3+S1*C3; RO54 = RO53*C4+RO82*S4; RO64 = RO63*C4+RO92*S4;
RO84 = -RO53*S4+RO82*C4; RO94 = -RO63*S4+RO92*C4; RO25 = RO23*C5-RO84*S5;
RO35 = RO33*C5-RO94*S5; RO85 = RO23*S5+RO84*C5; RO26 = RO25*C6+RO54*S6;
RO36 = RO35*C6+RO64*S6; RO27 = RO26*C7-RO85*S7; RO87 = RO26*S7+RO85*C7;
RL23 =RO22*D13 ; RL24 = RO23*D14; JT34_1 = RL23+RL24;
RL25 = RO23*D15 ; JT35_1 = JT34_1+RL25; RL26 = RO25*D16;
JT36_1 = JT35_1+RL26; RL27 = RO26*D17; RL28 = RO27*D18+RO87*q(8);
JT37_1 = JT36_1+RL27; JT38_1 = JT37_1+RL28; RL29 = RO27*D19;
J(3,1) = JT38_1+RL29;
```
(5.25)

For larger models, and when all the elements of a tangent matrix are needed, the advantage of the recursive differentiation is quite significant. In fact, the *explosive* increase in the size of the classical differentiation technique is quite understandable since it amounts to destroying the recursivity of the original scheme, leading to an *in extenso* formulation. Although the differentiation process is very consuming in terms of both memory storage and symbolic CPU time[6], these drawbacks are negligible in ROBOTRAN since the storage requirement is drastically controlled during the process, as explained in section 5.3.

5.7 Performance evaluation

5.7.1 Introduction

For the numerical simulation of multibody systems, a high speed code is essential since, despite the sustained increase in modern computer performances, models are becoming more and more complex (or refined) and the numerical analyses are also pushed further (e.g. asked to perform control and optimization rather than pure time simulation).
This section aims to compare the performances of various multibody formalisms. The present comparison deals with the number of arithmetical operations ($+, *, -, /$) required to obtain the accelerations $\ddot{q}$ of multibody systems *once*. This computation represents the *common denominator* of most of the numerical processes and is thus a good indicator of relative performance.

Only formalisms for tree-like systems will be compared here. These formalisms are systematically used for closed loop systems, since such systems are first reduced to trees (see section 4.1). Secondly, comparing formalisms for closed loop systems can be quite inconsistent since the comparison depends strongly on the way the constraints are solved (ex.: Newton/Raphson method, LU or QR decomposition of the Jacobian matrix, direct integration, etc.) and the way the dynamical equations are solved (reduction technique, DAE solvers, ...).

5.7.2 Performance comparison

Whereas the Order-N formalism [81] explicitly computes the accelerations $\ddot{q}$, the formalisms developed in chapter 3 lead to a final form 3.38 which is linear in terms of these accelerations $\ddot{q}$:

$$M(q)\,\ddot{q} + c(q,\,\dot{q}, F_{ext}, L_{ext}, g) = \mathcal{Q} \tag{5.26}$$

To compute the accelerations rigorously, we must[7] (see [87] for instance):

- perform a Cholesly decomposition of the (symmetric, positive definite) mass matrix ($M = G\,G^T$),
- solve the linear system $(G\,G^T)\,\ddot{q} = (\mathcal{Q} - c)$ by a forward elimination followed by a backward substitution.

[6] because thousands of "potential" partial derivatives are computed.
[7] Mass matrix inversion is not really advisable for both efficency and accuracy reasons.

For the present comparison, these two steps have been performed numerically and the associated arithmetical operations have been numbered and added to those included in the symbolic formulation of system 3.38: this makes the comparison with the Order N formalism fair.

Rather than dealing with specific applications like vehicles, robots etc. to confront the formalisms, we base the comparison on the following two classes of systems which represent, for a relative coordinates approach, *extreme* cases in terms of structure topology:

1. the *linear tree* system illustrated in figure 5.11_a, which appears in robot manipulator applications,
2. the *binary tree* system (figure 5.11_b) in which each body has two children.

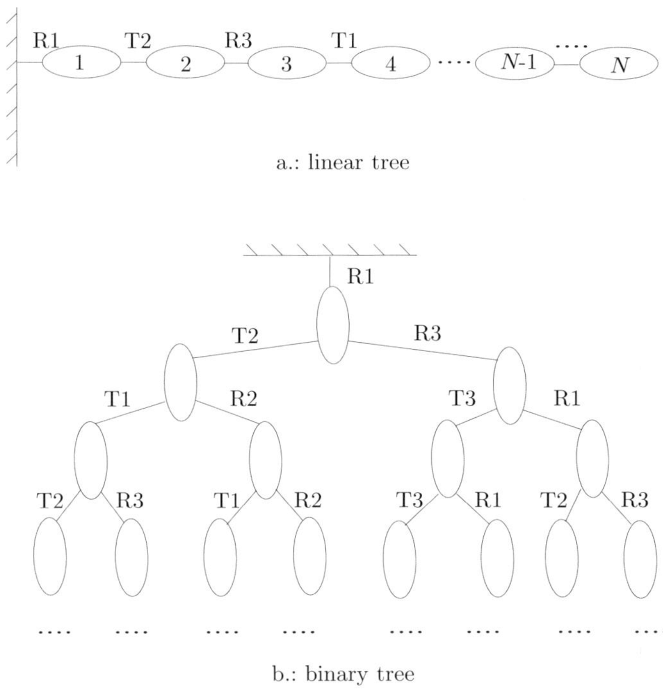

Figure 5.11: Analysed topologies

Figures 5.12 and 5.13 compare the number of operations $(+, -, *, /)$ required by the following formalisms for these two classes:

- PPP: virtual power principle *in extenso* form (section 3.2),
- NER: Newton/Euler recursive scheme with classical parameters (section 3.3),

- BAR: Newton/Euler recursive scheme with minimal set of barycentric parameters (section 3.4),
- ODN: Order-N formalism [81], [12].

Since it is not the main goal of the present comparison to point out the advantages of the symbolic approach, all the components of the vectors $\mathbf{d}^{ij}, \mathbf{d}^{ii}, \mathbf{F}^{i}_{ext}$, $\mathbf{L}^{i}_{ext}$ and tensors $\mathbf{I}^{i}$ have been represented by *non-zero* symbols in the data file. In fact, in practical applications, most of them are equal to zero and contribute to an additional efficiency of the symbolic approach. As already mentioned in the introduction, comparison tests [12] showed that, all other things being equal, a symbolic multibody model performs a time simulation between five and ten times faster than a purely numerical one.

5.7.3 Discussion

The *explosive* growth of the number of operations in the PPP formalism is a natural consequence of its *in extenso* formulation for obtaining the equations of motion (see section 3.2). Indeed, triple Σ summation symbols[8], covering the kinematic chains, appear in the final form of the equations of motion (see equations 3.34 - 3.35) and this makes the formalism[9] inappropriate for medium and large sized systems. The effect is less significant for the binary tree example (figure 5.13) since, for the same number of bodies, the length of the kinematic chains is shorter than for a linear tree.
Therefore, while leading to the most compact generation for very small systems (up to 4 d.o.f.), the advantage of this formalism is clearly not related to its computer efficiency but to the *legibility* of the symbolic equations. Indeed, being generated without any auxiliary variables, they are fully expressed in terms of the user symbolic data, ordered by ROBOTRAN in an elegant way as explained in section 5.2 (see also equation 5.7). Equation legibility is an attractive feature. Indeed, it allows us to visualize the impact of a given parameter (ex.: a force component) on a given equation, just as if the latter had been written analytically by hand. This feature is exploited in both educational (student projects) and scientific domains [71].

Turning now to the recursive formulations which clearly aim at minimizing the computational cost of multibody models, we can first observe and compare the two Newton/Euler recursive formalisms (NER and BAR) and conclude that:

- The arithmetical cost of both of these grows according to $O(N^2) + O(N^3)$ where $N = N^{dof}$, the number of d.o.f. of the system. This results from the mass matrix computation (Order N^2) and its Cholesky decomposition (Order N^3).

[8] against four if one uses classical rather than barycentric parameters.
[9] in this form

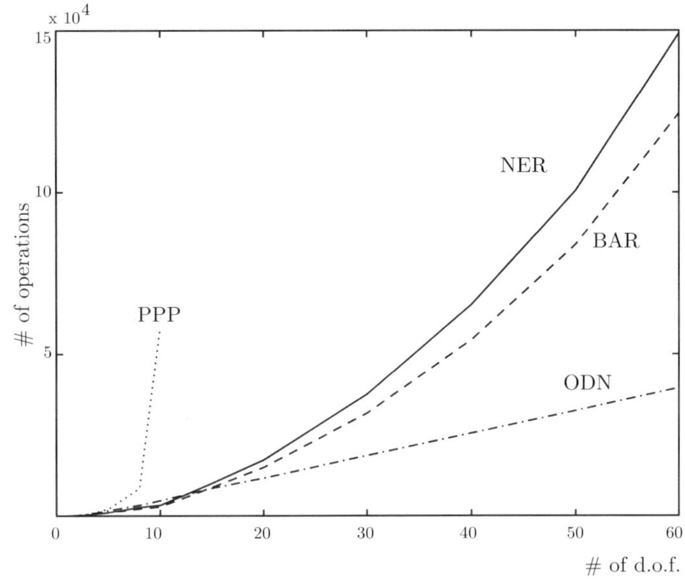

Figure 5.12: Linear tree

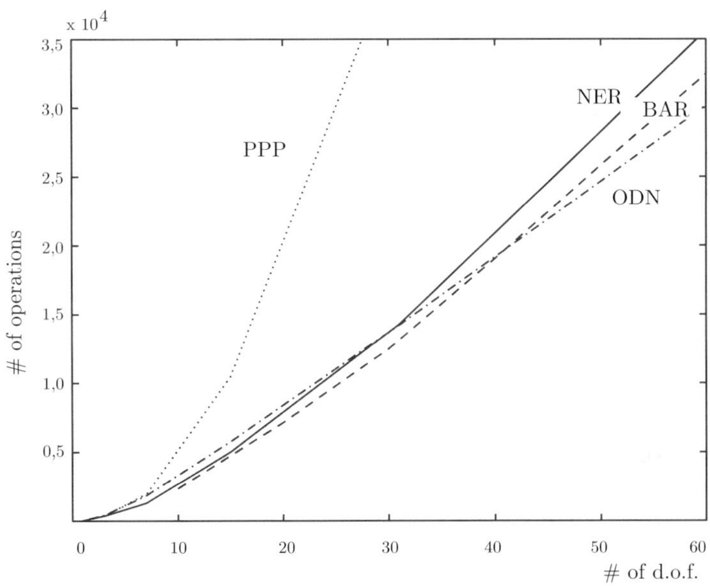

Figure 5.13: Binary tree

- Although the number of terms in the vector form is slightly larger for the formalism using the barycentric parameters (equations 3.110-3.115), the latter is slightly less costly than the classical formulation (NER): this is due to the fact that we used the combination rules of the barycentric parameters detailed in [18].
  On the other hand, classical parameters are more user-friendly for most people.

If we now compare the Newton/Euler schemes performance with that of the Order-N formalism (whose linearity is clearly shown in figures 5.12 and 5.13), we see that the latter only becomes the favorite for systems containing more than 13 d.o.f. ...(for a linear tree) or ... 30 d.o.f. (for a binary tree). This explains our preference for the Newton/Euler recursive schemes (NER, BAR). Two additional reasons reinforce this choice with respect to Order N techniques:

1. From a theoretical point of view, the introduction of the constraints for closed loop systems, using the Lagrange multipliers technique (see sections 2.4 and 4.1), is straightforward in the Newton/Euler approach since the form of the resulting equations (equations 3.118) corresponds exactly to those obtained by a virtual approach[10]. The Order N method, in which accelerations $\ddot{q}$ of the restored tree system are computed explicitly, is far more difficult to generalize to closed loop systems [21], when a reduced ODE system is desired.

2. From a practical point of view, experiments show that large multibody applications are generally made up of assemblies of distinct mechanical entities for which the sub-system segmentation technique (see section 4.4) is quite appropriate. This is the case for:

   - a railway vehicle consisting of a carbody structure and two bogies,
   - a car composed of four suspensions and a chassis (or 4 suspensions hanging on a $\frac{1}{4}$ chassis),
   - an earth-mover in which a rolling structure carries an articulated manipulator,
   - etc.

   Since the number of arithmetical operations is then related to the number of d.o.f. of the sub-systems and not of the whole structure, the Newton/Euler recursive approaches are generally the most efficient, according to the quantitative comparison of figures 5.12 and 5.13.

[10] which is needed to justify the Lagrange multipliers technique (see chapter 2)

5.8 Computer implementation

5.8.1 Joint modeling hypothesis

As stated in section 3.1.3, we have developed multibody formalisms using only 1 d.o.f. joints (prismatic or revolute) as depicted in figure 5.14. We recall that this modeling option does not represent a restriction for multi-d.o.f. joints encountered in applications: indeed, any of these can be modeled by properly combining and/or constraining revolute and prismatic joints which connect fictitious bodies[11] (see chapter 4).

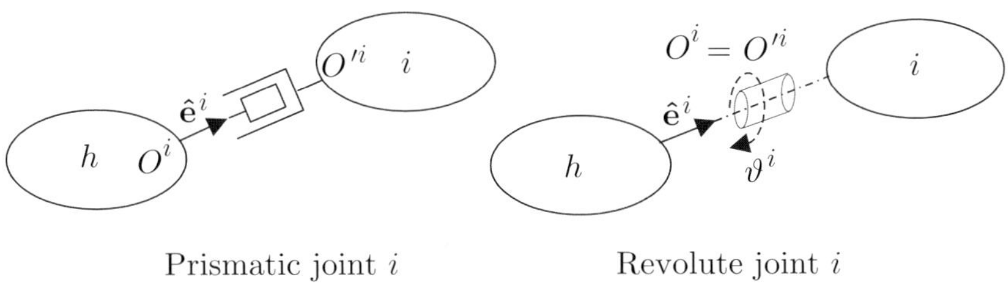

Figure 5.14: Elementary joint

The restriction to two kinds of joints was of course a practical option for the ROBOTRAN implementation. It is also attractive for the user. Indeed, experiments revealed that, in practice, combining elementary joints to match the internal degrees of freedom of a given physical device appears to be more user-friendly than manipulating an *exhaustive* library containing all the possible combinations.

An additional modeling hypothesis related to the joint definition has been adopted in ROBOTRAN and, once again, aims at simplifying the set of required data for a given application[12]. It does not relate to the nature of the joints, but to the direction of their axis (i.e. the direction of the $\hat{\mathbf{e}}^i$ vector represented in figure 5.14).
To avoid the proliferation of frames and rotation matrices in the model, the hypothesis states that the axis of each prismatic or revolute joint i, connecting body $h = inbody(i)$ to its child body i, must be aligned with one of the unit vectors $\hat{\mathbf{X}}_1$, $\hat{\mathbf{X}}_2$ or $\hat{\mathbf{X}}_3$ of body frames $\{\hat{\mathbf{X}}^h\}$, $\{\hat{\mathbf{X}}^i\}$. In other words, for unit vector $\hat{\mathbf{e}}^i$ (see figure 5.14) aligned with the joint i axis (pointing from the parent h to its child i), this vector must be such that:

$$\hat{\mathbf{e}}^i = \hat{\mathbf{X}}_1^h = \hat{\mathbf{X}}_1^i \quad \text{or} \quad \hat{\mathbf{e}}^i = \hat{\mathbf{X}}_2^h = \hat{\mathbf{X}}_2^i \quad \text{or} \quad \hat{\mathbf{e}}^i = \hat{\mathbf{X}}_3^h = \hat{\mathbf{X}}_3^i, \quad \forall i \tag{5.27}$$

Once these hypotheses have been established, we can introduce the notion

[11] fictitious means dimensionless and massless

[12] The present joint modeling hypothesis is introduced at this level of the book since it results only from the ROBOTRAN implementation, and not from any theoretical developments of the previous chapters.

of *reference configuration* of the tree-like multibody system (see figure 5.15), where:

1. all the body-fixed frames $\{\hat{\mathbf{X}}^i\}$ are aligned, in particular with the inertial frame $\{\hat{\mathbf{I}}\}$,

2. $O'^i = O^i$ for each prismatic joint i.

Hence, in the reference configuration: $q^i = 0\,,\ \forall\, i$.

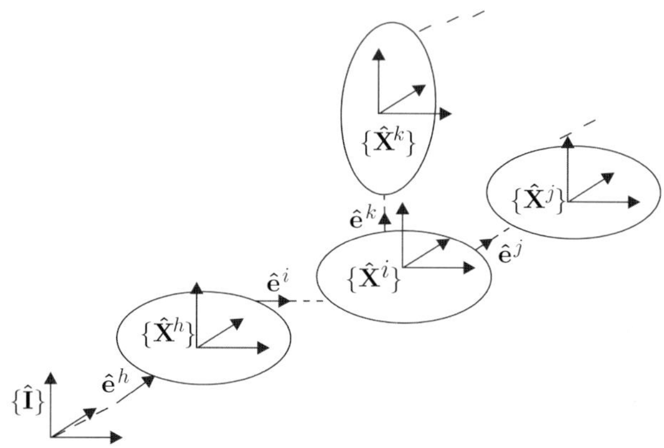

Figure 5.15: Reference configuration

This modeling hypothesis could appear restrictive insofar as, for practical applications, a given body can contain several joints whose relative directions are neither parallel nor perpendicular.

Fortunately, most applications can cope with this hypothesis without resorting to any trick regarding the orientation of the body frames. On the contrary, this hypothesis simplifies the set of data since only one frame per body needs to be defined[13].

However, in some specific cases[14], a so-called "skewed" joint must be introduced. Figure 5.16 illustrates how to take this into account in ROBOTRAN. Consider in figure 5.16_{left} a revolute joint i (angular variable q^i) whose axis makes a constant angle α (around the x direction), with the third axis $\hat{\mathbf{X}}_3^h$ of the body h fixed frame. Inserting a fictitious body and a revolute joint ($q^i = \alpha$) allows us to solve the problem (figure 5.16_{right}):

- the reference configuration is obtained for $q^i = 0$, $\forall\, i$

[13] in case of rigid bodies

[14] e.g., the inclined axis of a bicycle steering axle.

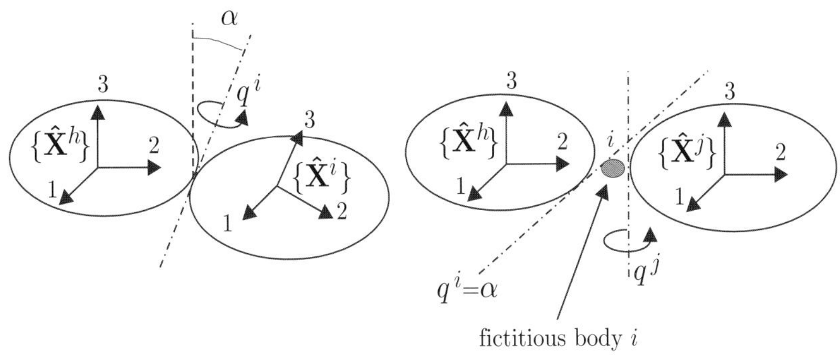

Figure 5.16: Modelling of a skewed joint

- the presence of this constant angle α in the real system will be taken into account during simulation by locking (i.e. constraining) this joint, i.e. $q^i = \alpha$, $\forall t$ (see figure 5.16_{right}).

By adding the variable q^i, we artificially increase the number of generalized coordinates as well as the number of equations of motion, by one. However, the relation $q^i = \alpha$ is clearly a kinematic constraint involving a Lagrange multiplier λ which appears in the equations of motion, multiplied by the Jacobian matrix transpose (see section 4.2.2, equation 4.21). Since only the ith joint equation of motion depends on this Lagrange multiplier λ, this equation can be disregarded in any numerical simulation[15].

5.8.2 Program overview

This section gives a short overview of the ROBOTRAN (for "ROBOt TRANslator") symbolic program possibilities[16].

ROBOTRAN is a stand-alone C program entirely dedicated to the *modeling of multibody systems.* Its purpose is to symbolically generate various inverse and direct formalisms (including those detailed in chapters 3, 4, 6 and 9) in *in extenso* or *recursive* forms. ROBOTRAN constitutes the computer version of the theoretical formalisms developed in this book.

The symbolic approach gives ROBOTRAN a "toolbox" rather than a "black-box" nature. Indeed, for any given application, the user builds, with ROBOTRAN, his or her own model by using the symbolic results to solve the specificities of the application at hand, e.g., a typical constraint, a particular joint, a specific force, etc.

ROBOTRAN is restricted to the generation of symbolic equations (ex. inverse or direct dynamics, direct kinematics, identification matrix, ...) related to a given multibody system. These equations which represent the *multibody model*

[15] However, this equation can be usefull to compute λ, which is the constrained generalized joint force/torque.

[16] The reader will find a user's guide and a demo version on the ROBOTRAN Web page.

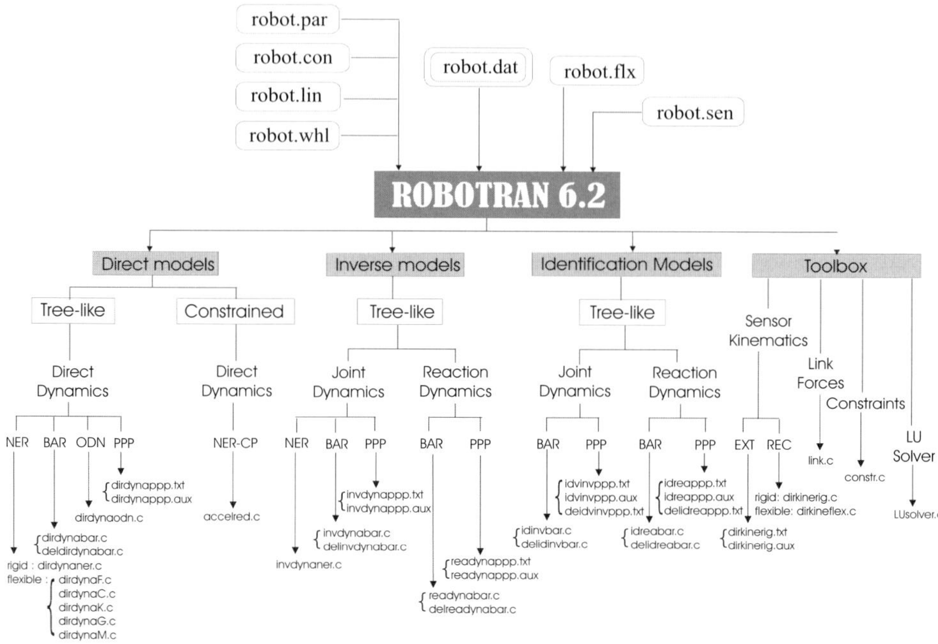

Figure 5.17: ROBOTRAN: main menu

are recorded in an output file with a specified syntax (Fortran, MATLAB or C). To exploit these symbolic equations, they must obviously be introduced into a numerical program for the *analysis* phase, and this is the user's responsibility. In other words, ROBOTRAN can be seen as a "multibody model writer" which ensures for the numerical phase:

- a high efficiency (because the symbolic ROBOTRAN process does the utmost to simplify the generated equations),
- a high portability (because the symbolic equations are generated with the desired syntax and the suitable interface).

Figure 5.17 depicts the main menu of ROBOTRAN (release 6.2) illustrated by a hypothetical application named "robot" (which can represent a vehicle, a mechanism, a manipulator, ...). The ROBOTRAN data files[17] contain *symbolic* parameters to respectively describe:

- joints and bodies: in *robot.dat*
- constraints (from loops or user defined, see sections 4.1 and 4.2): in *robot.con*

[17] can be generated via specific GUI (Graphical User Interface)

- point-to-point links (see section 4.3): in *robot.lin*
- wheels (car applications, see chapter 6): in *robot.whl*
- flexible beams (see chapter 9): in *robot.flx*
- sub-chain kinematics (see section 4.5): in *robot.sen*
- coordinate partitioning (see section 4.6.1), i.e., indices of the independent coordinates u and kinematically driven variables: $q = f(t)$: in *robot.par*.

In generating the symbolic equations for his/her project *robot*, the user is successively invited to choose:

1. The desired *model*: direct models, inverse models, identification models, toolbox models (sub-chain kinematics, constraints, link forces, ...),
2. The system *topology*: open-loop system, constrained system,
3. The desired *formalism:* Newton/Euler recursive scheme, barycentric approach, order-N method, virtual power principle,
4. The desired *output language*: FORTRAN, C, MATLAB, JAVA, ... (C language has been arbitrarily chosen in figure 5.17, in view of the output file extension[18]).

5.8.3 Description of the symbolic models

This section shortly describes each possible run of ROBOTRAN by following, from left to right, the diagram of figure 5.17.

Direct models

The objective is to compute the generalized accelerations of the MBS subjected to external or internal forces. The possible formalisms are:

- NER: recursive Newton/Euler formalism with classical parameters, for tree-like systems composed of rigid bodies (equation 3.43, section 3.3) or containing flexible beams (equation 9.94, chapter 9)
- BAR: recursive Newton/Euler formalism with minimal barycentric parameters, for tree-like systems (equation 3.118, section 3.4)
- ODN: Order-N formalism with classical parameters, for tree-like systems (from [81], [21])
- PPP: virtual power principle with barycentric parameters (equation 3.38, section 3.2)

[18]".c" stands for a C function; ".txt" and ".aux" relate to *in extenso* formalisms which are generated by ROBOTRAN in purely text format.

- NER-CP: recursive Newton/Euler formalism with classical parameters and coordinate partitioning method for closed-loop systems (equations 4.44 to 4.49, section 4.6.1)

Inverse models

The objective is to compute the generalized joint forces and torques ("joint dynamics") or the reaction forces and torques[19] ("reaction dynamics") of the MBS for a given trajectory.
Joint dynamics is based on the following formalisms:

- NER: recursive Newton/Euler formalism with classical parameters, for tree-like systems (equation 3.42, [52])
- BAR: recursive Newton/Euler formalism with minimal barycentric parameters, for tree-like systems (equation 3.42 using the barycentric formulation of section 3.4.2)
- PPP: virtual power principle with barycentric parameters (equation 3.37, section 3.2)

Reaction dynamics is obtained from the inverse model too, but considers only the forces and torques evaluated at an "imaginary" joint 0, i.e., located at the bedplate of the system (generally a robot). Formalisms BAR and PPP are implemented.

Identification models

Inverse models (joint and reaction dynamics) are linear with respect to barycentric parameters from which a minimal set δ can be found (see [18]). For dynamical identification purposes, it is necessary to extract the associated tangent matrix $\Psi = \frac{\partial \Phi}{\partial \delta^T}$. To obtain this matrix in ROBOTRAN, the formalisms to which the symbolic differentiation is applied are:

- BAR: recursive Newton/Euler formalism with minimal barycentric parameters, for tree-like systems (equation 3.42, section 3.4)
- PPP: virtual power principle with barycentric parameters (equation 3.37, section 3.2)

Toolboxes

Beside all those well-established formalisms, some interesting symbolic computations can be performed by ROBOTRAN to help the user in building his multibody model. This is the job of the "Toolbox" module which contains:

[19] mainly dedicated to robotics, reaction forces (and torques) are those evaluated at the bedplate of the robot, as detailed in [71].

- Sensor kinematics: symbolic computation of the absolute direct kinematics of a given body or point of the system (denoted as "sensors"): absolute position, orientation, velocities, accelerations and Jacobian matrix (section 4.5)

- Link forces: this relates to the symbolic computation of the point-to-point links (section 4.3)

- Constraints: symbolic computation of the loop and user constraints at position, velocity and acceleration levels (equations 4.1_b , 4.2, 4.3, section 4.1)

- LU solver: symbolic LU factorization and solution of a linear system $A\,x = b$.

5.9 A short example: the four-bar mechanism

To illustrate the symbolic inputs/outputs of a ROBOTRAN application, consider a simple planar four bar mechanism (figure 5.18_a) in which a link force $\mathbf{F}_{spring}$ is acting and there is a kinematic point sensor S.

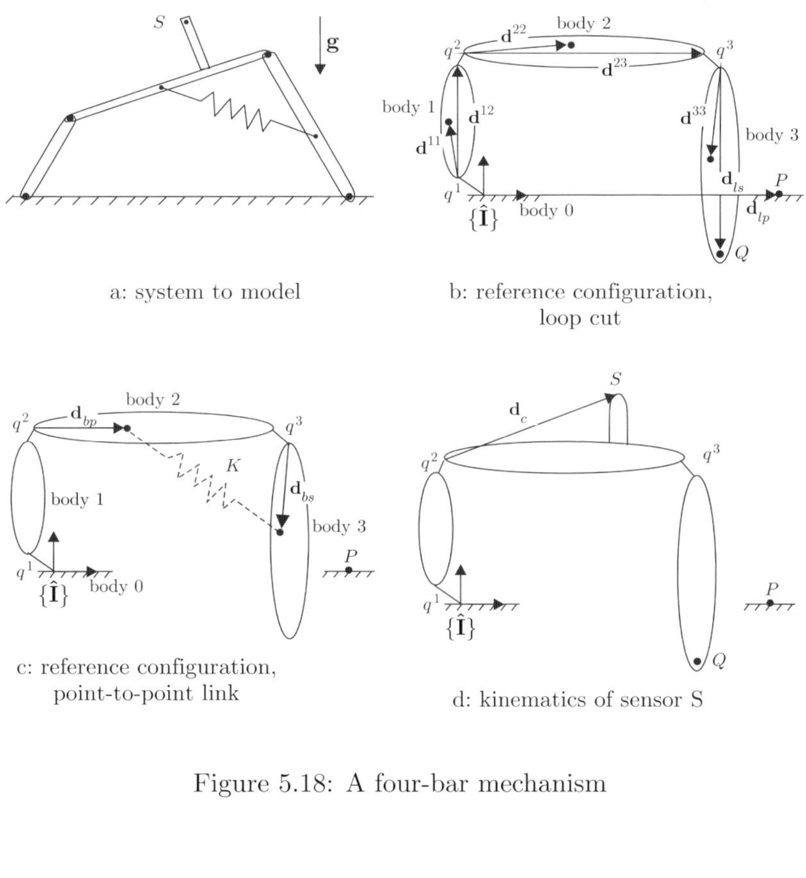

Figure 5.18: A four-bar mechanism

We first create a tree-like structure. Since we assume that the revolute joint connecting body 3 to the ground is ideal, it is cut as described in section 4.1.2. Figure 5.18$_b$ shows the *reference configuration* of the open(ed) structure: the positions of the two lips of the cut, points P and Q, are defined by the position vectors $\mathbf{d}_{lp}$ (on body 0) and $\mathbf{d}_{ls}$ (on body 3). The connection points of the spring are represented in figure 5.18$_c$ and located via vectors $\mathbf{d}_{bp}$ (on body 2) and $\mathbf{d}_{bs}$ (on body 3). Finally, the sensor S is located on body 2 by vector $\mathbf{d}_c$ (figure 5.18$_d$). All these position vectors have constant components in the frame attached to the associated body.

5.9.1 Symbolic input files

The symbolic input files related to this system are gathered in figure 5.19.

```
fbm.dat
3
0 1 2
--- Description of the joints ---
R3    q1    qd1    qdd1
R3    q2    qd2    qdd2
R3    q3    qd3    qdd3
--- Positions of the joints ---
0     0     0
0     D22   0
D13   0     0
--- Positions of mass centres ---
L11   L21   0
L12   L22   0
L13   L23   0
--- Masses ---
m1
m2
m3
--- Inertia matrices ---
0     0     0
      0     0
            I331
0     0     0
      0     0
            I332
0     0     0
      0     0
            I333
--- External forces resultant ---
0     0     0
frc12 frc22 0
frc13 frc23 0
--- External torques resultant ---
0     0     cp132
0     0     cp133
0     0     0
--- Gravity vector ---
0     0     g
```

```
fbm.con
0                             % Type 1
------------------------------
1                             % Type 2
--- Loop nbr 1 ---
0      3                      % Bodies
DLPx   0      0               % dlp
0      DLSy   0               % dls
--------------------------------
0                             % Type 3
--------------------------------
```

```
fbm.lin
1                             % # of links
--- Link nbr 1 ---
2      3                      % Bodies
DBPx   0      0               % dbp vector
0      DBSy   0               % dbs vector
K                             % Spring
```

```
fbm.sen
1                             % # of sensors
--- Sensor nbr 1 ---
2      0                      % Bodies
DCx    DCy    0               % dc vector
P      V      A               % Pos/Vel/Acc
```

Figure 5.19: Symbolic input files

The main symbolic data file, *fbm.dat,* first indicates the number of bodies and the *inbody* vector. The following sections successively describe:

- the joints (R for revolute, P for prismatic; 1, 2 or 3 for a x, y, z, axis respectively and $q(i), qd(i), qdd(i)$ for the generalized coordinates $q^i, \dot{q}^i, \ddot{q}^i$),

- the components of the joint position vectors $\mathbf{d}^{ij}$ (expressed in the $\{\hat{\mathbf{X}}^i\}$ frame),
- the components of the center of mass position vectors $\mathbf{d}^{ii}$ (expressed in the $\{\hat{\mathbf{X}}^i\}$ frame),
- the masses m^i and the components of the inertia tensors $\mathbf{I}^i$ (expressed in the $\{\hat{\mathbf{X}}^i\}$ frame),
- the components in body i-fixed frames of the external forces and torques $\mathbf{F}^i_{ext}, \mathbf{L}^i_{ext}$ and finally of the gravity vector $\mathbf{g}$, expressed in the inertial frame$\{\hat{\mathbf{I}}\}$.

For this four-bar mechanism, more than 50% of the components are zero; this will significantly reduce the number of equations.

The next data file, *fbm.con,* is related to the loop constraints, and indicates the number of loops for each type of cut (type 1 = cut of a body, type 2 = cut of a ball joint, type 3 = removal of a connecting rod) and, for each cut, the two adjacent bodies and the component of the closing vectors $\mathbf{d}_{lp}, \mathbf{d}_{ls}$.

The third data file, *fbm.lin,* deals with the point-to-point link forces and contains the number of links, and for each of them, the two connected bodies, the position vector $\mathbf{d}_{bp}, \mathbf{d}_{bs}$ of the anchor points and the nature of the constitutive equation (a linear spring of stiffness K in the example).

Finally, data file *fbm.sen* contains the number, the location (carrying body and position vector $\mathbf{d}_c$) of the sensors as well as the desired computation (position and/or velocity and/or acceleration for the point sensor S).

5.9.2 Symbolic output files

For illustration purposes only, the following three symbolic files, generated in MATLAB syntax [56], are shown in the next figures.

In figure 5.20, the MATLAB file *dirdynaner.m* contains the direct dynamics computed by the Newton/Euler recursive scheme (NER). Input arguments gather the variables $q^i, \dot{q}^i$ and the system parameters $\mathbf{d}^{ij}$, $\mathbf{d}^{ii}$, m^i, $\mathbf{I}^i$, $\mathbf{F}^i_{ext}$, $\mathbf{L}^i_{ext}$ and $\mathbf{g}$. Outputs are the mass matrix $M(q)$ and the dynamical vector $c(q, \dot{q}, ...)$ of equation 3.43. The *dirdynaner* function can be used as a *black box* for a MATLAB simulation program.

Figure 5.21 illustrates the contents of the *dirdynappp.txt* file, i.e. the same direct dynamics, now generated using the formulation based on the virtual power principle of section 3.2. Let us note that the barycentric parameters do not depend on the generalized coordinates q^i. They can thus be computed once, i.e. outside the integration scheme. We can also point out that this *in extenso* formulation does not contain any auxiliary variable and presents the equations in a very legible way, thanks to the lexicographical rules used to order the expressions.

Finally, in figure 5.22, the algebraic constraints at position level $h(q)$ and the Jacobian matrix $J(q)$ are computed recursively ("Toolbox" module) in the *constr.m* output file. Input $(q^i, \mathbf{d}^{ij})$ and output arguments (h, J) makes the *constr* function very portable towards the user numerical program, as it is the case for any file generated by ROBOTRAN.

```
function [M,c] = dirdynaner(q,qd,d,l,m,In,frc,trq,g)

% Trigonometric variables
S1 = sin(q(1));    C2 = cos(q(2));
C1 = cos(q(1));    S3 = sin(q(3));
S2 = sin(q(2));    C3 = cos(q(3));
% Forward Kinematics
BS51 = -qd(1)*qd(1);
AF11 = -g(2)*S1;
AF21 = -g(2)*C1;
OM32 = qd(1)+qd(2);
BS12 = -OM32*OM32;
AF12 = AF11*C2+S2*(AF21+BS51*d(2,2));
AF22 = -AF11*S2+C2*(AF21+BS51*d(2,2));
AM12_1 = -d(2,2)*C2;
AM22_1 = d(2,2)*S2;
OM33 = qd(3)+OM32;
BS53 = -OM33*OM33;
AF13 = AF22*S3+C3*(AF12+BS12*d(1,3));
AF23 = AF22*C3-S3*(AF12+BS12*d(1,3));
AM13_1 = AM12_1*C3+S3*(AM22_1+d(1,3));
AM23_1 = -AM12_1*S3+C3*(AM22_1+d(1,3));
AM13_2 = d(1,3)*S3;
AM23_2 = d(1,3)*C3;
% Backward Dynamics
FA13 = -frc(1,3)+AF13*m(3);
FA23 = -frc(2,3)+m(3)*(AF23+BS53*l(2,3));
CF33 = -trq(3,3)-FA13*l(2,3);
FB13_1 = m(3)*(AM13_1-l(2,3));
FB23_1 = AM23_1*m(3);
CM33_1 = In(9,3)-FB13_1*l(2,3);
FB13_2 = m(3)*(AM13_2-l(2,3));
FB23_2 = AM23_2*m(3);
CM33_2 = In(9,3)-FB13_2*l(2,3);
FB13_3 = -m(3)*l(2,3);
CM33_3 = In(9,3)-FB13_3*l(2,3);
FA12 = -frc(1,2)+m(2)*(AF12+BS12*l(1,2));
FA22 = -frc(2,2)+AF22*m(2);
FF12 = FA12+FA13*C3-FA23*S3;
FF22 = FA22+FA13*S3+FA23*C3;
CF32 = CF33-trq(3,2)+FA22*l(1,2)+d(1,3)*(FA13*S3+FA23*C3);
FB12_1 = AM12_1*m(2);
FB22_1 = m(2)*(AM22_1+l(1,2));
FM12_1 = FB12_1+FB13_1*C3-FB23_1*S3;
FM22_1 = FB22_1+FB13_1*S3+FB23_1*C3;
CM32_1 = CM33_1+In(9,2)+FB22_1*l(1,2)+d(1,3)*(FB13_1*S3+FB23_1*C3);
FB22_2 = m(2)*l(1,2);
CM32_2 = CM33_2+In(9,2)+FB22_2*l(1,2)+d(1,3)*(FB13_2*S3+FB23_2*C3);
FA11 = AF11*m(1);
CF31 = CF32-FA11*l(2,1)-d(2,2)*(FF12*C2-FF22*S2);
FB11_1 = -m(1)*l(2,1);
CM31_1 = CM32_1+In(9,1)-FB11_1*l(2,1)-d(2,2)*(FM12_1*C2-FM22_1*S2);
% Symbolic Outputs
c(1) = CF31;        c(2) = CF32;        c(3) = CF33;
M(1,1) = CM31_1; M(1,2) = CM32_1; M(1,3) = CM33_1;
M(2,2) = CM32_2; M(2,3) = CM33_2; M(3,3) = CM33_3;
```

Figure 5.20: ROBOTRAN: recursive direct dynamics (MATLAB syntax)

```
!
!    dirdynappp: Potential Power Principle - direct dynamics
!
!                          - M(i,j)  : symmetric mass matrix
!                          - c(j)    : dynamic terms
!
!           Trigonometric variables
!
S1 = sin(q1)
C1 = cos(q1)
S2 = sin(q2)
C2 = cos(q2)
q1p2 = q1+q2
S1p2 = sin(q1p2);
C1p2 = cos(q1p2);
S3 = sin(q3);
C3 = cos(q3)
q2p3 = q2+q3
S2p3 = sin(q2p3)
C2p3 = cos(q2p3)
q1p2p3 = q1+q2+q3
S1p2p3 = sin(q1p2p3)
C1p2p3 = cos(q1p2p3)
!
!           Barycentric parameters
!
mb1 = m1+m2+m3
mb2 = m2+m3
!
b11 = L11*m1
b12 = D22*mb2+L21*m1
b21 = D13*m3+L12*m2
b22 = L22*m2
b31 = L13*m3
b32 = L23*m3
!
K111 = D22*D22*mb2+L21*L21*m1
K112 = -L11*L21*m1
K122 = L11*L11*m1
K133 = I331+D22*D22*mb2+L11*L11*m1+L21*L21*m1
K211 = L22*L22*m2
K212 = -L12*L22*m2
K222 = D13*D13*m3+L12*L12*m2
K233 = I332+D13*D13*m3+L12*L12*m2+L22*L22*m2
K311 = L23*L23*m3
K312 = -L13*L23*m3
K322 = L13*L13*m3
K333 = I333+L13*L13*m3+L23*L23*m3
!
!      Dynamic equations
!
M(1,3) = K333+D13*(b31*C3-b32*S3)+D22*(b31*S2p3+b32*C2p3)
M(2,3) = K333+D13*(b31*C3-b32*S3)
M(3,3) = K333
c(3) = qd1*qd1*(D13*b31*S3+D13*b32*C3-D22*b31*C2p3+D22*b32*S2p3)
c(3) = c(3)+qd2*qd2*(D13*b31*S3+D13*b32*C3)
c(3) = c(3)+qd1*qd2*(2*D13*b31*S3+2*D13*b32*C3)
c(3) = c(3)-cp133-L13*frc23+L23*frc13
M(1,2) = M(1,3)+K233+D13*(b31*C3-b32*S3)+D22*(b21*S2+b22*C2)
M(2,2) = M(2,3)+K233+D13*(b31*C3-b32*S3)
c(2) = c(3)+qd1*qd1*(-D13*b31*S3-D13*b32*C3-D22*b21*C2+D22*b22*S2)
c(2) = c(2)+qd2*qd2*(-D13*b31*S3-D13*b32*C3)
c(2) = c(2)+qd1*qd2*(-2*D13*b31*S3-2*D13*b32*C3)
c(2) = c(2)+qd3*qd3*(-D13*b31*S3-D13*b32*C3)
c(2) = c(2)+qd1*qd3*(-2*D13*b31*S3-2*D13*b32*C3)
c(2) = c(2)+qd2*qd3*(-2*D13*b31*S3-2*D13*b32*C3)
c(2) = c(2)-cp132-D13*(frc13*S3+frc23*C3)-L12*frc22+L22*frc12
M(1,1) = M(1,2)+K133+D22*(b21*S2+b22*C2+b31*S2p3+b32*C2p3)
c(1) = c(2)+qd1*qd1*(D22*b21*C2-D22*b22*S2+D22*b31*C2p3-D22*b32*S2p3)
c(1) = c(1)+qd2*qd2*(D22*b21*C2-D22*b22*S2+D22*b31*C2p3-D22*b32*S2p3)
c(1) = c(1)+qd1*qd2*(2*D22*b21*C2-2*D22*b22*S2+2*D22*b31*C2p3-2*D22*b32*S2p3)
c(1) = c(1)+qd3*qd3*(D22*b31*C2p3-D22*b32*S2p3)
c(1) = c(1)+qd1*qd3*(2*D22*b31*C2p3-2*D22*b32*S2p3)
c(1) = c(1)+qd2*qd3*(2*D22*b31*C2p3-2*D22*b32*S2p3)
c(1) = c(1)+D22*(frc12*C2+frc13*C2p3-frc22*S2-frc23*S2p3)
!
```

Figure 5.21: ROBOTRAN: *in extenso* direct dynamics

```
function [h,J] = constr(q,D,DLP,DLS)
%
S1 = sin(q(1));                             C1 = cos(q(1));
S2 = sin(q(2));                             C2 = cos(q(2));
C1p2 = C1*C2-S1*S2;                   S1p2 = C1*S2+S1*C2;
S3 = sin(q(3));                             C3 = cos(q(3));
C3p1p2 = C3*C1p2-S3*S1p2;     S3p1p2 = C3*S1p2+S3*C1p2;
%
RL12 = -D(2,2)*S1;
RL22 = D(2,2)*C1;
RL13 = D(1,3)*C1p2;
RL23 = D(1,3)*S1p2;
PO13 = RL12+RL13;
PO23 = RL22+RL23;
JT13_1 = -RL22-RL23;
JT23_1 = RL12+RL13;
RL14 = -DLS(2,1)*S3p1p2;
RL24 = DLS(2,1)*C3p1p2;
PO14 = PO13+RL14;
PO24 = PO23+RL24;
JT14_1 = JT13_1-RL24;
JT24_1 = JT23_1+RL14;
JT14_2 = -RL23-RL24;
JT24_2 = RL13+RL14;
%
h(1) = DLP(1,1)-PO14;
h(2) = -PO24;
h(3) = 0.0;
%
J = zeros(3,3);
J(1,1) = -JT14_1;
J(2,1) = -JT24_1;
J(1,2) = -JT14_2;
J(2,2) = -JT24_2;
J(1,3) = RL24;
J(2,3) = -RL14;
%
```

Figure 5.22: ROBOTRAN: constraints and Jacobian matrix (MATLAB syntax)

Part II

Special topics

Chapter 6

Road vehicles: wheel/ground model

6.1 Introduction

Road vehicles represent an important class of multibody systems, all the more interesting since recent developments in this field rely on dynamical properties (stability, riding comfort, etc...). From the mechanical point of view, a vehicle is intrinsically of a multibody nature. Depending on the application, the model refinement can vary noticeably, while being rigorous with respect to the envisaged analysis. For instance, one could wish to check for vehicle stability, or to improve the under/over steering behavior in a curve: for such analyses, the model (including data) must be sufficiently refined as regards all the suspension elements, wheel/ground properties etc... yet can disregard, partially or even totally, the engine and transmission internal motions. Reciprocally, the latter would be at the heart of the model if vibration and noise were the main issues of the study.
Another extreme situation would be the case of a large truck equipped with a multi-wheelset trailer: in this case, rather than refining the engine or the suspension models, the use of the sub-system segmentation technique (see section 4.4) for assembling the cabin, the trailer, the wheelsets, etc... will yield a "coarse" but sufficiently reliable model for modal analysis or load stability.

Nevertheless, vehicle dynamics always relies on wheel/ground contact models. This does not mean that the tire tread laminations must be modeled, but to deal with the major issues in vehicle dynamics (lateral and entry curving stability, handling performance, passenger comfort, ...), a certain level of generality and accuracy is mandatory, both for the geometrical and dynamical points of view. In particular, and this obviously conditions the mathematical multibody model, the main tire properties (elasticity, damping, road-holding limitation, ...) must be known in the various directions.

The present chapter is organized as follows: section 6.2 describes the main definition and hypotheses underlying the wheel/ground model. Section 6.3 deals with the geometrical contact problem which is a key point for any rolling system. In section 6.4, two dynamical models of contact forces will be discussed; however, since wheel/ground models have been the subject of many publications (ex. [63], [1], [32], ...), the models proposed in section 6.4 will mainly illustrate the way constitutive equations of forces and torques can be introduced in a manner compatible with the previous hypotheses within the multibody context. Finally, section 6.5 will present some examples drawn from well-known international benchmarks [49] and joint researches.

6.2 Definitions and hypotheses

In the present multibody approach, we assume that each wheel is a *leaf* body of the multibody structure (which is for instance a whole vehicle or a single car suspension) as shown in figure 6.1 which illustrates the suspension of the ILTIS military vehicle (see section 6.5.2 for additional details). The corresponding ROBOTRAN multibody model is depicted in figure 6.2. This system was selected

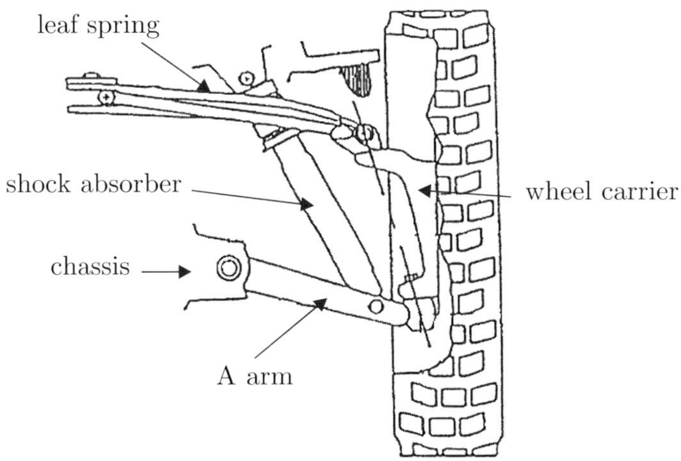

Figure 6.1: The front right ILTIS suspension

at the beginning of the nineties as an international benchmark for multibody software [49]. It is an interesting multibody application because it involves 3–dimensional loops of bodies (induced by the suspension rods and the steering rack), point-to-point links (for the spring, damper and bump stop elements) and is a good candidate for sub-system segmentation (see section 4.4).

As regards the wheel/ground contact, we could consider it as an additional loop of bodies (see figure 6.2) formed by the ground, the chassis, the suspension, the wheel and finally the ground; we shall see that, as a consequence of the tire's elasticity (always present in the real system), it is better to consider a wheel as

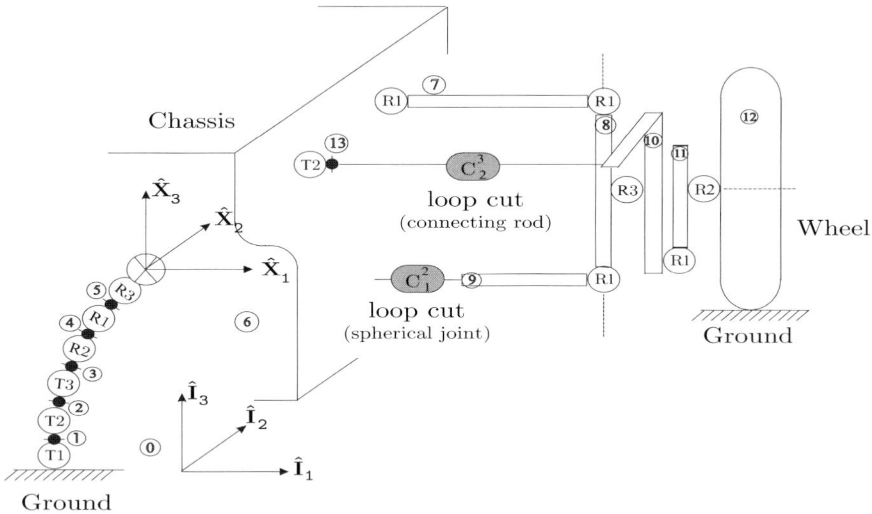

Figure 6.2: ILTIS suspension - multibody model

a *leaf* body i of the MBS on which ground contact forces/torques are *external,* i.e. computed via vectors $\mathbf{F}^i_{ext}$ and $\mathbf{L}^i_{ext}$ (see section 3.1.4). In this way, we avoid unnecessary loops of bodies. Moreover, the 6 relative degrees of freedom between each wheel and the ground, although they are effectively present in the model, are not treated as generalized coordinates q: this considerably reduces the size of the model, without any mathematical approximation.

The hypotheses underlying the present wheel/ground model are the following:

- From the *geometrical* point of view, the wheel has no thickness: the wheel (+tire) is simulated by a flat circular disc with a variable radius. Moreover, the contact between the tire and the ground is assumed to be at a point: the *equivalent* point of contact Q is unique on the disc periphery (see figure 6.3).

- From the *dynamical* point of view, the normal and shear stresses occurring over the real contact surface will be replaced by equivalent force and torque vectors applied to the wheel at the contact point Q. Their values will be computed via constitutive equations, drawn from tire/ground models compatible with the two previous hypotheses.

- The ground (see figure 6.3) is assumed to be infinitely rigid, and such that its class-$\mathcal{C}^1$ profile μ is known analytically in the form:

$$z = \mu(x, y) \tag{6.1}$$

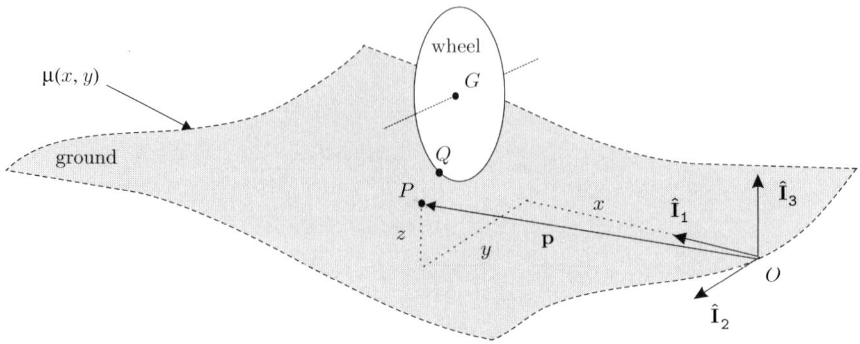

Figure 6.3: Ground profile and contact points

A point $P(x, y)$ on the ground surface (see figure 6.3) is located by the position vector $\mathbf{p}(x,y) = \overrightarrow{OP} = [\hat{\mathbf{I}}]^T \begin{pmatrix} x \\ y \\ \mu(x,y) \end{pmatrix}$

As regards the conventions we have chosen for road vehicle models, we first assume, according to the previous definition of the ground, that the inertial frame is such that $\hat{\mathbf{I}}_3$ is vertical and points upwards. As regards vectors $\hat{\mathbf{I}}_1$ and $\hat{\mathbf{I}}_2$, a well-established convention in road vehicle dynamics stipulates that $\hat{\mathbf{I}}_1$ and $\hat{\mathbf{I}}_2$ respectively correspond to the forward and left direction of the car in its nominal configuration[1] (see figure 6.4).

Finally, for the geometrical contact model, we assume that the axis of rotation of each wheel coincides with the y axis of the wheel body-fixed frame denoted $\{\hat{\mathbf{Y}}\}$ (see figure 6.4).

Figure 6.4: Road vehicle reference configuration

[1] The nominal configuration of a road vehicle denotes its static position on a flat ground ($\mu = 0$). It differs from the *reference* configuration of a multibody system defined in section 5.8.1.

6.3 Wheel/ground geometrical contact

6.3.1 Point and vector definitions

Let us consider in figure 6.5 a wheel as being a leaf body of a MBS which represents a vehicle or a part of it. The following frames located at the wheel center G are now introduced:

- $[\hat{\mathbf{Y}}] = R^M(q)\,[\hat{\mathbf{I}}]$ is the *material wheel* attached frame,
- $[\hat{\mathbf{X}}] = R^G(q)\,[\hat{\mathbf{I}}]$ is the *geometrical wheel* attached frame.

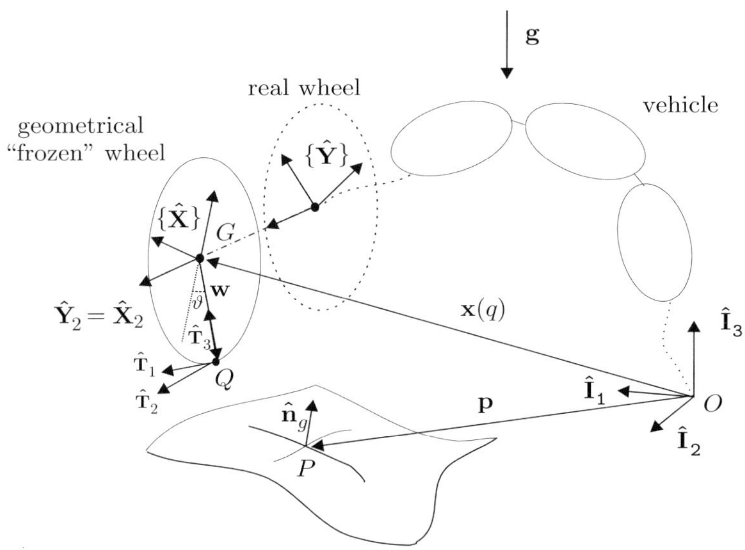

Figure 6.5: Wheel kinematics

The difference between these two frames is fundamental. The first one $\{\hat{\mathbf{Y}}\}$ is attached to the physical wheel (represented as a dotted circle in figure 6.5) which rotates around its axis of symmetry. The second one $\{\hat{\mathbf{X}}\}$ does not take this rotational velocity into account and is attached to a so-called *frozen* or *geometrical* wheel (continuous circle in figure 6.5) located at the same origin as the physical one but rigidly attached to the wheel carrier.

For the geometrical problem which consists in finding the contact point Q on the wheel for each configuration, it is more judicious to drop the wheel rotation and thus to deal with the *geometrical* wheel. In order to locate any potential contact point Q on its circumference, the following two quantities are introduced (see figure 6.5):

- a frame $\{\hat{\mathbf{T}}\}$ located at Q such that $[\hat{\mathbf{T}}] = R_2(\vartheta)[\hat{\mathbf{X}}]$ and $\hat{\mathbf{T}}_3$ points towards point G

$$R_2(\vartheta) \triangleq \begin{pmatrix} c\vartheta & 0 & -s\vartheta \\ 0 & 1 & 0 \\ s\vartheta & 0 & c\vartheta \end{pmatrix} \text{ where } c\vartheta \triangleq \cos\vartheta \text{ and } s\vartheta \triangleq \sin\vartheta.$$

- a vector $\mathbf{w} = \overrightarrow{GQ} = [\hat{\mathbf{T}}]^T \begin{pmatrix} 0 \\ 0 \\ -r_w \end{pmatrix}$ or equivalently: $\mathbf{w} = -r_w \hat{\mathbf{T}}_3$

As shown in figure 6.6, two auxiliary variables have been introduced in the previous expressions to locate Q: ϑ which allows us to define frame $\{\hat{\mathbf{T}}\}$, and r_w which represents the – time-varying – radius of the wheel. When the wheel is not in contact with the ground, $r_w = r_{nom}$, its nominal radius (see figure 6.6). However, for the geometrical solution only, we shall see that r_w can become larger than r_{nom}.

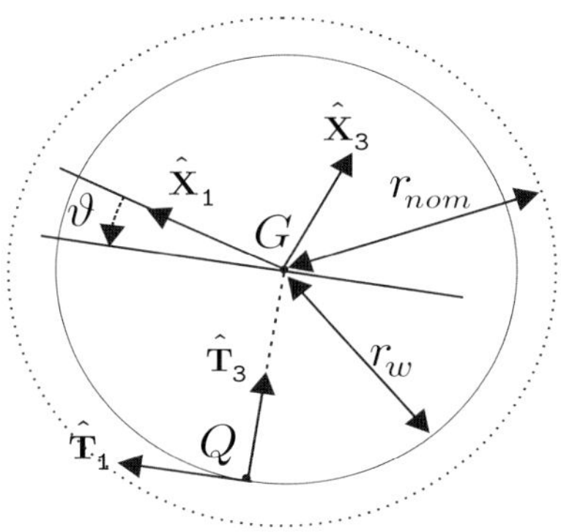

Figure 6.6: Auxiliary variables ϑ and r_w

In order to set up the geometrical model, we also introduce the following quantities (see figure 6.6):

- for the wheel:

- $\mathbf{x}(q) = \overrightarrow{OG} = [\hat{\mathbf{I}}]^T \begin{pmatrix} x_1 \\ x_2 \\ x_3 \end{pmatrix}$, the absolute position vector of the geometrical wheel center G which depends only on the generalized coordinates q appearing in the chain of bodies connecting the wheel to the inertial body.

- $\mathbf{u}(q, r_w, \vartheta) = \mathbf{x}(q) + \mathbf{w}(q, r_w, \vartheta) = [\hat{\mathbf{I}}]^T \begin{pmatrix} u_1 \\ u_2 \\ u_3 \end{pmatrix}$, the absolute position vector of the contact point Q on the geometrical wheel.

- for the ground:

- $\hat{\mathbf{n}}_g = \left(\frac{\partial \mathbf{p}}{\partial x} \times \frac{\partial \mathbf{p}}{\partial y}\right) / \left\|\frac{\partial \mathbf{p}}{\partial x} \times \frac{\partial \mathbf{p}}{\partial y}\right\|$, the unit vector normal to the ground surface at point $P(u_1, u_2)$ (see figure 6.5),

  where $\frac{\partial \mathbf{p}}{\partial x} = [\hat{\mathbf{I}}]^T \begin{pmatrix} 1 \\ 0 \\ \mu'_x(u_1, u_2) \end{pmatrix}$ with $\mu'_x \triangleq \frac{\partial \mu}{\partial x}$

  and $\frac{\partial \mathbf{p}}{\partial y} = [\hat{\mathbf{I}}]^T \begin{pmatrix} 0 \\ 1 \\ \mu'_y(u_1, u_2) \end{pmatrix}$ with $\mu'_y \triangleq \frac{\partial \mu}{\partial y}$

6.3.2 Contact point: geometrical solution

Although the wheel/ground contact will not be considered as a set of kinematic constraints in the sense of chapter 4, the problem at the algebraic level is of the same nature since it consists in solving — but with respect to the auxiliary variables ϑ and r_w in this case — the following geometrical problem: find a point and tangent contact between the *geometrical* wheel and the profiled ground.

This is ensured by the two following equations:

1. The contact point Q belongs to the ground profile:

$$h_1^w(q, r_w, \vartheta) \triangleq u_3 - \mu(u_1, u_2) = 0 \tag{6.2}$$

2. At the contact point Q, the geometrical wheel circumference is tangent to the ground:

$$h_2^w(q, r_w, \vartheta) \triangleq \hat{\mathbf{T}}_1 \,.\, \hat{\mathbf{n}}_g = 0 \tag{6.3}$$

To solve this nonlinear algebraic system with respect to the auxiliary variables ϑ and r_w, let us first consider equation 6.3 which can be expanded as follows.

Since $[\hat{\mathbf{T}}] = R_2(\vartheta) R^G(q)\, [\hat{\mathbf{I}}]$, one obtains:

$$\hat{\mathbf{T}}_1 = [\hat{\mathbf{I}}]^T \begin{pmatrix} c\vartheta\, R^G_{11} - s\vartheta\, R^G_{31} \\ c\vartheta\, R^G_{12} - s\vartheta\, R^G_{32} \\ c\vartheta\, R^G_{13} - s\vartheta\, R^G_{33} \end{pmatrix} \tag{6.4}$$

and from the previous definition of $\hat{\mathbf{n}}_g$,

$$\hat{\mathbf{n}}_g = \frac{1}{\sqrt{1 + \mu'^2_x(u_1, u_2) + \mu'^2_y(u_1, u_2)}} [\hat{\mathbf{I}}]^T \begin{pmatrix} -\mu'_x(u_1, u_2) \\ -\mu'_y(u_1, u_2) \\ 1 \end{pmatrix} \tag{6.5}$$

and finally,

$$\begin{aligned} h_2^w(q, r_w, \vartheta) = \left(c\vartheta\, R^G_{13} - s\vartheta\, R^G_{33}\right) - \mu'_x(u_1, u_2) \left(c\vartheta\, R^G_{11} - s\vartheta\, R^G_{31}\right) \\ -\mu'_y(u_1, u_2) \left(c\vartheta\, R^G_{12} - s\vartheta\, R^G_{32}\right) = 0 \end{aligned} \tag{6.6}$$

in which, from the above definition of vector **u**,

$$u_1 = x_1 - r_w \left(s\vartheta\, R^G_{11} + c\vartheta\, R^G_{31}\right) \text{ and } u_2 = x_2 - r_w \left(s\vartheta\, R^G_{12} + c\vartheta\, R^G_{32}\right) \qquad (6.7)$$

Similarly, we can expand equation 6.2 leading to

$$h_1^w(q, r_w, \vartheta) = x_3 - r_w(s\vartheta\, R^G_{13} + c\vartheta\, R^G_{33}) - \mu(u_1, u_2) = 0 \qquad (6.8)$$

with u_1, u_2 given by 6.7.

System (6.8, 6.6) must be solved, for each wheel and for any vehicle configuration, with respect to r_w and ϑ. Being nonlinear in these variables, the problem is solved by using a Newton-Raphson algorithm, as for the loop constraints solution: starting from a good initial guess $(r_w,\, \vartheta)_0$,

$$\begin{pmatrix} r_w \\ \vartheta \end{pmatrix}_{i+1} = \begin{pmatrix} r_w \\ \vartheta \end{pmatrix}_{i} - \begin{pmatrix} \frac{\partial h_1^w}{\partial r_w} & \frac{\partial h_1^w}{\partial \vartheta} \\ \frac{\partial h_2^w}{\partial r_w} & \frac{\partial h_2^w}{\partial \vartheta} \end{pmatrix}^{-1} \begin{pmatrix} h_1^w \\ h_2^w \end{pmatrix}_{i} \qquad (6.9)$$

should converge towards the correct solution.

We should stress the following three points:

1. The vehicle configuration (in particular the initial one) must be such that each wheel center G is above the ground surface: although being "obvious" from a physical point of view, this condition must also be satisfied for the present contact problem and thus for each wheel:

$$x_3 > \mu(x_1, x_2)$$

2. The geometrical problem can lead to a spurious solution in ϑ, i.e. $\pi + \vartheta$ rather than ϑ, for which frame $\{\hat{\mathbf{T}}\}$ would be inverted (dotted line frame in figure 6.7): this situation is not compatible with the conventions of the dynamical models of section 6.4. When using a numerical iterative procedure (ex.: Newton/Raphson), this spurious solution can be avoided by starting with a good guess for ϑ and by checking — *a posteriori* — the validity of the solution. Fortunately, the latter has a very simple mathematical expression using the value found for r_w. Indeed, the correct solution implies

$$r_w > 0,$$

and thus a negative value for r_w, associated with the spurious solution, must be rejected.

3. To satisfy the first two hypotheses of section 6.2 (point contact - no wheel thickness), we need to check, for each configuration of a wheel in contact, that:

   - the ground profile curvature at the contact point — measured on the curve defined by the intersection between the ground surface and the wheel plane — is less than $\frac{1}{r_w}$,the curvature of the wheel.

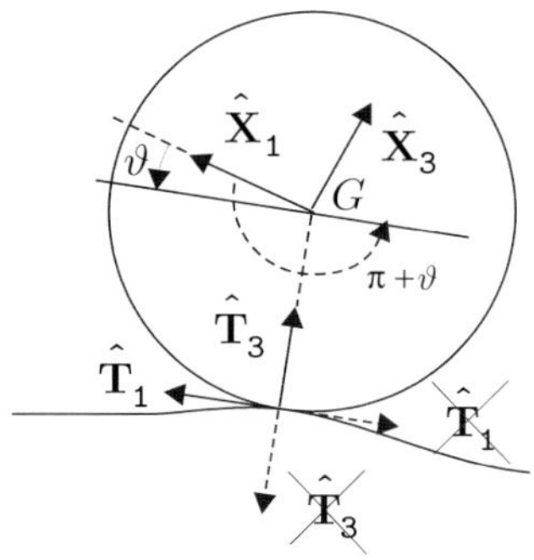

Figure 6.7: Spurious geometrical solution

- the wheel inclination is compatible with the profile slope at the contact point: in other words, the absolute value of the angle between the ground normal vector $\hat{\mathbf{n}}_g$ and the wheel plane defined by $\{\hat{\mathbf{X}}_1, \hat{\mathbf{X}}_3\}$ must be less than $\frac{\pi}{2}$.

In practice, these two tests are generally not performed in this way. Indeed, first of all, since the geometrical problem is solved via an iterative procedure, both tests must be valid in a "certain region" of the ground profile around the contact point solution, in order to ensure a robust convergence towards the correct point. Secondly, in most situations, the ground profile is known in advance and is such that, for any point (x, y) on the surface, the maximum curvature is smaller than the wheel nominal curvature $1/r_{nom}$: the previous hypotheses are thus automatically satisfied for any configuration.

In the particular situation of a horizontal flat ground (i.e.: $\mu = \mu'_x = \mu'_y = 0$, $\forall x, y$), a numerical procedure is not necessary to solve the geometrical problem; indeed, equations 6.8 and 6.6 reduce to

$$h_1^w \triangleq x_3 - r_w(s\vartheta\, R_{13}^G + c\vartheta\, R_{33}^G) = 0$$
$$h_2^w \triangleq c\vartheta\, R_{13}^G - s\vartheta\, R_{33}^G = 0$$

or equivalently,

$$\begin{aligned} \cos\vartheta\, R_{13}^G - \sin\vartheta\, R_{33}^G &= 0 \\ r_w &= \frac{x_3}{\sin\vartheta\, R_{13}^G + \cos\vartheta R_{33}^G} \end{aligned} \tag{6.10}$$

The first equation has two well-known solutions in $\sin\vartheta, \cos\vartheta$ one of which is irrelevant and corresponds to the spurious solution (frame $\{\hat{\mathbf{T}}\}$ inverted). The solution $(\sin\vartheta, \cos\vartheta)$ can then be introduced into the second equation to obtain r_w.

6.4 Wheel/ground forces and torques

6.4.1 Wheel/ground contact kinematics

Any wheel/ground contact force model which can be used in the present approach relies on specific geometrical and kinematic quantities which must be evaluated for each wheel configuration. Of course, the previous geometrical problem must first have been solved.
These quantities are:

- The *inclination* or *"camber" angle* φ_w which represents the angle between the wheel plane and the normal vector $\hat{\mathbf{n}}_g$. Since this vector belongs to the $\{\hat{\mathbf{T}}_2, \hat{\mathbf{T}}_3\}$ plane, the "camber" angle φ_w can be expressed as follows (see figure 6.8):

$$\varphi_w = \tan^{-1}\left(\frac{\hat{\mathbf{n}}_g \,.\, \hat{\mathbf{T}}_2}{\hat{\mathbf{n}}_g \,.\, \hat{\mathbf{T}}_3}\right) \qquad (6.11)$$

The camber angle φ_w is positive when the wheel, observed from the rear,

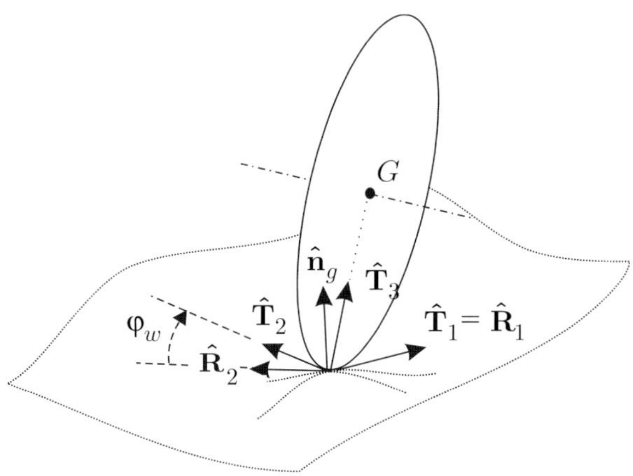

Figure 6.8: Wheel camber angle φ_w

tilts to the right. φ_w allows us to introduce the so-called ground contact frame $\{\hat{\mathbf{R}}\}$ which is such that (see figure 6.8)

$$\hat{\mathbf{R}}_1 = \hat{\mathbf{T}}_1 \ ; \hat{\mathbf{R}}_3 = \hat{\mathbf{n}}_g \ \ ; \hat{\mathbf{R}}_2 = \hat{\mathbf{R}}_3 \times \hat{\mathbf{R}}_1$$

and, using definition 6.11,

$$[\hat{\mathbf{R}}] = R_1^T(\varphi_w)[\hat{\mathbf{T}}], \quad \text{with} \quad R_1(\varphi_w) \triangleq \begin{pmatrix} 1 & 0 & 0 \\ 0 & c\varphi_w & s\varphi_w \\ 0 & \text{-}s\varphi_w & c\varphi_w \end{pmatrix} \qquad (6.12)$$

- The *sideslip angle* α_w which represents the angle – measured in the contact plane $\{\hat{\mathbf{R}}_1, \hat{\mathbf{R}}_2\}$ – between $\dot{\mathbf{u}}$, the absolute velocity of the *geometrical* contact point Q, and the unit vector $\hat{\mathbf{R}}_1$ which reflects the wheel orientation (see figure 6.9). The velocity vector $\dot{\mathbf{u}}$ can be computed as follows:

$$\dot{\mathbf{u}} = \dot{\mathbf{x}} + \dot{\mathbf{w}} = \dot{\mathbf{x}} + \overset{\circ}{\mathbf{w}} + \boldsymbol{\omega}^T \times \mathbf{w}$$

in which:

$$\begin{aligned} \dot{\mathbf{x}} &= [\hat{\mathbf{R}}]^T \, R_1^T(\varphi_w) \, R_2(\vartheta) \, R^G(q) \begin{pmatrix} \dot{x}_1 \\ \dot{x}_2 \\ \dot{x}_3 \end{pmatrix} \\ \overset{\circ}{\mathbf{w}} &= -\dot{r}_w \hat{\mathbf{T}}_3 \quad \text{and} \quad \boldsymbol{\omega}^T = \boldsymbol{\omega}^X + \dot{\vartheta} \hat{\mathbf{T}}_2 \end{aligned} \tag{6.13}$$

and where $\boldsymbol{\omega}^T$ and $\boldsymbol{\omega}^X$ denote the absolute angular velocities of frames $\{\hat{\mathbf{T}}\}$ and $\{\hat{\mathbf{X}}\}$ respectively. According to its definition and to figure 6.9, α_w can be computed as[2]

$$\alpha_w = \tan^{-1} \left(\frac{\dot{\mathbf{u}} \,.\, \hat{\mathbf{R}}_2}{\dot{\mathbf{u}} \,.\, \hat{\mathbf{R}}_1} \right) \tag{6.14}$$

α_w is positive when the vehicle turns to the right (see figure 6.9).

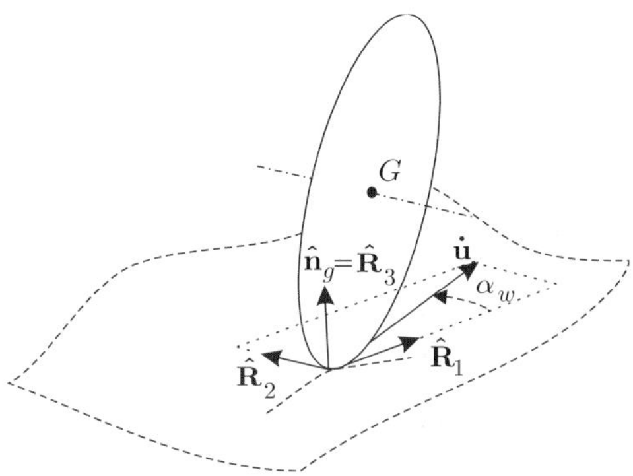

Figure 6.9: Wheel sideslip angle α_w

- The *longitudinal slip velocity* κ_w which is the projection onto $\hat{\mathbf{R}}_1$ of the velocity of the *material* point of contact of the wheel coinciding with Q :

$$\kappa_w = (\dot{\mathbf{x}} + \boldsymbol{\omega}^Y \times \mathbf{w} + \overset{\circ}{\mathbf{w}}) \,.\, \hat{\mathbf{R}}_1 \tag{6.15}$$

[2] In most situation (i.e. small $\dot{\varphi}_w$ and $\dot{\vartheta}$), the computation of α_w can be approximated in this formula by replacing $\dot{\mathbf{u}}$ by $\dot{\mathbf{x}}$, the velocity of the wheel center.

in which $\boldsymbol{\omega}^Y(q,\dot{q})$ denotes the angular velocity of frame $\{\hat{\mathbf{Y}}\}$, i.e. of the physical wheel. Since $\overset{\circ}{\mathbf{w}} = -\dot{r}_w \hat{\mathbf{T}}_3$ at the contact point and since $\hat{\mathbf{T}}_3 \,.\, \hat{\mathbf{R}}_1 = 0$, the latter expression reduces to

$$\kappa_w = (\dot{\mathbf{x}} + \boldsymbol{\omega}^Y \times \mathbf{w}) \,.\, \hat{\mathbf{R}}_1 \tag{6.16}$$

For tire dynamical models, the *longitudinal slip ratio* is generally used instead of κ_w, and is equal to κ_w divided by the longitudinal component of the wheel center velocity $\dot{\mathbf{x}}$:

$$\gamma_w \stackrel{\Delta}{=} \frac{\kappa_w}{\dot{\mathbf{x}} \,.\, \hat{\mathbf{R}}_1} \tag{6.17}$$

- The *tire penetration* p_w and penetration rate $\dot{p}_w$ which are measured along the normal vector $\hat{\mathbf{n}}_g$.
  From simple geometrical considerations (see figure 6.6 and 6.8), we find

  $$p_w = \max\{0, \cos\varphi_w\,(r_{nom} - r_w)\} \tag{6.18}$$

  p_w is thus positive when the wheel is in contact.
  The penetration rate $\dot{p}_w$ corresponds to the tire radial dilation rate and is obtained by projecting vector $\dot{\mathbf{w}}$ onto the normal vector $\hat{\mathbf{n}}_g$:

  $$\dot{p}_w = \dot{\mathbf{w}} \,.\, \hat{\mathbf{n}}_g = (\boldsymbol{\omega}^T \times \mathbf{w} - \dot{r}_w \hat{\mathbf{T}}_3) \,.\, \hat{\mathbf{n}}_g$$

  Using the definition of vector $\mathbf{w}$ and equation 6.12, we can write

  $$\dot{p}_w = r_w\,(\hat{\mathbf{n}}_g \times \hat{\mathbf{T}}_3) \,.\, \boldsymbol{\omega}^T - \dot{r}_w \cos\varphi_w$$

  in which the first term is negligible in most situations. Indeed

  $$r_w\,(\hat{\mathbf{n}}_g \times \hat{\mathbf{T}}_3) \,.\, \boldsymbol{\omega}^T = r_w \sin\varphi_w\,(\hat{\mathbf{T}}_1 \,.\, \boldsymbol{\omega}^T) \simeq r_w \sin\varphi_w\,\dot{\varphi}_w \quad ,$$

  thus $\dot{p}_w \simeq -\dot{r}_w \cos\varphi_w$. Of course, since a penetration rate only makes sense when the wheel is in contact, we write:

  $$\text{if } |p_w| > 0 \text{ then } \dot{p}_w \simeq -\dot{r}_w \cos\varphi_w \quad \text{else } \dot{p}_w = 0 \tag{6.19}$$

  $\dot{p}_w$ is thus positive when the tire sinks into the ground ($\dot{r}_w < 0$).

6.4.2 Contact force model

As stated in the introduction, a wide range of ground/tire contact models can be found in the scientific literature. From our point of view, two criteria can be used to "classify" these models:

1. Their *spectrum* with respect to various vehicle behaviors (lateral stability, braking/traction, vertical vibration, cornering, ...): most models are dedicated to one or more of these behaviors and it is often useless to use a full 3D model to analyze specific applications. Moreover, 3D models can represent a serious drawback in terms of CPU time or even numerical convergence and accuracy during the numerical process (ex. time integration, modal analysis, optimization).

2. Their *level of refinement*: let us illustrate this point via the following example. In order to compute the lateral ground/tire force F_{lat} for instance, a "lateral model" in which this force only depends linearly on the *sideslip angle* α_w is sufficient to perform modal analyses of the vehicle around its quasi-static equilibrium[3].
   When performing time simulations with steering manoeuvers, we must take saturation effects into account in the $F_{lat}(\alpha_w)$ relation for large slip angles.
   As soon as the roll motion of the vehicle becomes important, the influence of the camber angle φ_w is no longer negligible and must be included in the expression of F_{lat}.

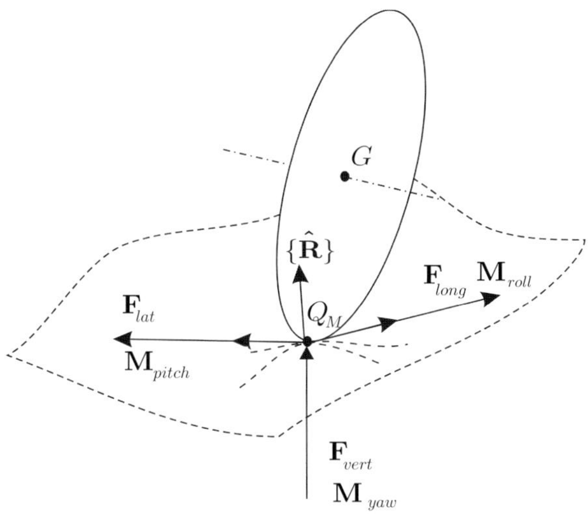

Figure 6.10: Wheel/ground contact forces and torques

Output of ground/tire models are a force resultant vector and a pure torque resultant vector acting on the real wheel at the material point Q_M coinciding with point Q (see figure 6.10). These vectors are expressed in the ground contact frame $\{\hat{\mathbf{R}}\}$, so as to be consistent with the standard models found in the literature.
The force vector is defined as:

$$\mathbf{F}_w = [\hat{\mathbf{R}}]^T \begin{pmatrix} F_{long} \\ F_{lat} \\ F_{vert} \end{pmatrix} \tag{6.20}$$

---

[3]Typically for a vehicle, quasi-static equilibrium means the equilibrium of the vehicle at forward constant speed V on a flat ground.

in which components F_{long}, F_{lat} and F_{vert} respectively represent the longitudinal, lateral and vertical forces. The pure torque vector reads

$$\mathbf{M}_w = [\hat{\mathbf{R}}]^T \begin{pmatrix} M_{roll} \\ M_{pitch} \\ M_{yaw} \end{pmatrix} \tag{6.21}$$

in which components M_{roll}, M_{pitch} and M_{yaw} are respectively the overturning, the rolling resistance and the aligning torques.

For additional details on tire models, the reader is referred to more specialized publications or textbooks (ex.: [63], [1], [32], ...). Two models of tire are proposed below. Before detailing these, let us first consider the vertical force F_{vert} which we do not actually associate with any particular model but is in fact considered rather as an "input" for the other force/torque components.

Vertical force

The vertical ground/tire force $\mathbf{F}_{vert}$ depends on the tire penetration p_w and penetration rate $\dot{p}_w$. The most popular constitutive law is linear and reads:

$$\mathbf{F}_{vert} = F_{vert}\,\hat{\mathbf{R}}_3 \quad \text{with } F_{vert} = K_{rad}\, p_w + D_{rad}\,\dot{p}_w \tag{6.22}$$

in which K_{rad} and D_{rad} denote the tire radial equivalent stiffness and damping coefficient respectively.

Besides this characteristic of a tire in the radial direction, for which equation 6.22 provides a quite good approximation, the tire is also in itself a dynamical system with resonances which affect the transmission of vibrations to the vehicle [32]: for instance, a first mode which is characterized by the tire eccentricity with respect to the wheel center, occurs – if excited – somewhere around $60Hz$ for a passenger car tire. Since the entire tread band moves up and down in unison, the force associated with the resonance is transmitted vertically to the wheel and axle: this phenomenon is obviously absent from equation 6.22 since it would involve the "local" tire dynamics. Thus, a model based on 6.22 should be used with caution for high speeds over very rough road profiles.

Calspan model

The first model, denoted here as the "Calspan" model, was proposed within the framework of the IAVSD road benchmark [49] dealing with the modeling and simulation of the "ILTIS" military vehicle (see chapter 4). The Calspan model is restricted to the computation of the lateral force $\mathbf{F}_{lat}$ and the aligning torque $\mathbf{M}_{yaw}$ of a moving vehicle: it is concerned mainly with the vehicle lateral stability, in the linear domain (modal analysis) and nonlinear domain (handling performances under various steering behaviors). The coefficients needed in the Calspan model are given in Table 6.1.
The lateral force F_{lat} is given by [49]

$$\mathbf{F}_{lat} = F_{lat}\,\hat{\mathbf{R}}_2 \quad \text{with} \quad F_{lat} = f\, F_{vert}\, g(\bar{\alpha})\, F_{slip} \tag{6.23}$$

Coefficient	Value	Description	Units
A_0	2625	force coefficient	N
A_1	14.47	force coefficient	none
A_2	12930	force coefficient	N
A_3	2.29	force coefficient	none
A_4	18175	force coefficient	N
B_1	$-0.464 \,.\, 10^{-4}$	friction coefficient	N^{-1}
B_2	1.216	friction coefficient	none
B_3	$0.218 \,.\, 10^{-10}$	friction coefficient	N^{-2}
S_N	1.0274	Skid number ratio	none
F_1	$-0.171 \,.\, 10^{-4}$	torque coefficient	$\frac{m}{N}$
F_2	$0.171 \,.\, 10^{-4}$	torque coefficient	$\frac{m}{N}$

Table 6.1: Calspan coefficients

where:

- f is the tire sideforce friction coefficient given by

$$f = (B_1 F_{vert} + B_2 + B_3 (F_{vert})^2) S_N \tag{6.24}$$

- $g(\bar{\alpha})$ is the dimensionless sideforce shaping function for combined sideslip angle α_w and camber angle φ_w :

$$\begin{aligned} g(\bar{\alpha}) &= \bar{\alpha} - \frac{1}{3}\bar{\alpha}\,|\bar{\alpha}| + \frac{1}{27}\,\bar{\alpha}^3 \quad \text{for} \quad |\bar{\alpha}| < 3 \\ g(\bar{\alpha}) &= \frac{\bar{\alpha}}{|\bar{\alpha}|} \quad \text{for} \quad |\bar{\alpha}| \geq 3 \end{aligned} \tag{6.25}$$

where $\bar{\alpha}$ is the non-dimensional equivalent sideslip angle defined as follows (refer to table 6.1):
- for $F_{vert} \leq A_2$

$$\begin{aligned} \bar{\alpha} &= \frac{A_1 F_{vert}(F_{vert} - A_2) - A_0\, A_2}{A_2\, f\, F_{vert}} (\alpha_w + \alpha_\varphi) \\ \text{with } \alpha_\varphi &= \frac{-A_2\, A_3 (A_4 - F_{vert}) F_{vert}}{A_4 (A_1\, F_{vert}(F_{vert} - A_2) - A_0\, A_2)}\, \varphi_w \end{aligned}$$

- else

$$\begin{aligned} \bar{\alpha} &= \frac{-A_0}{f\, F_{vert}} (\alpha_w + \alpha_\varphi) \\ \alpha_\varphi &= \frac{-A_2\, A_3 (A_4 - A_2) A_2}{-A_0\, A_4}\, \varphi_w \end{aligned}$$

Figure 6.11 illustrates function $g(\bar{\alpha})$ whose main goal is to take the *contact saturation* into account for an increasing equivalent sideslip angle.

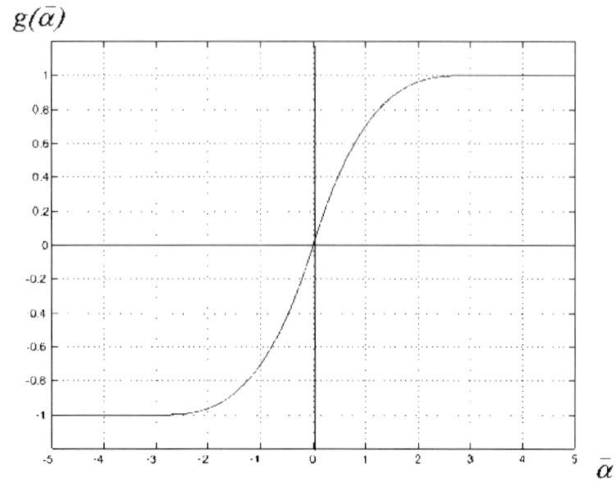

Figure 6.11: Saturation function $g(\bar{\alpha})$

- F_{slip} is a dimensionless shaping coefficient accounting for longitudinal slip, with a value close to unity except for cases of strong traction or braking. For the IAVSD benchmark, $F_{slip} = 1$.

The aligning torque $\mathbf{M}_{yaw}$ in the Calspan model is given by:

$$\mathbf{M}_{yaw} = M_{yaw}\hat{\mathbf{R}}_3 \text{ with } M_{yaw} = (F_1\, F_{vert} + F_2\, |F_{lat}|)\, F_{lat} \tag{6.26}$$

The physical interpretation of the aligning torque M_{yaw} is illustrated by figure 6.12 issued from [32]. The figure illustrates the top view of a rolling tire

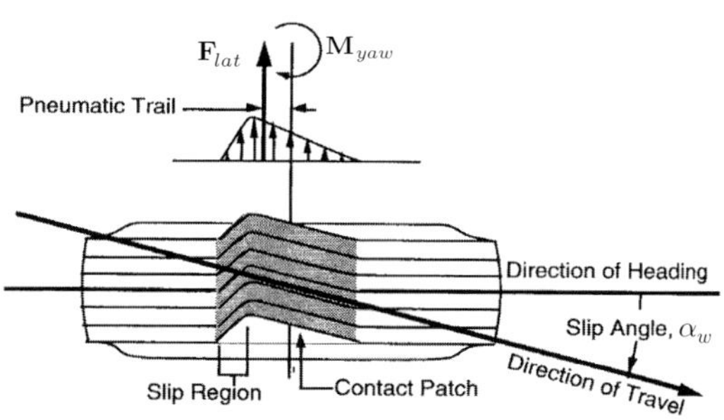

Figure 6.12: Lateral deformation of a rolling tire (frorm [32])

deformed under the action of a lateral force: the lateral stress distribution developed throughout the contact patch takes the form shown in figure 6.12. The integration of the lateral stress over the contact patch yields the resultant lateral force F_{lat} whose point of action is offset towards the rear of the contact point by a distance known as the *pneumatic trail*: this induces the aligning torque computed by equation 6.26 in the Calspan model.

Bakker model [1]

This model is based on numerous experimental measurements, to which non-linear functions have been fitted. Their authors propose the following formulae for the longitudinal and lateral forces and for the aligning torque, in terms of the longitudinal slip ratio expressed as a percentage $\gamma'_w = 100 * \gamma_w$, the sideslip and the camber angles both being expressed in degrees, $\alpha^o_w \stackrel{\Delta}{=} \frac{180}{\pi}\alpha_w$ and $\varphi^o_w \stackrel{\Delta}{=} \frac{180}{\pi}\varphi_w$ and the vertical force F_{vert} expressed in [kNewton].
For the longitudinal and lateral forces, the model at first considers the particular case for which these forces are uncoupled and thus computed independently: a longitudinal force denoted F^*_{long} and a lateral force denoted F^*_{lat} respectively.

- Longitudinal force (in the absence of any lateral force):

$$\begin{aligned} F^*_{long} &= -D\sin(C\ \tan^{-1}(B\phi)) \\ \text{with } \phi &= (1-E)\,\gamma'_w + \frac{E}{B}\tan^{-1}(B\,\gamma'_w) \end{aligned} \tag{6.27}$$

  where:

  . $D = a_1\,(F_{vert})^2 + a_2\,F_{vert}$
  . $C = 1.65$
  . $B = \frac{a_3\,(F_{vert})^2 + a_4\,F_{vert}}{C\,D\,e^{a_5 F_{vert}}}$
  . $E = a_6\,(F_{vert})^2 + a_7\,F_{vert} + a_8$

- Lateral force (in the absence of any longitudinal force):

$$\begin{aligned} F^*_{lat} &= -D\sin(C\ \tan^{-1}(B\,\phi)) + \Delta S_v \\ \text{with } \phi &= (1-E)(\alpha^o_w + \Delta S_h) + \frac{E}{B}\tan^{-1}(B\,(\alpha^o_w + \Delta S_h)) \end{aligned} \tag{6.28}$$

  where:

  . $D = a_1\,(F_{vert})^2 + a_2\,F_{vert}$
  . $C = 1.30$
  . $B = \frac{a_3\sin(a_4\ \tan^{-1}(a_5\,F_{vert}))}{C\,D}(1 - a_{12}|\varphi^o_w|)$
  . $E = a_6\,(F_{vert})^2 + a_7\,F_{vert} + a_8$
  . $\Delta S_h = a_9\,\varphi^o_w$
  . $\Delta S_v = (a_{10}\,(F_{vert})^2 + a_{11}\,F_{vert})\,\varphi^o_w$

- Aligning torque:

$$\begin{aligned} M_{yaw} &= D\sin(C\ \tan^{-1}(B\,\phi)) + \Delta S_v \\ \text{with } \phi &= (1-E)(\alpha^o_w + \Delta S_h) + \frac{E}{B}\tan^{-1}(B\,(\alpha^o_w + \Delta S_h)) \end{aligned} \tag{6.29}$$

  where:

. $D = a_1\,(F_{vert})^2 + a_2\,F_{vert}$

. $C = 2.40$

. $B = \frac{a_3\,(F_{vert})^2 + a_4\,F_{vert}}{C\,D\,e^{a_5 F_{vert}}}(1 - a_{12}|\varphi_w^o|)$

. $E = (a_6\,(F_{vert})^2 + a_7\,F_{vert} + a_8)(1 - a_{13}|\varphi_w^o|)$

. $\Delta S_h = a_9\,\varphi_w^o$

. $\Delta S_v = (a_{10}\,(F_{vert})^2 + a_{11}\,F_{vert})\,\varphi_w^o$

The coefficients a_i for the three components appear in the following table.

	a_1	a_2	a_3	a_4	a_5	a_6	a_7	a_8
F^*_{long}	−21.3	1144	49.6	226	0.069	−0.006	0.056	0.486
F^*_{lat}	−22.1	1011	1078	1.82	0.208	0	−0.354	0.707
M_{yaw}	−2.72	−2.28	−1.86	−2.73	0.11	−0.07	0.643	−4.04
	a_9	a_{10}	a_{11}	a_{12}	a_{13}			
F^*_{lat}	0.028	0	14.8	0.022	0			
M_{yaw}	0.015	−0.066	0.945	0.03	0.07			

Finally, we must take into account the general case where the vehicle undergoes lateral and longitudinal slip simultaneously. For this situation, Bakker computes the effective forces F_{long} and F_{lat} on the basis of the slip ratios in the longitudinal (γ_w) and lateral ($\tan\alpha_w$) directions:

$$F_{long} = \left|\frac{\sigma_x}{\sigma}\right| F^*_{long} \tag{6.30}$$

$$F_{lat} = \left|\frac{\sigma_y}{\sigma}\right| F^*_{lat} \tag{6.31}$$

where:

$$\sigma_x = \frac{\gamma'_w}{1+\gamma'_w}$$

$$\sigma_y = \frac{\tan\alpha_w^o}{1+\gamma'_w}$$

$$\sigma = \sqrt{\sigma_x^2 + \sigma_y^2}$$

For instance, figure 6.13 shows the evolution of the longitudinal force F_{long} with respect to the longitudinal slip ratio γ'_w (expressed as a percentage) for different values of the sideslip angle α_w^o (expressed in degrees); the vertical load and the camber angle have been fixed as follows: $F_{vert} = 1000\,N$ and $\varphi_w^o = 0$.

Equivalently, figure 6.14 describes the evolution of the lateral force F_{lat} with respect to the sideslip angle α_w^o (in degrees) for different values of γ'_w (with $F_{vert} = 1000\,N$ and $\varphi_w^o = 0$).

These figures lead to the following comments:

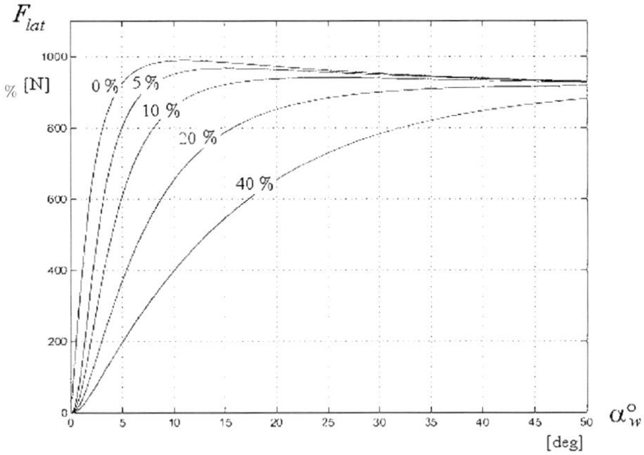

Figure 6.13: F_{long} versus γ'_w ($F_{vert} = 1000\,N$, $\varphi^o_w = 0$)

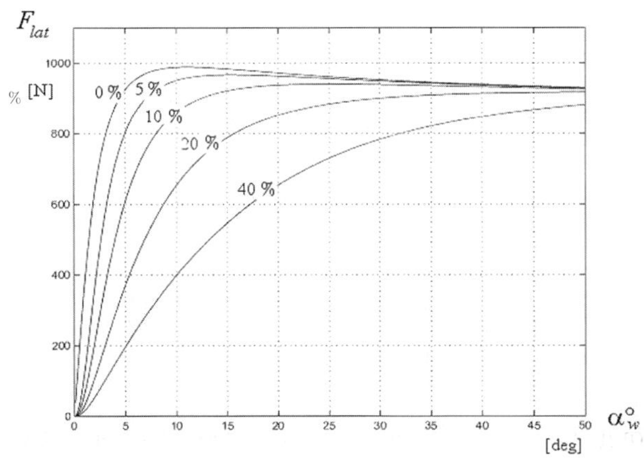

Figure 6.14: F_{lat} versus α^o_w ($F_{vert} = 1000\,N$, $\varphi^o_w = 0$)

1. Both F_{long} and F_{lat} obey a "natural" saturation effect for large slip, all the more reduced since the slip in the other direction increases: this simply illustrates the fact that the reserve of longitudinal force – for instance – is larger when braking along a straight line than in a curve.

2. The overshoot[4] in the curve $F_{long}(\gamma'_w)$ is an important feature which is at the root of the Anti-Lock Brake Systems ("ABS"). We clearly observe that a larger braking force is "available" at lower longitudinal slip: one of the objectives of an ABS is to keep each tire of the vehicle operating near to the peak point of the $F_{long}(\gamma'_w)$ curve. We note that this effect vanishes as soon as the lateral slip grows: this can be understood intuitively since the overshoot is directly related to the difference between the so-called "static" and "dynamical" friction coefficients. Since the wheel is slipping laterally, the potential of the static friction is lost for the longitudinal force and conversely.

6.5 Numerical examples

6.5.1 Introduction

In the previous sections, a geometrical and kinematic model for the wheel/ground contact was developed within the general framework of multibody dynamics, i.e. to model any type of vehicle equipped with wheels, whatever the multibody topology (which can represent a simplified vehicle, a Mc-Pherson suspension, a complete car or a multi-axle truck).

Using the "sensor kinematics" menu of ROBOTRAN (see chapter 5), we can generate the full wheel kinematics symbolically, whatever the complexity of the multibody topology. Thus, as soon as the symbolic equations are generated, the only specific ingredients that the user must provide to complete the model are:

1. the ground profile $\mu(x, y)$ and its spatial derivatives,

2. the wheel/ground dynamical model, i.e., the constitutive equations of the force $\mathbf{F}_w$ and the torque $\mathbf{M}_w$ expressed in the $\{\hat{\mathbf{R}}\}$ frame.

This section presents three examples modeled using ROBOTRAN and simulated using our MATLAB simulation program MBSOFT [23], each providing the model of a realistic vehicle, namely:

- The ILTIS military vehicle (from the IAVSD benchmark) [49],
- An off-road vehicle studied within a student thesis [51],
- A complete modern car equipped with a 4-bar suspension and a semi-active anti-roll system, studied within a student thesis in collaboration with an industrial partner [4].

[4] also present to a lesser degree in the other curve $F_{lat}(\alpha^o_w)$

The models and results will not be discussed in detail (just the data of the third example would cover more than ten pages !) but their role is to illustrate the possibilities of the multibody approach.

6.5.2 The ILTIS vehicle benchmark

Figure 6.15: The ILTIS military vehicle

Depicted in figure 6.15, the Bombardier ILTIS vehicle must be defined "*in model form, suitable for computing specified modal analysis and response properties*" (extract of the IAVSD benchmark description [49]).

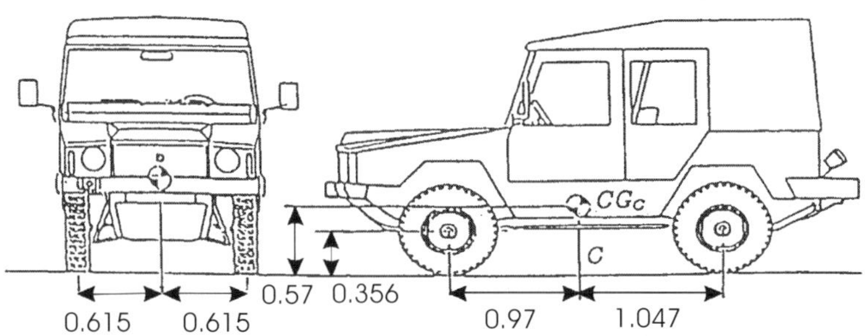

Figure 6.16: ILTIS vehicle : reference configuration

Figure 6.16 shows the vehicle in its nominal configuration, which does not necessarily correspond to its static equilibrium. The absolute positions of the bodies and joints within a suspension are depicted in figure 6.17 and are given in the nominal configuration. The four suspensions have the same topology and thus yield similar symbolic models. The details of the model are the following:

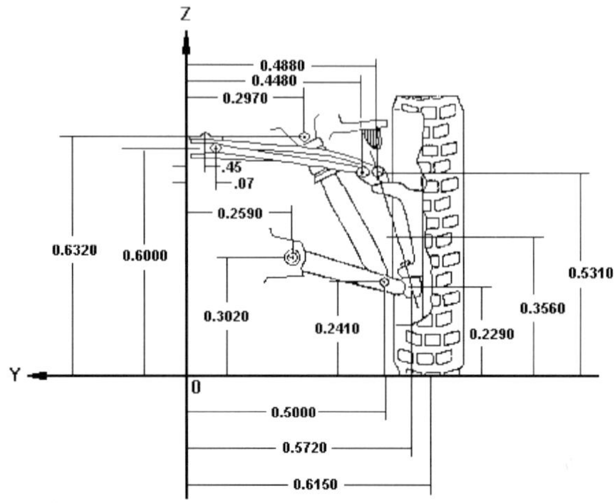

Figure 6.17: ILTIS suspension - nominal configuration (units : meters)

- The carbody has 6 d.o.f. with respect to the ground.
- Each suspension contains 2 loops of bodies. The first is planar and formed by the leaf spring, the wheel carrier and the "A-arm" (see figures 6.1 and 6.2). The other involves the tie rod and allows the steering of the wheel (also exists for the rear suspensions but with a locked tie rod).
- As regards stiffness and damping, each suspension is equipped with a shock absorber (modeled on the basis of an experimental force law), a leaf spring (modeled as an equivalent nonlinear massless spring between the suspension and the chassis) and a rubber bump stop ("bumper") which limits the wheel hop for large road disturbances.
- The wheel/ground forces are computed using the Calspan model (see section 6.4.2). The vertical tire force is assimilated to a linear spring with no damping.

Due to the obvious symmetry of the suspensions, the multibody model of the ILTIS was set up using the sub-system segmentation technique (see chapter 4). It contains:

- $4 \times 13 = 52$ generalized coordinates q
- 18 elementary constraints between sub-systems $h(u)$ (see section 4.4)
- $4 \times 3 = 12$ loop constraints (within sub-systems) $h(q)$
- 1 constraint for the left/right tie rod connection

- 6 kinematically driven joints ($q = f(t)$ or $q = cst$)

  This leads to $52 - 37 = 15$ degrees of freedom: 6 for the chassis, 1 for each suspension wheel hop, 1 for each wheel rotation and 1 for the steering.

The tests required by the benchmark were the following [49]:

- *Static equilibrium*: this simply requires us to determine the vertical equilibrium of the vehicle at rest. For this purpose, neither lateral nor longitudinal forces are considered at the wheel/ground contact. On the basis of the ROBOTRAN symbolic model of the ILTIS, the MBSOFT program iteratively converges towards the static equilibrium using the Newton/Raphson method, with a numerical evaluation of the Jacobian. Detailed results are presented in Part III (problems 7 and 8).

- *Modal analysis* at constant forward speed (20 m/sec): this requires the Calspan model. The MBSOFT program first finds the quasi-static equilibrium of the vehicle (i.e. at constant speed $V = 20\frac{m}{\text{sec}}$), then linearizes these equations on the basis of a parabolic fitting technique. The linearized model can then be used as an input to the modal analysis, which provides and sorts the various eigenmodes (eigenvalues and eigenvectors) of the vehicle: stable, unstable, damped, undamped modes are presented and can be animated if necessary.

- *Time simulation*: three situations were proposed in the benchmark for two forward velocities: $5\frac{m}{\text{sec}}$ and $10\frac{m}{\text{sec}}$.

  1. straight line, driving on a in-phase cosine bump (length $= 5\,m$, height $= 0.1\,m$.), i.e. for which the right and left wheel undergo the same excitation ("sleeping policeman"),
  2. straight line, driving on a anti-phase cosine bump (length $= 5\,m$, height $= +0.1\,m/-0.1\,m$.), i.e. for which the right and left wheels undergo excitations which are equal in magnitude but with opposite signs (when one wheel goes up, the other goes down),
  3. flat ground, sinusoidal steer: The steering manoeuvre starts at the beginning of the simulation with the vehicle in its equilibrium state. The steering rack displacement is given by:

$$r(t) = R\sin(2\pi t) \quad [mm] \quad \text{with} \quad R = 1, 2 \text{ and } 4\,mm$$

For the time simulation of the anti-phase cosine bump at $V = 10\frac{m}{\text{sec}}$, figure 6.18 illustrates the roll and pitch angles of the carbody while figures 6.19 and 6.20 show the wheel/ground normal and lateral forces respectively: losses of contact are obvious in these figures. The high peaks in the normal force curves result from the very stiff bump stops which induce high frequency transient responses when hit by the suspension. Finally, figure 6.21 shows some snapshots of the animation of the vehicle on the in-phase cosine bump.

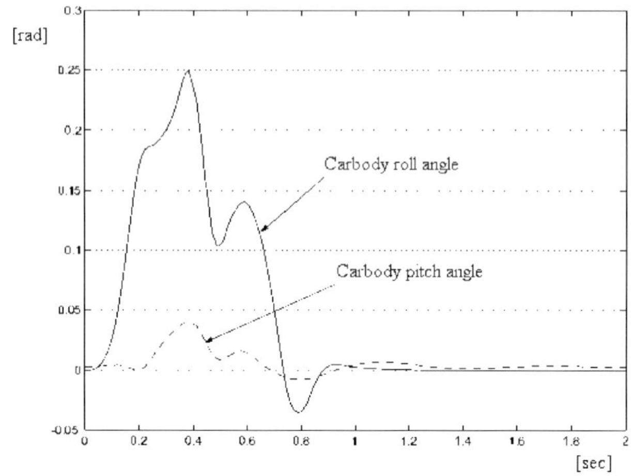

Figure 6.18: Carbody roll and pitch - anti-phase cosine obstacle

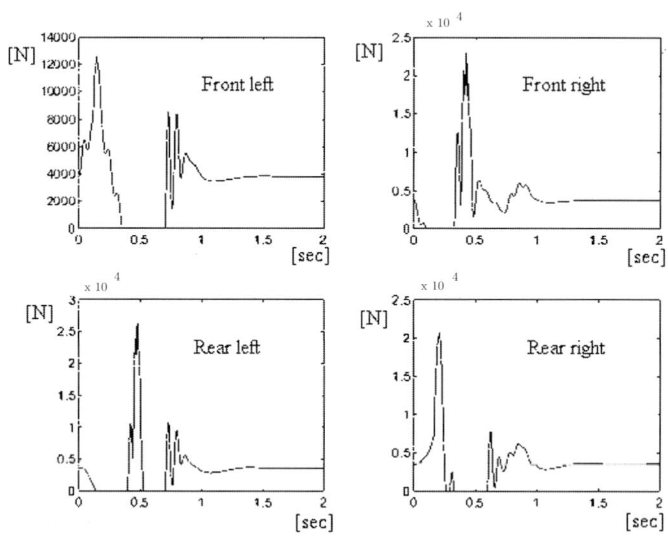

Figure 6.19: Wheel/ground normal forces: F_{vert}

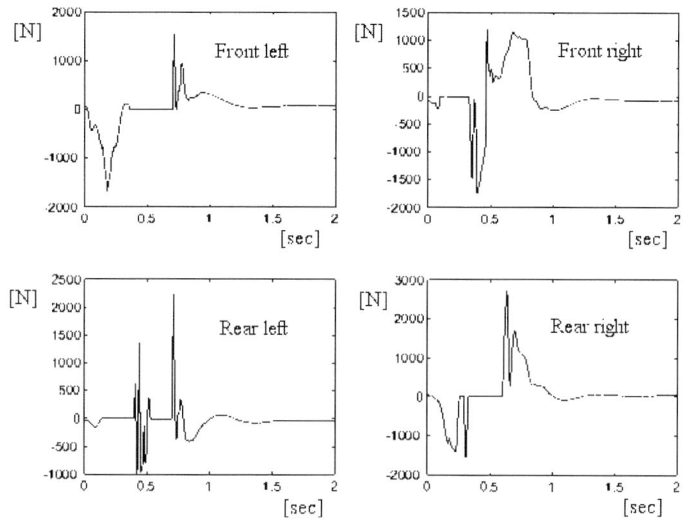

Figure 6.20: Wheel/ground lateral forces (Calspan model): F_{lat}

Figure 6.21: Animation of the ILTIS jeep over the in-phase cosine bump (snapshots)

6.5.3 An off-road vehicle

This example (studied in [51]) deals with the stability analysis of an off-road vehicle with 4 steering wheels, depicted in figures 6.22 and 6.23.

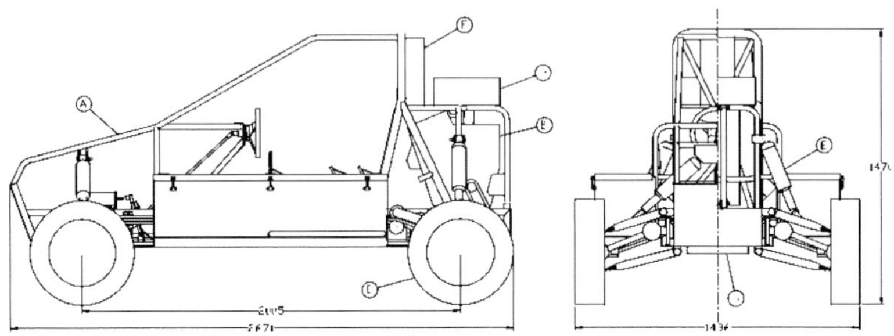

Figure 6.22: An original off-road vehicle (from [51])

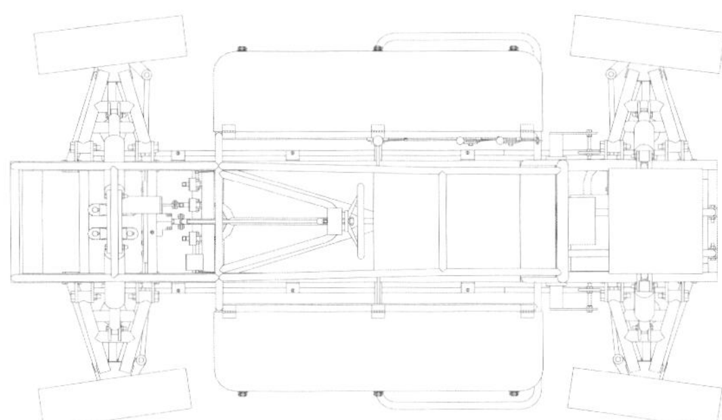

Figure 6.23: 4 steering wheel suspensions (from [51])

Since the general design of the vehicle follows the customer specifications (size, weight, wheel gauge, ...), the goal of the study was to improve the vehicle suspensions in terms of lateral stability.

For this purpose, the "line-change" test depicted in figure 6.24 was selected and the multibody model of the vehicle using the Bakker wheel/ground model enabled us to considerably improve the dynamical performances thanks to a new suspension design (see figure 6.25, right). The main improvements with respect to the original suspension (figure 6.25, left) are:

- the raising (towards the car center of mass) of the car roll center [32], in order to reduce the longitudinal torque on the chassis— and thus the chassis roll — in a curve,
- the limitation of the camber angle φ_w of the wheels in a curve (i.e. when the vehicle is subjected to rolling motions), which affects the lateral wheel/ground force (see section 6.4.2, equation 6.28).

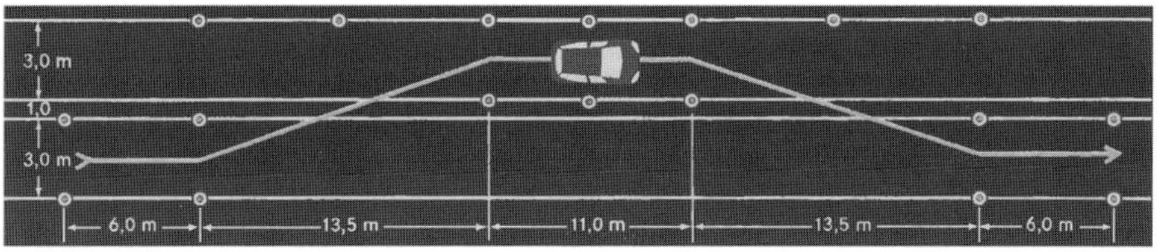

Figure 6.24: The "line-change" test ($V = 60$ km/h)

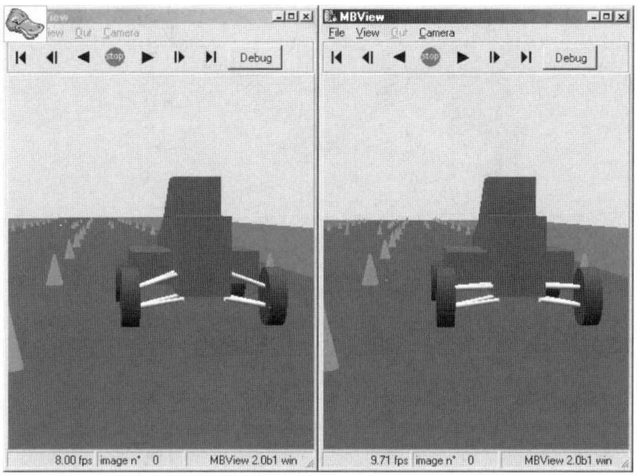

Figure 6.25: Animation snapshots - left : original suspension — right : optimized suspension [51]

6.5.4 A complete modern car

This last example (studied in [4]) deals with the modeling of a complete modern car equipped with a semi-active anti-roll system (front and rear anti-roll bars are connected via a hydraulic system which induces a rear/front coupling). Figure 6.26 shows the front wheel-axle unit consisting of a part of the chassis, the suspension rods, the dampers, the anti-roll bar, the tie rod and the rubber joint bushings.
From the multibody point of view, this vehicle is somewhat more complex than the previous ones since it involves:

- more than 100 generalized coordinates;
- multi-link suspensions (rear and front), which induce several 3D loop constraints;

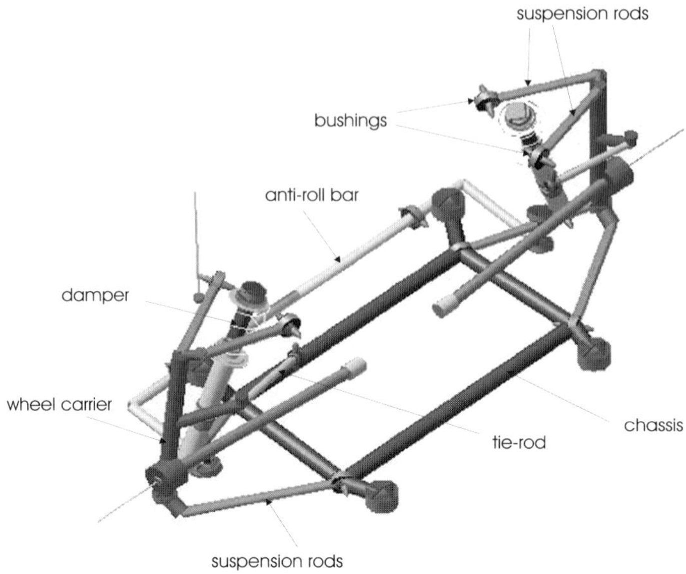

Figure 6.26: Front wheel-axle unit

- compliant bushing elements between the various rods and the chassis, the damper and the wheel carrier: these bushings can deform in translation and rotation in the three directions, inducing internal forces and torques which need to be characterized in terms of these deformations and deformation rates;
- a hydraulic model of the dampers (governed by differential equations) and of an anti-roll semi-active system (not detailed here).

Figure 6.27 illustrates the ROBOTRAN multibody representation of one suspension (front left) in which each circle denotes a joint involving a generalized coordinate.

Two representative results are shown in figures 6.28 and 6.29. In obtaining those of figure 6.28, we performed a static parameterization and computed the carbody roll when subjected to an increasing centrifugal effect (quantified in the plot by an equivalent lateral force [N]). The study was carried out in three situations: the car without anti-roll bars, with standard anti-roll bars, and with ten-fold stiffer anti-roll bars. Note that each point of the curves represents a quasi-static equilibrium of the vehicle, which can be obtained classically by time simulation, or more straightforwardly by using a Newton/Raphson iterative method applied to vector $(c_{red} - \mathcal{Q}_{red})$ of the equations of motion 4.48.

The second figure 6.29 is a time simulation of curve entry (forward velocity $V = 20 \frac{m}{\sec}$, average curve radius $= 50\,m$). After a transient response, the vehicle reaches its quasi-static equilibrium in accordance with the results of figure 6.28.

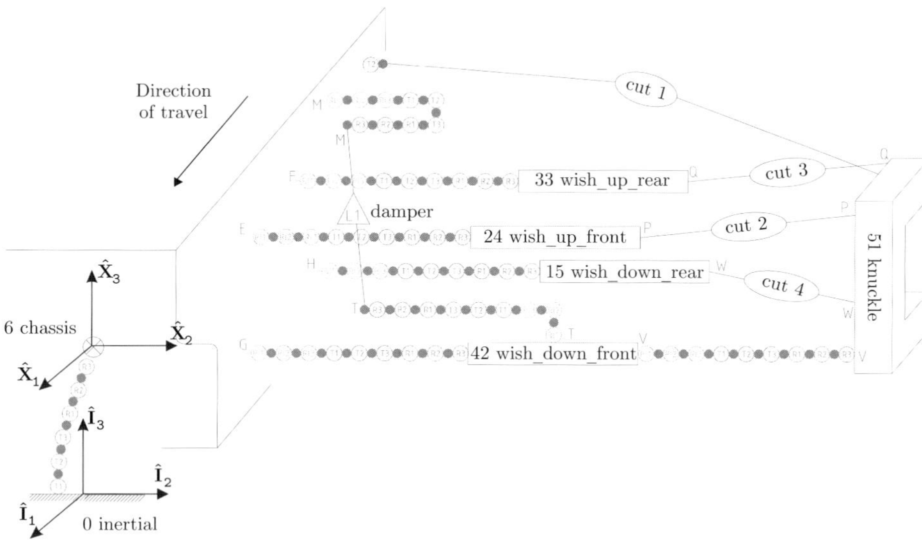

Figure 6.27: Multibody model of the front left suspension

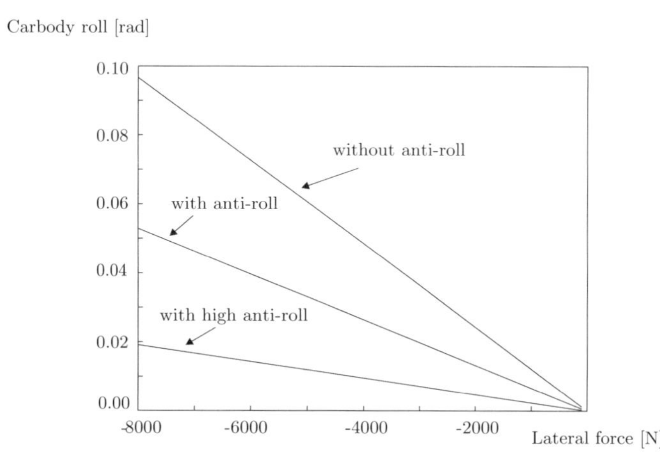

Figure 6.28: Steady curve equilibrium

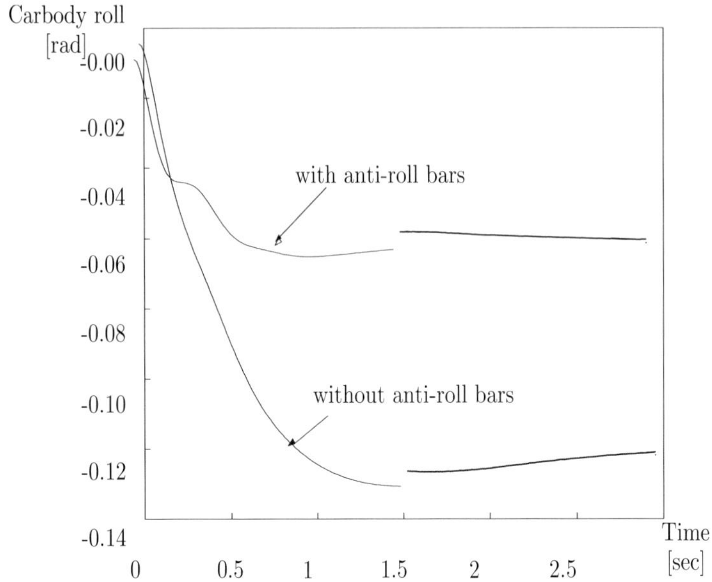

Figure 6.29: Curve entry at constant speed

Chapter 7

Railway vehicles: wheel/rail model

7.1 Introduction

Vehicles - on road or on track - certainly represent one of the most important types of application of the multibody approach. In the case of railway vehicles, an arduous aspect of the modeling phase results from the contact between the wheels and the track. Indeed, from a purely geometrical point of view, locating the contact point between a wheel and a rail becomes complicated since both are profiled, and from a dynamical point of view, the large number of parameters (shape of the profiles in contact, contact pressure, relative contact velocity, physical properties of the materials, ...) leads to complex theories such as those developed by Kalker [45].

The classical approach for the geometrical problem is dedicated to conventional railway vehicles whose "rolling elements" consist of wheelsets. The contact between a wheelset and a track has been abundantly developed in the literature (see for instance [27], [9]); the corresponding models take advantage of the fact that the left and right wheels are rigidly linked by a common axle and have, generally, identical profiles. From a dynamical point of view, the classical approach often splits the whole vehicle (and the corresponding mathematical model) into several sub-systems, the wheelset being one of them. The coupling between the latter and the rest of the vehicle is then modeled by linear springs and dampers (primary suspension) which allow relative motions with up to 6 degrees of freedom.

This classical approach is no longer suitable for new bogie designs, such as for instance the so-called "BAS 2000" bogie developed by the Belgian company B.N.- Eurorail (see figure 7.1); indeed this bogie consists of an articulated frame with independent wheels, meaning that the left and right wheels are no longer mounted on a common axle; moreover, the front and rear wheels of the bogie may have different geometrical dimensions and profiles.

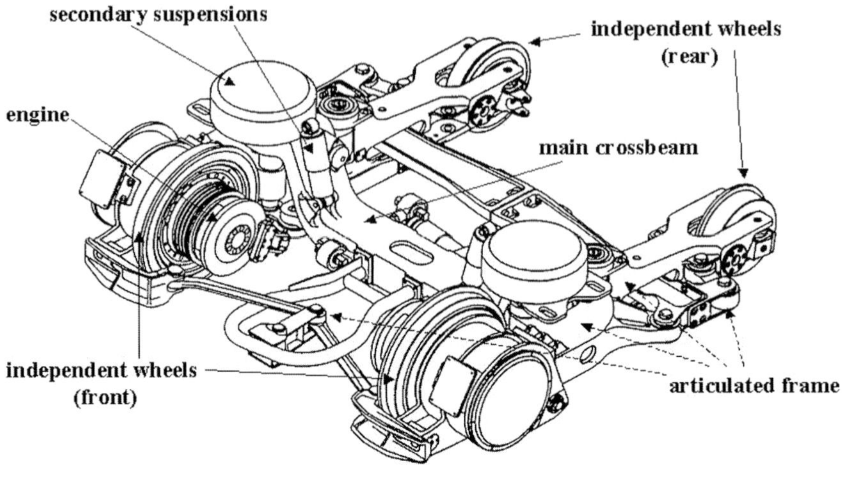

Figure 7.1: Bogie "BAS 2000"

For that reason, the classical approach is unsuitable because in the present case, each wheel of the vehicle should be considered separately. This section therefore presents an appropriate model for a single wheel/rail contact (which, being more general than a wheelset model, can deal with any railway application).

In order to take the wheel/rail contact constraints into account, we may consider as in [7] that the track belongs to the multibody system, the wheel/rail contact then being modeled as an *internal* joint. Since this approach increases the size of the model, a better efficiency is obtained by restricting the multibody system to the vehicle itself and by considering the kinematic contact constraints and the corresponding contact forces and torques as *external*, for each individual wheel [20]. One of the advantages of this method is that the auxiliary variables, which are needed to solve the geometrical problem, are easily eliminated from the constraint equations and from their time derivatives. In this way, the dynamics of the vehicle is described by a set of equations of motion which depend only on the generalized coordinates relating to the multibody representation of the vehicle: this considerably reduces the CPU time requirements.

When relative coordinates are used, most of the constraints encountered in multibody systems result from kinematic loops (for instance, the BAS 2000 articulated bogie contains five independent loops closed by connecting rods). As we shall see, each wheel/rail contact also defines a kinematic loop which is closed by the track and must also be taken into account when formulating the equations.

7.2 Wheel/rail kinematic model

Considering the previous definition of the "independent wheel", it is clear that each wheel/rail contact must be analyzed separately. An isolated wheel moving along on a straight track has five d.o.f. which could be naturally described by:

- the lateral and longitudinal displacements of the wheel center of mass G,
- the yaw and roll angles of the wheel,
- the rotation angle of the wheel around its axis of symmetry.

This set of coordinates would however require solving a preliminary nonlinear geometrical problem, because the position of the wheel/rail contact point cannot be known in advance since the rail and the wheel are profiled. This is the reason why a set of auxiliary geometrical variables is used, which explicitly refer to the position of the contact point.

7.2.1 Contact model of a wheel on a straight track

Definition of the kinematic quantities

Considering the wheel and the straight track illustrated in figure 7.2, let us first denote Q the contact point on the wheel side and P the contact point on the rail side respectively. Denoting by $\{\hat{\mathbf{I}}\}$ the inertial frame located at the fixed point O, we define the frames $\{\hat{\mathbf{Y}}\}$ and $\{\hat{\mathbf{X}}\}$ as follows[1]:

$[\hat{\mathbf{Y}}] = R^M[\hat{\mathbf{I}}]$ is the *material wheel* attached frame,

$[\hat{\mathbf{X}}] = R^G[\hat{\mathbf{I}}]$ is the *geometrical wheel* attached frame.

Note the difference between the latter two frames. The first, $\{\hat{\mathbf{Y}}\}$, is attached to the physical wheel which rotates around its axis of symmetry according to the speed of the vehicle. The second, $\{\hat{\mathbf{X}}\}$, does not take this rotational speed into account and is attached to a so-called *frozen* or *geometrical* wheel which would be rigidly attached to its bearing.

Further, we define

$$[\hat{\mathbf{S}}] = T^{\vartheta}[\hat{\mathbf{X}}], \qquad \text{with} \quad T^{\vartheta}(\vartheta) = R_2(\vartheta) \triangleq \begin{pmatrix} \cos\vartheta & 0 & -\sin\vartheta \\ 0 & 1 & 0 \\ \sin\vartheta & 0 & \cos\vartheta \end{pmatrix} \tag{7.1}$$

which is such that the wheel point of contact Q belongs to the $\{\hat{\mathbf{S}}_2, \hat{\mathbf{S}}_3\}$ plane;

$$[\hat{\mathbf{T}}] = T^{\beta}[\hat{\mathbf{S}}], \qquad \text{with} \quad T^{\beta}(\beta) = R_1(\beta) \triangleq \begin{pmatrix} 1 & 0 & 0 \\ 0 & \cos\beta & \sin\beta \\ 0 & -\sin\beta & \cos\beta \end{pmatrix} \tag{7.2}$$

[1] as we did for the wheel/ground model in chapter 6

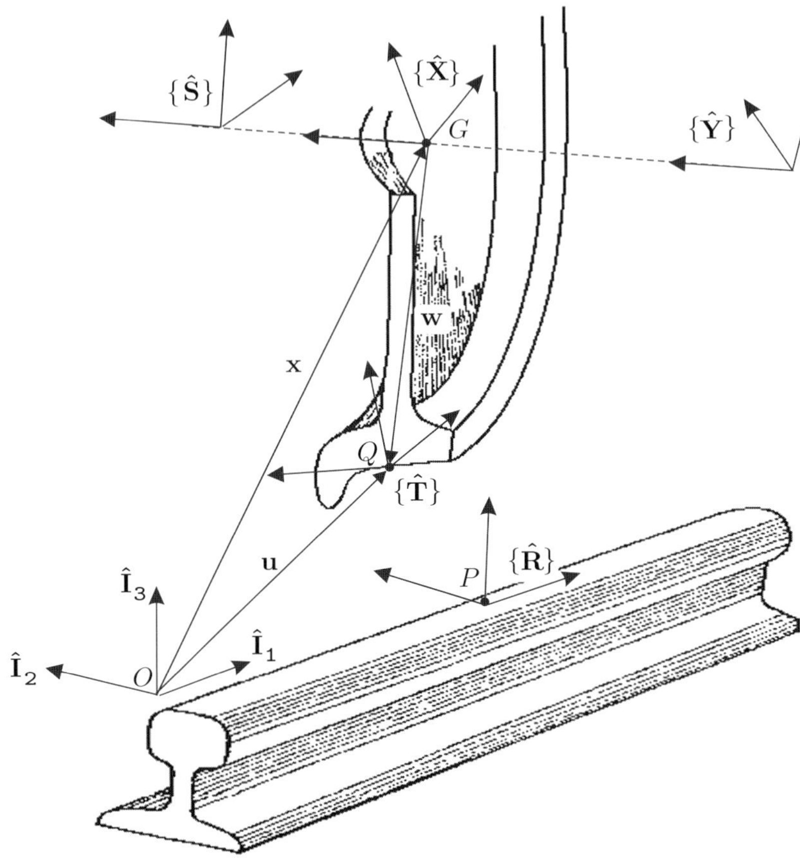

Figure 7.2: Wheel and straight track

which is such that the unit vectors $\hat{\mathbf{T}}_1$ and $\hat{\mathbf{T}}_2$ belong to the wheel tangent plane at the contact point Q.

The position of the mass center and of the wheel contact point are given by the following vectors:

$$\mathbf{x} = \overrightarrow{OG} \qquad \text{and} \qquad \mathbf{w} = \overrightarrow{GQ} = [\hat{\mathbf{S}}]^T \begin{pmatrix} 0 \\ w \\ \rho(w) \end{pmatrix} \tag{7.3}$$

where $\rho(w)$ is the *wheel profile* function with[2] $\beta(w) = \arctan(\rho'(w))$. Since the *frozen* wheel is considered to be a part of the multibody system, the position of point G and the orientation of the $\{\hat{\mathbf{X}}\}$ frame depend only on the multibody generalized coordinates q (see figure 7.3). On the other hand, two auxiliary vari-

[2] where $\rho'(w)$ stands for $\frac{d\rho}{dw}$.

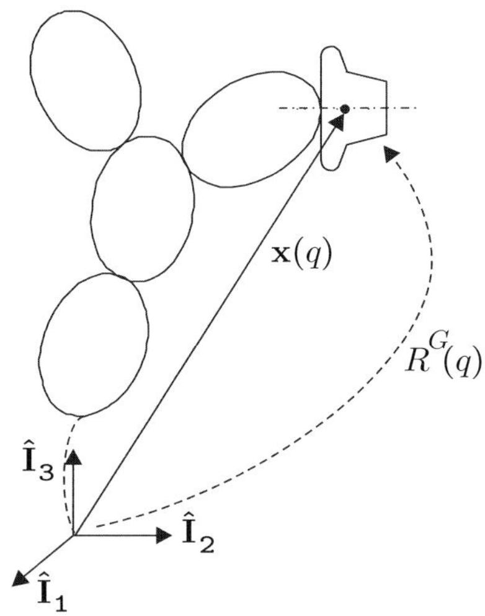

Figure 7.3: Location of the *geometrical* wheel in the multibody system.

ables, w and ϑ, have been used to locate point Q and to specify the orientation of its related frame $\{\hat{\mathbf{T}}\}$:

$$
\begin{aligned}
[\hat{\mathbf{T}}] &= T^{\beta} T^{\vartheta} R^{G}(q)[\hat{\mathbf{I}}] \\
\mathbf{u} \stackrel{\Delta}{=} \mathbf{x} + \mathbf{w} &= [\hat{\mathbf{I}}]^{T} \begin{pmatrix} u_1 \\ u_2 \\ u_3 \end{pmatrix} \\
&= [\hat{\mathbf{I}}]^{T} \left(x(q) + \left(R^{G}(q)\right)^{T} \left(T^{\vartheta}(\vartheta)\right)^{T} \begin{pmatrix} 0 \\ w \\ \rho(w) \end{pmatrix} \right)
\end{aligned}
$$

Geometrical constraints

In order to ensure the geometrical contact between the profiled wheel and rail:

- point Q must belong to the rail surface:

$$h_1(q, w, \vartheta) \stackrel{\Delta}{=} u_3 - \mu(u_2) = 0 \tag{7.4}$$

  where μ is the describing function of the *rail profile* in the $\{\hat{\mathbf{I}}_2, \hat{\mathbf{I}}_3\}$ plane.

- unit vector $\hat{\mathbf{T}}_3$ must be normal to the rail surface:

$$h_2(q, w, \vartheta) \stackrel{\Delta}{=} \hat{\mathbf{T}}_3 \,.\, \hat{\mathbf{R}}_1 = 0 \tag{7.5}$$

$$h_3(q, w, \vartheta) \stackrel{\Delta}{=} \hat{\mathbf{T}}_3 \,.\, \hat{\mathbf{R}}_2 = 0 \tag{7.6}$$

where we have introduced a new frame $\{\hat{\mathbf{R}}\}$, defined as follows:

$$[\hat{\mathbf{R}}] = T^{\alpha}[\hat{\mathbf{I}}] \qquad \text{with} \quad T^{\alpha} = R_1(\alpha) \triangleq \begin{pmatrix} 1 & 0 & 0 \\ 0 & \cos\alpha & \sin\alpha \\ 0 & -\sin\alpha & \cos\alpha \end{pmatrix} \tag{7.7}$$

and which is such that the unit vectors $\hat{\mathbf{R}}_1$ and $\hat{\mathbf{R}}_2$ belong to the tangent plane to the rail surface at contact point P. The angle α must be consistent with the rail profile at the contact point and therefore satisfy the relation

$$\tan\alpha = \mu'(u_2) \qquad \text{with} \quad \mu'(u_2) \triangleq \frac{d\mu}{du_2} \tag{7.8}$$

The constraints 7.5 and 7.6 are orientation constraints for the wheel, and it can be shown (see [74]) that the associated Lagrange multipliers are two pure torques (along the tangent directions) applied by the rail at contact point Q. However, the wheel and rail being considered as rigid, and assuming that their respective profiles are not conformal (i.e. the contact patch area reduces to point Q), the rail is physically unable to apply such pure torques to the wheel. As a consequence, it can be shown (see [74]) that these Lagrange multipliers are identically equal to zero.

Nevertheless, these orientation constraints will allow us to find, in an explicit way, the value of the auxiliary variables w and ϑ corresponding to a given value of the generalized coordinates q. Equation 7.5 can be developed by expressing the scalar product in the inertial frame; it leads to

$$\hat{\mathbf{T}}_3 \,.\, \hat{\mathbf{R}}_1 = \begin{pmatrix} 0 & 0 & 1 \end{pmatrix} T^{\beta}(\beta)\, T^{\vartheta}(\vartheta)\, R^{G}(q) \begin{pmatrix} 1 \\ 0 \\ 0 \end{pmatrix} = 0$$

Therefore

$$R^{G}_{(3,1)}\cos\vartheta + R^{G}_{(1,1)}\sin\vartheta = R^{G}_{(2,1)}\tan\beta(w) \tag{7.9}$$

This typical equation has two well-known solutions in $(\sin\vartheta, \cos\vartheta)$ one of which $(\cos\vartheta < 0)$ is irrelevant in our case because it corresponds to a contact point located on the upper part of the wheel.

In the same way, equation 7.6 can be developed as follows:

$$\hat{\mathbf{T}}_3 \,.\, \hat{\mathbf{R}}_2 = \begin{pmatrix} 0 & 0 & 1 \end{pmatrix} T^{\beta}(\beta)\, T^{\vartheta}(\vartheta)\, R^{G}(q) \begin{pmatrix} 0 \\ \cos\alpha \\ \sin\alpha \end{pmatrix} = 0$$

Assuming that $\cos\alpha$ and $\cos\beta$ are non zero (zero would correspond to a contact on a purely vertical tangent plane), this leads to

$$\tan\alpha = -\frac{R^{G}_{(1,2)}\sin\vartheta - R^{G}_{(2,2)}\tan\beta(w) + R^{G}_{(3,2)}\cos\vartheta}{R^{G}_{(1,3)}\sin\vartheta - R^{G}_{(2,3)}\tan\beta(w) + R^{G}_{(3,3)}\cos\vartheta} \tag{7.10}$$

where angle α must also satisfy relation 7.8.

In order to solve the geometrical constraints 7.4, 7.9 and 7.10, we could blindly resort to the Newton/Raphson algorithm as explained in section 4.6.1. By experience, for some practical wheel/rail pairs, the nonlinearity of the orientation constraints 7.9 and 7.10 can lead to a numerical divergence of the Newton/Raphson formula 4.49.

To circumvent this problem, the following numerical procedure is thus carried out for each wheel:

1. computation of the rotation matrix $R^G(q)$ for a given multibody configuration by means of the recursive kinematics of section 4.5 (see figure 7.3),
2. choice of a circle line on the wheel profile: $w, \rho(w), \beta(w)$,
3. computation of ϑ from 7.9,
4. computation of $\tan\alpha$ from 7.10 and computation of $\tan\alpha$ from 7.8
5. choice of a new circle (step 2) until $\left|\tan\alpha^{(7.10)} - \tan\alpha^{(7.8)}\right| < \epsilon$, a numerical threshold[3], using a *dichotomic* method, which takes advantage of the fact that on one side of the "assumed" contact point, the slope of the wheel profile 7.10 is larger than that of the rail 7.8, contrary to the other side (see figure 7.4).

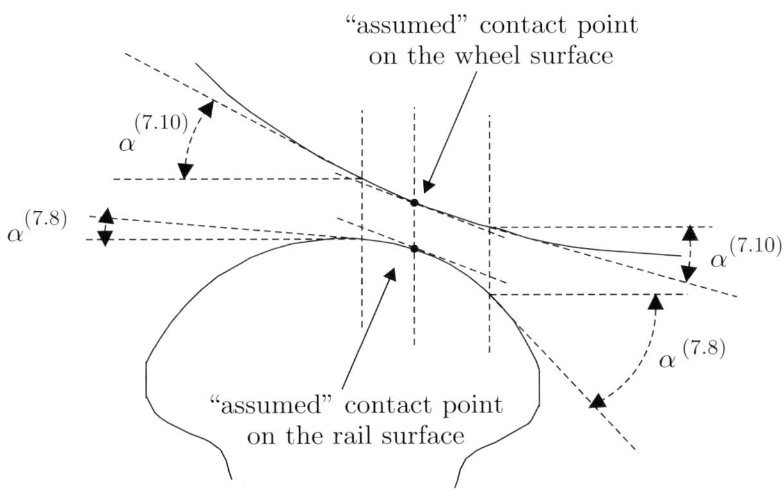

Figure 7.4: Geometry of the wheel and rail contact point.

After computing the auxiliary variables (w, ϑ), we may consider equation 7.4 as an external constraint linking the generalized coordinates q of the multibody

[3] $\epsilon = 10^{-8}...10^{-9}$

system. This constraint, which expresses that the wheel has only five d.o.f. with respect to inertial space, is implicit and nonlinear. It will thus be taken into account by means of the Lagrange multipliers technique and the coordinate partitioning method (see section 4.6.1) for which the generalized coordinates q must be partitioned into dependent (v) and independent (u) variables.

Figure 7.5 clearly makes the distinction between two interwoven iterative process:

1. the dichotomic procedure explained above, applied to each wheel separately, which provides a new estimate for the auxiliary variables w, ϑ for a given configuration of the system (u, v);

2. the Newton/Raphson procedure, applied to the complete set of multibody constraints (i.e. including loop and user constraints), which provides a new estimate for the dependent variables v.

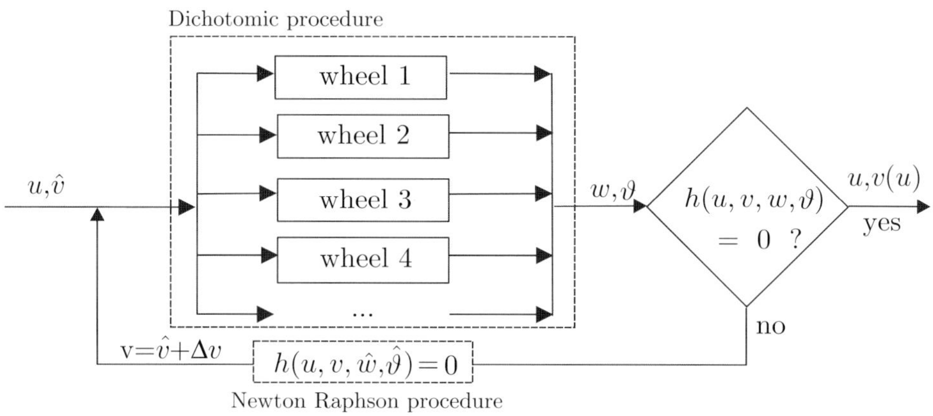

Figure 7.5: General scheme for the geometrical solution of the contact.

As shown in figure 7.5, the set of constraints 7.4 associated with each wheel of the vehicle – denoted h – is solved by a Newton-Raphson procedure, for which the Jacobian matrix $\frac{dh}{dv^T}$ is needed:

$$\frac{dh}{dv^T} = \frac{\partial h}{\partial v^T} + \frac{\partial h}{\partial w}\frac{\partial w}{\partial v^T} + \frac{\partial h}{\partial \vartheta}\frac{\partial \vartheta}{\partial v^T} \tag{7.11}$$

As will be seen later, the quantities $\frac{\partial h}{\partial w}$ and $\frac{\partial h}{\partial \vartheta}$ are equal to zero *when the constraints are satisfied.* Thus, insofar as the initial conditions of the Newton-Raphson procedure are close to the final solution, a good convergence is obtained by "freezing" the auxiliary variables $(\hat{w}, \hat{\vartheta})$ within this procedure and thus restricting the Jacobian to the first term of 7.11: $\frac{\partial h}{\partial v^T}$.

Although it is more costly than a global Newton/Raphson method which would also iterate on the orientation constraints 7.9, 7.10, the procedure summarized in figure 7.5 is far more reliable in terms of convergence for the wheel/rail contact problem, especially when highly nonlinear profiles are considered (see example of section 7.4.1).

Constraint derivatives

In order to obtain the Jacobian matrix corresponding to the geometrical constraint 7.4 associated with the wheel/rail contact, the partial derivatives of the constraint function $h_1(q, w, \vartheta)$ with respect to the generalized coordinates q are needed. Let us calculate the time derivative of 7.4:

$$\dot{h}_1 = \frac{d}{dt}\left(\mathbf{u}\,.\,\hat{\mathbf{I}}_3\right) - \frac{d}{dt}\left(\mu(\mathbf{u}\,.\,\hat{\mathbf{I}}_2)\right) = \dot{\mathbf{u}}\,.\,\hat{\mathbf{I}}_3 - \mu'\left(u_2\right)\dot{\mathbf{u}}\,.\,\hat{\mathbf{I}}_2 = 0$$

Using equations 7.7 and 7.8 and assuming, as previously explained, that α is different from $\pm\frac{\pi}{2}$, it can be written

$$\dot{h}_1 = \frac{1}{\cos\alpha}\left(\dot{\mathbf{u}}\,.\,\hat{\mathbf{R}}_3\right) = 0$$

This finally leads to the relation

$$\dot{\mathbf{u}}\,.\,\hat{\mathbf{R}}_3 = 0 \tag{7.12}$$

which simply expresses that the geometrical contact point has no velocity component along the direction normal to the contact plane. Since this velocity constraint is equivalent to the contact constraint 7.4 (assuming compatible initial conditions), we can now extract from 7.12 the Jacobian matrix associated with the Lagrange multipliers in the dynamical equations.
The velocity vector $\dot{\mathbf{u}}$ can be further developed as follows:

$$\dot{\mathbf{u}} = \dot{\mathbf{x}} + \dot{\mathbf{w}} = \dot{\mathbf{x}} + \overset{\circ}{\mathbf{w}} + \boldsymbol{\omega}^S \times \mathbf{w} \tag{7.13}$$

where

$$\begin{aligned} \overset{\circ}{\mathbf{w}} &= [\hat{\mathbf{S}}]^T \begin{pmatrix} 0 \\ 1 \\ \rho'(w) \end{pmatrix} \dot{w} = \left(\hat{\mathbf{S}}_2 + \rho'(w)\hat{\mathbf{S}}_3\right)\dot{w} \\ \text{with} \quad \rho'(w) &= \tan\beta \quad \text{and} \quad \boldsymbol{\omega}^S = \boldsymbol{\omega}^X + \dot{\vartheta}\hat{\mathbf{X}}_2 \end{aligned}$$

and where $\boldsymbol{\omega}^S$ and $\boldsymbol{\omega}^X$ are the angular velocities of the $\{\hat{\mathbf{S}}\}$ and $\{\hat{\mathbf{X}}\}$ frames respectively.
By using equations 7.1, 7.2 and 7.3, we can show that

$$\hat{\mathbf{S}}_2 + \rho'(w)\hat{\mathbf{S}}_3 = \frac{1}{\cos\beta}\hat{\mathbf{T}}_2 \quad \text{and} \quad \dot{\vartheta}\,\hat{\mathbf{X}}_2 \times \mathbf{w} = \dot{\vartheta}\,\rho(w)\hat{\mathbf{T}}_1$$

Equation 7.12 can then be rewritten:

$$\left(\dot{\mathbf{x}} + \frac{\dot{w}}{\cos\beta}\hat{\mathbf{T}}_2 + \boldsymbol{\omega}^X \times \mathbf{w} + \dot{\vartheta}\,\rho(w)\hat{\mathbf{T}}_1\right).\,\hat{\mathbf{R}}_3 = 0$$

Since the constraints 7.5 and 7.6 imply that $\hat{\mathbf{R}}_3$ is normal to the plane $\{\hat{\mathbf{T}}_1, \hat{\mathbf{T}}_2\}$, this expression reduces to:

$$\left(\dot{\mathbf{x}} + \boldsymbol{\omega}^X \times \mathbf{w}\right) . \hat{\mathbf{R}}_3 = 0 \tag{7.14}$$

Since vector $\boldsymbol{\omega}^X$ is the angular velocity of the "frozen" wheel, we observe that the term between brackets in the left hand side of equation 7.14 represents the velocity of a point attached to the "frozen" wheel when it is located at Q. Through linearity of $\dot{\mathbf{x}}$ and $\boldsymbol{\omega}^X$ with respect to the multibody generalized velocities $\dot{q}$, equation 7.14 has the desired form:

$$J\left(q, w(q), \vartheta(q)\right) \dot{q} = 0 \tag{7.15}$$

since it does not depend on the auxiliary velocities $\dot{w}$ and $\dot{\vartheta}$. As mentioned before, $\frac{\partial h}{\partial w}$ and $\frac{\partial h}{\partial \vartheta}$ are thus equal to zero in equation 7.11 when the constraints are satisfied.

Therefore, this matrix J can be used as the Jacobian matrix associated with the constraint 7.4 of the multibody system. Furthermore, it can easily be shown that the corresponding Lagrange multiplier is equal to the normal component of the contact force (see [74]) whose value is indispensable for the calculation of the tangent forces.

Since the differential-algebraic system 4.1 formed by the dynamical and constraint equations will be solved using the coordinate partitioning method (section 4.6.1), the second time derivatives of the constraints are also needed. Let us thus calculate the time derivative of 7.12:

$$\frac{d}{dt}\left(\dot{\mathbf{u}} . \hat{\mathbf{R}}_3\right) = \ddot{\mathbf{u}} . \hat{\mathbf{R}}_3 + \dot{\mathbf{u}} . \left(\boldsymbol{\omega}^R \times \hat{\mathbf{R}}_3\right) \tag{7.16}$$

where $\boldsymbol{\omega}^R$ is the angular velocity of the $\{\hat{\mathbf{R}}\}$ frame. As mentioned above, the velocity vector $\dot{\mathbf{u}}$ can be written as:

$$\dot{\mathbf{u}} = \dot{\mathbf{x}} + \dot{\mathbf{w}} \qquad \text{with} \quad \dot{\mathbf{w}} = \tilde{\omega}^X . \mathbf{w} + \dot{\vartheta}\, \rho(w) \hat{\mathbf{T}}_1 + \frac{\dot{w}}{\cos\beta} \hat{\mathbf{T}}_2 \tag{7.17}$$

The acceleration vector $\ddot{\mathbf{u}}$ can then be developed as follows:

$$\begin{aligned} \ddot{\mathbf{u}} &= \ddot{\mathbf{x}} + \dot{\tilde{\omega}}^X . \mathbf{w} + \tilde{\omega}^X . \dot{\mathbf{w}} + \tilde{\omega}^T . \left(\dot{\vartheta}\, \rho(w) \hat{\mathbf{T}}_1 + \frac{\dot{w}}{\cos\beta} \hat{\mathbf{T}}_2\right) \\ &\quad + \frac{d}{dt}\left(\dot{\vartheta}\, \rho(w)\right) \hat{\mathbf{T}}_1 + \frac{d}{dt}\left(\frac{\dot{w}}{\cos\beta}\right) \hat{\mathbf{T}}_2 \end{aligned}$$

where $\boldsymbol{\omega}^T$ is the angular velocity vector of the $\{\hat{\mathbf{T}}\}$ frame.

Since, from the constraints 7.5 and 7.6, $\hat{\mathbf{T}}_3$ and $\hat{\mathbf{R}}_3$ are continuously aligned, the following property holds:

$$\tilde{\omega}^T . \hat{\mathbf{R}}_3 = \tilde{\omega}^R . \hat{\mathbf{R}}_3$$

Using the latter and substituting 7.17 into 7.16, we obtain

$$\left(\ddot{\mathbf{x}} + \dot{\tilde{\boldsymbol{\omega}}}^X . \mathbf{w} + \tilde{\boldsymbol{\omega}}^X . \dot{\mathbf{w}} - \tilde{\boldsymbol{\omega}}^R . \left(\dot{\mathbf{x}} + \tilde{\boldsymbol{\omega}}^X . \mathbf{w}\right)\right) . \hat{\mathbf{R}}_3 = 0 \tag{7.18}$$

which does not depend on the second time derivatives $\ddot{\vartheta}$, $\ddot{w}$ of the auxiliary variables: their computation is thus superfluous[4].

Developing $\dot{\mathbf{w}}$ as in 7.17, the final form of 7.16 becomes

$$\left(\ddot{\mathbf{x}} + \left(\dot{\tilde{\boldsymbol{\omega}}}^X + \tilde{\boldsymbol{\omega}}^X . \tilde{\boldsymbol{\omega}}^X\right) . \mathbf{w} + \tilde{\boldsymbol{\omega}}^X . \left(\dot{\vartheta}\,\rho(w)\hat{\mathbf{T}}_1 + \frac{\dot{w}}{\cos\beta}\hat{\mathbf{T}}_2\right)\right) . \hat{\mathbf{R}}_3$$
$$- \left(\tilde{\boldsymbol{\omega}}^R . \left(\dot{\mathbf{x}} + \tilde{\boldsymbol{\omega}}^X . \mathbf{w}\right)\right) . \hat{\mathbf{R}}_3 = 0 \tag{7.19}$$

where $\boldsymbol{\omega}^R = \dot{\alpha}\,\hat{\mathbf{I}}_1$ from 7.7.

This last equation 7.19 corresponds to the general form

$$J\ddot{q} + \left(\frac{\partial J}{\partial q^T}\dot{q} + \frac{\partial J}{\partial w}\dot{w} + \frac{\partial J}{\partial \vartheta}\dot{\vartheta}\right) = 0 \tag{7.20}$$

from which the velocities $\dot{w}$ and $\dot{\vartheta}$ of the auxiliary variables should now be eliminated. For this purpose, by differentiating the auxiliary constraints 7.5 and 7.6 with respect to time, we obtain

$$\boldsymbol{\omega}^T . \hat{\mathbf{R}}_2 = 0 \quad \text{and} \quad \boldsymbol{\omega}^T . \hat{\mathbf{R}}_1 = \dot{\alpha} \tag{7.21}$$

where:

$$\boldsymbol{\omega}^T = \boldsymbol{\omega}^X + \dot{\vartheta}\hat{\mathbf{X}}_2 + \dot{\beta}\hat{\mathbf{S}}_1 = \boldsymbol{\omega}^X + \dot{\vartheta}\hat{\mathbf{X}}_2 + \frac{\rho''(w)}{1 + \left(\rho'(w)\right)^2}\,\dot{w}\,\hat{\mathbf{S}}_1$$

and, from 7.8,

$$\dot{\alpha} = \frac{d}{dt}\left(\arctan\left(\mu'(u_2)\right)\right) = \left(\frac{\mu''(u_2)}{1 + \left(\mu'(u_2)\right)^2}\right)\dot{\mathbf{u}} . \hat{\mathbf{I}}_2$$

where $\dot{\mathbf{u}}$ is given by 7.17 as a linear function of $\dot{\vartheta}$ and $\dot{w}$. By substituting the latter two expressions into 7.21, we finally obtain a 2 by 2 linear system of equations with the form

$$\begin{pmatrix} a_{11} & a_{12} \\ a_{21} & a_{22} \end{pmatrix}\begin{pmatrix} \dot{w} \\ \dot{\vartheta} \end{pmatrix} = \begin{pmatrix} b_1 \\ b_2 \end{pmatrix}\dot{q} \tag{7.22}$$

Since this system is invertible analytically, the auxiliary velocities $\dot{w}$ and $\dot{\vartheta}$ can be eliminated from 7.20 in order to obtain the second time derivative associated with the constraint 7.4:

$$J\left(q, w(q), \vartheta(q)\right)\ddot{q} + \dot{J}\left(\dot{q}, q, \dot{w}(q,\dot{q}), \dot{\vartheta}(q,\dot{q}), w(q), \vartheta(q)\right)\dot{q} = 0 \tag{7.23}$$

[4] contrary to what it is suposed in [82].

7.2.2 Contact of a wheel on a curved track (with constant radius)

These developments can be extended to curved track by considering for each wheel (see figure 7.6) a *local* straight track whose center line, tangent to the curved track at point C, is perpendicular to the vertical plane Π which contains the track center of curvature O and the center of mass G of the wheel.

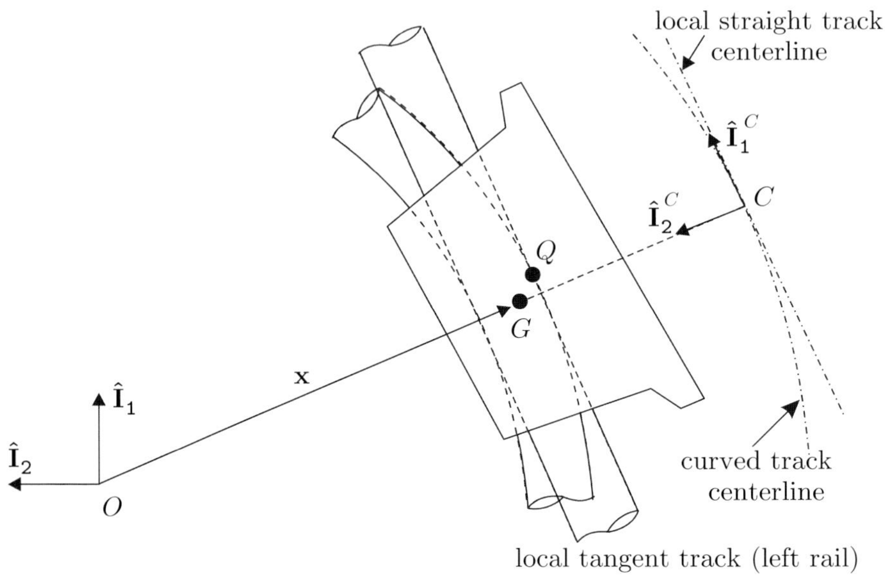

Figure 7.6: Wheel on a curved track with constant radius.

Let us define the vector $\mathbf{u}^* \triangleq \overrightarrow{CQ}$. The geometrical constraints express that:

1. point Q belongs to the (straight) rail surface:

$$h_1^*(q, w, \vartheta) = \mathbf{u}^* \,.\, \hat{\mathbf{I}}_3^C - \mu\left(\mathbf{u}^* \,.\, \hat{\mathbf{I}}_2^C\right) = 0 \tag{7.24}$$

   where μ is the describing function of the local straight track rail profile, defined in the $\{\hat{\mathbf{I}}^C\}$ frame,

2. the surfaces of the wheel and rail must be tangent at the contact point:

$$h_2^*(q, w, \vartheta) = h_2 \quad \text{with } h_2 \text{ given by 7.5} \tag{7.25}$$

$$h_3^*(q, w, \vartheta) = h_3 \quad \text{with } h_3 \text{ given by 7.6} \tag{7.26}$$

   where w, ϑ are the auxiliary variables already defined for the straight track.

When evaluated in extreme cases (high curvature, important yaw angle of the wheel, high inclined wheel profiles, ...), the geometrical error due to the assumption of a local straight track is less than 5.10^{-6} meters. The developments related to the time derivatives of these constraints are thus similar to those presented above for a straight track, all calculations essentially being made with respect to the $\{\hat{\mathbf{I}}^C\}$ moving frame.

7.3 Wheel/rail contact forces and torques

7.3.1 Wheel/rail contact kinematics

As for the wheel/ground contact forces presented in section 6.4, wheel/rail contact forces rely on specific kinematic quantities measured at the *material* point of the wheel which coincides with the *geometrical* contact point Q, previously computed by solving the geometrical constraints 7.4, 7.5, 7.6. This contact point is in fact a simplified geometrical representation of a small elliptical contact surface which can be divided into a *rolling* and a *slipping* region. The relative size and the shape of these regions in the ellipse strongly depend on the linear and angular velocities of contact: a detailed discussion of this very complex phenomenon can be found in the literature (ex. [45], [27]). As our model relies on a point contact – which coincides with the ellipse center – the above-mentioned rolling and slipping phenomena must be condensed in a unique kinematic concept, denoted as *creepages*. They thus represent the deviations from a pure rolling motion of the wheel on the rail and they are at the root of contact force models, such as those proposed by Kalker in [45].
Referring to figure 7.2, the creepages of the *material* contact point which coincides with the *geometrical* contact point Q are defined with respect to the contact plane $\{\hat{\mathbf{T}}_1, \hat{\mathbf{T}}_2\}$ as follows:

- The *longitudinal creepage* ξ_x is the longitudinal velocity of the material contact point divided by the longitudinal velocity[5] of the wheel center:

$$\xi_x \triangleq \frac{\left(\dot{\mathbf{x}} + \boldsymbol{\omega}^Y \times \mathbf{w}\right) . \hat{\mathbf{T}}_1}{\dot{\mathbf{x}} . \hat{\mathbf{R}}_1} \tag{7.27}$$

- The *lateral creepage* ξ_y is the lateral velocity of the material contact point divided by the longitudinal velocity of the wheel center:

$$\xi_y \triangleq \frac{\left(\dot{\mathbf{x}} + \boldsymbol{\omega}^Y \times \mathbf{w}\right) . \hat{\mathbf{T}}_2}{\dot{\mathbf{x}} . \hat{\mathbf{R}}_1} \tag{7.28}$$

- The *spin creepage* ξ_{sp} is the angular velocity of the wheel in the direction normal to the contact plane, divided by the longitudinal velocity of the

[5] Depending on the authors, the denominator can slighty differs, being for instance the rolling circumferential velocity or the mean of the latter and of the forward wheel velocity, ...

wheel center:

$$\xi_{sp} \triangleq \frac{\boldsymbol{\omega}^Y \,.\, \hat{\mathbf{T}}_3}{\dot{\mathbf{x}} \,.\, \hat{\mathbf{R}}_1} \tag{7.29}$$

The two first are dimensionless while the spin has the dimension of length^{-1}.

7.3.2 Wheel/rail contact forces

Let us first write the contact forces and torques in vector form as follows:

$$\mathbf{F}_w = F_{long}\,\hat{\mathbf{T}}_1 + F_{lat}\,\hat{\mathbf{T}}_2 + F_{vert}\,\hat{\mathbf{T}}_3 \quad \text{and} \quad \mathbf{M}_w = M_{spin}\,\hat{\mathbf{T}}_3$$

According to Kalker's linear theory ([45], [27]), for small creepages, the following linear relationship holds:

$$\begin{pmatrix} F_{long} \\ F_{lat} \\ M_{spin} \end{pmatrix} = \begin{pmatrix} -k_{11} & 0 & 0 \\ 0 & -k_{22} & -k_{23} \\ 0 & k_{23} & -k_{33} \end{pmatrix} \begin{pmatrix} \xi_x \\ \xi_y \\ \xi_{sp} \end{pmatrix} \tag{7.30}$$

where:

$$\begin{aligned} k_{11} &= (ab)G^{sh}c_{11} \\ k_{22} &= (ab)G^{sh}c_{22} \\ k_{33} &= (ab)^2\,G^{sh}c_{33} \\ k_{23} &= (ab)^{3/2}\,G^{sh}c_{23} \end{aligned}$$

in which:

- G^{sh} is the combined shear modulus of rigidity of wheel and rail materials;
- a, b respectively denote the longitudinal and lateral semi-axes of the contact ellipse. The product (ab) depends – nonlinearly – on the normal force F_{vert} and on the contact curvatures, the Poisson coefficients σ^{Pois} and the Young moduli of elasticity E^Y of the materials in contact;
- c_{ij} are the Kalker's coefficients, which he tabulated as a function of the Poisson coefficients σ^{Pois} and the a/b ratio.

Let us point out that the normal force F_{vert} is a constraint force which is already taken into account via the Lagrange multiplier λ associated with the normal constraint 7.14. For the numerical simulation its value, which is required for the computation of the Kalker's coefficients, can be reasonably picked up from the previous time integration step. Indeed, computing the current value of F_{vert} $(= \lambda)$ would require a global iterative procedure on the whole set of equations, since creep forces and torque in relation (7.30) nonlinearly depend on F_{vert}.

Finally, as for the wheel/ground contact, creep forces and torque values must be saturated when large creepages occur. Various models can be found in the literature (see [27]): the most elementary ones only deal with the saturation of the forces with respect to the longitudinal and lateral creepages. Refinements can be found in models (as developed by Kalker for instance) which carefully take the spin creepage influence into account, especially when the contact occurs in the wheel flange region where ξ_{sp} can become very large.

7.4 Applications in railway dynamics

This section first presents some simulation results for conventional vehicles which, by comparison with those obtained by means of classical methods, validate the single wheel/rail model of the previous section. The model of a non conventional vehicle, equipped with BAS 2000 articulated bogies (depicted in figure 7.1), is then described and illustrated with some typical simulation results.

7.4.1 Geometrical contact between a S1002 wheelset and UIC60 rails

The geometrical particularity of the standardized "S1002" wheel profile (see figure 7.7) on a standardized "UIC60" rail profile (see figure 7.8) with a $\frac{1}{40}$ cant slope is the following: when the wheelset is progressively displaced laterally from

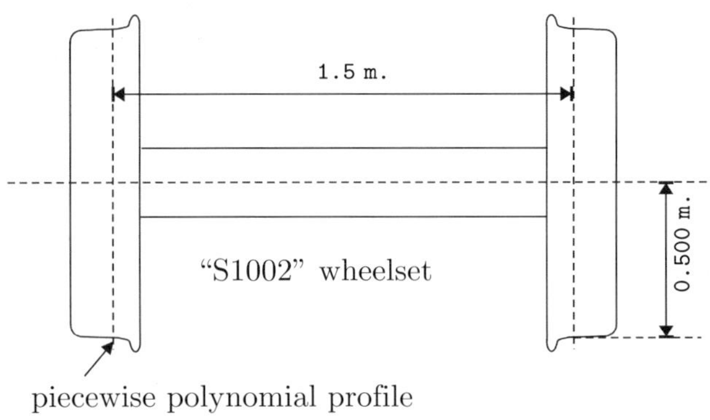

Figure 7.7: S1002 wheel profile

its central position on the track to the extreme position corresponding to the derailment limit, the location of the contact point on the wheel and rail profiles does not change continuously, but three *jumps* occur[6].

[6]assuming new profiles which perfectly fit the standardized norms, and assuming the wheel and rail are perfectly rigid.

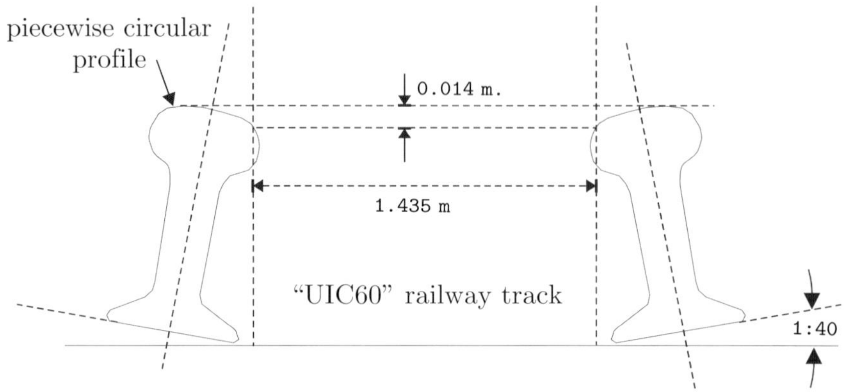

Figure 7.8: UIC60 rail profile

These discontinuities obviously affect the rolling radii (see figure 7.9) and cause modeling difficulties (for the contact geometry, the computation of contact forces and the numerical analysis). In the literature (see [28] for instance), the locations of these jumps (versus the lateral wheelset displacement) vary, depending on the geometrical modeling approach which is used. In some extreme cases, the first contact jump simply disappears because of the smoothing techniques used in describing the wheel and rail profiles. A rigorous approach to this geometrical problem is presented in [95]. Figure 7.9 compares the results obtained in [95] for a rigid wheelset with those obtained with the present single wheel/rail contact model, in terms of variations of the difference between left and right rolling radii, given as a function of the lateral displacement of the wheelset. One may observe on this figure the three discontinuities which clearly illustrate the contact jump phenomenon, typical of the S1002 wheelset / UIC60 rail pair.

7.4.2 Limit cycle of a rigid wheelset at constant speed

Reference [43] makes an extensive study of the limit cycles and chaotic motions of (rigid) wheelsets and (classical) bogies. Such behaviors are of great practical importance. Indeed, one essential characteristic of a railway vehicle is its *critical speed.* The latter, denoted v^{cr}, is generally determined by means of a modal analysis of the vehicle running at constant speed on a straight track. It corresponds to the speed for which an eigenmode involving lateral and yaw motions of the wheelsets (or the bogies) becomes unstable and could provoke the derailment of the vehicle. However, as it results from a linear analysis, the critical speed v^{cr} does not provide any information on the occurrence of *limit cycles* which, even if they are stable, are unacceptable from a practical point of view. Such limit cycles appear for speeds above a certain limit velocity denoted v^{lim}. Since $v^{lim} < v^{cr}$, the limit speed v^{lim} should be considered in practice

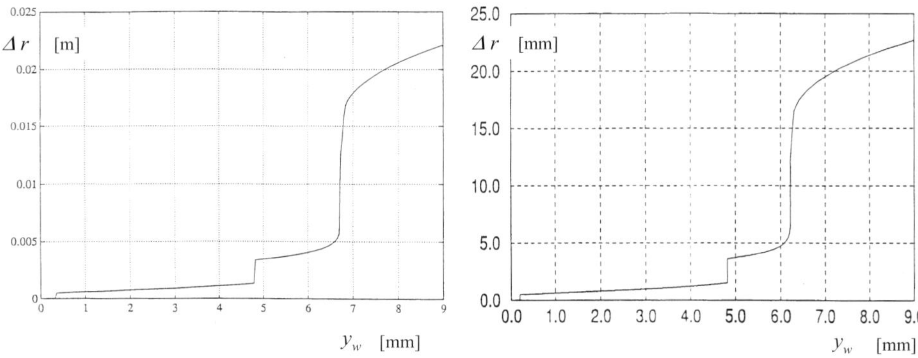

Figure 7.9: Rolling radius difference Δr between left and right wheels versus wheelset lateral displacement y_w (left : results from [20] - right : results from [95]

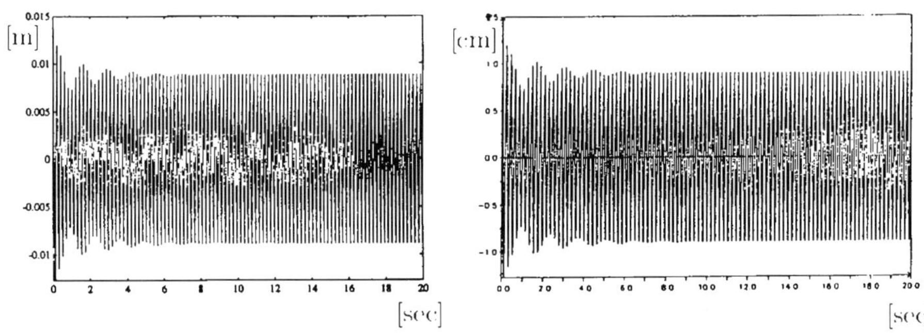

Figure 7.10: Wheelset limit cycle (left: from [20] - right: from [43])

as the effective critical speed. A nonlinear dynamical analysis (using numerical integration) of the vehicle at constant speed is one of the means to detect the presence of limit cycles. Figure 7.10 shows an illustrative example of such an analysis, performed on the right via a classical wheelset approach [43] and on the left by using the present wheel/rail contact model imbedded in a multibody representation of the wheelset [20]. By looking at this figure, we see that such a behavior cannot be accepted during an actual vehicle ride!

7.4.3 BAS 2000 bogie

The development of a single-wheel contact model – instead of a wheelset model – within the context of a multibody approach is motivated by applications which involve articulated bogies like the BAS 2000 represented in figure 7.1. The open-loop multibody model of this bogie, sketched in figure 7.11, contains 24 elementary joints (either revolute or prismatic) distributed as follows:

- 6 (fictitious) joints connecting the ground and the crossbeam of the bogie,
- 10 revolute joints within the articulated structure of the bogie itself,
- 4×2 joints (one prismatic and one revolute) connecting the wheels to the bogie.

The open-loop structure of the bogie was obtained by cutting six independent kinematic loops: five of these must be closed by connecting rods and the last one by a ball joint (see section 4.1). These kinematic loops imply a set of 8 independent closure constraints. In addition, each wheel induces an additional constraint due to its contact with the rail. The constrained system therefore has $24 - 8 - 4 = 12$ degrees of freedom:

- 6 d.o.f. (translations + rotations) of the bogie with respect to the ground (conferred via the 6 first joints),
- 2 d.o.f. associated with the chassis deformation (around a vertical and a transverse axis respectively),
- 4×1 d.o.f. for the wheel suspensions (between each wheel bearing and its carrying beam).

Joint and externally applied forces (and torques) are modeled as follows.

1. The (external) interactions with the carbody through the secondary suspension are modeled by six force/torque components applied to the main crossbar. This allows us to model the complete vehicle by means of the sub-system segmentation technique explained in section 4.4.

2. Stiffness is introduced in the revolute joints of the bogie. It corresponds to the elasticity introduced into some joints in order to prevent excessive yaw deformation of the bogie.

3. The tangent contact forces on the wheels are computed according to Kalker's theory [45].

4. Since this particular bogie is designed to run with cylindrical wheels, the guidance along the track is ensured mainly by *second* contacts occurring between the wheel flange and the rail. Due to the lateral clearance between the wheels and the rail gauges (3.9 mm for the BAS 2000), these contacts are intermittent. The kinematic multibody tools of section 4.5 are thus used to detect this contact which is approximated as indicated in figure 7.12.

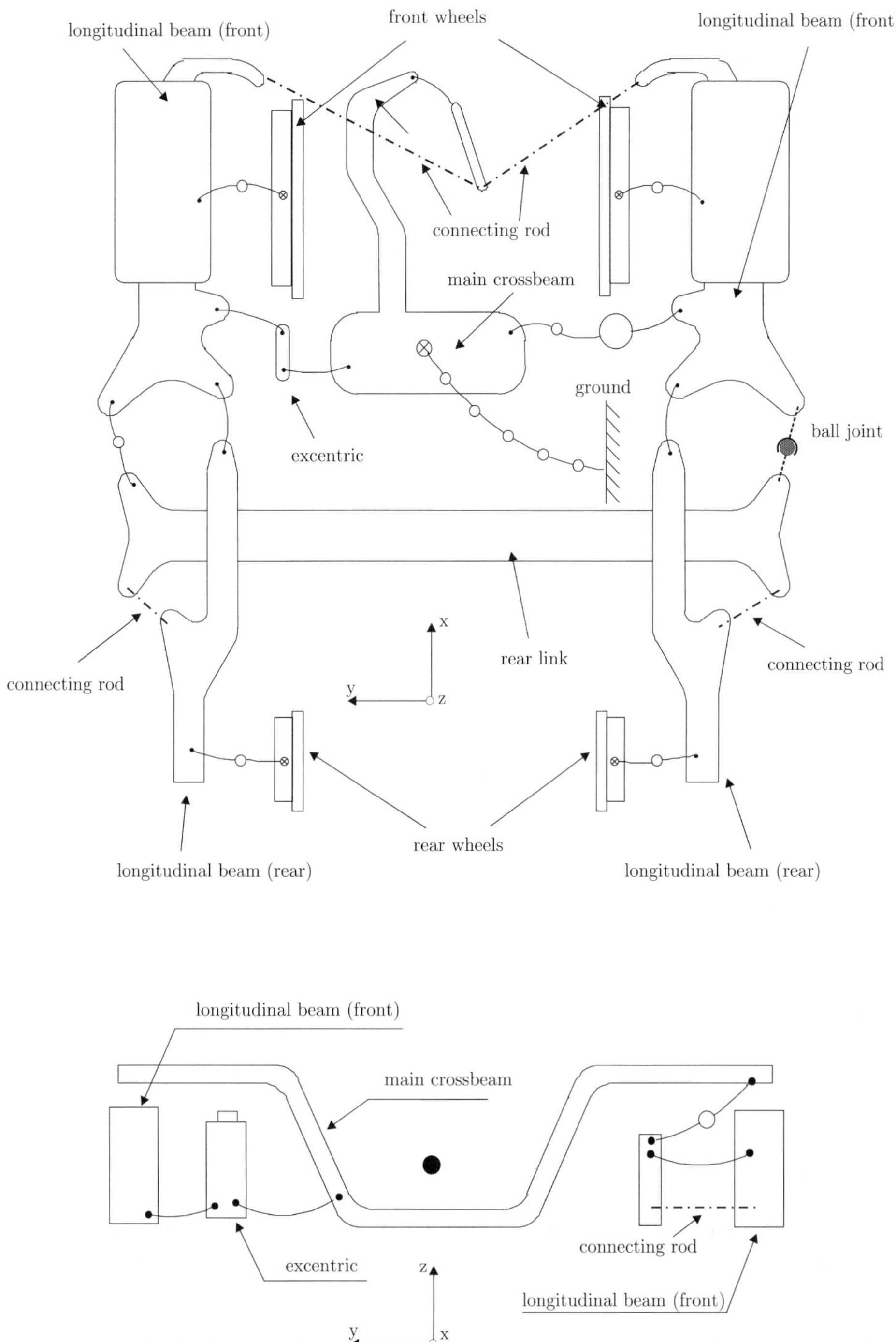

Figure 7.11: Multibody model of BAS 2000 bogie

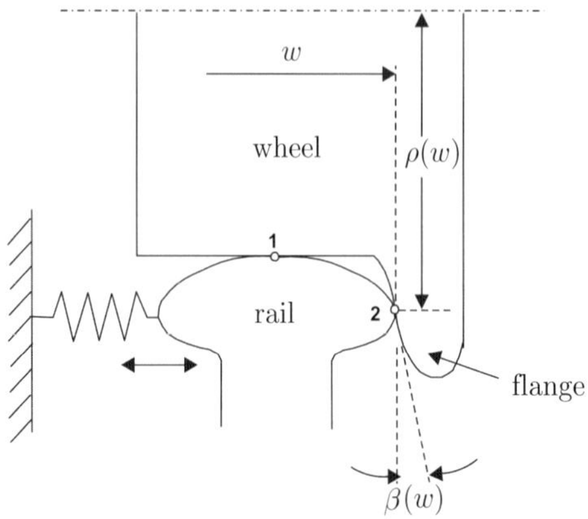

Figure 7.12: Second wheel-rail contact.

Since this second contact occurs on the wheel flange, we assume that its lateral position w on the wheel profile is constant. In the wheel-rail contact model of section 7.2, the values $w, \rho(w)$ and $\beta(w)$ are thus kept constant while the so-called *shift angle* ϑ is still evaluated from 7.9. Indeed, due to the large value of β in the flange area (more than one radian), the shift angle rapidly becomes significant in the presence of yaw.

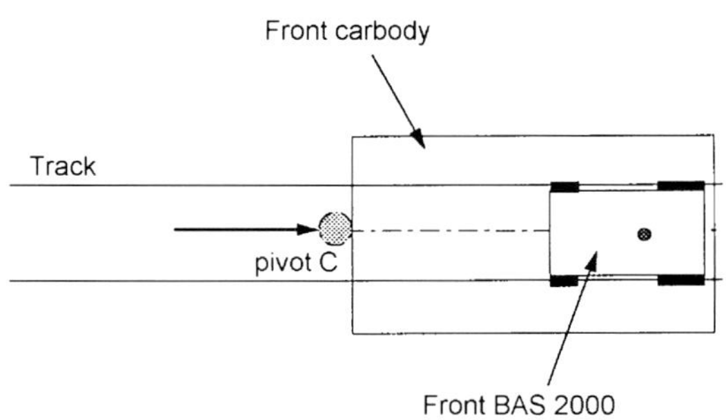

Figure 7.13: Simulation of the front BAS 2000 bogie

The normal contact force F_N at the second contact on the rail is evaluated by considering an elastically restrained rail (as shown in figure 7.12). Such a contact model does not introduce a constraint, since the wheel is allowed to "penetrate"[7] into the rail. Considering the high level of creepage occurring at this contact point, the model for the contact tangent force applied to the wheel

[7] The lateral stiffness of a rail results mainly from its roll deformation, or, in some situations, from the global transverse motion of the roadbed with respect to the inertial frame.

is dry friction: $F_T = \mu F_N$, oppositely aligned with the direction of the velocity of the material point of the wheel which is in contact with the rail. Note that for a large *shift angle* ϑ, the vector of the contact point velocity (and thus of the tangent force) has a large vertical component which can significantly load or unload the wheel (see [14]).

Some simulation results obtained with this model are illustrated here. The front carbody of a tramway, supported by a single BAS 2000 bogie (see figure 7.13) is assumed to be driven at constant speed along a straight track by the rest of the vehicle: this means that the pivot C between the first and second carbodies travels at constant speed along the track-centerline. Under these conditions, the resulting behavior of the yaw angle (measured at point C) between the front carbody and the direction of the track is then observed. As could be expected with cylindrical wheels, the curves plotted in figure 7.14 show that the vehicle bounces laterally between the flanges of the left and right rails. The amplitude of this bouncing phenomenon increases with the clearance between the wheel and rail gauges. For small clearances (ex.: 2 mm), this bouncing behavior disappears and the vehicle tends to remain in permanent contact on the left or right side, depending on the initial perturbation.

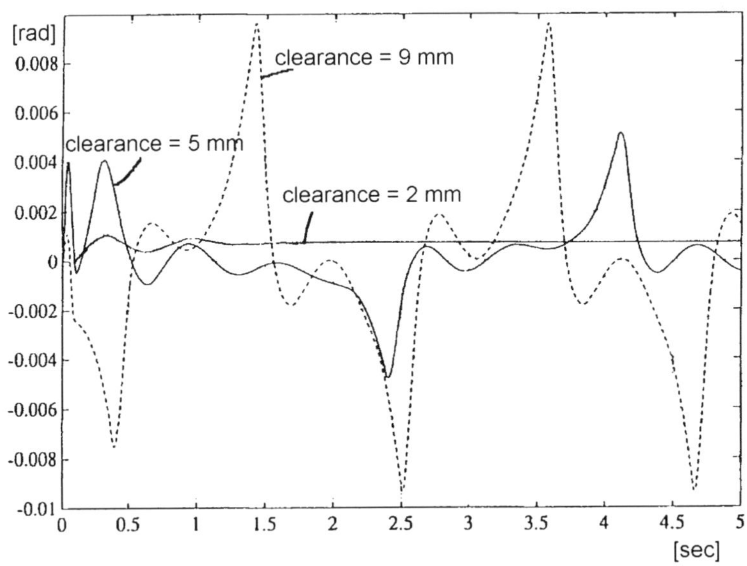

Figure 7.14: Unstable behavior of BAS 2000 bogie: wheel/rail bouncing

7.4.4 Tramway 2000

The complete vehicle (represented in figure 7.15) consists of three articulated carbodies, two BAS 2000 bogies (mounted in opposite directions) at the ends and

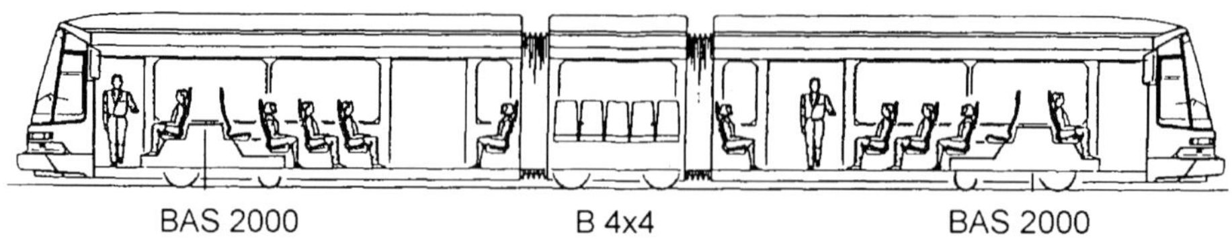

Figure 7.15: The tramway 2000

one conventional bogie (denoted B4×4) carrying the central carbody. According to the subsystem segmentation technique presented in section 4.4, the multibody model of this complex system involves:
- 4 sub-systems (one for the carbodies and one for each bogie),
- 74 joint variables in the open-loop multibody representation,
- 33 constraints due to 17 kinematic loops and 12 wheel-rail contacts.

The complete system thus has 41 degrees of freedom. Since it is designed for urban transportation, the vehicle must be able to travel over tracks featuring tight curves. During curving, the yaw angle formed by a wheel and the rail (denoted attack angle, see figure 7.16) is a critical parameter: the smaller this *attack angle*, the lower the creep friction at the contact points. Reducing the angle of attack therefore reduces energy dissipation and wear. Figure 7.18 indicates that the operational behavior of the BAS 2000 bogies during the curve entry depicted in figure 7.17 is much better (nearly one order of magnitude) than that of the conventional B4×4 bogie in terms of this angle of attack.

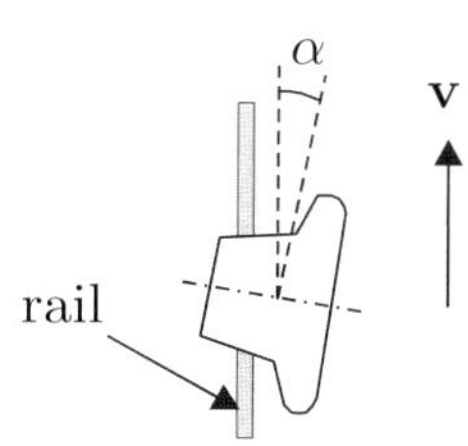

Figure 7.16: Wheel/rail attack angle

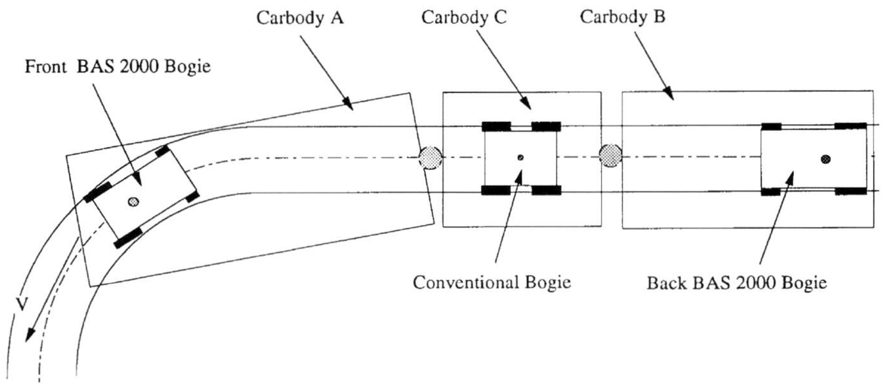

Figure 7.17: Tramway 2000: curve entry simulation

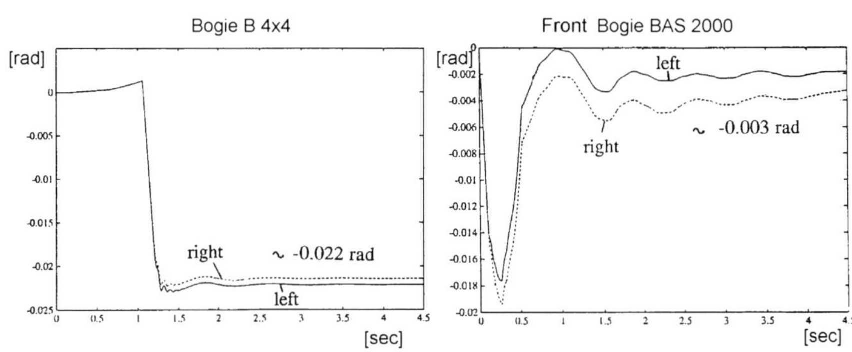

Figure 7.18: Comparaison of wheel attack angles during curving, for a conventional and a BAS 2000 bogie (from [14])

Chapter 8

Mechanisms: cam/follower model

8.1 Introduction

Cam/follower devices are used in a wide spectrum of applications, especially in automatic machines and instruments, textile machinery, printing presses, food-processing equipment, internal combustion engines, control systems, ... They constitute a simple and economic device with few movable parts. It is possible to design a cam such that the follower has a specific motion type. In particular, cam optimization [34] refers to determining the cam surface producing a given follower displacement that minimizes a certain cost (such as material volume or inertia of moving parts, ...) or maximizes a certain profit (force transmission, life span, ...).

This chapter is more concerned with the modeling of cam/follower devices in a multibody context, in which cams and followers are treated indistinctly (as in [5]), leading to a quite general formulation (see [16]). Starting from the postulate that the permanent contact between a cam and a follower is the most natural and frequent situation, we consider loop kinematic constraints to ensure the tangent and normal contact between the bodies. The benefits of such an approach is to avoid numerical instabilities which could take place when considering the normal contact force as a contributive one, modeled for instance by a spring and a damper in parallel, as pointed out in [5]. On the other hand, this option to kinematically constrain the bodies in their "contact phase" leads to a change in the number of degrees of freedom when the contact disappears or reappears. In that situation, since the multibody equations of motion are obtained on the basis of the coordinate partitioning method (see section 4.6.1), a permanent control of the variable partitioning is absolutely necessary during time simulation.

The chapter is organized as follows: section 8.2 presents the main definitions, notations and hypotheses underlying the cam/follower model. Section 8.3 details the establishment and the solution of the contact constraints, at position, velocity and acceleration levels. Section 8.4 is then devoted to the computation of the contact forces in the normal and tangent directions, and to the transient phases related to contact loss and recovery. Finally, section 8.5 presents some simulation examples:

- a comparison with another multibody formalism for which results are available in the literature [5]
- an experimental validation based on a simple cam/follower set-up
- the modeling of universal wheels of mobile robots, which constitutes a rather original application of the proposed formulation and demonstrates the degree of generality underlying the multibody approach.

8.2 Description of cam/follower systems

8.2.1 Hypotheses and general notations

The hypotheses underlying the mathematical model are:

- The cam and follower are rigid bodies and have relative planar motion.
- Contact (if it exists) is unique and occurs at a point[1] at all times.
- The profiles of the cam and of the follower are described by piecewise functions (line, circular arcs, cubic splines). Both open and closed profiles can be considered, but the curvature of any possible contact point on the profiles must be bounded and thus profiles cannot contain cusps. However, the specific "knife-edge" follower will be considered as a particular case of the method.
- The contact surfaces have relative sliding motion and intermittent contact is allowed. Tangent friction or viscous forces due to sliding can be taken into account.

A multibody sub-structure containing a cam/follower pair is represented in figure 8.1.

As previously mentioned, the resulting model does not include any assumption regarding which body is the cam and which is the follower[2]. Each of the two bodies involved is a *leaf* body of a kinematic chain: we make a distinction in figure 8.1 between a *primary* chain from body i to body p (the cam, for instance) and a *secondary* chain from body i to body s (the follower, for instance). Body i is their common *ancestor*. By hypothesis, the joints belonging to the chains $\{i, ..., p\}$ and $\{i, ..., s\}$ kinematically ensure a relative planar motion between bodies p and s.

[1] or along a straight line if one endows bodies with a given thickness.
[2] except when the latter is a knife-edge.

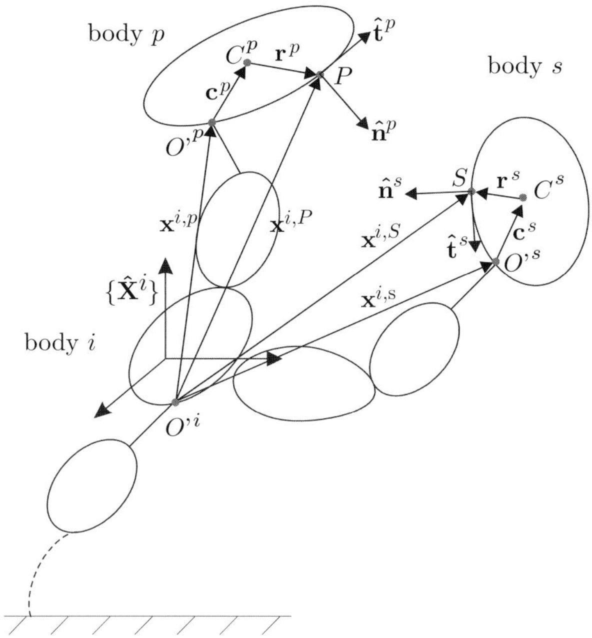

Figure 8.1: Cam/follower : general notations

In addition to the multibody elements introduced in Part 1, let us now define the following quantities:

- $\{\hat{\mathbf{Y}}^p\}$ (resp. $\{\hat{\mathbf{Y}}^s\}$), a frame attached to body p (resp. body s) and located at point C^p (resp. C^s), with respect to which the profile of the body is defined (see figure 8.2). By convention, unit vectors $\hat{\mathbf{Y}}_1$ and $\hat{\mathbf{Y}}_2$ lie in the plane of the body. In particular, a constant rotation matrix relates frame $\{\hat{\mathbf{Y}}^p\}$ to the body-fixed frame $\{\hat{\mathbf{X}}^p\}$, i.e.: $[\hat{\mathbf{Y}}^p] = R_y^p\,[\hat{\mathbf{X}}^p]$, and similarly for body s : $[\hat{\mathbf{Y}}^s] = R_y^s\,[\hat{\mathbf{X}}^s]$,

- vectors $\mathbf{x}^{i,p}(q) = \overrightarrow{O'^iO'^p}$ and $\mathbf{x}^{i,s}(q) = \overrightarrow{O'^iO'^s}$ which depend on the generalized joint coordinates q,

- constant vectors $\mathbf{c}^p = \overrightarrow{O'^pC^p}$ and $\mathbf{c}^s = \overrightarrow{O'^sC^s}$: these result from the fact that, generally speaking, the cam and follower profiles are defined with respect to a point C which does not necessarily coincide with the joint reference point O',

- vectors $\mathbf{r}^p = \overrightarrow{C^p\,P}$ and $\mathbf{r}^s = \overrightarrow{C^s\,S}$, the position vectors of the candidate geometrical contact points[3] P and S respectively,

[3] $\neq$ material points !

- η^p and η^s, the arc-length parameters along the profiles of bodies p and s respectively (see figure 8.2).

8.2.2 Cam and follower profiles

As already mentioned in the introduction, the definition of the cam and follower profiles is a wide and quite complex subject which is the topic of entire books (see [34] for instance), since it involves optimization methods and covers a large variety of applications. Within the framework of the present model — in which the cam/follower mechanism represents a particular case of two bodies p and s in planar contact — we will assume that the profiles are available (as inputs) in the form of open or closed smooth curves in polar coordinates. Considering body p (or s) in figure 8.2, let us assume for instance that its profile is known via a set of points in polar coordinates $\{\rho_\alpha, \vartheta_\alpha\}$.
The proposed formalism requires the knowledge of the following *analytical* functions: $\rho^p(\eta^p)\,,\vartheta^p(\eta^p)$ for body p and $\rho^s(\eta^s)\,,\vartheta^s(\eta^s)$ for body s. The arc-length parameter η is – arbitrarily – counted from the point defined by the intersection of the profile and the axis which coincides with unit vector $\hat{\mathbf{Y}}_1$ (see figure 8.2).

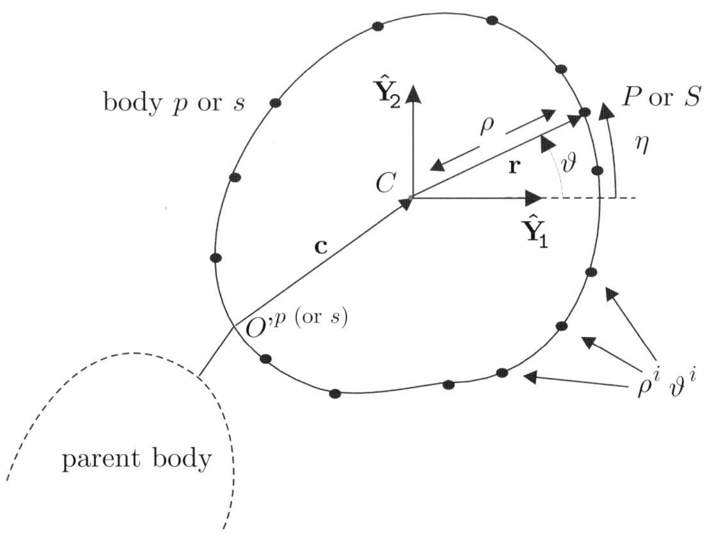

Figure 8.2: Cam/follower profile description

Starting from a set of data points $\{\rho_\alpha, \vartheta_\alpha\}$ defining the cam (or the follower) profile, we perform the following pre-process, based in this illustrative example on cubic spline interpolation:

N.B.: For a given function $f(x)$, the following notations are adopted: $f'(x) \triangleq \frac{df}{dx}$, $f''(x) \triangleq \frac{d^2 f}{dx^2}$

$$M \text{ points } \{\rho_\alpha, \vartheta_\alpha\}$$
$$\Downarrow$$
$$M \text{ periodic cubic splines}$$
$$\Downarrow$$
$$\rho(\vartheta)\ ,\ \rho'(\vartheta)$$
$$\Downarrow$$

$\theta_i = 0\ ; \Delta\,\theta = 2\pi/M$

for $i = 1\ :\ M$

$\eta_i = \int_0^{\theta_i} \sqrt{\rho^2(\theta) + \rho'^2(\theta)}\,d\theta$

store $\{\rho_i, \theta_i, \eta_i\}$

$\theta_i := \theta_i + \Delta\theta$

end

$$\Downarrow$$
$$M \text{ points } \{\rho_i(\eta_i), \vartheta_i(\eta_i)\}$$
$$\Downarrow$$
$$M \text{ cubic splines } \rho(\eta)\ ,\ \vartheta(\eta)$$
$$\Downarrow$$
$$\text{Analytical derivatives: } \rho'(\eta)\ ,\ \rho''(\eta)\ ;\ \vartheta'(\eta)\ ,\ \vartheta''(\eta)$$

For a knife-edge follower (= body s by convention), the contact point S is unique and the existence of η^s is superfluous ; ρ^s and ϑ^s become constant parameters.

Once this pre-process is performed, the following functions can be generated for the profile of each pair of bodies p/s of the multibody system:

$$\begin{aligned}(\rho^p, \rho'^p, \rho''^p, \vartheta^p, \vartheta'^p, \vartheta''^p) &= \text{profil\_p}\,(\eta^p),\\ (\rho^s, \rho'^s, \rho''^s, \vartheta^s, \vartheta'^s, \vartheta''^s) &= \text{profil\_s}\,(\eta^s);\end{aligned}$$

and they provide all the necessary functions analytically for given values of the input parameters η^p and η^s.
Under the assumption of unicity of contact between bodies p and s (see above), an additional geometrical condition involving the profile curvatures κ^p and κ^s must be satisfied to ensure the unicity of the solution for any possible relative position of bodies p and s (see [34]):

$$\min\,\kappa^p + \min\,\kappa^s > 0 \tag{8.1}$$

with $\kappa^p \triangleq \frac{\rho^{p\,2} + 2\rho^{p\prime 2} - \rho^p \rho^{p\prime\prime}}{(\rho^{p\,2} + \rho^{p\prime 2})^{3/2}}$ and $\kappa^s \triangleq \frac{\rho^{s\,2} + 2\rho^{s\prime 2} - \rho^s \rho^{s\prime\prime}}{(\rho^{s\,2} + \rho^{s\prime 2})^{3/2}}$

Piecewise interpolation functions combining lines, circles and cubic splines are available in the MBSOFT program [23] but must be such that no "cusp" exist at the transient points as mentioned in the hypotheses; in other words, functions $\rho(\eta)$ and $\vartheta(\eta)$ must be of class $\mathcal{C}^1$.

8.3 Kinematic constraints

The first step in the formalism consists in deriving the rigorous set of kinematic constraints, as for the wheel/rail contact problem (see chapter 7), which correspond to a permanent point contact between body p and body s. In this way, we consider a configuration of the system in which bodies p and s lean against each other: this is of course an *a priori* assumption which must be checked from a dynamical point of view as we shall see later.

8.3.1 Preliminary computations

Let us first express the vectors $\mathbf{r}^p$ and $\mathbf{r}^s$. On the basis of figure 8.2 and using the pre-process results, we can write

$$\mathbf{r}^p = [\hat{\mathbf{Y}}^p]^T \; r^p \tag{8.2}$$

with $r^p = \begin{pmatrix} \rho^p \cos\vartheta^p \\ \rho^p \sin\vartheta^p \\ 0 \end{pmatrix}$ and where $\rho^p = \rho^p(\eta^p), \vartheta^p = \vartheta^p(\eta^p)$

$$\mathbf{r}^s = [\hat{\mathbf{Y}}^s]^T \; r^s \tag{8.3}$$

with $r^s = \begin{pmatrix} \rho^s \cos\vartheta^s \\ \rho^s \sin\vartheta^s \\ 0 \end{pmatrix}$ and where $\rho^s = \rho^s(\eta^s), \vartheta^s = \vartheta^s(\eta^s)$

On the basis of figure 8.1, we can introduce position vectors — denoted $\mathbf{x}^{i,P}$ and $\mathbf{x}^{i,S}$ — which locate the candidate contact points P and S with respect to the *ancestor* body i:

$$\mathbf{x}^{i,P} \triangleq \overrightarrow{O'^iP} = \mathbf{x}^{i,p}(q) + \mathbf{c}^p + \mathbf{r}^p = [\hat{\mathbf{X}}^i]^T \, x^{i,p}(q) + [\hat{\mathbf{X}}^p]^T \, c^p + [\hat{\mathbf{Y}}^p]^T \, r^p \tag{8.4}$$

Using the rotation matrices between the previous frames,

$$[\hat{\mathbf{X}}^p] = R^{p,i}(q)[\hat{\mathbf{X}}^i] \quad \text{and} \quad [\hat{\mathbf{Y}}^p] = R^p_y[\hat{\mathbf{X}}^p] \tag{8.5}$$

we can write vector $\mathbf{x}^{i,P}$ in the $\{\hat{\mathbf{X}}^i\}$ frame as

$$\mathbf{x}^{i,P} = [\hat{\mathbf{X}}^i]^T \; \left\{ x^{i,p}(q) + R^{i,p} \left(c^p + R_y^{p^T} \, r^p\right)\right\}$$

Similarly for body s,

$$\mathbf{x}^{i,S} \triangleq \overrightarrow{O'^iS} = \mathbf{x}^{i,s}(q) + \mathbf{c}^s + \mathbf{r}^s = [\hat{\mathbf{X}}^i]^T \, x^{i,s}(q) + [\hat{\mathbf{X}}^s]^T \, c^s + [\hat{\mathbf{Y}}^s]^T \, r^s \tag{8.6}$$

in which the rotation matrices are defined as

$$[\hat{\mathbf{X}}^s] = R^{s,i}(q)[\hat{\mathbf{X}}^i] \quad \text{and} \quad [\hat{\mathbf{Y}}^s] = R^s_y[\hat{\mathbf{X}}^s]$$

and thus

$$\mathbf{x}^{i,S} = [\hat{\mathbf{X}}^i]^T \left\{ x^{i,s}(q) + R^{i,s} \, (c^s + R_y^{s^T} \, r^s) \right\}$$

Since the kinematic contact constraints will be expressed in the tangent and normal directions (with respect to the profiles at the contact point), we need to compute the corresponding vectors $\mathbf{t}^p$, $\mathbf{n}^p$ and $\mathbf{t}^s$, $\mathbf{n}^s$ (see figure 8.1).
For body p, we can write

$$\mathbf{t}^p \triangleq \frac{d\mathbf{r}^p}{d\eta^p} = [\hat{\mathbf{Y}}^p]^T \, t^p \quad \text{with} \quad t^p = \begin{pmatrix} \rho'^p \cos\vartheta^p - \rho^p \vartheta'^p \sin\vartheta^p \\ \rho'^p \sin\vartheta^p + \rho^p \vartheta'^p \cos\vartheta^p \\ 0 \end{pmatrix} \tag{8.7}$$

It is convenient for further developments to norm this vector:

$$\hat{\mathbf{t}}^p = \frac{1}{||\mathbf{t}^p||} \, \mathbf{t}^p \;\text{ with }\; ||\mathbf{t}^p|| = \sqrt{(t_1^p)^2 + (t_2^p)^2} \tag{8.8}$$

The normal vector $\mathbf{n}^p$ can be computed as follows (see figures 8.1 and 8.2):

$$\mathbf{n}^p = \mathbf{t}^p \times \hat{\mathbf{Y}}_3^p = [\hat{\mathbf{Y}}^p]^T \, n^p \;\text{ with } n^p = \begin{pmatrix} t_2^p \\ -t_1^p \\ 0 \end{pmatrix} \tag{8.9}$$

with a norm:

$$\hat{\mathbf{n}}^p = \frac{1}{||\mathbf{n}^p||} \, \mathbf{n}^p \quad \text{with} \quad ||\mathbf{n}^p|| = ||\mathbf{t}^p|| \tag{8.10}$$

Similarly for body s, we can write

$$\mathbf{t}^s \triangleq \frac{d\mathbf{r}^s}{d\eta^s} = [\hat{\mathbf{Y}}^s]^T \, t^s \quad \text{with } t^s = \begin{pmatrix} \rho'^s \cos\vartheta^s - \rho^s \vartheta'^s \sin\vartheta^s \\ \rho'^s \sin\vartheta^s + \rho^s \vartheta'^s \cos\vartheta^s \\ 0 \end{pmatrix} \tag{8.11}$$

$$\hat{\mathbf{t}}^s = \frac{1}{||\mathbf{t}^s||} \, \mathbf{t}^s \quad \text{with} \;\; ||\mathbf{t}^s|| = \sqrt{(t_1^s)^2 + (t_2^s)^2} \tag{8.12}$$

and

$$\mathbf{n}^s = \mathbf{t}^s \times \hat{\mathbf{Y}}_3^s = [\hat{\mathbf{Y}}^s]^T \, n^s \quad \text{with } n^s = \begin{pmatrix} t_2^s \\ -t_1^s \\ 0 \end{pmatrix} \tag{8.13}$$

$$\hat{\mathbf{n}}^s = \frac{1}{||\mathbf{n}^s||} \, \mathbf{n}^s \;\text{ with } ||\mathbf{n}^s|| = ||\mathbf{t}^s|| \tag{8.14}$$

For further developments, let us define the following scalar ratio:

$$\alpha \triangleq \frac{||\mathbf{t}^s||}{||\mathbf{t}^p||} = \frac{||\mathbf{n}^s||}{||\mathbf{n}^p||} \tag{8.15}$$

8.3.2 Contact constraints

As previously mentioned, we will first assume a multibody configuration in which bodies p and s are in point contact, i.e. in the presence of a contact force. Based on figure 8.1, three kinematic constraints are necessary to ensure the contact between bodies P and S at a "common" geometrical point P ($\in$ body p surface) $= S$ ($\in$ body s surface):

- An "alignment" constraint which specifies that tangent vectors $\mathbf{t}^p$ and $\mathbf{t}^s$ are parallel, or equivalently:

$$h_1 \triangleq \mathbf{n}^p \,.\, \mathbf{t}^s = 0 \tag{8.16}$$

  This first constraint is of course irrelevant (and superfluous) for a knife-edge follower s and will simply be omitted in that case.

- Two "contact" constraints which force points P and S to coincide in the tangent direction:

$$h_2 \triangleq (\mathbf{x}^{i,S} - \mathbf{x}^{i,P}) \,.\, \mathbf{t}^p = 0 \tag{8.17}$$

  and in the normal direction

$$h_3 \triangleq (\mathbf{x}^{i,S} - \mathbf{x}^{i,P}) \,.\, \mathbf{n}^p = 0 \tag{8.18}$$

  By using the preliminary computations (equations 8.2 - 8.14), we can easily compute these constraints as follows:

$$h_1(q, \eta^p, \eta^s) \equiv n^{p^T} \, R_y^p \, R^{p,i} \, R^{i,s} \, R_y^{s^T} \, t^s = 0 \tag{8.19}$$

$$\begin{aligned} h_2(q, \eta^p, \eta^s) \equiv t^{p^T} \, R_y^p \, R^{p,i} \, \{ & x^{i,s} - x^{i,p} + R^{i,s} \, (c^s + R_y^{s^T} \, r^s) \\ & - R^{i,p} \, (c^p + R_y^{p^T} \, r^p) \} = 0 \end{aligned} \tag{8.20}$$

$$\begin{aligned} h_3(q, \eta^p, \eta^s) \equiv n^{p^T} \, R_y^p \, R^{p,i} \, \{ & x^{i,s} - x^{i,p} + R^{i,s} \, (c^s + R_y^{s^T} \, r^s) \\ & - R^{i,p} \, (c^p + R_y^{p^T} \, r^p) \} = 0 \end{aligned} \tag{8.21}$$

Let us emphasize that the above kinematic constraints involve two auxiliary variables η^p and η^s — introduced to locate the *geometrical* points P and S on the profiles — which momentarily add to the set of joint coordinates q. Thus, by imposing these three independent constraints on the multibody system, the number of d.o.f. of the latter is reduced by only one (3 constraints - 2 auxiliary

variables), which is kinematically consistent for such a contact[4].
Thus, each kinematic loop induced by a cam/follower contact formally leads to selecting in vector q, an additional[5] dependent coordinate, denoted v^s hereafter. A "physical" reasoning will often lead to choosing the relative (angular or prismatic) displacement of the follower q^s (see figure 8.3) as dependent variable v^s, as long as this choice remains acceptable from a numerical point of view: this can easily be checked on line by analyzing the conditioning of the constraints Jacobian matrix.
If the joint variable v^s does not appear in the other constraints applied to the system[6], the solution of the cam/follower constraints (8.19 - 8.21) can be performed independently, downstream from the others. Thus the Newton/Raphson method for these two types of constraints (cam/follower and "others") can be decoupled and performed in series as shown in figure 8.3.

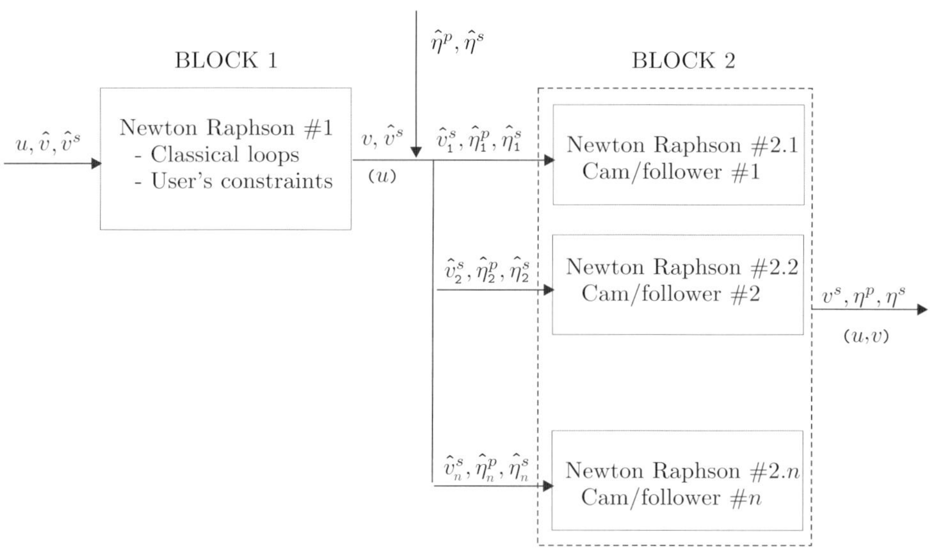

Figure 8.3: Numerical solution of the kinematic constraints

The first Newton/Raphson block ("BLOCK 1") is dedicated to solving the loop and user's constraints, and iteratively finds the correct values of the dependent variables v as functions of u (see equation 4.49 in section 4.6.1), starting from a good estimate $\hat{v}$. The second block ("BLOCK 2") is then devoted to the solution of the cam/follower constraints, i.e. the iterative solution of equations (8.19 - 8.21) for each cam/follower pair, solved for the variables v^s, η^p, η^s. The

[4] as far as one assumes a tangent slip between bodies p and s.

[5] with respect to those (v) resulting from other multibody constraints (loops of bodies, "user's constraints", ... see chapter 4).

[6] which is the case for most practical applications.

i^{th} iteration step can be written as:

$$\begin{pmatrix} \Delta v^s \\ \Delta \eta^p \\ \Delta \eta^s \end{pmatrix}_{(i)} = -(\mathcal{J})^{-1}_{(i)} \begin{pmatrix} h_1(u, v, v^s, \eta^p, \eta^s) \\ h_2(u, v, v^s, \eta^p, \eta^s) \\ h_3(u, v, v^s, \eta^p, \eta^s) \end{pmatrix}_{(i)} \quad \text{with} \quad i = 1, ..., iter_{max} \tag{8.22}$$

where $\mathcal{J}$ is the 3×3 Jacobian matrix:

$$\mathcal{J} \triangleq \begin{pmatrix} \frac{\partial h_1}{\partial v^s} & \frac{\partial h_1}{\partial \eta^p} & \frac{\partial h_1}{\partial \eta^s} \\ \frac{\partial h_2}{\partial v^s} & \frac{\partial h_2}{\partial \eta^p} & \frac{\partial h_2}{\partial \eta^s} \\ \frac{\partial h_3}{\partial v^s} & \frac{\partial h_3}{\partial \eta^p} & \frac{\partial h_3}{\partial \eta^s} \end{pmatrix},$$

which can be numerically evaluated.
Experiments have shown that the convergence of equation 8.22 was very often jeopardized by highly nonlinear profiles. This observation led us to modify the solution method in the following way, in a similar manner as for the wheel/rail contact model of chapter 7.
The idea is to decouple the solution of constraints h_1, h_2 from that of h_3 by splitting the numerical procedure (for each block 2.k of figure 8.3) as shown in figure 8.4.

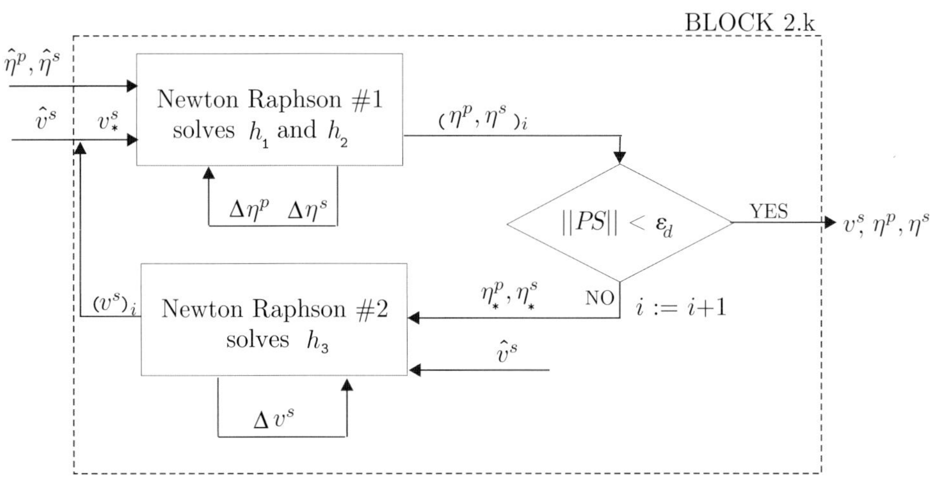

Figure 8.4: Numerical solution of the cam/follower constraints

The first step of the procedure (N.R. #1) solves constraints h_1 and h_2 with respect to η^p, η^s for a fixed value of the variable v^s, momentarily considered as a parameter: v^s_*. Physically, this consists in finding "good candidate" points P and S on the bodies, namely, the closest ones when for instance bodies p and s are in a "non-contact" configuration. The second "sub-Newton/Raphson" block (N.R. #2) in figure 8.4 is dedicated to constraint h_3 and inverts the roles of the

"momentary" parameters η_*^p, η_*^s and the variable which is now υ^s. This solution consists in fact in setting to zero the projection onto the normal vector $\mathbf{n}^p$, of vector $\overrightarrow{PS}$ resulting from the first step. The global procedure iterates until the distance between the candidate points P and S, provided by the first step, falls below a predefined threshold ϵ_d.
Experiences have shown that better convergence and robustness were obtained with this method which of course is more CPU time consuming but far more reliable for highly nonlinear profiles than the algorithm depicted in figure 8.3.

8.3.3 Constraint derivatives

Once the kinematic constraints are satisfied at position level, their first and second time derivatives must be generated and solved to set up the final form of the equations of motion and to reduce it using the coordinate partitioning technique (see section 4.6.1).
For a reason which will appear later, we will first focus on the third constraint (h_3) derivatives, by making the corresponding Jacobian matrix appear.
The first time derivative of constraint h_3 can be written as

$$\dot{h}_3 = \left(\dot{\mathbf{x}}^{i,S} - \dot{\mathbf{x}}^{i,P}\right) .\mathbf{n}^p + \left(\mathbf{x}^{i,S} - \mathbf{x}^{i,P}\right) .\dot{\mathbf{n}}^p = 0 \tag{8.23}$$

Since $\left(\mathbf{x}^{i,S} - \mathbf{x}^{i,P}\right) = 0$, once constraints h_2 and h_3 are satisfied at position level, the second term of equation 8.23 vanishes and thus

$$\dot{h}_3 = \left(\dot{\mathbf{x}}^{i,S} - \dot{\mathbf{x}}^{i,P}\right) .\mathbf{n}^p = 0 \tag{8.24}$$

On the basis of equations 8.6, 8.3, let us develop $\dot{\mathbf{x}}^{i,S}$ as follows:

$$\dot{\mathbf{x}}^{i,S} = \dot{\mathbf{x}}^{i,s} + \dot{\mathbf{c}}^s + \dot{\mathbf{r}}^s$$

with:

- $\dot{\mathbf{x}}^{i,s} = [\hat{\mathbf{X}}^i]^T \dot{x}^{i,s} + \boldsymbol{\omega}^i \times \mathbf{x}^{i,s}$ where:
  - $\dot{x}^{i,s} = J_t^{i,s}\,\dot{q}$ is easily computable from the topology of the kinematic chain $\{i, ..., s\}$ (see for instance section 4.5),
  - $\boldsymbol{\omega}^i = [\hat{\mathbf{X}}^i]^T\, \omega^i$ is the angular velocity vector of the $\{\hat{\mathbf{X}}^i\}$ frame.
- $\dot{\mathbf{c}}^s = \boldsymbol{\omega}^s \times \mathbf{c}^s = (\boldsymbol{\omega}^i + \boldsymbol{\omega}^{i,s}) \times \mathbf{c}^s$ where:
  - $\boldsymbol{\omega}^{i,s} = [\hat{\mathbf{X}}^i]^T\, \omega^{i,s}$ with $\omega^{i,s} = J_r^{i,s}\,\dot{q}$ is the angular velocity of $\{\hat{\mathbf{X}}^s\}$ relative to $\{\hat{\mathbf{X}}^i\}$, also available from the chain $\{i, ..., s\}$,
- $\dot{\mathbf{r}}^s = [\hat{\mathbf{Y}}^s]^T\, \dot{r}^s + \boldsymbol{\omega}^s \times \mathbf{r}^s = \dot{\eta}^s \mathbf{t}^s + (\boldsymbol{\omega}^i + \boldsymbol{\omega}^{i,s}) \times \mathbf{r}^s$
  In the latter expression, the term $\dot{\eta}^s \mathbf{t}^s$ clearly denotes the fact that the relative velocity of geometrical point S is physically tangent to the profile of the body.

Finally, we obtain

$$\dot{\mathbf{x}}^{i,S} = [\hat{\mathbf{X}}^i]^T \; J_t^{i,s}\,\dot{q} + \boldsymbol{\omega}^i \times \mathbf{x}^{i,S} + \boldsymbol{\omega}^{i,s} \times (\mathbf{c}^s + \mathbf{r}^s) + \dot{\eta}^s \mathbf{t}^s \tag{8.25}$$

A similar reasoning holds of course for body p, ending up with

$$\dot{\mathbf{x}}^{i,P} = [\hat{\mathbf{X}}^i]^T \; J_t^{i,p}\,\dot{q} + \boldsymbol{\omega}^i \times \mathbf{x}^{i,P} + \boldsymbol{\omega}^{i,p} \times (\mathbf{c}^p + \mathbf{r}^p) + \dot{\eta}^p \mathbf{t}^p$$

where $\boldsymbol{\omega}^{i,p} = [\hat{\mathbf{X}}^i]^T\,\omega^{i,p}$, with $\omega^{i,p} = J_r^{i,p}\,\dot{q}$, is the angular velocity of frame $\{\hat{\mathbf{X}}^p\}$ with respect to $\{\hat{\mathbf{X}}^i\}$.

Introducing these results into constraint 8.24 and using the fact that contact constraints are satisfied at position level, we find

$$\begin{aligned} \{[\hat{\mathbf{X}}^i]^T \left((J_t^{i,s} - J_t^{i,p})\,\dot{q}\right) + \dot{\eta}^s \mathbf{t}^s - \dot{\eta}^p \mathbf{t}^p \\ + \boldsymbol{\omega}^{i,s} \times (\mathbf{c}^s + \mathbf{r}^s) - \boldsymbol{\omega}^{i,p} \times (\mathbf{c}^p + \mathbf{r}^p)\}\,.\,\mathbf{n}^p = 0 \end{aligned}$$

Since $\mathbf{t}^s.\,\mathbf{n}^p = \mathbf{t}^p.\,\mathbf{n}^p = 0$ at all time, the previous expression reduces to

$$\left\{[\hat{\mathbf{X}}^i]^T \left((J_t^{i,s} - J_t^{i,p})\,\dot{q}\right) + \boldsymbol{\omega}^{i,s} \times (\mathbf{c}^s + \mathbf{r}^s) - \boldsymbol{\omega}^{i,p} \times (\mathbf{c}^p + \mathbf{r}^p)\right\}\,.\,\mathbf{n}^p = 0 \tag{8.26}$$

which no longer depends on $\dot{\eta}^s$ and $\dot{\eta}^p$.
The scalar form of equation 8.26 can be obtained in a straightforward manner by using the appropriate projections. Without going into details, this yields:

$$\begin{aligned} \{ \left[n^{p^T}\, R_y^p\, R^{p,i} (J_t^{i,s} - J_t^{i,p}) \right] - \\ \left[\frac{1}{\alpha}\,(r^s + R_y^s c^s)^T \tilde{n}^{\,s} R_y^s\, R^{s,i}\, J_r^{i,s} + (r^p + R_y^p c^p)^T\, \tilde{n}^{\,p} R_y^p\, R^{p,i}\, J_r^{i,p} \right] \}\ \dot{q} = 0 \end{aligned} \tag{8.27}$$

in which definition 8.15 has been used, leading to $\mathbf{n}^s = -\alpha\,\mathbf{n}^p$.
Equation 8.27 has the following generic form:

$$\dot{h}_3(q,\dot{q}) = J_3\,(q, \eta^p(q), \eta^s(q))\,\dot{q} = 0 \tag{8.28}$$

from which $\dot{\eta}^s$ and $\dot{\eta}^p$ have disappeared as mentioned above. The Jacobian matrix appearing in equation 8.28 – which is the term in brackets '{}' in equation 8.27 – will contribute to the right-hand-side of the dynamical equations 4.1. As for the wheel/rail contact model (see chapter 7), it can be shown that the Lagrange multiplier λ_3 associated with constraint h_3 (which imposes the contact in the *normal* direction – see equation 8.18) is proportional to the normal force F_n at the contact point $P \equiv S$, and more precisely in the present case:

$$|F_n| = ||\mathbf{n}^p||\,|\lambda_3| \tag{8.29}$$

Moreover, it is also shown in [74] that the Lagrange multipliers associated with kinematic constraints of types h_1 (equation 8.16) and h_2 (equation 8.17) are identically equal to zero at any time. This can be interpreted physically since these constraints consist in aligning the tangent vectors $\mathbf{t}^p, \mathbf{t}^s$ and in locating *geometrical* points (P and S) in the direction tangent to the profiles: such constraints do not require any force or torque, since these geometrical points are "massless" and free to move over the surface. Further to these considerations, we can simply disregard the dynamical contribution ($= 0!$) of constraints h_1 and h_2 on the right-hand-side of the dynamical equations 4.1.

Since the differential/algebraic equations are solved using the coordinate partitioning method (see section 4.6.1), the second time derivative of constraint h_3 is required. As we shall see later, this computation will require the knowledge of derivatives $\dot{\eta}^p$ and $\dot{\eta}^s$. The latter will be obtained by solving the auxiliary constraints h_1 and h_2 at velocity level.

The first time derivative of constraint h_1 can be written as

$$\dot{h}_1 = \dot{\mathbf{n}}^p \,.\, \mathbf{t}^s + \mathbf{n}^p \,.\, \dot{\mathbf{t}}^s = 0 \tag{8.30}$$

By expanding the time derivatives using the angular velocity vectors and equations 8.9 and 8.11, equation 8.30 becomes

$$[\hat{\mathbf{Y}}^p]^T \dot{n}^p \,.\, \mathbf{t}^s + \mathbf{n}^p \,.\, [\hat{\mathbf{Y}}^s]^T \dot{t}^s = (\mathbf{n}^p \times \mathbf{t}^s) \,.\, (\boldsymbol{\omega}^{i,s} - \boldsymbol{\omega}^{i,p}) \tag{8.31}$$

As regards constraint h_2, its time derivative can be expressed as

$$\dot{h}_2 = \left(\dot{\mathbf{x}}^{i,S} - \dot{\mathbf{x}}^{i,P}\right) \,.\, \mathbf{t}^p = 0$$

which, by expanding the velocity vectors as already done in obtaining equation 8.26, yields

$$\begin{aligned} \dot{\eta}^s\, \mathbf{t}^s \,.\, \mathbf{t}^p - \dot{\eta}^p\, \mathbf{t}^p \,.\, \mathbf{t}^p = & \\ & - \{([\hat{\mathbf{X}}^i]^T\, (J_t^{i,s} - J_t^{i,p})\dot{q} + \boldsymbol{\omega}^{i,s} \times (\mathbf{c}^s + \mathbf{r}^s) \\ & - \boldsymbol{\omega}^{i,p} \times (\mathbf{c}^p + \mathbf{r}^p)\} \,.\, \mathbf{t}^p \end{aligned} \tag{8.32}$$

After expansions and projections, equations 8.31 and 8.32 give rise to the following 2×2 linear system:

$$\begin{pmatrix} \alpha\left(t_2^p t_1^{p\prime} - t_1^p t_2^{p\prime}\right) & \frac{1}{\alpha}\left(t_1^s t_2^{s\prime} - t_2^s t_1^{s\prime}\right) \\ ||\mathbf{t}^p||^2 & \frac{1}{\alpha}||\mathbf{t}^s||^2 \end{pmatrix} \begin{pmatrix} \dot{\eta}^p \\ \dot{\eta}^s \end{pmatrix} =$$

$$\begin{pmatrix} ||\mathbf{t}^p||\,||\mathbf{t}^s||(\omega_3^{i,p} - \omega_3^{i,s}) \\ t^{p^T} R_y^p\, R^{p,i}\left((J_t^{i,s} - J_t^{i,p})\dot{q} + \tilde{\omega}^{i,s}\, R^{i,s}(c^s + R_y^{s^T} r^s) - \tilde{\omega}^{i,p}\, R^{i,p}(c^p + R_y^{p^T} r^p)\right) \end{pmatrix} \tag{8.33}$$

in which definition 8.15 is used, and where $t_1^{p\prime}$ and $t_2^{p\prime}$ (resp. $t_1^{s\prime}$ and $t_2^{s\prime}$) denote the derivative of the components of t^p (8.7) (resp. t^s (8.11)) with respect to η^p (resp. η^s), viz.:

$$t_1^{p\prime} = \rho^{p\prime\prime} \cos\vartheta^p - 2\rho^{p\prime}\vartheta^{p\prime} \sin\vartheta^p - \rho^p\left(\vartheta^{p\prime^2} \cos\vartheta^p + \vartheta^{p\prime\prime} \sin\vartheta^p\right) \tag{8.34}$$

$$t_2^{p\prime} = \rho^{p\prime\prime} \sin\vartheta^p - 2\rho^{p\prime}\vartheta^{p\prime} \cos\vartheta^p - \rho^p\left(\vartheta^{p\prime^2} \sin\vartheta^p + \vartheta^{p\prime\prime} \cos\vartheta^p\right) \tag{8.35}$$

A similar result holds for $t_1^{s\prime}$ and $t_2^{s\prime}$.
The solution of 8.33 is of course straightforward. Let us now focus on the second time derivative of equation 8.18:

$$\ddot{h}_3 = \left(\ddot{\mathbf{x}}^{i,S} - \ddot{\mathbf{x}}^{i,P}\right) . \mathbf{n}^p + 2\left(\dot{\mathbf{x}}^{i,S} - \dot{\mathbf{x}}^{i,P}\right) . \dot{\mathbf{n}}^p + \left(\mathbf{x}^{i,S} - \mathbf{x}^{i,P}\right) . \ddot{\mathbf{n}}^p = 0$$

Since, at this stage, the three contact constraints h_1, h_2 and h_3 are satisfied at position and velocity levels, the previous expression reduces to

$$\ddot{h}_3 = \left(\ddot{\mathbf{x}}^{i,S} - \ddot{\mathbf{x}}^{i,P}\right) . \mathbf{n}^p = 0, \tag{8.36}$$

which physically means that the geometrical points P and S have the same acceleration in the normal direction. By expanding $\ddot{\mathbf{x}}^{i,S}$ and $\ddot{\mathbf{x}}^{i,P}$ as done previously for the velocity vectors, we find that equation 8.36 becomes

$$\begin{aligned}
&\{[\hat{\mathbf{X}}^i]^T\left((J_t^{i,s}\,\ddot{q} + \dot{J}_t^{i,s}\dot{q}) - (J_t^{i,p}\,\ddot{q} + \dot{J}_t^{i,p}\dot{q})\right)\\
&+\left(\ddot{\eta}^s\mathbf{t}^s + \dot{\eta}^s[\hat{\mathbf{Y}}^s]\dot{t}^s\right) - \left(\ddot{\eta}^p\mathbf{t}^p + \dot{\eta}^p[\hat{\mathbf{Y}}^p]\dot{t}^p\right)\\
&+2\left(\dot{\eta}^s\boldsymbol{\omega}^{i,s}\times\mathbf{t}^s - \dot{\eta}^p\boldsymbol{\omega}^{i,p}\times\mathbf{t}^p\right)\\
&+(\tilde{\dot{\boldsymbol{\omega}}}^{i,s} + \tilde{\boldsymbol{\omega}}^{i,s}\,\tilde{\boldsymbol{\omega}}^{i,s}) . (\mathbf{r}^s + \mathbf{c}^s)\\
&-(\tilde{\dot{\boldsymbol{\omega}}}^{i,p} + \tilde{\boldsymbol{\omega}}^{i,p}\,\tilde{\boldsymbol{\omega}}^{i,p}) . (\mathbf{r}^p + \mathbf{c}^p)\} . \mathbf{n}^p = 0
\end{aligned} \tag{8.37}$$

in which we can compute the components of $\dot{\boldsymbol{\omega}}^{i,s}$ and $\dot{\boldsymbol{\omega}}^{i,p}$ in terms of the generalized coordinates $q, \dot{q}, \ddot{q}$ as follows[7] :

$$\dot{\omega}^{i,s} = J_r^{i,s}\,\ddot{q} + \dot{J}_r^{i,s}\dot{q} \quad\text{and}\quad \dot{\omega}^{i,p} = J_r^{i,p}\,\ddot{q} + \dot{J}_r^{i,p}\dot{q}$$

In equation 8.37, the prior resolution of the constraints at position and velocity levels has not yet been thoroughly exploited. Indeed, owing to the position constraint h_1, the terms $(...\ddot{\eta}^s\mathbf{t}^s...) . \mathbf{n}^p$ and $(...\ddot{\eta}^p\mathbf{t}^p...) . \mathbf{n}^p$ are identically equal to zero: the second time derivative of constraint h_3 consequently does not depend on the quantities $\ddot{\eta}^s, \ddot{\eta}^p$ whose computation is thus superfluous.
After some term rearrangements, equation 8.36 takes the following generic form:

$$\ddot{h}_3 = J_3(q, \eta^p, \eta^s)\,\ddot{q} + (\dot{J}\dot{q})_3(q, \dot{q}, \eta^p, \dot{\eta}^p, \eta^s, \dot{\eta}^s) = 0 \tag{8.38}$$

in which only the second term is needed for the coordinate partitioning reduction.
This term can easily be computed from 8.37 using the appropriate projections and finally leads to the following expression:

$$\begin{aligned}
(\dot{J}\dot{q})_3 \;=\; & n^{p^T} R_*^p\,\{\, \dot{J}_t^{i,s}\dot{q} - \dot{J}_t^{i,p}\dot{q} + \dot{\eta}^s R_*^{s^T}\dot{t}^s - \dot{\eta}^p R_*^{p^T}\dot{t}^p\\
&+2\,\dot{\eta}^s\tilde{\omega}^{i,s}\,R_*^{s^T}t^s - 2\,\dot{\eta}^p\tilde{\omega}^{i,p}\,R_*^{p^T}t^p\\
&+\left(\left(\dot{J}_r^{i,s}\dot{q}\right)^{\sim} + \tilde{\omega}^{i,s}\,\tilde{\omega}^{i,s}\right)(R^{s^T}c^s + R_*^{s^T}r^s)\\
&-\left(\left(\dot{J}_r^{i,p}\dot{q}\right)^{\sim} + \tilde{\omega}^{i,p}\,\tilde{\omega}^{i,p}\right)(R^{p^T}c^p + R_*^{p^T}r^p)\}
\end{aligned} \tag{8.39}$$

[7] In these relations, the term $\dot{J}\dot{q}$ vanishes if all the revolute joint axes of the sub-chains $\{i...p\}$, $\{i...s\}$ are aligned, as is generally the case.

in which $R_*^p \triangleq R_y^p \, R^{p,i}$ and $R_*^s \triangleq R_y^s \, R^{s,i}$ have been introduced for conciseness. In equation 8.39, $\dot{\eta}^p$ and $\dot{\eta}^s$ are available from the solution of the linear system 8.33 and $\dot{t}^p$ (resp. $\dot{t}^s$) can be obtained from equations 8.34, 8.35, i.e.: $\dot{t}^p = t^{p\prime} \, \dot{\eta}^p$ (resp. $\dot{t}^s = t^{s\prime} \, \dot{\eta}^s$).

8.4 Contact forces

Once the geometrical contact problem has been solved, the next step will be the introduction of the contact forces at the contact point $P \equiv S$. This part of the modeling is subject to numerous hypotheses, e.g. that the contact occurs at a point, which follows from the hypothesis of body rigidity. It is a well-known fact that contact force modeling is a difficult problem, and all the more complex when collisions are involved. In this connection, the models to be found in the literature are of two types (see [31]):

- *discrete models* which are based on the notion of *impact* between rigid bodies, and for which body motion is a function of the coefficient of restitution which defines a relationship between the normal components of the velocities before and after the impact at the contact point.

- *continuous models* which are based on the *compliance* of the contact: a contact force is computed from the local deformation and deformation rate of the bodies in contact, using specific constitutive laws.

It is not our purpose here to compare these two approaches which both present specific advantages and drawbacks [31]. For practical reasons, we opt for the second one when intermittent contact takes place [16].

8.4.1 Permanent contact

Normal contact force

In section 8.3, kinematic constraints were derived by assuming a permanent contact between bodies p and s. The easiest way to check this *a priori* assumption is to compute the normal force of the current configuration. The norm of this normal force is given by equation 8.29 through the Lagrange multiplier associated with constraint h_3. As regards the sign of this force — which is indispensable at this stage — a virtual approach can be used to show that

$$F_n = ||\mathbf{n}^p|| \, \lambda_3 \tag{8.40}$$

represents the component along $\hat{\mathbf{n}}^p$ of the normal force $\mathbf{F}_n^{ps}$ of body p on body s. Since this component is positive as long as contact lasts, the existence of the contact is governed by the following test:

$$F_n > 0 \quad \text{or equivalently} \quad \lambda_3 > 0 \tag{8.41}$$

Although the reduction procedure (see section 4.6.1) aims at eliminating the Lagrange multipliers λ to set up the final reduced form 4.48, the λ can be computed during the reduction via equation 4.50 (see section 4.6.1):

$$\lambda = \left((J_v)^T\right)^{-1} \{ (M_{vu} + M_{vv} B_{vu})\, \ddot{u} + c_v - \mathcal{Q}_v - M_{vv} J_v^{-1} b\} \tag{8.42}$$

So, as long as the contact is ensured[8], the model takes the normal contact forces into account via the Lagrange multipliers, without resorting to the contact compliance. Numerical instabilities which can occur with compliant models [5] are completely absent here.

Tangent forces

Tangent forces can take place between bodies p and s and can be of different natures. Indeed, depending on the application, Coulomb forces arising from dry friction or viscous forces arising from lubricated contact — or both — can act at the contact point between the bodies. In other circumstances, a pure rolling contact model could also be suitable, and one could decide to switch from one model ("stick" phase) to another ("slip" phase) in the same way as for the intermittent normal contact (see below).
Whatever the model, it will require, in addition to the instantaneous value of the normal force, the computation of the relative velocities between bodies p and s at the contact point.

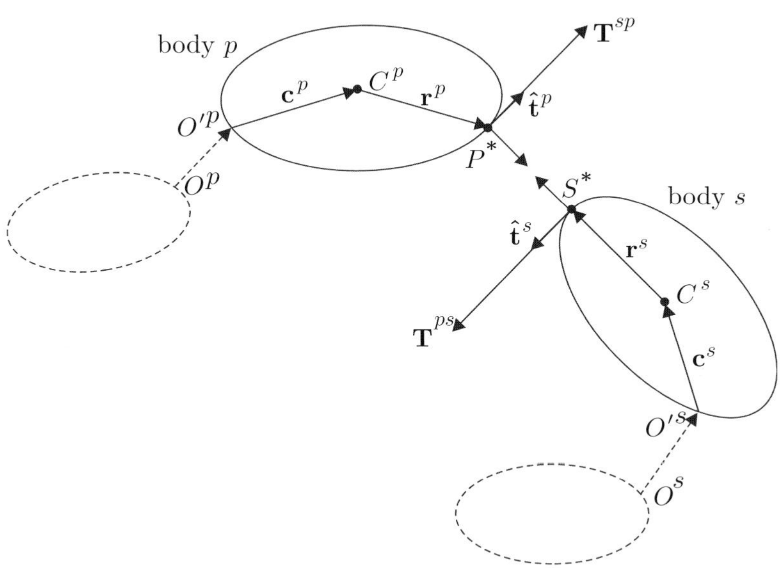

Figure 8.5: Tangent contact forces

[8]permanent contact is the most frequent situation for cam/follower devices

In order to compute this velocity, we must consider in figure 8.5 the velocity difference between *material* points S^* and P^* which coincide with the geometrical points S and P respectively[9], and then project this vector along the tangent direction. Hence,

$$v^{s^*p^*} = (\dot{\mathbf{x}}^{s^*} - \dot{\mathbf{x}}^{p^*}) \,.\, \hat{\mathbf{t}}^p \tag{8.43}$$

denotes the relative velocity of material point S^* with respect to P^* along $\hat{\mathbf{t}}^p$. Using the kinematic vector relations of section 8.3, we can develop equation 8.43 as

$$v^{s^*p^*} = \left([\hat{\mathbf{X}}^i]^T \left(J_t^{i,s}\,\dot{q} - J_t^{i,p}\,\dot{q} \right) + \boldsymbol{\omega}^{i,s} \times (\mathbf{c}^s + \mathbf{r}^s) - \boldsymbol{\omega}^{i,p} \times (\mathbf{c}^p + \mathbf{r}^p) \right) .\, \hat{\mathbf{t}}^p \tag{8.44}$$

which can easily be computed using the appropriate projections, as before.

Coulomb forces For dry friction (with coefficient μ), Coulomb forces appear at the contact point and the following scheme holds:

if $\lambda_3 > 0$

$$\mathbf{T}_c^{sp} = \mu\, F_n \,\mathrm{sign}\,(v^{s^*p^*})\, \hat{\mathbf{t}}^p \tag{8.45}$$

and

$$\mathbf{T}_c^{ps} = -\mathbf{T}_c^{sp} = \mu\, F_n \,\mathrm{sign}\,(v^{s^*p^*})\, \hat{\mathbf{t}}^s$$

else

$$\mathbf{T}_c^{sp} = \mathbf{T}_c^{ps} = 0$$

This scheme checks for the presence of the contact ($\lambda_3 \overset{?}{>} 0$) otherwise the computation of the tangent forces is of course irrelevant.
Let us stress the following points:

- This Coulomb friction law contains a well-known singularity around the origin (at zero speed) whose analytical smoothing (using for instance a magic formula or a hyperbolic tangent-based function with an adjustable slope around the origin) is recommended[10], unless one chooses to investigate a more general approach in which a "stick" model and a "slip" model can follow one another [31].

- The normal force (introduced in the previous expressions via λ_3) is of course the instantaneous value of this force. However, the computation of the latter requires the knowledge of the Lagrange multiplier associated with constraint h_3, i.e. the computation of equation 8.42, which can be performed only at the end of the equation reduction phase. In order to solve this problem, the following two solutions are proposed:

[9] At this stage, $P \equiv S$, but they have been separated in figure 8.5 for clarity.
[10] if reached during simulation

1. Simultaneously compute the accelerations $\ddot{u}$ and the Lagrange multipliers λ on the basis of the whole system {4.48, 8.42, 8.45): this will produce an exact result, but at the expense of a higher CPU time.
2. In equation 8.45, consider the value of the normal force computed at the previous valid integration step. Numerous experiments in railway simulation [20] have shown that this trick is quite reasonable as long as the time step remains very small, as it is anyway required for systems with contact.

Viscous forces A tangent viscous force, arising for instance from the contact lubrification, can be introduced to the model by means of a constitutive equation.
For linear viscous damping e.g.,
if $\lambda_3 > 0$

$$\mathbf{T}_v^{sp} = D_v\left(v^{s^*p^*}\right)\hat{\mathbf{t}}^p \text{ and } \mathbf{T}_v^{ps} = -\mathbf{T}_v^{sp} = D_v\left(v^{s^*p^*}\right)\hat{\mathbf{t}}^s$$

else

$$\mathbf{T}_v^{sp} = \mathbf{T}_v^{ps} = 0$$

in which D_v is a damping contact coefficient to be identified. On the basis of figure 8.5, one can thus consider a general case where both Coulomb friction and viscous contact forces arise and write

$$\mathbf{T}^{sp} = \mathbf{T}_c^{sp} + \mathbf{T}_v^{sp} \quad \text{and} \quad \mathbf{T}^{ps} = -\mathbf{T}^{sp} \tag{8.46}$$

Pure rolling contact: constraint forces As mentioned in the introduction, the proposed formalism deals with a contact model for which the cam/follower device represents a particular case. This assertion justifies the fact that the present model could be used for applications in which bodies p and s undergo a "pure rolling" motion : in that case a pure rolling model would be more advisable in the tangent direction.
Thus, in the same way as for normal contact, the approach consists in reducing the number of degrees of freedom of this system by one, by introducing a non-slip constraint between bodies p and s. This constraint simply imposes the condition that the material points P^* and S^* have the same tangent velocity, i.e.:

$$\dot{h}_4(q, \eta^p(q), \eta^s(q), \dot{q}) \equiv v^{s^*p^*} = 0 \tag{8.47}$$

where $v^{s^*p^*}$ is given by 8.44.
Note that equation 8.47 does not depend on $\dot{\eta}^p$, $\dot{\eta}^s$ since it involves the velocity of *material* points on the surfaces. Thus, equation 8.47 has the following form:

$$\dot{h}_4(q, \eta^p(q), \eta^s(q), \dot{q}) = J_4(q)\,\dot{q} = 0$$

which can be introduced into the dynamical equations via the Lagrange multipliers technique, as for the normal contact constraint. Similarly, the tangent force is directly deduced from the Lagrange multiplier associated with constraint 8.47: its value could thus also be used to check the validity of the pure rolling assumption, otherwise the analysis must be interrupted, or one must shift to a "slip" model.

8.4.2 Intermittent contact

In the present section, we will shortly present the way we deal with intermittent contact, i.e. when the normal contact condition 8.41 is not necessarily satisfied during the motion.
In order to clarify this, we show in figure 8.6, the three distinct phases of intermittent contact between bodies p and s.

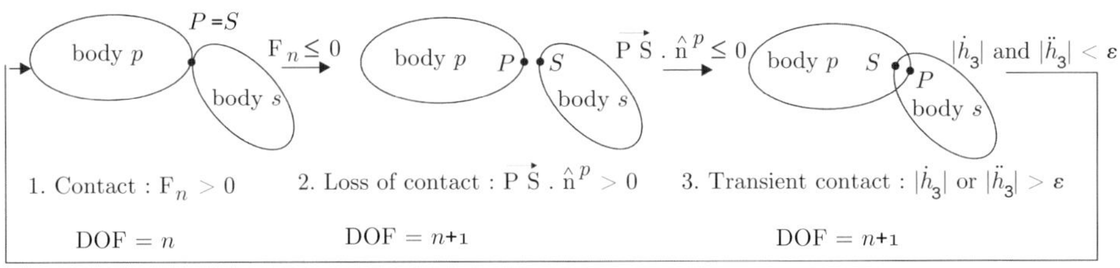

Figure 8.6: Intermittent contact

Phase 1: Permanent contact (arbitrarily considered as the reference configuration)

Bodies p and s are (momentarily) in permanent contact. Test 8.41 is satisfied and constraints h_1, h_2, h_3 are solved. The loop formed by the contact reduces the number of d.o.f of the MBS by 1 with respect to an "open configuration". Practically, the coordinate partitioning procedure (see section 4.6.1) considers v^s as a dependent variable in the reduction phase.

Phase 2: Loss of contact

As soon as the contact test 8.41 is no longer satisfied, bodies p and s lose contact and the multibody structure opens, with an increase in d.o.f. Formally, the configuration of the system directly preceding the test violation[11] is then stored to serve as initial condition for the sequel: the dependent variable v^s becomes *independent* and will be denoted u^s according to the previous notations. During this new phase, the minimal distance between bodies p and s is computed on

[11] The exact transition time is difficult to determine; practically, the previous (small!) time step is chosen. However, in some high level numerical integrators, a more precise point can be found.

line as follows: constraints h_1, h_2 precisely locate the candidate points P and S, while the l.h.s. of h_3 (equation 8.18) in which $\mathbf{n}^p$ is normed and thus replaced by $\hat{\mathbf{n}}^p$, computes the distance between the points, i.e. the projection e^{ps} of vector $(\mathbf{x}^{i,s} - \mathbf{x}^{i,p})$ onto the normal unit vector $\hat{\mathbf{n}}^p$. As long as this projection remains positive, bodies p and s are not in contact.

Phase 3: Transient contact

As soon as this projection h_3 becomes negative, a so-called "transient contact" phase takes place since the bodies — momentarily — interpenetrate. From that time onwards, a compliant model using an elastic and dissipative normal force is introduced between points P and S.
Denoting e^{ps} the penetration and v^{ps} the penetration rate:

$$e^{ps} \triangleq (\mathbf{x}^{i,S} - \mathbf{x}^{i,P}) \,.\, \hat{\mathbf{n}}^p \quad , \quad v^{ps} \triangleq (\dot{\mathbf{x}}^{i,S} - \dot{\mathbf{x}}^{i,P}) \,.\, \hat{\mathbf{n}}^p$$

and introducing the elastic and damping coefficients[12] K^m and D^m,

$$\mathbf{F}_n^{ps} = -(K^m e^{ps} + D^m v^{ps})\, \hat{\mathbf{n}}^p \quad \text{and} \quad \mathbf{F}_n^{sp} = -\mathbf{F}_n^{ps} \tag{8.48}$$

represent respectively the normal transient force of body p on body s, and conversely.
Despite the penetration, the determination of points P and S is similar to that described above (during phase 2) and the quantities e^{ps}, v^{ps} can be directly computed from the l.h.s. of equations 8.18, 8.24.

A final test now needs to be established to decide whether or not a transient contact (phase 3) can evolve and again become a permanent contact (phase 1 in figure 8.6). For that purpose, we continuously examine during phase 3 the l.h.s. of the normal constraint h_3 at both velocity and acceleration levels (equations 8.23, 8.36). As soon as both these quantities $|\dot{h}_3|$ and $|\ddot{h}_3|$ fall below a given threshold, we decide to return to the permanent contact model (phase 1). A reasonable threshold for $|\dot{h}_3|$ and $|\ddot{h}_3|$ can be for instance a certain percentage (1...2 %) of their initial values, i.e. those computed (and stored !) at the first time integration step of phase 3. The normal contact is then stabilized and it is quite reasonable to "close" the loop again: the independent variable u^s again becomes dependent ($u^s \to v^s$) and the additional d.o.f is lost. Note that a very small "jump" is unavoidable between the last value of u^s and the first value of v^s, since the bodies still interpenetrate slightly at the end of phase 3 and are in point contact at the beginning of phase 1.
Note that this "stabilization" test does not imply that bodies p and s interpenetrate continuously during the transient phase 3. In other words, once phase 2 is finished (first contact occurrence), the test is active, whether the bodies remain in contact or not. Of course, in the latter situation, the normal force given by equation 8.48 disappears, as well as the possible tangent forces.

---

[12] which should be chosen with caution [5], [31] and are not necessarily constant.

Discussion

This methodology implies various nonlinearities[13] (in the constitutive equations, at the "switching" points before and after phase 1, ...) and ill-conditioning (high stiffness coefficient) which are well-known obstacles for integration schemes and which, in any case, require very small time steps and robust integrators. It is not the purpose of the present work to discuss integration methods; we are aware of the difficulties resulting both from nonlinearities and high frequencies. For the latter issue, stiff integrators have been developed and are available in high level mathematical libraries. In particular, we will present in chapter 10 a time integration method for closed multibody systems containing high frequencies, combining the Newmark scheme and the coordinate partitioning technique. As regards the stop and restart of the integration scheme, we extensively use the MATLAB® features for such a process, which is all the more complex in the present situation since the number of variables changes each time a cam/follower pair loses or retrieves contact.

More important than these high nonlinearities, one of the major issues of the transient phase relates to the way the collision of bodies is modeled. In a compliant approach for instance, the constitutive equation of the normal force between two bodies in contact is extremely difficult to model and, as is the case in many other contributions, we considered only a linear behavior with constant parameters (see equation 8.48), keeping in mind that this remains a rather simplified model.

Fortunately, in most applications of cam/follower systems, the *permanent* contact phase (phase 1) is preponderant, or even the only one. Thus, considering the normal contact as a kinematic constraint (see section 8.3) is a great advantage from both the efficiency and reliability points of view since we do not resort to "stiff elements" in the normal direction as done in various multibody softwares.

Finally, let us emphasize that the present model takes into account intermittent contact only in the *normal* direction. As regards the *tangent* contact, one could imagine modeling in a similar manner the shift between a "slip" model in which tangent forces are introduced via a specific constitutive equation (Coulomb, viscous force, ...), and a "stick" model which would constrain bodies to be in pure rolling contact (according to equation 8.47).

8.5 Numerical examples

8.5.1 Introduction

The present model was implemented in the computer program MBSOFT [23] and its symbolic core ROBOTRAN [55]. The MATLAB® environment is well-suited to implement the present model. Indeed, in addition to variable partitioning and matrix permutations, the "switch" from one model to another must be

[13] which are of course present in the real system

performed: this operation requires not only a control of the number of d.o.f. and the role switches of the corresponding variables but also the interruption (and restart) of the integrator scheme, which is not a trivial task.

8.5.2 Cam/follower model: numerical validation

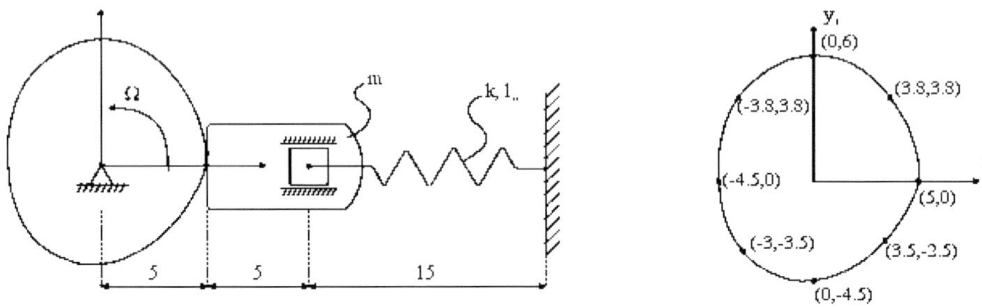

Figure 8.7: A simple cam/follower device (from [5])

This first example comes from [5] and consists of a cam/follower pair with a constant angular speed at the input shaft (see figure 8.7). The geometry of the cam was defined by giving the eight points indicated in figure 8.7 and using a cubic spline interpolation between these. As detailed in [5], the spring constant k equals 500, the unstressed length l_0 of the spring is 22.5 and the mass m of the follower equals 1. The angular speed of the input shaft is $\Omega = 1.75\,\text{rev/s}$ and tangent friction is neglected (see [5] for more details). Figure 8.8 compares the time history of the displacement of the follower. Despite a small difference observed around 0.2 sec, 0.8 sec, etc. ... due to analytical interpolations of the profiles, results are quite similar.

Figure 8.9 represents the time history of the normal contact force F_n between the cam and the follower and compares the results from [5] and those obtained by the present approach. Of course, the small "geometrical" deviation observed previously in figure 8.8 has noticeable repercussions here at the same places. Note that in our approach, since the current cam/follower loop sets the number of d.o.f to zero, the problem reduces to computing an inverse dynamical problem via equation 8.42. The result is thus straightforward since it does not require any numerical integrator. Finally, figure 8.10 compares the required input torque at the driving shaft M_{in} with (continuous line) and without (dashed line) the presence of a tangent friction force (friction coefficient $\mu = 0.2$).

8.5.3 Cam/follower model: experimental validation

As a second application, we propose a comparison with an experimental set-up in our laboratory described in fig. 8.11, which consists of a motor-driven cam and a follower mounted on a rotating arm (whose free rotation axis is located on

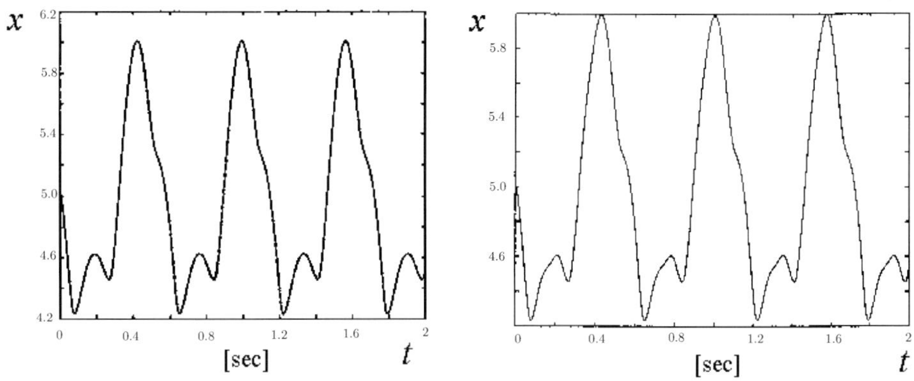

Figure 8.8: Rectilinear displacement of the follower (left : results from MBSOFT, right : results from [5])

the left in figure 8.11). In addition to the gravity force, a spring acts vertically on the arm to preload the system. A small part was inserted between the arm and the follower (namely, an "opened" square element) to fix the strain gauges in such a way that their output reflects the normal contact force as accurately as possible. The main numerical characteristics of the system are the following:

- arm : length $= 0.75\,m$; mass $= 1.8\,kg$
- cam : profile :
  - 2 arcs (circular) + 2 straight segments (tangent)
  - size : large radius $= 40\,mm$; small radius $= 15\,mm$; total vertical length $= 110\,mm$
- follower 1 (depicted in figure 8.11) :
  - profile : opened arc (circle)
  - size : radius $= 400\,mm$; horizontal length $= 100\,mm$
- follower 2 (not shown) :
  - profile : circular roller (closed circle)
  - size : radius $= 10\,mm$

The cam is driven by a D.C. motor, at a constant speed of $5.2\,\frac{rad}{\sec}$. The normal force is recorded via the previously calibrated strain gauges. The friction force between the cam and the follower is neglected. Figures 8.12 and 8.13 show a comparison between experiment (continuous line) and simulation (dotted line). Figure 8.12 considers the case where the contact between the cam and the

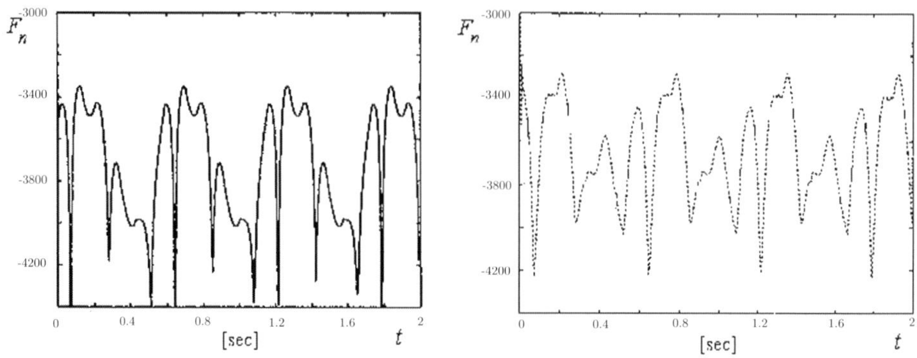

Figure 8.9: Normal contact force between the cam and the follower (left : results from MBSOFT, right : results from [5])

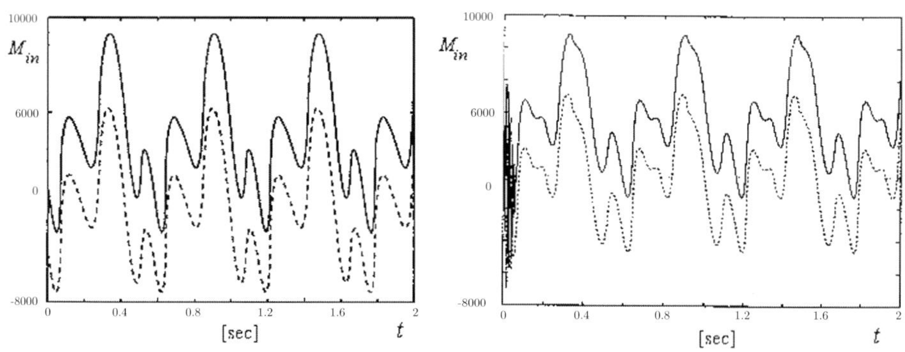

Figure 8.10: Input driving torque (left: results from MBSOFT, right: from [5])

follower 1 is permanent. In the second experiment (figure 8.13), the follower 2 is used and the spring preload has been increased in such a way that loss of contact appears twice, namely before and after the contact on the small arc of the cam profile (N.B.: the high frequencies in the experimental curve result from some difficulties in tuning the low-pass filter of the strain gauges in this experiment).

8.5.4 Modeling of universal wheels

Most existing mobile robots are equipped with classical wheels. Depending on the design of the various wheels, the mobility of these robots is restricted. One way of achieving perfect mobility is to use universal wheels allowing independent rotation and translation in any direction (see figure 8.14). The treads of such wheels are not tires but consist of several freely rotating rollers. When a wheel

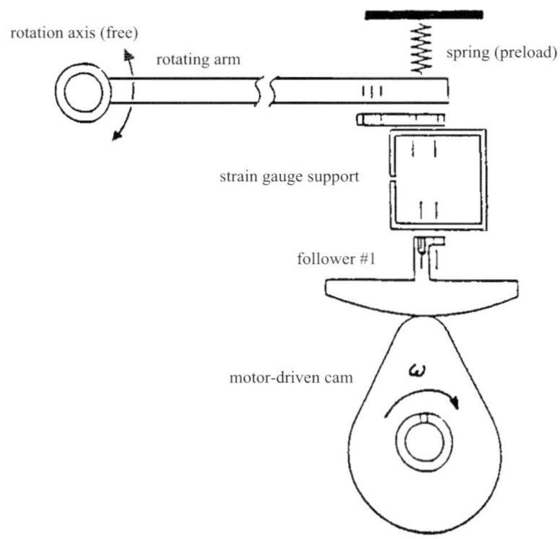

Figure 8.11: Sketch of the experimental set-up

axis rotates, the wheel is driven as usual but, at the same time, the free rotation of the roller in contact with the ground allows motion of the wheel parallel to the driven axis.

The main limitations of such wheels are their load capacity and the surmountable bump height (due to the small radius of the rollers) and the vertical vibrations resulting from the successive shocks occurring each time a new roller contacts the ground. In order to circumvent the first two limitations, the ROLLMOBS omnimobile robot [11] shown in figure 8.15 was developed in our laboratory: in this robot, three compliant spheres carry the platform, each being driven by a small universal wheel.

Since the latter is no longer in direct contact with the ground, this considerably improves the robot's performances in terms of load capacity and surmountable bump height.

As regards the vibration phenomenon, which is also reduced but still present in this new design, various concepts of universal wheels, namely the "classical", the "alternate" and the "half roller" (see figure 8.16), have been investigated in order to minimize the gap between two successive rollers.

In order to properly quantify and compare the vibration amplitudes of these designs, the present model of contact between profiled bodies has been used, taking the intermittent contact into account both from the geometrical and dynamical points of view. For the three universal wheels (see figure 8.16), we note the fact that such wheels actually match the previous multibody model, since each roller can be assimilated to a "cam" and the ground to a "follower". For instance, figure 8.17 compares the vertical displacement of the three wheels moving over flat ground at constant velocity. As expected, the smaller the gap

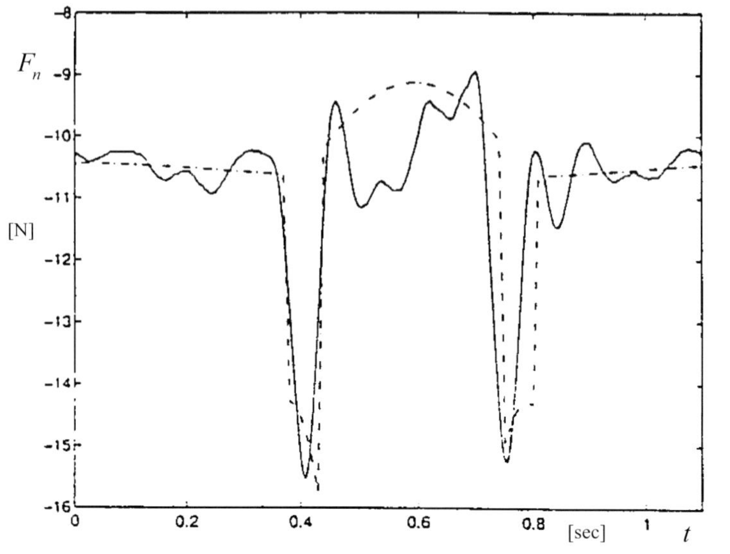

Figure 8.12: Simulation versus experiment: permanent contact

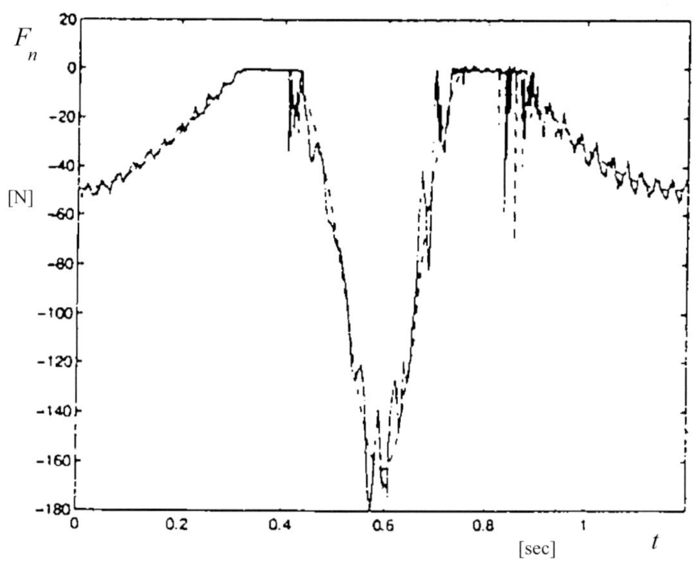

Figure 8.13: Simulation versus experiment: intermittent contact

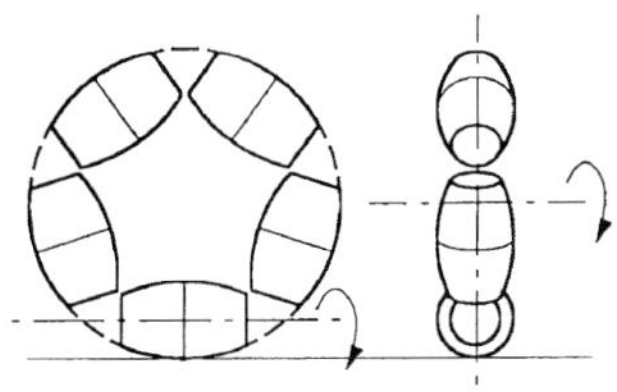

Figure 8.14: The universal wheel principle

Figure 8.15: ROLLMOBS: prototype

between rollers, the lower the vibration amplitude: of course the latter also depends on several parameters such as the roller elasticity, size and type.

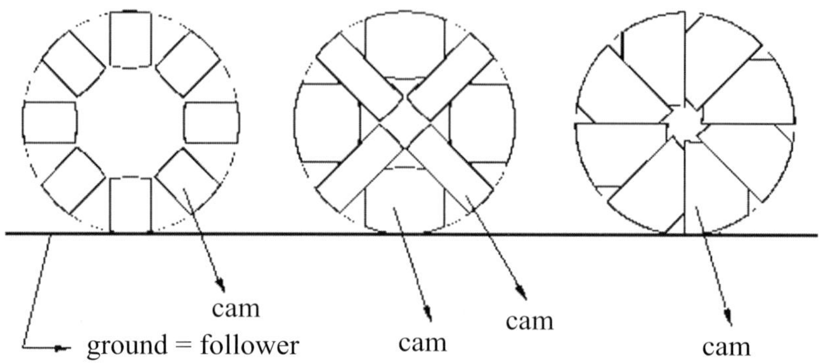

Figure 8.16: Universal wheel designs : classical - alternate - half rollers

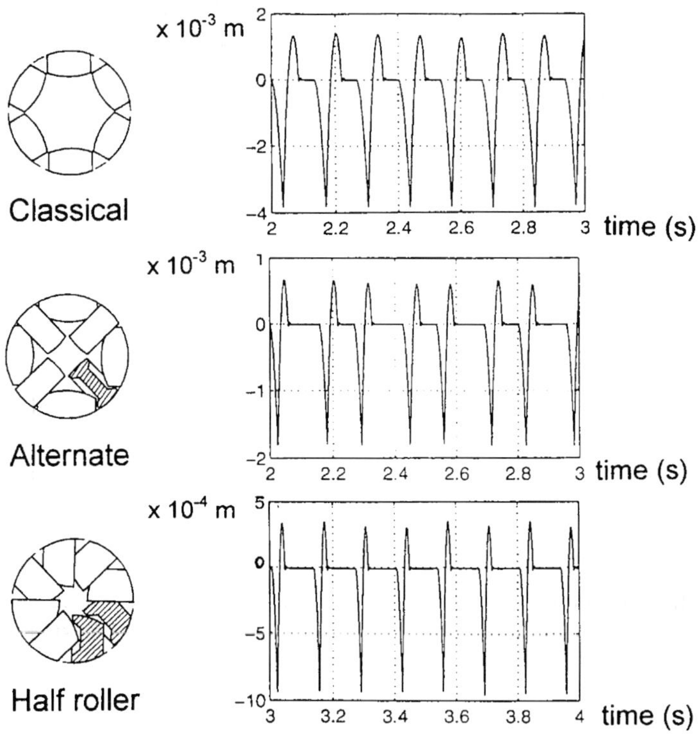

Figure 8.17: Vertical vibration of three universal wheel designs

Chapter 9

Multibody systems with flexible beams

9.1 Introduction

Flexibility effects in multibody dynamics certainly represent one of the most arduous aspects of the modeling of mechanical systems. There are several reasons for this. First of all, contrary to the rigid case, flexible bodies require both physical and geometrical hypotheses (e.g.: elastic linear behavior, small flexible motion, ...) whose validity must be questioned *a posteriori* as regards the accuracy or even the meaningfulness of the simulation results.

Next, since deformable bodies theoretically have an infinite number of degrees of freedom, their spatial discretization is generally indispensable (except for some analytically solvable cases) which naturally induces another source of approximation. Unfortunately, the latter is not easy to control, in particular due to the fact that the type of spatial discretization and its refinement are often problem-dependent whatever the method implemented (Rayleigh-Ritz, Finite Segment or Finite Element methods, ...) (see [57] for instance).

Finally, as a consequence of the discretization, the increase in the number of degrees of freedom of flexible multibody systems has a severe impact on the cost of the numerical analysis. Despite the power of modern computers and the use of parallel computation, one still needs to improve the mathematical model of flexible multibody systems in terms of size, matrix conditioning and numerical treatment (especially the time integration of stiff systems). These difficulties have stimulated a vast variety of approaches (see [86], [83], [29], [47], [91], [85], [13], [68] for instance) which in fact stem either from classical rigid multibody dynamics or structural mechanics: indeed, the inclusion of flexibility effects involves a blend of these two disciplines.

One way to deal with flexible multibody systems emerges from researches in the field of structural mechanics. This represents a marked departure from the

classical multibody approach presented in this book. As a matter of fact, Finite Element Methods (FEM) may be adopted in a general sense to represent any type of components appearing in a multibody system: rigid and elastic bodies, mechanical joints, interactions between bodies or with the external world. In such formulations, the global motion (including rigid motion and elastic deformation) of a given body is generally referred directly to the inertial frame, using absolute coordinates. As regards flexible bodies, an appropriate finite element discretization is developed and introduced in the kinematic quantities of the dynamical equations of motion. A joint between two bodies is generally modeled via kinematic constraints which, in a finite element context, express relations (often trivial identification) between nodal degrees of freedom.
With respect to the classical multibody approach which incorporates flexibility effects into its dynamical formalism, one could say that the FEM approach focuses mainly on the flexible bodies and uses the global multibody motion as a source of external and dynamical loading. Since all the geometrical nonlinearities are generally taken into account in a FEM approach, the space integrals which appear in the equations of motion must be evaluated at each step of the numerical treatment [78]: this is unavoidably costly in terms of computational time.

In a purely multibody context, one generally assumes that the global motion of a flexible body is decomposed into a rigid body motion and a superimposed small deformation. It has been shown [47], [91] that this restriction, if one does not take care, can lead to completely erroneous results for flexible bodies submitted to external or inertial loads. Although this so-called *geometric stiffening* problem can be solved by keeping second order terms in the expression of the deformation of the material, the amplitude of the flexible motions (e.g. the tip angle of a rotating beam) must remain small in those formalisms which are generally based on kinematic restrictions as regards flexibility. Moreover, the space discretization is generally based on an *assumed mode* technique. In this context, it is apparent that the selection of an appropriate set of shape functions for a given multibody application is quite a delicate task and could penalize this approach probably more than the above-mentioned kinematic assumptions. Indeed, there is no systematic method or rule to set up an *a priori* suitable set of shape functions since this choice is strongly problem-dependent ([57], [88]). Moreover, it is shown in [57] that even admissible[1] shape functions can be poor or inadequate to satisfy the natural boundary conditions of a system, whatever the envisaged refinement.

Assumed modes are nevertheless selected in this chapter to discretize the beam deformations. As regards the shape functions, we propose (as in [44]) to use monomials as an alternative to more dedicated functions. Monomials offer two important advantages. First they are problem-independent and since their combination approximates a Taylor series, they should be able to approximate any plausible deformation. Secondly, their "trivial" differentiation and integration is most suitable to symbolic computation as detailed in section 9.3.5.

[1] in a Rayleigh-Ritz sense

From a practical point of view, this approach offers the following two advantages:

- The use of an invariable set of shape functions (monomials in our case) considerably alleviates the user's task since there is no need for a prior modal analysis to determine a set of appropriate shape functions.

- Thanks to the trivial derivation and integration of monomials, the *fully symbolic*[2] *generation* of the dynamical equations is performed with a high level of simplification thanks to the ROBOTRAN symbolic capabilities [13], [17]. As we shall see later, the resulting benefits are clearly revealed by the CPU time performances of the numerical simulations, especially in comparison with a fully nonlinear finite element method [88].

The chapter is organized as follows. First the finite segment method is briefly described [42]. While simple in concept, such an approach gives satisfactory and efficient results, particularly for planar applications. Then, based on the Timoshenko theory, a more elaborate formalism is developed, in which beam flexibility is modeled via a 3 D approach including shear deformation and cross-section rotary inertia. The deformation equations are obtained from the virtual power principle, using approaches proposed in [86], [29] and taking into account the previously mentioned stiffening effects. The Newton-Euler recursive formalism in relative coordinates of section 3.3 is then adapted in order to take into account the flexibility of the beam. The whole model is implemented in ROBOTRAN, including the fully symbolic differentiation of the dynamical equations to provide the tangent matrices with respect to the generalized coordinates q, velocities $\dot{q}$ and accelerations $\ddot{q}$. Finally, with the help of ROBOTRAN, several experiments and numerical simulations were performed on various multibody systems with flexible beams. For each of these, the discretization of the beam deformations was systematically achieved by using one of the following techniques:

- a finite segment approach in which the beam is replaced by a sequence of rigid segments interconnected by equivalent springs [42],

- an assumed modes method using the "power series" monomials mentioned above [44], [13],

- an assumed modes method using shape functions resulting from a preliminary modal analysis [77],

- a finite element model, using an appropriate FEM code [78].

[2]contrary to a semi-symbolic approach ([58] for instance), in which symbolic terms cannot be traced back to input variables

9.2 The finite segment approach

A detailed description of the so-called *finite segment* modeling of slender bodies can be found in [42], in particular for the case of rectilinear beams. The method consists in replacing each flexible beam of the physical system by a chain of N rigid bodies – or finite segments – connected by springs as illustrated in figure 9.1. Basic principles of structural mechanics are then used to establish the

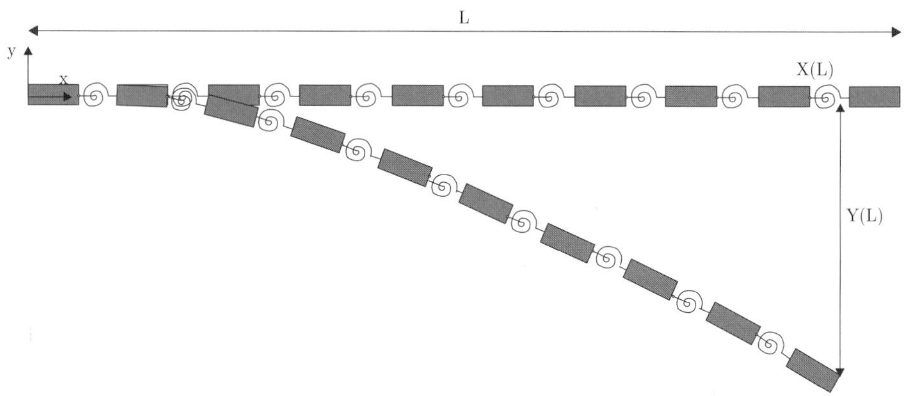

Figure 9.1: Finite Segment Method

relations between the force and moment components and the displacements and rotations between two adjacent segments, in order to identify *equivalent* stiffness coefficients. Various configurations are treated in [42] (extension, torsion, bending of straight or tapered segments, ...) leading to systematic rules for computing the equivalent stiffnesses of a given chain of finite segments. The use of this method is illustrated in a specific problem of the tutorial (Part III, section 12.9).

From a multibody point of view, the method offers the advantage of dealing with flexible bodies in a purely rigid multibody context, i.e. formally, of using a conventional multibody code. Indeed, as shown in figure 9.1, the chain of finite segments (and intermediate joints) is part of the multibody topology, and treated in the same way as the other elements. However, and this probably represents the main limitation of the method, experiments show that a minimum of ...10... finite segments per beam is generally required: this strongly increases the length of the kinematic chains and thus the number of d.o.f., especially when dealing with 3D flexible motions. This is the reason why, from an efficiency point of view, this technique is certainly attractive for planar applications but becomes less adequate in other situations; this will be quantitatively apparent in the examples of section 9.4.

9.3 The assumed mode approach

9.3.1 Description of the flexible beam

The modeling of any flexible body results from various hypotheses. These are listed and detailed below in the present case:

. *Geometry*: Only prismatic beams are considered, and the centroidal axis is assumed to be rectilinear in the undeformed configuration. The elastic axis is assumed to coincide with the centroidal axis.

. *Material*: The beam consists of a homogeneous isotropic material which conforms to linear elasticity.

. *Deformation model* [29]: 3D-model, conservation of plane cross-sections, shear deformation and cross-section rotary inertia included.

. *Kinematics*: The angular deformations and the curvatures of the beam are assumed to be small. As regards the angular deformations, the present hypothesis will lead to the linearization of the 3×3 rotation matrix between two cross-section fixed-frames (as in [47]).
Since this assumption may be restrictive for some applications, a subdivision of the complete beam into several sub-bodies will sometimes be applied in the present multibody approach, at the expense of course of a higher computational cost.

. *Topology*: From this point of view, the beam has the same status as a rigid body but it is assumed that the joint connection points (for the parent and for the children bodies) are located on the centroidal axis[3]. The *reference body frame* $\{\hat{\mathbf{X}}^i\}$ is thus defined in such a way that its plane $\{\hat{\mathbf{X}}_2^i, \hat{\mathbf{X}}_3^i\}$ coincides with the beam cross-section $S \triangleq S_i$ (see figure 9.2) which is attached to parent body h through joint i.

Figure 9.2 summarizes the various kinematic quantities related to the beam[4]. The beam is characterized by its natural length L and by a cross-sectional area A. Let s be the current abscissa of the beam centroidal axis, i.e. the distance from point O^E, located at the root of the beam, to the plane of a generic cross-section S of the beam in its undeformed configuration. A frame $\{\hat{\mathbf{S}}(s)\}$ is fixed to the plane of this cross-section S, oriented so that $\hat{\mathbf{S}}_1$ is perpendicular to the plane, while $\hat{\mathbf{S}}_2$ and $\hat{\mathbf{S}}_3$ are parallel to the principal axes of the cross-section. Some particular sections need to be identified:

. S_E, the root section at $s = 0$ ($\ni O^E$) and the beam reference frame $\{\hat{\mathbf{E}}\} \triangleq \{\hat{\mathbf{S}}(0)\}$,

[3] This restrictive hypothesis can be circumvented in practical situations by using intermediate massless bodies and fictitious locked joints.

[4] For clarity, the superscript i has been deliberately omitted for the various vectors and variables relating to beam i.

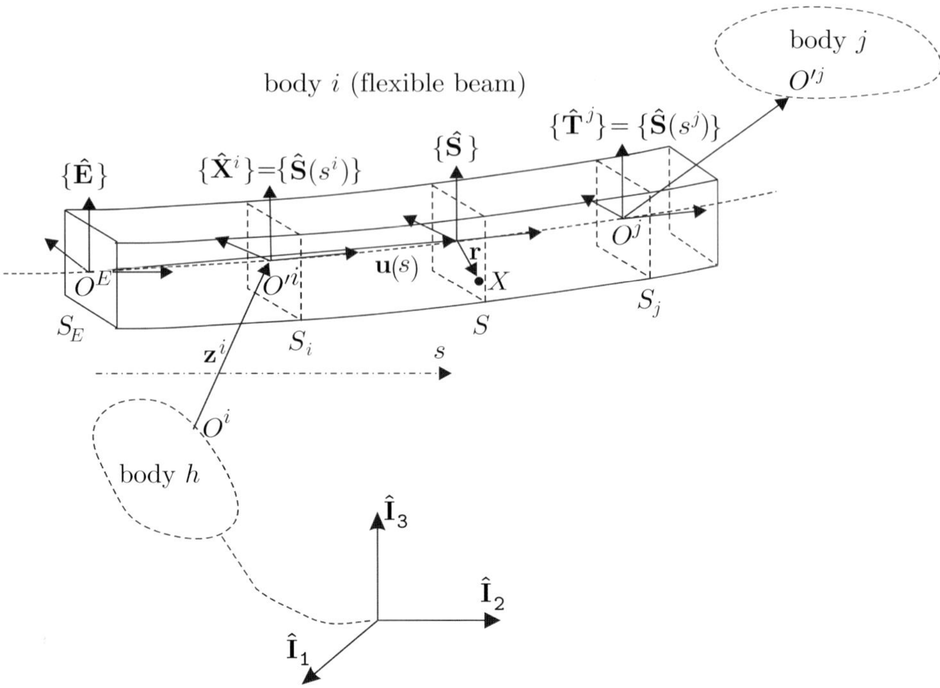

Figure 9.2: A flexible beam i in the multibody system

. S_i, the section containing the parent connection point O'^i at $s = s^i$ (frame $\{\hat{\mathbf{X}}^i\} \triangleq \{\hat{\mathbf{S}}(s^i)\}$),

. S_j, the section containing the connection point O^j of joint $j \in \bar{\imath}$ at $s = s^j$ (frame $\{\hat{\mathbf{T}}^j\} \triangleq \{\hat{\mathbf{S}}(s^j)\}$).

On the basis of the previous kinematic assumption, a linearized rotation matrix $R(s)$ is used to express the orientation of a given cross-section frame $\{\hat{\mathbf{S}}(s)\}$ with respect to $\{\hat{\mathbf{E}}\}$:

$$[\hat{\mathbf{S}}(s)] = R(s)[\hat{\mathbf{E}}] \tag{9.1}$$

Defining $\vartheta \triangleq (\vartheta_1, \vartheta_2, \vartheta_3)^T$, the three Tait-Bryan angles around the 1, 2 and 3 directions successively, we can write the linear approximation of the general expression 1.42 of the rotation matrix $R(s)$ as follows:

$$R(s) = \begin{pmatrix} 1 & \vartheta_3(s) & -\vartheta_2(s) \\ -\vartheta_3(s) & 1 & \vartheta_1(s) \\ \vartheta_2(s) & -\vartheta_1(s) & 1 \end{pmatrix} = E - \tilde{\vartheta}(s) \tag{9.2}$$

In particular for sections S_i and $S_{j\in\bar{\imath}}$,

$$[\hat{\mathbf{X}}^i] = R(s^i)[\hat{\mathbf{E}}] \quad \text{with} \quad R(s^i) = E - \tilde{\vartheta}^i \tag{9.3}$$
$$[\hat{\mathbf{T}}^j] = R(s^j)[\hat{\mathbf{E}}] \quad \text{with} \quad R(s^j) = E - \tilde{\vartheta}^j \tag{9.4}$$

In the undeformed state, the reference position with respect to O^E of a point P which belongs to the centroidal axis is specified by its abscissa s:

$$\overrightarrow{O^E P}_{ref} = [\hat{\mathbf{E}}]^T \begin{pmatrix} s \\ 0 \\ 0 \end{pmatrix}$$

and when the beam deforms, the current position vector of this point becomes

$$\mathbf{u}(s) \triangleq \overrightarrow{O^E P} = [\hat{\mathbf{E}}]^T u(s) \tag{9.5}$$

The displacement field on the centroidal axis is thus given by the function

$$v(s) = u(s) - \begin{pmatrix} s \\ 0 \\ 0 \end{pmatrix} \tag{9.6}$$

Similarly, the reference position $\mathbf{u}_r(X)$ of a arbitrary point X of a cross-section S with respect to point O^E (see figure 9.2) is given in the *undeformed state* by

$$\begin{aligned} \mathbf{u}_r(X) \triangleq \overrightarrow{O^E X}_{ref} &= \overrightarrow{O^E P}_{ref} + \mathbf{r}(X) \\ &= [\hat{\mathbf{E}}]^T \begin{pmatrix} s \\ 0 \\ 0 \end{pmatrix} + [\hat{\mathbf{S}}(s)]^T r(X) \\ \text{with} \quad [\hat{\mathbf{S}}(s)] &= [\hat{\mathbf{E}}] \quad \text{and} \quad r(X) = \begin{pmatrix} 0 & r_2(X) & r_3(X) \end{pmatrix}^T \end{aligned} \tag{9.7}$$

In the *current* (deformed) configuration, this position vector becomes

$$\begin{aligned} \mathbf{u}(X) \triangleq \overrightarrow{O^E X} &= [\hat{\mathbf{E}}]^T u(s) + [\hat{\mathbf{S}}(s)]^T r(X) \\ &= [\hat{\mathbf{E}}]^T u(s) + [\hat{\mathbf{E}}]^T R^T(s) r(X) \end{aligned} \tag{9.8}$$

and can be written under the form

$$\mathbf{u}(X) = \mathbf{u}_r(X) + \mathbf{u}_d(X) \tag{9.9}$$

Using 9.6 and 9.2, the displacement vector $\mathbf{u}_d(X)$ then reads

$$\mathbf{u}_d(X) = [\hat{\mathbf{E}}]^T \left(v(s) + \left(R^T(s) - E\right) r(X)\right) = [\hat{\mathbf{E}}]^T \Big(v(s) + \tilde{\vartheta}(s) r(X)\Big) \tag{9.10}$$

The functional discretization of the beam is then achieved by means of admissible shape functions

$$\{\ \tau_1^\alpha(s),\quad \tau_2^\alpha(s),\quad \tau_3^\alpha(s)\ \} \text{ and } \{\ \rho_1^\alpha(s),\quad \rho_2^\alpha(s),\quad \rho_3^\alpha(s)\ \}$$

respectively approximating the displacements $v(s)$ on the centroidal axis and the three angles $\vartheta(s)$ characterizing the rotational deflections of the cross sections. In a synthetic matrix form, we write

$$v(s) = \tau(s) q_t \tag{9.11}$$

where

$$\tau(s) \triangleq \begin{pmatrix} \tau_1(s) \\ \tau_2(s) \\ \tau_3(s) \end{pmatrix} = \begin{pmatrix} \tau_1^1 & \cdots & \tau_1^{n_{t1}} & 0 & \cdots & \cdots & \cdots & \cdots & 0 \\ 0 & \cdots & 0 & \tau_2^1 & \cdots & \tau_2^{n_{t2}} & 0 & \cdots & 0 \\ 0 & \cdots & \cdots & \cdots & \cdots & 0 & \tau_3^1 & \cdots & \tau_3^{n_{t3}} \end{pmatrix} \tag{9.12}$$

and

$$q_t \triangleq \begin{pmatrix} q_{t_1}^1 & \cdots & q_{t_1}^{n_{t1}} & q_{t_2}^1 & \cdots & q_{t_2}^{n_{t2}} & q_{t_3}^1 & \cdots & q_{t_3}^{n_{t3}} \end{pmatrix}^T \tag{9.13}$$

denotes the translational deformation coordinates.

For the angular displacements, the discretization is

$$\vartheta(s) = \rho(s) q_r \tag{9.14}$$

where

$$\rho(s) \triangleq \begin{pmatrix} \rho_1(s) \\ \rho_2(s) \\ \rho_3(s) \end{pmatrix} = \begin{pmatrix} \rho_1^1 & \cdots & \rho_1^{n_{r1}} & 0 & \cdots & \cdots & \cdots & \cdots & 0 \\ 0 & \cdots & 0 & \rho_2^1 & \cdots & \rho_2^{n_{r2}} & 0 & \cdots & 0 \\ 0 & \cdots & \cdots & \cdots & \cdots & 0 & \rho_3^1 & \cdots & \rho_3^{n_{r3}} \end{pmatrix} \tag{9.15}$$

and

$$q_r \triangleq \begin{pmatrix} q_{r_1}^1 & \cdots & q_{r_1}^{n_{r1}} & q_{r_2}^1 & \cdots & q_{r_2}^{n_{r2}} & q_{r_3}^1 & \cdots & q_{r_3}^{n_{r3}} \end{pmatrix}^T \tag{9.16}$$

denotes the rotational deformation coordinates.

The choice and the number of these shape functions are generally the user's responsibility, and depend on the application. If the user opts for monomials for $\tau_i(s)$ (or $\rho_i(s)$), ROBOTRAN (see section 9.3.5) will use the following set of shape functions

$$(\frac{s}{L}), (\frac{s}{L})^2, ...(\frac{s}{L})^n$$

of the normalized variable $\frac{s}{L}$, to represent the corresponding displacement $v_i(s)$ (or $\vartheta_i(s)$).

The beam is characterized by material and cross-sectional properties:

- A, the surface of the cross-section,
- Young's modulus E^Y, shear modulus G^{sh} and μ, the mass per unit length of the beam,
- $I_2^S = \int_S (r_3(X))^2\, dA$, $I_3^S = \int_S (r_2(X))^2\, dA$ and $J_p^S = I_2^S + I_3^S$, respectively the principal second moments of area of section S around unit vectors $\hat{\mathbf{S}}_2$ and $\hat{\mathbf{S}}_3$ and the section polar inertia around the centroid.

9.3.2 Kinematics

As mentioned in section 5.8.1, an intrinsic convention[5] of the ROBOTRAN program considers a *reference configuration* of the multibody structure in which all the body frames $\{\hat{\mathbf{X}}^i\}$ coincide with the inertial frame $\{\hat{\mathbf{I}}\}$, and in which each joint axis $\hat{\mathbf{e}}^i$ is aligned with one of the axes of the inertial frame. In addition, we assume that all the flexible beams are undeformed in this reference configuration.

Besides the computations relating to the flexible beams, we assume that the joint vectors $\boldsymbol{\varphi}^j$ and $\boldsymbol{\psi}^j$ as well as the joint rotation matrix $R^{i,j}(q^j)$ are computed beforehand for each joint j of the system. In accordance with section 3.1.3, one has for joint j:

$$\boldsymbol{\psi}^j \triangleq \hat{\mathbf{e}}^j \text{ and } \boldsymbol{\varphi}^j \triangleq 0 \quad \text{when } j \text{ is prismatic,} \tag{9.17}$$

$$\boldsymbol{\varphi}^j \triangleq \hat{\mathbf{e}}^j \text{ and } \boldsymbol{\psi}^j \triangleq 0 \quad \text{when } j \text{ is revolute.} \tag{9.18}$$

$$[\hat{\mathbf{X}}^j] = R^{j,i}(q^j)[\hat{\mathbf{X}}^i] \quad \text{when the parent body } i \text{ is rigid} \tag{9.19}$$

$$\text{and} \quad [\hat{\mathbf{X}}^j] = R^{j,i}(q^j)[\hat{\mathbf{T}}^j] \quad \text{when the parent body } i \text{ is flexible} \tag{9.20}$$

Before computing the complete multibody kinematics, we precompute quantities associated with each flexible beam i, for a given value of the deformation generalized coordinates $q_t, \dot{q}_t, \ddot{q}_t$ and $q_r, \dot{q}_r, \ddot{q}_r$. First of all, the relative angular velocity vector of a cross-section frame $\{\hat{\mathbf{S}}(s)\}$ with respect to $\{\hat{\mathbf{E}}\}$ can be expressed in linear approximation as (see equation 1.61)

$$\boldsymbol{\Omega}(s) = [\hat{\mathbf{S}}(s)]^T \Omega(s) = [\hat{\mathbf{S}}]^T B(s)\dot{\vartheta}(s) \quad \text{with} \quad B(s) = \begin{pmatrix} 1 & \vartheta_3(s) & 0 \\ -\vartheta_3(s) & 1 & 0 \\ \vartheta_2(s) & 0 & 1 \end{pmatrix} \tag{9.21}$$

The relative time derivative of this vector is given by

$$\overset{\circ}{\boldsymbol{\Omega}}(s) \triangleq [\hat{\mathbf{S}}(s)]^T \dot{\Omega}(s) \quad \text{with} \quad \dot{\Omega}(s) = B(s)\ddot{\vartheta}(s) + \dot{B}(s)\dot{\vartheta}(s)$$

---

[5] which is not restrictive

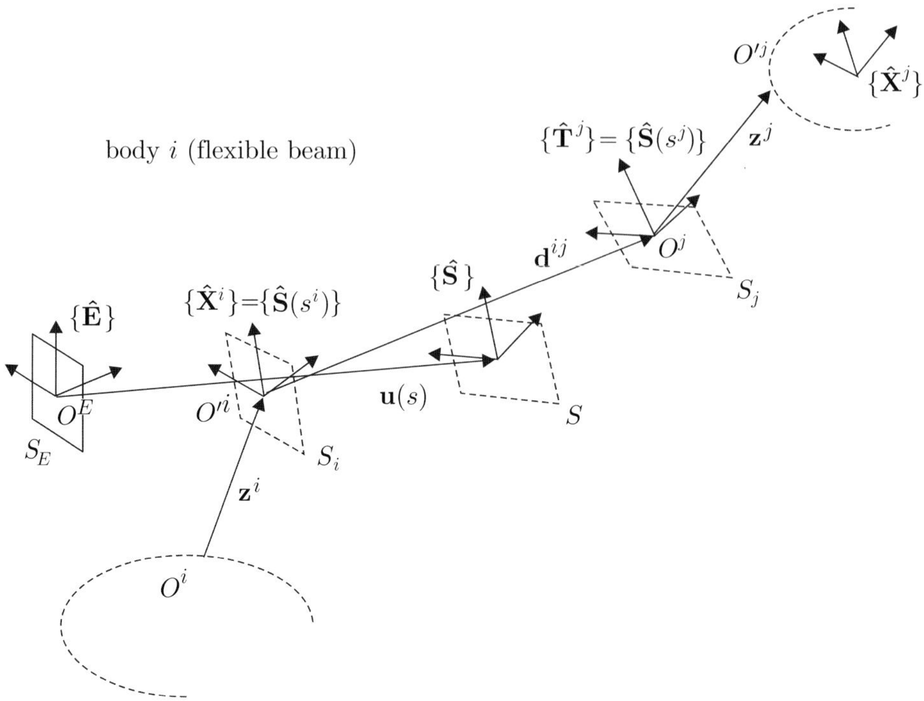

Figure 9.3: Flexible beam

On the basis of 9.5, 9.6 and 9.11, we can express the relative position vector $\mathbf{u}(s)$ of an arbitrary point P of the centroidal axis in frame $\{\hat{\mathbf{E}}\}$ as well as its corresponding relative time derivatives as

$$\begin{aligned}
\mathbf{u}(s) &= [\hat{\mathbf{E}}]^T u(s) \quad \text{with} \quad u(s) = \begin{pmatrix} s & 0 & 0 \end{pmatrix}^T + \tau(s) q_t \\
\overset{\circ}{\mathbf{u}}(s) &= [\hat{\mathbf{E}}]^T \dot{u}(s) \quad \text{with} \quad \dot{u}(s) = \tau(s) \dot{q}_t \\
\overset{\circ\circ}{\mathbf{u}}(s) &= [\hat{\mathbf{E}}]^T \ddot{u}(s) \quad \text{with} \quad \ddot{u}(s) = \tau(s) \ddot{q}_t
\end{aligned} \tag{9.22}$$

The computation of the joint position vector $\mathbf{d}^{ij} \triangleq \overrightarrow{O'^i O^j}$ in the current configuration is then straightforward since (see figure 9.3):

$$\mathbf{d}^{ij} = \mathbf{u}(s^j) - \mathbf{u}(s^i) \triangleq [\hat{\mathbf{E}}]^T d^{ij}_{[E]} \quad \text{with} \quad d^{ij}_{[E]} = \begin{pmatrix} s^j - s^i \\ 0 \\ 0 \end{pmatrix} + \left(\tau(s^j) - \tau(s^i)\right) q_t \tag{9.23}$$

Its relative time derivatives with respect to $\{\hat{\mathbf{E}}\}$ are easily deduced from 9.22:

$$\begin{aligned}\overset{\circ}{\mathbf{d}}{}^{ij} &= [\hat{\mathbf{E}}]^T \dot{d}^{ij}_{[E]} \quad \text{with} \quad \dot{d}^{ij}_{[E]} = \left(\tau(s^j) - \tau(s^i)\right) \dot{q}_t \\ \overset{\circ\circ}{\mathbf{d}}{}^{ij} &= [\hat{\mathbf{E}}]^T \ddot{d}^{ij}_{[E]} \quad \text{with} \quad \ddot{d}^{ij}_{[E]} = \left(\tau(s^j) - \tau(s^i)\right) \ddot{q}_t\end{aligned}$$

At this stage, all the "ingredients" are available to compute the kinematics of the tree structure in a forward recursive manner, i.e. by moving along the kinematic chains from the inertial body 0 to the terminal bodies. Recursive computation is the technique (see [72] [83] [22] for instance) we developed in section 3.3.2 and which consists in computing a quantity relating to a given body as a function of those relating to its parent. Auxiliary variables are then used to store the corresponding recursive results so that the amount of arithmetical operations is minimized.

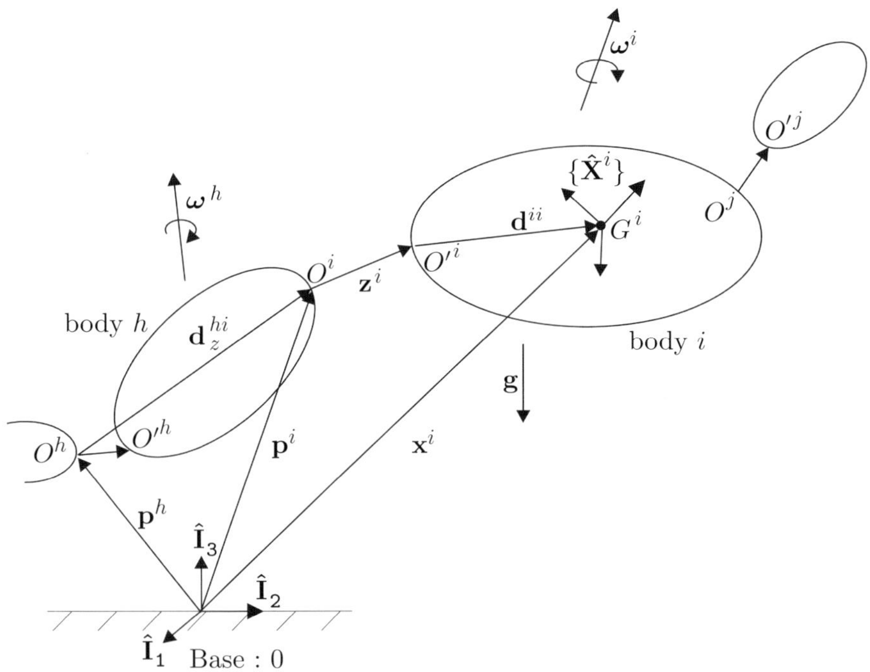

Figure 9.4: Forward kinematics

Vector / tensor formulation

This forward kinematic recursive scheme (see figures 9.4 and 9.3) is first presented in vector/tensor form for legibility.

- Initialization: $\boldsymbol{\omega}^0 = \dot{\boldsymbol{\omega}}^0 = 0 \quad \text{and} \quad \boldsymbol{\alpha}^0 = -\mathbf{g}$
  For $j = 1 : N^{body}$

$$i = \text{index of the parent of body } j \tag{9.24}$$

- Absolute angular velocities $\boldsymbol{\omega}^j$ and $\boldsymbol{\omega}^E$ of frames $\{\hat{\mathbf{X}}^j\}$ and $\{\hat{\mathbf{E}}\}$:
  - if body i is rigid:

$$\boldsymbol{\omega}^j = \boldsymbol{\omega}^i + \boldsymbol{\varphi}^j\,\dot{q}^j \tag{9.25}$$

  end.
  - if body i is flexible:

$$\begin{aligned} \boldsymbol{\omega}^E &= \boldsymbol{\omega}^i - \boldsymbol{\Omega}^{\hat{\mathbf{X}}^i} \\ \boldsymbol{\omega}^{\hat{\mathbf{T}}^j} &= \boldsymbol{\omega}^E + \boldsymbol{\Omega}^{\hat{\mathbf{T}}^j} \\ \boldsymbol{\omega}^j &= \boldsymbol{\omega}^{\hat{\mathbf{T}}^j} + \boldsymbol{\varphi}^j\dot{q}^j \end{aligned} \tag{9.26}$$

  (where $\boldsymbol{\Omega}^{\hat{\mathbf{X}}^i} \triangleq \boldsymbol{\Omega}(s^i)$ and $\boldsymbol{\Omega}^{\hat{\mathbf{T}}^j} \triangleq \boldsymbol{\Omega}(s^j)$ are computed from 9.21)
  end.

- Absolute angular accelerations $\dot{\boldsymbol{\omega}}^j$ and $\dot{\boldsymbol{\omega}}^E$ of frames $\{\hat{\mathbf{X}}^j\}$ and $\{\hat{\mathbf{E}}\}$:
  - if body i is rigid:

$$\dot{\boldsymbol{\omega}}^j = \dot{\boldsymbol{\omega}}^i + \tilde{\boldsymbol{\omega}}^j\,.\,\boldsymbol{\varphi}^j\dot{q}^j + \boldsymbol{\varphi}^j\ddot{q}^j \tag{9.27}$$

  end.
  - if body i is flexible:

$$\dot{\boldsymbol{\omega}}^E = \dot{\boldsymbol{\omega}}^i - \overset{\circ}{\boldsymbol{\Omega}}{}^{\hat{\mathbf{X}}^i} - \tilde{\boldsymbol{\omega}}^i\,.\,\boldsymbol{\Omega}^{\hat{\mathbf{X}}^i} \tag{9.28}$$

$$\boldsymbol{\beta}^E \triangleq \tilde{\dot{\boldsymbol{\omega}}}^E + \tilde{\boldsymbol{\omega}}^E\,.\,\tilde{\boldsymbol{\omega}}^E \tag{9.29}$$

$$\dot{\boldsymbol{\omega}}^{\hat{\mathbf{T}}^j} = \dot{\boldsymbol{\omega}}^E + \overset{\circ}{\boldsymbol{\Omega}}{}^{\hat{\mathbf{T}}^j} + \tilde{\boldsymbol{\omega}}^{\hat{\mathbf{T}}^j}\,.\,\boldsymbol{\Omega}^{\hat{\mathbf{T}}^j} \tag{9.30}$$

$$\dot{\boldsymbol{\omega}}^j = \dot{\boldsymbol{\omega}}^{\hat{\mathbf{T}}^j} + \tilde{\boldsymbol{\omega}}^j\,.\,\boldsymbol{\varphi}^j\dot{q}^j + \boldsymbol{\varphi}^j\ddot{q}^j \tag{9.31}$$

  end.

$$\boldsymbol{\beta}^j \triangleq \tilde{\dot{\boldsymbol{\omega}}}^j + \tilde{\boldsymbol{\omega}}^j\,.\,\tilde{\boldsymbol{\omega}}^j \tag{9.32}$$

- Absolute position vectors (see equations 3.44 and 3.45):

$$\begin{aligned} \mathbf{p}^j &= \mathbf{p}^i + \mathbf{z}^i + \mathbf{d}^{ij}, \text{ for attachment point } O^j \text{ on body } i, \\ \mathbf{x}^i &= \mathbf{p}^i + \mathbf{z}^i + \mathbf{d}^{ii}, \text{ for the center of mass of body } i. \end{aligned} \tag{9.33}$$

- Absolute velocity vectors:

$$\begin{aligned} \dot{\mathbf{p}}^j &= \dot{\mathbf{p}}^i + \dot{\mathbf{z}}^i + \dot{\mathbf{d}}^{ij} \\ &= \dot{\mathbf{p}}^i + \overset{\circ}{\mathbf{z}}{}^i + \tilde{\omega}^i \,.\, \mathbf{z}^i + \dot{\mathbf{d}}^{ij} \end{aligned} \tag{9.34}$$

- Absolute accelerations:

$$\ddot{\mathbf{p}}^j = \ddot{\mathbf{p}}^i + (\overset{\circ\circ}{\mathbf{z}}{}^i + 2\,\tilde{\omega}^i \,.\, \overset{\circ}{\mathbf{z}}{}^i + \tilde{\omega}^i \,.\, \tilde{\omega}^i \,.\, \mathbf{z}^i + \dot{\tilde{\omega}}^i \,.\, \mathbf{z}^i) + \ddot{\mathbf{d}}^{ij}$$

Defining (as in 3.53)

$$\boldsymbol{\alpha}^j \triangleq \ddot{\mathbf{p}}^j + \overset{\circ\circ}{\mathbf{z}}{}^j + 2\,\tilde{\omega}^j \,.\, \overset{\circ}{\mathbf{z}}{}^j - \mathbf{g} \tag{9.35}$$

we find the recursive formula:

$$\boldsymbol{\alpha}^j = \boldsymbol{\alpha}^i + \boldsymbol{\beta}^i \,.\, \mathbf{z}^i + 2\,\tilde{\omega}^j \,.\, \overset{\circ}{\mathbf{z}}{}^j + \overset{\circ\circ}{\mathbf{z}}{}^j + \ddot{\mathbf{d}}^{ij} \tag{9.36}$$

which becomes

  - if body i is rigid:

$$\boldsymbol{\alpha}^j = \boldsymbol{\alpha}^i + \boldsymbol{\beta}^i \,.\, (\mathbf{z}^i + \mathbf{d}^{ij}) + 2\,\tilde{\omega}^j \,.\, \overset{\circ}{\mathbf{z}}{}^j + \overset{\circ\circ}{\mathbf{z}}{}^j \tag{9.37}$$

  end.

  - if body i is flexible:

$$\boldsymbol{\alpha}^{*i} \triangleq \boldsymbol{\alpha}^i + \boldsymbol{\beta}^i \,.\, \mathbf{z}^i \tag{9.38}$$

$$\boldsymbol{\alpha}^j = \boldsymbol{\alpha}^{*i} + 2\,\tilde{\omega}^j \,.\, \overset{\circ}{\mathbf{z}}{}^j + \overset{\circ\circ}{\mathbf{z}}{}^j + \boldsymbol{\beta}^E \,.\, \mathbf{d}^{ij}_{[E]} + 2\,\tilde{\omega}^E \,.\, \overset{\circ}{\mathbf{d}}{}^{ij}_{[E]} + \overset{\circ\circ}{\mathbf{d}}{}^{ij}_{[E]} \tag{9.39}$$

  end.

- end.

Matrix formulation

Written in its final *matrix form*, this recursive algorithm becomes:

- Initialization: $\omega^0 = \dot{\omega}^0 = 0$ and $\alpha^0 = -g$

  For $j = 1 : N^{body}$

$$i = \text{index of the parent of body } j \tag{9.40}$$

- Absolute angular velocities of frames $\{\hat{\mathbf{X}}^j\}$ and $\{\hat{\mathbf{E}}\}$:
  - if body i is rigid:

$$\omega^j = R^{j,i}(q^j)\omega^i + \varphi^j\,\dot{q}^j \tag{9.41}$$

end.

  - if body i is flexible:

$$\begin{aligned} \omega^E &= R^T(s^i)\left(\omega^i - \Omega(s^i)\right) \\ \omega^{\hat{\mathbf{T}}^j} &= R(s^j)\omega^E + \Omega(s^j) \\ \omega^j &= R^{j,i}(q^j)\omega^{\hat{\mathbf{T}}^j} + \varphi^j\,\dot{q}^j \end{aligned} \tag{9.42}$$

end.

- Absolute angular accelerations $\dot{\omega}^j$ and $\dot{\omega}^E$ of frames $\{\hat{\mathbf{X}}^j\}$ and $\{\hat{\mathbf{E}}\}$:
  - if body i is rigid:

$$\dot{\omega}^j = R^{j,i}(q^j)\dot{\omega}^i + \tilde{\omega}^j\varphi^j\dot{q}^j + \varphi^j\,\ddot{q}^j$$

end.

  - if body i is flexible:

$$\dot{\omega}^E = R^T(s^i)\left(\dot{\omega}^i - \dot{\Omega}(s^i) - \tilde{\omega}^i\,\Omega(s^i)\right) \tag{9.43}$$

$$\beta^E = \tilde{\dot{\omega}}^E + \tilde{\omega}^E\tilde{\omega}^E \tag{9.44}$$

$$\dot{\omega}^{\hat{\mathbf{T}}^j} = R(s^j)\dot{\omega}^E + \dot{\Omega}(s^j) + \tilde{\omega}^{\hat{\mathbf{T}}^j}\,\Omega(s^j) \tag{9.45}$$

$$\dot{\omega}^j = R^{j,i}(q^j)\dot{\omega}^{\hat{\mathbf{T}}^j} + \tilde{\omega}^j\varphi^j\dot{q}^j + \varphi^j\,\ddot{q}^j \tag{9.46}$$

end.

$$\beta^j = \tilde{\dot{\omega}}^j + \tilde{\omega}^j\tilde{\omega}^j \tag{9.47}$$

- Absolute linear accelerations:
  - if body i is rigid:

$$\alpha^j = R^{j,i}(q^j)\left(\alpha^i + \beta^i(z^i + d^{ij})\right) + 2\,\tilde{\omega}^j\psi^j\dot{q}^j + \psi^j\,\ddot{q}^j \tag{9.48}$$

end.

  - if body i is flexible:

$$\alpha^{*i} = \alpha^i + \beta^i\psi^i q^i \tag{9.49}$$

$$\begin{aligned} \alpha^j = {} & R^{j,i}(q^j)R(s^j)\left(R^T(s^i)\alpha^{*i} + \ddot{d}^{ij}_{[E]} + 2\,\tilde{\omega}^E\dot{d}^{ij}_{[E]} + \beta^E d^{ij}_{[E]}\right) \\ & +2\,\tilde{\omega}^j\psi^j\dot{q}^j + \psi^j\,\ddot{q}^j \end{aligned} \tag{9.50}$$

end.

- end.

9.3.3 Joint equations

Introduction

Particular care is taken to obtain the various equations corresponding to joint displacements and body deformations in a unified manner as in [25] by applying the virtual power principle to the tree-like structure. This principle can be written in the following general form [93] :

$$\sum_{k=1}^{N^{body}} \int_{X \in k} (\ddot{\mathbf{x}}(X) - \mathbf{f}) \, . \, \Delta\dot{\mathbf{x}}(X) \, dm = 0 \tag{9.51}$$

where:

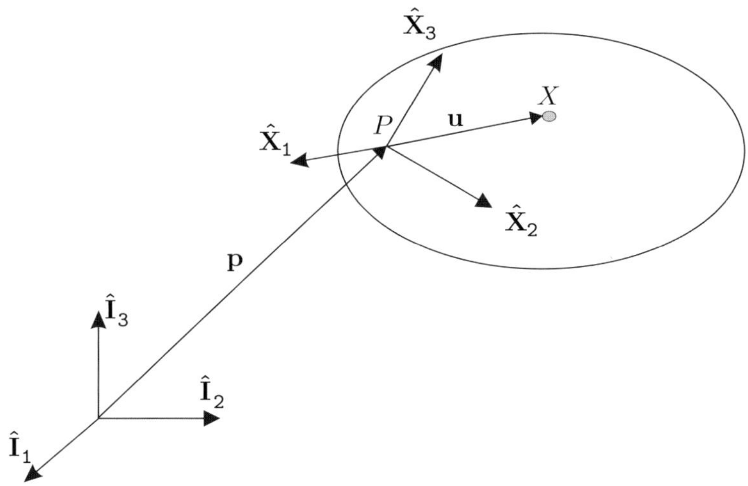

Figure 9.5: Virtual velocity field

. $\ddot{\mathbf{x}}(X)$ denotes the absolute acceleration of a material point X (of body k),

. $\mathbf{f}$ is the local force density.

In order to generate the first set of equations of motion related to the N^{body} joint degrees of freedom, we may choose a field of virtual velocity changes $\Delta\dot{\mathbf{x}}(X)$ such that (see figure 9.5):

. bodies $k \neq i$ are virtually frozen $\Leftrightarrow \Delta\dot{\mathbf{x}}(X) = 0, \ \forall$ body $k \neq i$,

. body i is virtually rigidified $\Leftrightarrow \Delta\dot{\mathbf{x}}(X) = \Delta\dot{\mathbf{p}} + \Delta\boldsymbol{\omega}^i \times \mathbf{u}(X)$, if body $k = i$, (P being an arbitrary reference point of body i, see figure 9.5).

It then proves useful to separate the internal and external interactions by writing: $\mathbf{f} = \mathbf{f}_{int} + \mathbf{f}_{ext}$. Since the chosen virtual velocity field is rigidifying for body i, the virtual power of the internal interactions $\mathbf{f}_{int}$ vanishes and hence, introducing this field into (9.51),

$$\left(\int_i \ddot{\mathbf{x}}(X)dm - \mathbf{F}_{ext}\right) . \Delta\dot{\mathbf{p}} + \left(\int_i \mathbf{u}(X)\, dm \times \ddot{\mathbf{p}} + \int_i \mathbf{u}(X) \times \ddot{\mathbf{u}}(X)\, dm - \mathbf{L}^P_{ext}\right) . \Delta\boldsymbol{\omega}^i = 0$$

where

. $\mathbf{F}_{ext} \triangleq \int_i \mathbf{f}_{ext}\, dm$ is the resultant of the external forces applied to body i,

. $\mathbf{L}^P_{ext} \triangleq \int_i \mathbf{u} \times \mathbf{f}_{ext}\, dm$ is the external torque resultant on body i (sum of the moments of external forces and pure torques) with respect to point P.

Since this equation must be satisfied for any choice of the virtual velocity changes $\Delta\dot{\mathbf{p}}$ and $\Delta\boldsymbol{\omega}^i$, we finally obtain Newton's translational equation and the rotational equation with respect to the arbitrary point P, i.e.:

$$m^i \ddot{\mathbf{x}}^i = \mathbf{F}_{ext} \quad (9.52)$$

$${}^P\dot{\mathbf{H}}^i + m^i \mathbf{u}^i \times \ddot{\mathbf{p}} = \mathbf{L}^P_{ext} \quad (9.53)$$

where:

. $\ddot{\mathbf{x}}^i$ is the absolute acceleration vector of the center of mass G^i,

. $\mathbf{u}^i$ is the relative position vector[6] of the center of mass G^i with respect to point P

$$m^i \mathbf{u}^i \triangleq \int_i \mathbf{u}(X)\, dm \quad (9.54)$$

. ${}^P\dot{\mathbf{H}}^i \triangleq \int_i \mathbf{u}(X) \times \ddot{\mathbf{u}}(X)\, dm$ is the time derivative of the angular momentum of body i with respect to the arbitrary point P.

Backward dynamics

With the forward kinematic computation having been performed, the backward dynamical recursion (from body N^{body} to body 1) can now take place to generate the first set of equations of motion relating to the N^{body} joint degrees of freedom. In figure 9.6, the following vector quantities have been introduced:

. $\mathbf{F}^i = [\hat{\mathbf{X}}^i]^T F^i$ and $\mathbf{L}^i = [\hat{\mathbf{X}}^i]^T L^i$, respectively the force and torque acting on body i through joint i, evaluated at point O^i,

[6] which should not be confused with $\mathbf{u}(s^i)$.

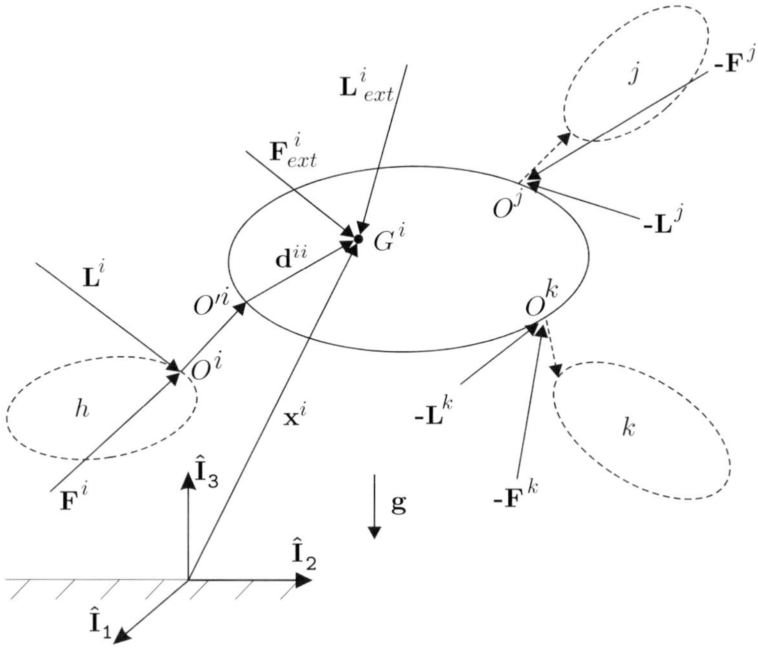

Figure 9.6: Dynamical notations for rigid bodies

. $\mathbf{F}^i_{ext} = [\hat{\mathbf{X}}^i]^T F^i_{ext}$ and $\mathbf{L}^i_{ext} = [\hat{\mathbf{X}}^i]^T L^i_{ext}$, the external interactions on body i (barring gravity) under the form of an equivalent resultant force applied to the center of mass G^i and a pure torque. These external forces and torques will not be considered when body i is a beam since it turns out that their introduction is more user-friendly in ROBOTRAN by adding to the beam a fictitious joint and a massless child-body to which the desired external load is applied.

During the backward recursion, the formulation of the joint equations of motion depends on the type of body encountered: rigid or flexible. To clarify the present section, the two cases are treated separately, while in the computational algorithm, a simple test is of course performed first for each body, allowing a switch to the appropriate equations.

Rigid body On the basis of figure 9.6, we can write the equations of motion 9.52 and 9.53 for body i, the reference point for the rotational equation in this case being the center of mass G^i of the body. Then, using the previous kinematic computations (see section 9.3.2) as well as the definition of the joint rotation matrices, we can project these equations into the body-fixed frame $\{\hat{\mathbf{X}}^i\}$ and obtain the Newton-Euler equations in recursive form:

$$
\begin{aligned}
& W^i \triangleq m^i\,(\alpha^i + \beta^i\, d_z^{ii}) - F_{ext}^i \\
F^i &= \sum_{j \in \bar{i}} R^{i,j}(q^j) F^j + W^i \qquad (9.55)\\
L^i &= \sum_{j \in \bar{i}} \left(R^{i,j}(q^j) L^j + \tilde{d}_z^{ij}\, R^{i,j}(q^j) F^j \right) + \tilde{d}_z^{ii}\, W^i - L_{ext}^i \\
& + I^i\, \dot{\omega}^i + \tilde{\omega}^i\, I^i\, \omega^i \qquad (9.56)
\end{aligned}
$$

Apart from the fact that in the present case we do not recursively extract the mass matrix, these equations are strictly equivalent to those obtained in section 3.3 (equations 3.67 and 3.69).

Flexible beam On the basis of figure 9.7 and now choosing point O^E as the reference point for the equation of rotation, we can project 9.52 onto frame $\{\hat{\mathbf{X}}^i\}$ $(\triangleq \{\hat{\mathbf{S}}(s^i)\}\,)$.

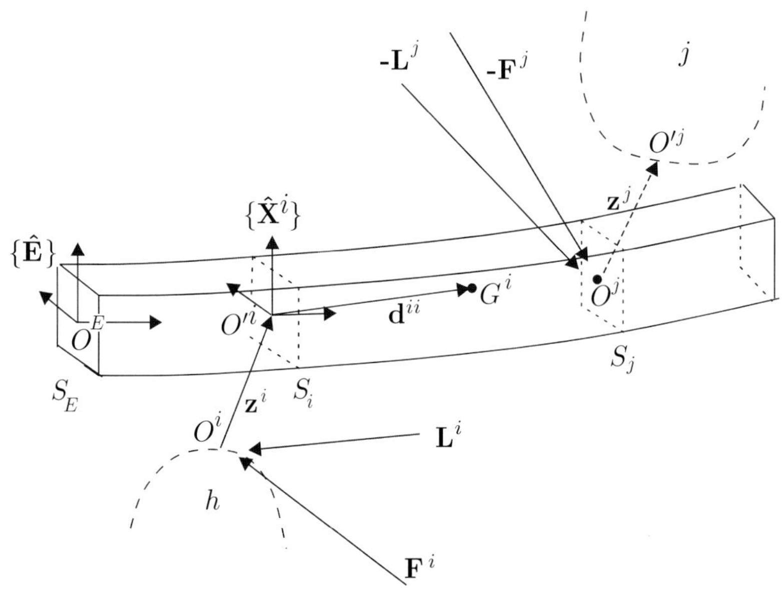

Figure 9.7: Dynamical notations for flexible beams

Let us first describe the position vector (and its "relative" derivatives) of the center of mass of beam i (see figure 9.7):

$$
\mathbf{d}^{ii} = \overrightarrow{O'^i\, G^i} = \mathbf{u}^i - \mathbf{u}(s^i) = [\hat{\mathbf{E}}]^T d_{[E]}^{ii} \qquad (9.57)
$$

$$
\overset{\circ}{\mathbf{d}}{}^{ii} = [\hat{\mathbf{E}}]^T \dot{d}_{[E]}^{ii} \text{ and } \overset{\circ\circ}{\mathbf{d}}{}^{ii} = [\hat{\mathbf{E}}]^T \ddot{d}_{[E]}^{ii} \qquad (9.58)
$$

The vector translational equation can thus be written as[7] :

$$\mathbf{F}^i - \sum_{j \in \bar{\imath}} \mathbf{F}^j + m^i \mathbf{g} = m^i \ddot{\mathbf{x}}^i \tag{9.59}$$

and, thanks to definition 9.35:

$$\boldsymbol{\alpha}^j \triangleq \ddot{\mathbf{p}}^j + \overset{\circ\circ}{\mathbf{z}}^j + 2\,\tilde{\boldsymbol{\omega}}^j \,.\, \overset{\circ}{\mathbf{z}}^j - \mathbf{g}$$

and its consequence (resulting from 9.33):

$$\ddot{\mathbf{x}}^i - \mathbf{g} = \boldsymbol{\alpha}^i + \boldsymbol{\beta}^i \,.\, \mathbf{z}^i + \ddot{\mathbf{d}}^{ii}$$

this vector equation becomes

$$\mathbf{F}^i = \sum_{j \in \bar{\imath}} \mathbf{F}^j + m^i (\boldsymbol{\alpha}^i + \boldsymbol{\beta}^i \,.\, \mathbf{z}^i + \overset{\circ\circ}{\mathbf{d}}^{ii} + 2\,\tilde{\boldsymbol{\omega}}^E \,\overset{\circ}{\mathbf{d}}^{ii} + \boldsymbol{\beta}^E \,.\, \mathbf{d}^{ii})$$

In matrix form, and using 9.49, the joint translational equation can thus be written as

$$F^i = m^i \alpha^{*i} + R(s^i) \left(\sum_{j \in \bar{\imath}} R^T(s^j) R^{i,j}(q^j) F^j + m^i \left(\ddot{d}^{ii}_{[E]} + 2\,\tilde{\omega}^E \dot{d}^{ii}_{[E]} + \beta^E d^{ii}_{[E]} \right) \right) \tag{9.60}$$

The vector rotational equation 9.53 with respect to reference point O^E reads

$$\begin{aligned} {}^E\dot{\mathbf{H}}^i + m^i \mathbf{u}^i \times \ddot{\mathbf{x}}^E &= \mathbf{L}^E_{ext} \\ &= \mathbf{L}^i + \overrightarrow{O^E O^i} \times \mathbf{F}^i + m^i \mathbf{u}^i \times \mathbf{g} - \sum_{j \in \bar{\imath}} \left(\mathbf{L}^j + \overrightarrow{O^E O^j} \times \mathbf{F}^j \right) \end{aligned}$$

where the angular momentum ${}^E\mathbf{H}^i$ is given by

$${}^E\mathbf{H}^i = \int_i \mathbf{u} \times \dot{\mathbf{u}}\, dm = \int_i \tilde{\mathbf{u}} \,.\, \overset{\circ}{\mathbf{u}}\; dm - \int_i \tilde{\mathbf{u}} \,.\, \tilde{\mathbf{u}}\, dm \,.\, \boldsymbol{\omega}^E$$

Denoting by $\mathbf{I}^E \triangleq -\int_i \tilde{\mathbf{u}} \,.\, \tilde{\mathbf{u}}\, dm = -[\hat{\mathbf{E}}]^T \int_i \tilde{u} \,.\, \tilde{u}\, dm [\hat{\mathbf{E}}]$, the inertia tensor of the flexible beam i with respect to O^E, the rotational equation becomes

$$\begin{aligned} \mathbf{L}^i \;=\; & \mathbf{I}^E \,.\, \dot{\boldsymbol{\omega}}^E + \dot{\mathbf{I}}^E \,.\, \boldsymbol{\omega}^E + \tilde{\boldsymbol{\omega}}^E \,.\, \mathbf{I}^E \,.\, \boldsymbol{\omega}^E + \int_i \tilde{\mathbf{u}} \,.\, \overset{\circ\circ}{\mathbf{u}}\; dm + \tilde{\boldsymbol{\omega}}^E \,.\, \int_i \tilde{\mathbf{u}} \,.\, \overset{\circ}{\mathbf{u}}\; dm \\ & + m^i \tilde{\mathbf{u}}^i \,.\, \left(\boldsymbol{\alpha}^i + \boldsymbol{\beta}^i \,.\, \mathbf{z}^i - \overset{\circ\circ}{\mathbf{u}}\,(s^i) - 2\,\tilde{\boldsymbol{\omega}}^E \,.\, \overset{\circ}{\mathbf{u}}\,(s^i) - \boldsymbol{\beta}^E \,.\, \mathbf{u}(s^i) \right) \\ & - (\tilde{\mathbf{u}}(s^i) - \tilde{\mathbf{z}}^i) \,.\, \mathbf{F}^i + \sum_{j \in \bar{\imath}} (\mathbf{L}^j + \tilde{\mathbf{u}}(s^j) \,.\, \mathbf{F}^j) \end{aligned}$$

---

[7] according to our previous assumptions, there are no external forces acting directly on the flexible elements.

The rotational equation with respect to reference point O^E is then given in matrix form by

$$\begin{aligned} L^i \; = \; & R(s^i)\Big[I^E \dot{\omega}^E + \dot{I}^E \omega^E + \tilde{\omega}^E I^E \omega^E + \int_i \tilde{u}\,\ddot{u}\,dm + \tilde{\omega}^E \int_i \tilde{u}\,\dot{u}\,dm \\ & + m^i \tilde{u}^i \left(R^T(s^i)\alpha^{*i} - \ddot{u}(s^i) - 2\,\tilde{\omega}^E\,\dot{u}(s^i) - \beta^E\,u(s^i) \right) \\ & - \tilde{u}(s^i) R^T(s^i)\,F^i \\ & + \sum_{j \in \bar{i}} \left(R^T(s^j) R^{i,j}(q^j) L^j + \tilde{u}(s^j)\,R^T(s^j) R^{i,j}(q^j) F^j \right) \Big] + q^i\,\tilde{\psi}^i\,F^i \end{aligned} \tag{9.61}$$

Quantities relating to the beam mass distribution appearing in 9.60 and 9.61 must be precomputed, at each time step, concurrently with the kinematic computations detailed in section 9.3.2. These quantities are:

1. The mass center relative position vector $\mathbf{u}^i$ defined in 9.54, which can be computed by integrating over the entire beam, using 9.9:

$$\begin{aligned} m^i \mathbf{u}^i \triangleq \int_i \mathbf{u}(X)\,dm \; &= \; [\hat{\mathbf{E}}]^T \int_i u_r(X)\,dm + [\hat{\mathbf{E}}]^T \int_i u_d(X)\,dm \\ &= \; m^i [\hat{\mathbf{E}}]^T u_r^i + [\hat{\mathbf{E}}]^T \int_i u_d(X)\,dm \end{aligned} \tag{9.62}$$

2. The body i-inertial quantities (with respect to reference point O^E):

$$I^E \triangleq - \int_i \tilde{u}(X)\,\tilde{u}(X)\,dm, \text{ its time derivative } \dot{I}^E,$$

$$\text{and the integrals: } \int_i \tilde{u}(X)\,\ddot{u}(X)\,dm \;\text{ and }\; \int_i \tilde{u}(X)\,\dot{u}(X)\,dm \tag{9.63}$$

Quantities 9.62 and 9.63 still need to be expanded using 9.9 and 9.10 with the discretizations 9.11 and 9.14, first to make the dynamical data $\mathbf{u}_r^i$ and $\mathbf{I}_r^i$ appear, respectively the position vector of the center of mass and the inertia tensor with respect to this point of the undeformed beam, and secondly to isolate integrals over (products of) shape functions. Since the latter are constant with respect to time, they can be computed once and for all and considered as parameters (i.e. unequivocal symbols) within the equations of motion. These substitutions and the corresponding developments are lengthy and tedious, and they will not be detailed here: the corresponding integrals over shape functions are given in Appendix B.

Joint equations of motion The joint equations of motion are finally obtained from (9.55 - 9.56) and (9.60 - 9.61) in a straightforward manner, by projection onto the joint axes, using the joint vector definitions 9.17 and 9.18. The final backward recursion can thus be summarized as follows:

For $i = N^{body} : 1$

- if body i is rigid:

- if joint i is prismatic, the equation of motion is the product: $\left(\psi^i\right)^T$ (9.55$_{rhs}$)

- if joint i is revolute, the equation of motion is the product: $\left(\varphi^i\right)^T$ (9.56$_{rhs}$)

- if body i is flexible (beam):

- if joint i is prismatic, the equation of motion is the product: $\left(\psi^i\right)^T$ (9.60$_{rhs}$)

- if joint i is revolute, the equation of motion is the product: $\left(\varphi^i\right)^T$ (9.61$_{rhs}$)

end.

One should note at this stage that these joint equations of motion are fully implicit with respect to the generalized accelerations $\ddot{q}^i, \{\ddot{q}_t^i\}$ and $\{\ddot{q}_r^i\}$. The problem of time integration of flexible multibody system (for simulation purposes) will be addressed in chapter 10.

9.3.4 Deformation equations

In the multibody system being investigated, let us assume that $N^{beam}(\leq N^{body})$ bodies represent flexible beams whose main modeling hypotheses were set out in section 9.3.1. The present section will now focus on the corresponding deformation equations of motion. The following developments are largely inspired by [86], [29] and applied to each beam i of the tree-structure which undergoes arbitrary motions ($\ddot{\mathbf{p}}^i, \boldsymbol{\omega}^i, \dot{\boldsymbol{\omega}}^i$ — see section 9.3.1) and is submitted to external forces and torques ($\mathbf{g}, \mathbf{F}^i, \mathbf{L}^i, \mathbf{F}^{j\in\bar{\imath}}, \mathbf{L}^{j\in\bar{\imath}}$).

Beam deformation

Let us focus (figure 9.8) on an arbitrary section S (frame $[\hat{\mathbf{S}}(s)] = R(s)[\hat{\mathbf{E}}]$) and more precisely on the material just beyond this section. In particular a section $S^\star$ located at $s^\star = s + ds$ is considered:

$$[\hat{\mathbf{S}}(s^*)] = (E - \tilde{\vartheta}^\star)[\hat{\mathbf{S}}(s)] \tag{9.64}$$

In order to express the local deformation in the material, we first express the position vector $d\mathbf{u}$ of an arbitrary point $X^\star$ of section $S^\star$ with respect to the position $\mathbf{u}(X_r^\star)$ it would have *if the material between sections S and $S^\star$ were undeformed.* In that case, $S^\star = S_r^\star$, $X^\star = X_r^\star$ and $[\hat{\mathbf{S}}(s^*)] = [\hat{\mathbf{S}}(s)]$ in figure 9.8, and according to equation 9.8, we can write

$$\mathbf{u}(X_r^\star) \triangleq \overrightarrow{O^E\, X_r^\star} = \mathbf{u}(s) + [\hat{\mathbf{S}}(s)]^T \begin{pmatrix} 1 \\ 0 \\ 0 \end{pmatrix} ds + [\hat{\mathbf{S}}(s)]^T r(X^\star) \tag{9.65}$$

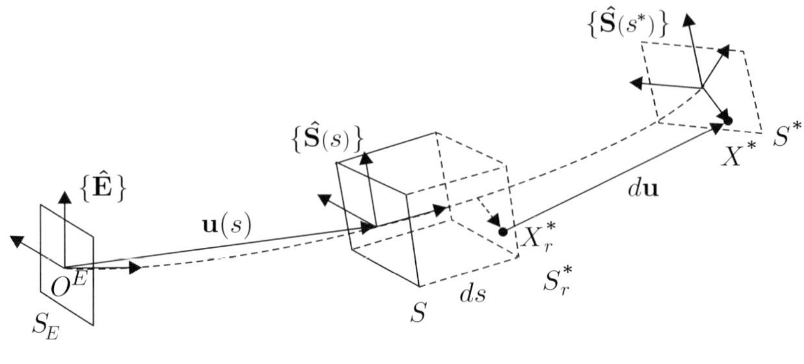

Figure 9.8: Description of the beam deformation

In a limiting process, the *actual position* of $X^\star$ when the material between S and $S^\star$ is deformed is given by

$$\begin{aligned}\mathbf{u}(X^\star) \triangleq \overrightarrow{O^E\, X^\star} &= \mathbf{u}(s) + \frac{\partial \mathbf{u}}{\partial s}\, ds + [\hat{\mathbf{S}}(s^*)]^T r(X^\star) \\ &= \mathbf{u}(s) + \frac{\partial \mathbf{u}}{\partial s}\, ds + [\hat{\mathbf{S}}(s)]^T (E + \tilde{\vartheta}^\star) r(X^\star)\end{aligned}$$

where

$$\frac{\partial \mathbf{u}}{\partial s} \triangleq [\hat{\mathbf{E}}]^T \frac{\partial u}{\partial s}$$

Using the definitions 9.5 of $u(s)$ and of the transformation matrices, we obtain

$$d\mathbf{u} \triangleq \mathbf{u}(X^\star) - \mathbf{u}(X_r^\star) = [\hat{\mathbf{S}}(s)]^T \left[R(s)\frac{\partial u}{\partial s} - \begin{pmatrix} 1 \\ 0 \\ 0 \end{pmatrix} \right] ds + [\hat{\mathbf{S}}(s)]^T \left[\tilde{\vartheta}^\star\, r(X^\star) \right]$$

In order to obtain the displacement gradient $\frac{\partial \mathbf{u}}{\partial s}$ expressed in the local frame $\{\hat{\mathbf{S}}(s)\}$, we note that

$$[\hat{\mathbf{S}}(s^*)] = (R(s) + dR)\, [\hat{\mathbf{E}}] = (E - \tilde{\vartheta}^\star) R(s) [\hat{\mathbf{E}}]$$

whence

$$\tilde{\vartheta}^\star = \tilde{K}(s)\, ds \quad \text{with} \quad \tilde{K}(s) \triangleq -\frac{\partial R(s)}{\partial s} R^T(s) \tag{9.66}$$

and finally

$$\frac{\partial \mathbf{u}}{\partial s} = [\hat{\mathbf{S}}(s)]^T \left[\Gamma(s) + \tilde{K}(s)\, r(X^\star) \right] \tag{9.67}$$

where:

. $\mathbf{\Gamma}(s) = [\hat{\mathbf{S}}(s)]^T \Gamma(s)$, with $\Gamma(s) \triangleq R(s)\frac{\partial u}{\partial s} - \begin{pmatrix} 1 \\ 0 \\ 0 \end{pmatrix}$, is the strain vector of the centroidal axis (Γ_1: extension; Γ_2 and Γ_3 : shear),

. $\mathbf{K}(s) = [\hat{\mathbf{S}}(s)]^T K(s)$, is the beam curvature vector (K_1: twist ; K_2 and K_3 : bending).

Deformation velocities

In the context of a virtual principle, the virtual power of the internal elastic forces involves the expression of "local" velocities when formulating a virtual velocity field compatible with the beam deformation. For that reason, it is useful to establish the following time derivatives:

1. The relative derivative of the strain vector $\mathbf{\Gamma}$:

$$\overset{\circ}{\mathbf{\Gamma}}(s) \triangleq [\hat{\mathbf{S}}(s)]^T \dot{\Gamma}(s) = \frac{\partial \overset{\circ}{\mathbf{u}}}{\partial s} - \tilde{\mathbf{\Omega}}(s) \, . \, \frac{\partial \mathbf{u}}{\partial s} \tag{9.68}$$

where:

$$\frac{\partial \overset{\circ}{\mathbf{u}}}{\partial s} \triangleq [\hat{\mathbf{E}}]^T \frac{\partial \dot{u}}{\partial s} \quad \text{and} \quad \mathbf{\Omega}(s) = [\hat{\mathbf{S}}]^T B(s) \rho(s) \dot{q}_r$$

This result is straightforward from the previous definition of $\mathbf{\Gamma}$ and the relation $\tilde{\Omega} = R(s)\dot{R}^T(s)$.

2. The relative derivative of the curvature vector $\mathbf{K}(s)$:

$$\overset{\circ}{\mathbf{K}}(s) \triangleq [\hat{\mathbf{S}}(s)]^T \dot{K}(s) = \frac{\partial \mathbf{\Omega}}{\partial s} \tag{9.69}$$

This results from

$$\tilde{\mathbf{\Omega}}(s^*) = \tilde{\mathbf{\Omega}} + \frac{\partial \tilde{\mathbf{\Omega}}}{\partial s} ds = [\hat{\mathbf{S}}(s^*)]^T \, R(s^*)\dot{R}^T(s^*) \, [\hat{\mathbf{S}}(s^*)] \tag{9.70}$$

which, with $R(s^*) = (E - \tilde{\vartheta}^\star)R(s)$, leads to

$$\begin{aligned}
\tilde{\mathbf{\Omega}}(s^*) &= \tilde{\mathbf{\Omega}} + \frac{\partial \tilde{\mathbf{\Omega}}}{\partial s} ds \\
&= [\hat{\mathbf{S}}(s^*)]^T \, (E - \tilde{\vartheta}^\star) R(s) \dot{R}^T(s) (E + \tilde{\vartheta}^\star) \, [\hat{\mathbf{S}}(s^*)] \\
&\quad + [\hat{\mathbf{S}}(s^*)]^T \, (E - \tilde{\vartheta}^\star) R(s) R^T(s) \, \tilde{\dot{\vartheta}}^\star \, [\hat{\mathbf{S}}(s^*)] \\
&= [\hat{\mathbf{S}}(s)]^T \, R(s)\dot{R}^T(s) \, [\hat{\mathbf{S}}(s)] \\
&\quad + [\hat{\mathbf{S}}(s)]^T \, \tilde{\dot{\vartheta}}^\star \, [\hat{\mathbf{S}}(s^*)] \\
&= \tilde{\mathbf{\Omega}}(s) + [\hat{\mathbf{S}}(s)]^T \, \tilde{\dot{\vartheta}}^\star \, (E - \tilde{\vartheta}^\star) \, [\hat{\mathbf{S}}(s)]
\end{aligned}$$

and, in view of the differential nature of $\vartheta^\star$, implies

$$\frac{\partial \tilde{\mathbf{\Omega}}}{\partial s} ds = [\hat{\mathbf{S}}(s)]^T \, \tilde{\dot{\vartheta}}^\star \, [\hat{\mathbf{S}}(s)] = [\hat{\mathbf{S}}(s)]^T \, \tilde{\dot{K}}(s) ds \, [\hat{\mathbf{S}}(s)] \tag{9.71}$$

Internal stress resultant

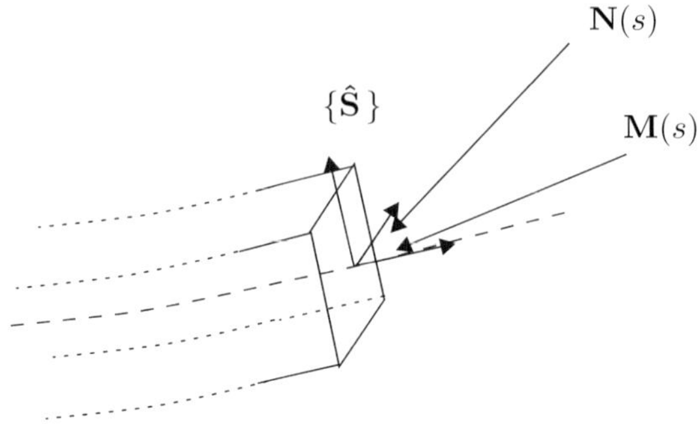

Figure 9.9: Internal stress resultant

In a context of linear elasticity, the internal stress resultants[8] (viz. a force **N** and a pure torque **M**) are related linearly to the beam strain and curvatures:

$$\mathbf{N} \triangleq [\hat{\mathbf{S}}(s)]^T N \tag{9.72}$$

$$\mathbf{M} \triangleq [\hat{\mathbf{S}}(s)]^T M \tag{9.73}$$

with:

$$\begin{array}{rcl} N & = & C^N \Gamma \quad \text{and} \quad C^N = diag(\ E^Y A\,, \quad k_2\, G^{sh} A\,, \quad k_3\, G^{sh} A\ ) \\ M & = & C^M K \quad \text{and} \quad C^M = diag(\ G^{sh} J_p^S\,, \quad E^Y I_2^S\,, \quad E^Y I_3^S\ ) \end{array}$$

where k_2, k_3 are the shear correction factors in the $\hat{\mathbf{S}}_2$ and $\hat{\mathbf{S}}_3$ directions.

Local equations of motion

Let us consider (figure 9.10) a portion of the beam on which the internal stress resultants and gravity are acting. In a limiting process, we may write the vector equations of motion of this element as

$$d\mathbf{N} + \mu\, \mathbf{g}\, ds = \mu\, \ddot{\mathbf{x}}(s)\, ds \tag{9.74}$$

$$d\mathbf{u} \times \mathbf{N} + d\mathbf{M} = \mathbf{I}\, ds\, \dot{\boldsymbol{\omega}} + \tilde{\boldsymbol{\omega}}\, .\, \mathbf{I}\, ds\, .\, \boldsymbol{\omega} \tag{9.75}$$

where:

- $\ddot{\mathbf{x}}(s) = \ddot{\mathbf{p}}^i + \ddot{\mathbf{z}}^i - \ddot{\mathbf{u}}(s^i) + \ddot{\mathbf{u}}(s)$ is the absolute acceleration of the origin of $\{\hat{\mathbf{S}}(s)\}$,
- $\mathbf{I} = [\hat{\mathbf{S}}(s)]^T I\, [\hat{\mathbf{S}}(s)]$ with $I = (\mu/A)\ \text{diag}\,(I_2^S + I_3^S, I_2^S, I_3^S)$,
- $\boldsymbol{\omega} = \boldsymbol{\omega}^E + \boldsymbol{\Omega}(s)$ is the absolute angular velocity of frame $\{\hat{\mathbf{S}}(s)\}$.

[8] at the section gravity center.

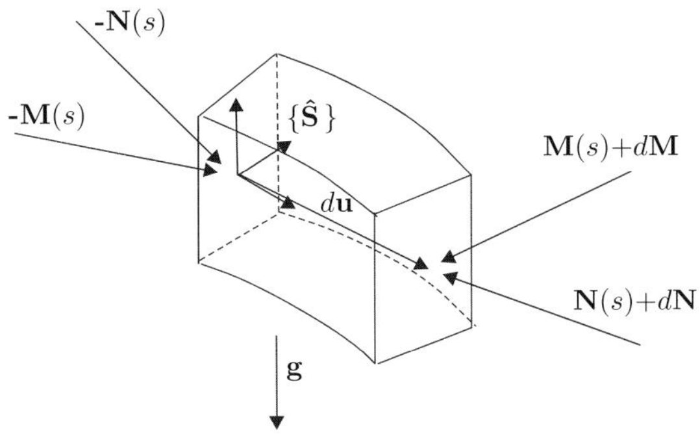

Figure 9.10: Local equations of motion

Virtual power principle

In section 9.3.3, the chosen virtual velocity field was such that each body $k \neq i$ was frozen and body i was rigidified: the resulting equation of motion reflected the overall motion of body i along the degree of freedom of the preceding joint. In the present case, the goal being to set up the deformation equations of the beam i, a virtual velocity field is chosen in such a way that:

- bodies $k \neq i$ are virtually frozen $\Leftrightarrow \Delta\dot{\mathbf{x}}(X) = 0, \ \forall$ body $k \neq i$,
- body (i)-frame $\{\hat{\mathbf{E}}\}$ is virtually frozen while virtual deformation velocities are considered $\Leftrightarrow \Delta\dot{\mathbf{x}}(X) = \Delta \overset{\circ}{\mathbf{u}}(s) + \Delta\mathbf{\Omega}(s) \times \mathbf{r}(X)$, if body $k = i$.

Introducing this velocity field into the virtual power principle 9.51 and integrating over the entire beam on the basis of equations of motion 9.74 and 9.75, we find (since external loads arise solely from the joints i and $j \in \bar{\imath}$ and are applied at O^i and O^j respectively)

$$
\begin{aligned}
&\int_0^L \left(\mu\,(\ddot{\mathbf{x}}(s) - \mathbf{g}) - \frac{\partial \mathbf{N}}{\partial s}\right) . \Delta \overset{\circ}{\mathbf{u}}(s)ds \\
&+ \int_0^L \left(\mathbf{I} . \dot{\boldsymbol{\omega}} + \tilde{\boldsymbol{\omega}} . \mathbf{I} . \boldsymbol{\omega} - \frac{\partial \mathbf{M}}{\partial s} - \frac{\partial \mathbf{u}}{\partial s} \times \mathbf{N}\right) . \Delta\mathbf{\Omega}(s)ds \\
&- \mathbf{F}^i . \Delta \overset{\circ}{\mathbf{u}}(O^i) - \mathbf{L}^i . \Delta\mathbf{\Omega}(s^i) + \sum_{j \in \bar{\imath}} \left(\mathbf{F}^j . \Delta \overset{\circ}{\mathbf{u}}(s^j) + \mathbf{L}^j . \Delta\mathbf{\Omega}(s^j)\right) = 0
\end{aligned}
\tag{9.76}
$$

Since, on the basis of 9.68 and 9.69,

$$
\Delta \overset{\circ}{\mathbf{\Gamma}}(s) = \frac{\partial \Delta \overset{\circ}{\mathbf{u}}}{\partial s} - \Delta\tilde{\mathbf{\Omega}}(s) . \frac{\partial \mathbf{u}}{\partial s} \quad \text{and} \quad \Delta \overset{\circ}{\mathbf{K}}(s) = \frac{\partial \Delta\mathbf{\Omega}}{\partial s} \tag{9.77}
$$

by integrating by parts the terms

$$\frac{\partial \mathbf{N}}{\partial s} \, . \, \Delta \overset{\circ}{\mathbf{u}}(s) \quad \text{and} \quad \frac{\partial \mathbf{M}}{\partial s} \, . \, \Delta \mathbf{\Omega}(s)$$

we obtain after rearranging terms

$$\begin{aligned} & \int_0^L \mu \, (\ddot{\mathbf{x}}(s) - \mathbf{g}) \, . \, \Delta \overset{\circ}{\mathbf{u}}(s) ds + \int_0^L \left(\mathbf{I} \, . \, \dot{\omega} + \tilde{\omega} \, . \, \mathbf{I} \, . \, \omega\right) \, . \, \Delta \mathbf{\Omega}(s) ds \\ = \; & \mathbf{F}^i \, . \, \Delta \overset{\circ}{\mathbf{u}}(O^i) + \mathbf{L}^i \, . \, \Delta \mathbf{\Omega}(s^i) - \sum_{j \in \bar{\imath}} \left(\mathbf{F}^j \, . \, \Delta \overset{\circ}{\mathbf{u}}(s^j) + \mathbf{L}^j \, . \, \Delta \mathbf{\Omega}(s^j)\right) \\ & - \int_0^L \mathbf{N} \, . \, \Delta \overset{\circ}{\mathbf{\Gamma}}(s) ds - \int_0^L \mathbf{M} \, . \, \Delta \overset{\circ}{\mathbf{K}}(s) ds \end{aligned} \tag{9.78}$$

Finally, the virtual power principle is expressed in terms of virtual velocity changes in the deformation coordinates $\Delta \dot{q}_t$ and $\Delta \dot{q}_r$ by substituting into 9.78 (and 9.77) the expressions:

- $\Delta \overset{\circ}{\mathbf{u}}(s) = [\hat{\mathbf{E}}]^T \tau(s) \Delta \dot{q}_t$ (from 9.22_b)
- $\Delta \mathbf{\Omega}(s) = [\hat{\mathbf{S}}(s)]^T B(s) \rho(s) \Delta \dot{q}_r = [\hat{\mathbf{E}}]^T R^T(s) B(s) \rho(s) \Delta \dot{q}_r$ (from 9.21 and 9.14)

The latter relations justify retaining terms in $\vartheta \dot{\vartheta}$ in the expression 9.21 of $\mathbf{\Omega}(s)$, since it appears clearly that they lead to first order terms (in ϑ) in the deformation equations: their presence is therefore indispensable as regards the hypotheses underlying the model.

Deformation equations

The final scalar form of the deformation equations of motion can be obtained by:

- expressing the kinematic quantities relating to beam i $(u, \dot{u}, \ddot{u}, R, \Omega, \dot{\Omega}, ...)$ as functions of the deformation coordinates q_t and q_r (see sections 9.3.1 and 9.3.2),
- performing the various vector operations in the appropriate frames,
- isolating in 9.78 the coefficients of each virtual generalized velocity $\Delta \dot{q}_t^k$, $\Delta \dot{q}_r^k$, the latter being assumed independent.

Further to this, the various terms appearing in 9.78 are expanded in such a way that the integrals to be evaluated involve only combinations of the shape functions $\tau_{1[2,3]}^{\alpha}, \rho_{1[2,3]}^{\beta}$ (or their spatial derivatives). These integrals are gathered in Appendix B.

In the present text, the developments[9] of 9.78 will be curtailed, since they are of little interest.

[9] as required by a multibody program

- *Term 1*: $\int_0^L \Delta \overset{\circ}{\mathbf{u}}(s) \,.\, (\ddot{\mathbf{x}}(s) - \mathbf{g})\, \mu ds$

First of all, we can write on the basis of the definition of vectors $\boldsymbol{\alpha}^i$ (see 9.35) and $\boldsymbol{\alpha}^{*i}$ (see 9.38)

$$\begin{aligned} \ddot{\mathbf{x}}(s) - \mathbf{g} &= \ddot{\mathbf{p}}^i + \ddot{\mathbf{z}}^i - \ddot{\mathbf{u}}(s^i) + \ddot{\mathbf{u}}(s) - \mathbf{g} \\ &= \boldsymbol{\alpha}^i + \boldsymbol{\beta}^i \,.\, \mathbf{z}^i - \ddot{\mathbf{u}}(s^i) + \ddot{\mathbf{u}}(s) \\ &= \boldsymbol{\alpha}^{*i} - \ddot{\mathbf{u}}(s^i) + \ddot{\mathbf{u}}(s) \end{aligned} \tag{9.79}$$

and compute the second time derivative $\ddot{\mathbf{u}}(s)$ on the basis of 9.22 (using 9.44)

$$\ddot{\mathbf{u}}(s) = \boldsymbol{\beta}^E \,.\, \mathbf{u}(s) + 2\,\tilde{\boldsymbol{\omega}}^E \,.\, \overset{\circ}{\mathbf{u}}(s) + \overset{\circ\circ}{\mathbf{u}}(s) \tag{9.80}$$

Term 1 can thus be written as follows:

$$\begin{aligned} &(\Delta\dot{q}_t)^T \int_0^L \tau^T(s)\mu ds \left(R^T(s^i)\alpha^{*i} - \left(\ddot{u}(s^i) + 2\,\tilde{\omega}^E\,\dot{u}(s^i) + \beta^E\, u(s^i) \right) \right) \\ &+ (\Delta\dot{q}_t)^T \int_0^L \tau^T(s) \left(\ddot{u}(s) + 2\,\tilde{\omega}^E\,\dot{u}(s) + \beta^E\, u(s) \right) \mu ds \end{aligned} \tag{9.81}$$

where:

. α^{*i} was computed by 9.49,

. ω^E and β^E were computed by 9.42 and 9.44 respectively.

On the basis of 9.11 and 9.14, $u(s), \dot{u}(s), \ddot{u}(s), R(s), \Omega(s)$ and $\dot{\Omega}(s)$ must now be expressed in terms of the shape functions in such a way that the latter contribute to the final form through constant integrals (see Appendix B). Expression 9.81 is formally of the form

$$(\Delta\dot{q}_t)^T \ (\ ... \ )$$

in which each element inside the parentheses contributes to the corresponding translational deformation equation.

- *Term 2*: $\int_0^L \Delta\boldsymbol{\Omega}(s) \,.\, \left(\mathbf{I} \,.\, \dot{\boldsymbol{\omega}} + \tilde{\boldsymbol{\omega}} \,.\, \mathbf{I} \,.\, \boldsymbol{\omega}\right) ds$

Expressing the absolute angular velocity of the frame $\{\hat{\mathbf{S}}(s)\}$ as

$$\boldsymbol{\omega}(s) = \boldsymbol{\omega}^E + \boldsymbol{\Omega}(s) = [\hat{\mathbf{S}}(s)]^T \left(R(s)\omega^E + B(s)\dot{\vartheta}(s) \right)$$

and the absolute angular acceleration successively as

$$\begin{aligned} \dot{\boldsymbol{\omega}}(s) &= \dot{\boldsymbol{\omega}}^E + \overset{\circ}{\boldsymbol{\Omega}}(s) + \tilde{\boldsymbol{\omega}}(s)\boldsymbol{\Omega}(s) \\ &= \dot{\boldsymbol{\omega}}^E + \overset{\circ}{\boldsymbol{\Omega}}(s) + \tilde{\boldsymbol{\omega}}^E(s)\boldsymbol{\Omega}(s) \\ &= [\hat{\mathbf{S}}(s)]^T \left(R(s)\dot{\omega}^E + B(s)\ddot{\vartheta}(s) + \dot{B}(s)\dot{\vartheta}(s) + \left(R(s)\omega^E\right)^{\sim} B(s)\dot{\vartheta}(s) \right) \end{aligned}$$

we can expand this second term as follows:

$$
\begin{aligned}
&(\Delta \dot{q}_r)^T \Big[\int_0^L \rho^T(s) B^T(s)\, I\, R(s)\, ds\, \dot{\omega}^E \\
&+ \int_0^L \rho^T(s) B^T(s)\, I \left(B(s)\ddot{\vartheta}(s) + \dot{B}(s)\dot{\vartheta}(s) + \left(R(s)\omega^E\right)^{\sim} B(s)\dot{\vartheta}(s) \right) ds \\
&+ \int_0^L \rho^T(s) B^T(s) \left(R(s)\omega^E + B(s)\dot{\vartheta}(s) \right)^{\sim} I \left(R(s)\omega^E + B(s)\dot{\vartheta}(s) \right) ds \Big]
\end{aligned}
\tag{9.82}
$$

where $\dot{\omega}^E$ and ω^E were computed by 9.43 and 9.42.
Each kinematic quantity $(\vartheta(s), \dot{\vartheta}(s), \ldots)$ must then be expressed as a function of the shape functions $\rho(s)$ and the generalized coordinates q_r on the basis of 9.14. The final resulting expression can be written in the following form:

$$
(\Delta \dot{q}_r)^T \ (\ \ldots\ ) \tag{9.83}
$$

in which each element inside the parenthesis contributes to the corresponding rotational deformation equation.

- *Terms 3 and 4*: $\mathbf{F}^i \,.\, \Delta\, \overset{\circ}{\mathbf{u}}\, (O^i) + \mathbf{L}^i \,.\, \Delta\boldsymbol{\Omega}(s^i)$

The joint force and torque $\mathbf{F}^i$ and $\mathbf{L}^i$ were computed during the backward dynamical recursion via equations 9.55, 9.60 and 9.56, 9.61 respectively, and expressed in the body-frame $\{\hat{\mathbf{X}}^i\}$. Their contribution to the virtual power principle 9.78 results from the virtual velocity changes of point O^i and frame $\{\hat{\mathbf{X}}^i\}$ induced by the virtual deformation velocities. The relative position of O^i with respect to O^E is given by

$$
\mathbf{u}(O^i) = \overrightarrow{O^E\, O^i} = [\hat{\mathbf{E}}]^T \left(u(s^i) - R^T(s^i) z^i \right)
$$

whence virtual velocity change reads

$$
\begin{aligned}
\Delta\, \overset{\circ}{\mathbf{u}}\, (O^i) &= [\hat{\mathbf{E}}]^T \left(\tau(s^i)\, \Delta\dot{q}_t - R^T(s^i) \left(B(s^i)\rho(s^i)\Delta\dot{q}_r \right)^{\sim} z^i \right) \\
&= [\hat{\mathbf{E}}]^T \left(\tau(s^i)\, \Delta\dot{q}_t + R^T(s^i)\, \tilde{z}^i B(s^i)\rho(s^i)\Delta\dot{q}_r \right)
\end{aligned}
$$

The computation of the virtual power is thus straightforward and can be written as

$$
\left[\left(F^i\right)^T R(s^i)\tau(s^i) \right] \Delta\dot{q}_t + \left[\left(L^i\right)^T B(s^i)\rho(s^i) + \left(F^i\right)^T \tilde{z}^i\, B(s^i)\rho(s^i) \right] \Delta\dot{q}_r
$$

or equivalently:

$$
(\Delta\dot{q}_t)^T \left[\tau^T(s^i) R^T(s^i) F^i \right] + (\Delta\dot{q}_r)^T \left[\rho^T(s^i) B^T(s^i) \left(L^i - q^i\, \tilde{\psi}^i F^i \right) \right] \tag{9.84}
$$

- *Term 5:* $-\sum_{j\in\bar{\imath}}\left(\mathbf{F}^j\,.\,\Delta\,\overset{\circ}{\mathbf{u}}\,(s^j)+\mathbf{L}^j\,.\,\Delta\mathbf{\Omega}(s^j)\right)$

The reasoning is similar for joints j such that $j\in\bar{\imath}$. The corresponding virtual power is given by

$$-\sum_{j\in\bar{\imath}}\left[\left(F^j\right)^T R^{j,i}(q^j)R(s^j)\tau(s^j)\right]\Delta\dot{q}_t-\sum_{j\in\bar{\imath}}\left[\left(L^j\right)^T R^{j,i}(q^j)B(s^j)\rho(s^j)\right]\Delta\dot{q}_r$$

or

$$-\sum_{j\in\bar{\imath}}\left((\Delta\dot{q}_t)^T\left[\tau^T(s^j)R^T(s^j)R^{i,j}(q^j)F^j\right]+(\Delta\dot{q}_r)^T\left[\rho^T(s^j)B^T(s^j)R^{i,j}(q^j)L^j\right]\right) \tag{9.85}$$

- *Term 6:* $-\int_0^L\mathbf{N}\,.\,\Delta\,\overset{\circ}{\mathbf{\Gamma}}\,(s)ds$

This term represents the virtual power of the internal elastic forces when the beam is subjected to axial or shear deformation.
In order to compute this term properly, it is convenient to momentarily disregard how the constitutive equations of $\mathbf{N}=[\hat{\mathbf{S}}]^TN$ are formulated (see equation 9.72), and consider the latter as a generic (internal) force which contributes to the virtual power through a projection along the virtual velocity field $\Delta\,\overset{\circ}{\mathbf{\Gamma}}$.
Starting from 9.77_a :

$$\Delta\,\overset{\circ}{\mathbf{\Gamma}}\,(s)=\frac{\partial\Delta\,\overset{\circ}{\mathbf{u}}}{\partial s}-\Delta\tilde{\mathbf{\Omega}}(s)\,.\,\frac{\partial\mathbf{u}}{\partial s}$$

$\Delta\,\overset{\circ}{\mathbf{\Gamma}}$ can be computed considering that, from 9.22,

$$\frac{\partial\Delta\,\overset{\circ}{\mathbf{u}}\,(s)}{\partial s}=[\hat{\mathbf{E}}]^T\tau'(s)\Delta\dot{q}_t\quad\text{with}\quad\tau'(s)\triangleq\frac{d\tau(s)}{ds}$$

and

$$\frac{\partial\mathbf{u}}{\partial s}=[\hat{\mathbf{E}}]^T\frac{\partial u(s)}{\partial s}=[\hat{\mathbf{E}}]^T\begin{pmatrix}1+\tau_1'(s)q_t\\ \tau_2'(s)q_t\\ \tau_3'(s)q_t\end{pmatrix}$$

resulting in

$$\begin{aligned}\Delta\,\overset{\circ}{\mathbf{\Gamma}}\;=\;&[\hat{\mathbf{S}}(s)]^T\left(R(s)\tau'(s)\right)\Delta\dot{q}_t\\ &+[\hat{\mathbf{S}}(s)]^T\left[R(s)\begin{pmatrix}1+\tau_1'(s)q_t\\ \tau_2'(s)q_t\\ \tau_3'(s)q_t\end{pmatrix}\right]^{\sim}B(s)\,\rho(s)\,\Delta\dot{q}_r\end{aligned}$$

Under the linear kinematic assumption of section 9.3.1, all terms having the form $\vartheta_\alpha \tau'_\beta(s) q_t$ are neglected and we can write *term 6* as:

$$\int_0^L \begin{pmatrix} N_1 & N_2 & N_3 \end{pmatrix} \begin{pmatrix} (\tau'_1 + \tau'_2\,\vartheta_3 - \tau'_3\,\vartheta_2) \\ (\tau'_2 - \tau'_1\,\vartheta_3 + \tau'_3\,\vartheta_1) \\ (\tau'_3 + \tau'_1\,\vartheta_2 - \tau'_2\,\vartheta_1) \end{pmatrix} ds\,\Delta\dot{q}_t$$
$$+ \int_0^L \begin{pmatrix} N_1 & N_2 & N_3 \end{pmatrix} \begin{pmatrix} -\rho_2\,(\tau'_3 q_t + \vartheta_2) + \rho_3\,(\tau'_2 q_t - \vartheta_3) \\ -\rho_3 + (\rho_1\,\tau'_3 q_t - \rho_3\,\tau'_1 q_t) \\ +\rho_2 + (\rho_2\,\tau'_1 q_t - \rho_1\,\tau'_2 q_t) \end{pmatrix} ds\,\Delta\dot{q}_r \tag{9.86}$$

This expression clearly reveals the coupling between the translational and rotational deformation equations. Indeed, with the help of figure 9.11, let us focus on the axial strain force component N_1 in a $2\,D$-context ($1-2$ plane). This figure highlights the presence of the term $N_1\,\vartheta_3$ in the lateral deformation equation $(...\tau'_2\Delta\dot{q}_t)$ and of the contribution $N_1\,(\tau'_2 q_t - \vartheta_3)$ to the rotational equation (in the third direction: $...\,\rho_3\Delta\dot{q}_r$).

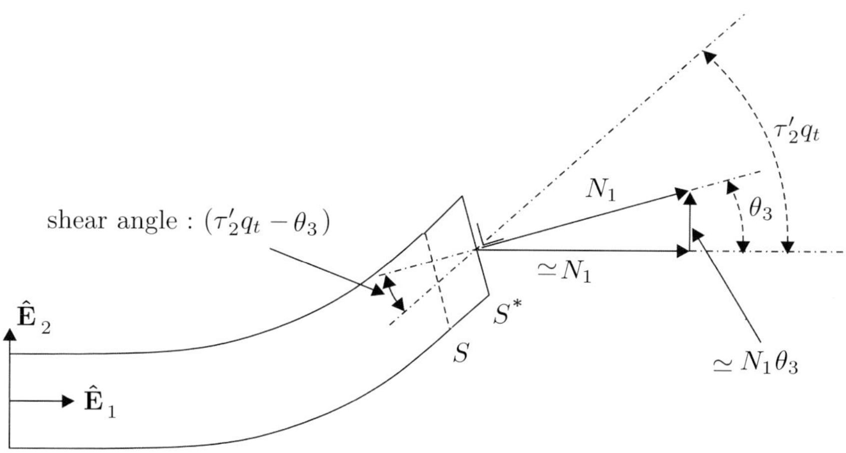

Figure 9.11: Coupling of the deformation equations

Now focus on the computation of the internal elastic force N in *term 6*, whose constitutive equation is given by 9.72:

$$\mathbf{N} \triangleq [\hat{\mathbf{S}}(s)]^T C^N \Gamma$$

where Γ denotes the components of the local strain vector in frame $\{\hat{\mathbf{S}}(s)\}$. Based on its definition (see 9.67), Γ can be expressed using 9.6 as:

$$\Gamma(s) = (R(s) - E)\begin{pmatrix} 1 \\ 0 \\ 0 \end{pmatrix} + R(s)\begin{pmatrix} \tau'_1(s) \\ \tau'_2(s) \\ \tau'_3(s) \end{pmatrix} q_t \tag{9.87}$$

- Γ_1 : the axial strain

When computing Γ_1, and considering only the first order terms while neglecting terms of the form $\vartheta_\alpha \tau'_\beta(s) q_t$, we would obtain the following result:

$$\Gamma_1(s) = \tau'_1(s) q_t \tag{9.88}$$

As explained below, this approximation is a poor and quite erroneous estimation of the axial strain for the most frequent situation, i.e., beam bending [47]. Indeed, when developing the axial strain Γ_1 up to the second order, we obtain

$$\Gamma_1(s) = \tau'_1(s) q_t + (\tau'_2(s) q_t\, \vartheta_3 - \frac{(\vartheta_3)^2}{2}) - (\tau'_3(s) q_t\, \vartheta_2 + \frac{(\vartheta_2)^2}{2}) \tag{9.89}$$

To emphasize the physical meaning of this result, consider in figure 9.12 a segment dl of the beam subjected to pure bending in the plane $\{1,2\}$ where the axial strain of the centroidal axis Γ_1 equals 0. We may write:

$$\tau'_1(s) q_t = \frac{x}{dl} \overset{(2^d order)}{=} -\frac{(\vartheta_3)^2}{2} \quad \text{and} \quad \tau'_2(s) q_t \simeq \vartheta_3$$

The first order computation 9.88 yields $\Gamma_1 = -\frac{(\vartheta_3)^2}{2}$, which is absolutely erroneous; expression 9.89 yields

$$\Gamma_1 \simeq -\frac{(\vartheta_3)^2}{2} + \frac{(\vartheta_3)^2}{2} = 0$$

as it should be. In other words, relation 9.89 expresses an important fact, namely, that a beam submitted to pure bending does not change its length [47], [91], [68].

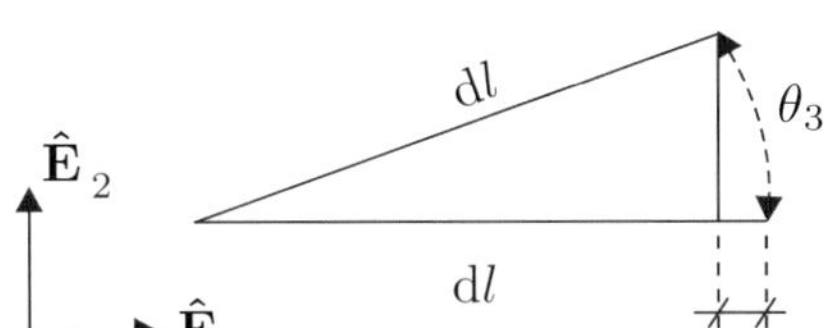

Figure 9.12: A beam in pure bending

The previous considerations as well as the kinematic assumption of section 9.3.1 lead us to write the following relation of order between the quantities $\tau'_\alpha(s) q_t$ and ϑ_β for a beam undergoing bending:

$$\text{order}\left(\tau'_1(s) q_t, \frac{(\vartheta_2)^2}{2}, \frac{(\vartheta_3)^2}{2}\right) \ll \text{order}\,(\tau'_2(s) q_t, \tau'_3(s) q_t, \vartheta_1, \vartheta_2, \vartheta_3) \tag{9.90}$$

- Γ_2 : The shear deformation (y direction)

Expanding the second component of Γ (see 9.87) and taking into account the relation of order 9.90, we finally obtain the following acceptable approximation for Γ_2 (as in [47]):

$$\Gamma_2 = \tau_2'(s)q_t - \vartheta_3$$

- Γ_3 : The shear deformation (z direction)

Similarly, Γ_3 can be expressed as

$$\Gamma_3 = \tau_3'(s)q_t - \vartheta_2$$

Term 6 is finally expanded by:

. introducing the strain components Γ_1, Γ_2 and Γ_3 into 9.72,

. introducing the latter into integral 9.86,

. expressing ϑ_1, ϑ_2, ϑ_3 in terms of the shape functions $\rho(s)$ and the generalized coordinates q_r,

. isolating the integrals over the shape function combinations in the various terms of 9.86. Note that integrals of products of 2, 3 and 4 shape functions are involved. They result from the necessity of keeping higher order terms in the computation of the strain vector Γ, as detailed above.

- *Term 7:* $-\int_0^L \mathbf{M} \,.\, \Delta \overset{\circ}{\mathbf{K}}(s)ds$

A similar reasoning is followed for this last term by momentarily disregarding the constitutive equation of the internal torque $\mathbf{M} = [\hat{\mathbf{S}}(s)]^T M$ (see equation 9.73) and by computing the virtual velocity field $\Delta \overset{\circ}{\mathbf{K}}(s)$ given by 9.77 on the basis of 9.21:

$$\begin{aligned}
\Delta \overset{\circ}{\mathbf{K}} &= \frac{\partial \Delta \mathbf{\Omega}(s)}{\partial s} = \frac{\partial}{\partial s}[\hat{\mathbf{S}}(s)]^T \left(B(s)\rho(s)\right) \Delta\dot{q}_r \\
&= [\hat{\mathbf{E}}]^T \frac{\partial}{\partial s} \left(R^T(s)B(s)\rho(s)\right) \Delta\dot{q}_r \\
&= [\hat{\mathbf{S}}(s)]^T \left[R(s)\frac{\partial \left(R^T(s)B(s)\right)}{\partial s}\rho(s) + B(s)\,\rho'(s)\right] \Delta\dot{q}_r \qquad (9.91)
\end{aligned}$$

The computation of 9.91 is performed under the kinematic hypotheses relating to the angular deformation and curvature of the beam (see section 9.3.1). In concrete terms, assuming both ϑ_α and $\vartheta'_\beta \triangleq \frac{\partial \vartheta_\beta}{\partial s}$ are small, we may express $\Delta \overset{\circ}{\mathbf{K}}$ by dropping terms of second order and higher, and introduce it into *term 7*, leading to

$$\int_0^L \left\{ \begin{pmatrix} M_1 & M_2 & M_3 \end{pmatrix} \begin{pmatrix} \rho_1' + \vartheta_3\rho_2' + \vartheta_2'\rho_3 \\ \rho_2' - \vartheta_3\rho_1' - \vartheta_1'\rho_3 \\ \rho_3' + \vartheta_2\rho_1' + \vartheta_1'\rho_2 \end{pmatrix} \Delta\dot{q}_r \right\} ds \qquad (9.92)$$

As regards the internal torque **M** itself (see equation 9.73) and in particular the curvature vector **K**, its three components can be computed under the kinematic assumptions and using the previously formulated relation of order, resulting in

$$\mathbf{K} = [\hat{\mathbf{S}}(s)]^T K \quad \text{with} \quad K = \begin{pmatrix} \vartheta'_1 \\ \vartheta'_2 \\ \vartheta'_3 \end{pmatrix} \tag{9.93}$$

As with the 6th term, *term 7* is finally obtained by:

. introducing 9.93 into 9.73,

. introducing the latter into the integral 9.92,

. expressing $\vartheta_1, \vartheta_2, \vartheta_3$ and $\vartheta'_1, \vartheta'_2, \vartheta'_3$ in terms of the shape functions and the generalized coordinates,

. isolating in 9.92 the integrals over the shape functions (see Appendix B).

These developments providing the beam deformation equations must be carried out for all the beams ($i = 1 : N^{beam}$) of the multibody system; the corresponding equations are then added to the joint equations of section 9.3.3 to produce the final system of equations of motion:

$$\mathcal{F}(q, \dot{q}, \ddot{q}) = 0 \tag{9.94}$$

with:

$$q = \left(\overbrace{q^1 ... q^{N^{body}}}^{joints}, \overbrace{(\{q_t\}, \{q_r\})}^{beam\ 1}, ... \overbrace{(\{q_t\}, \{q_r\})}^{beam\ N^{beam}} \right)$$

It should be noted that these equations of motion were set up without worrying about the computation of the mass matrix of the system. In other words, they are fully implicit with respect to the generalized accelerations $\ddot{q}$. Nevertheless, in a simulation context, the knowledge of this matrix is indispensable (or highly desirable) whatever is the integration method. To avoid a pure numerical derivation of the latter, which is quite costly, it is preferable to derive it analytically, as shown in chapter 5, section 5.6.

9.3.5 Symbolic computation of the equations of motion

Using its own symbolic library and taking advantage of the recursive nature of the equations of motion, a symbolic module has been added to ROBOTRAN to compute the multibody equations 9.94 as well as the associated tangent matrices required by numerical integrators (see chapter 10):

$$K = \frac{\partial \mathcal{F}}{\partial q^T}, G = \frac{\partial \mathcal{F}}{\partial \dot{q}^T} \text{ and } M = \frac{\partial \mathcal{F}}{\partial \ddot{q}^T}, \text{ the mass matrix}$$

These matrices are obtained by symbolic differentiation according to chapter 5, section 5.6.

To generate the multibody model, ROBOTRAN requires a specific symbolic data file (see chapter 5, figure 5.17) related to the flexible beams of the system. Such a file is illustrated in figure 9.13.

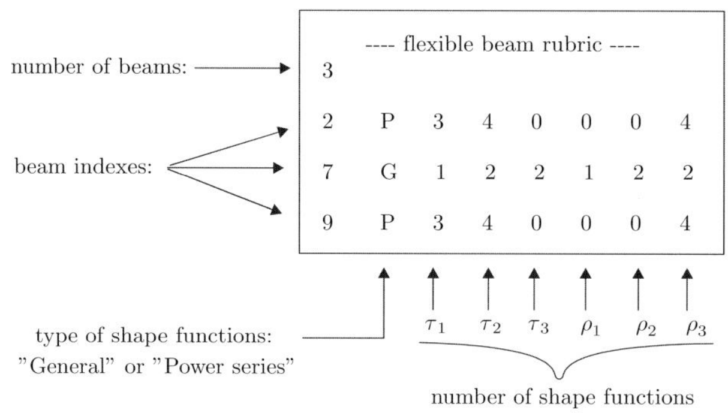

Figure 9.13: the ROBOTRAN data file (part)

For the shape functions, the user can choose between two possibilities:

. the so-called "general" case (flag G in the data file): ROBOTRAN then implements unequivocal strings to represent the shape functions and the corresponding integrals. The latter must be computed by the user (see Appendix B),

. the so-called "power-series" case (flag P in the data file) : ROBOTRAN then evaluates the monomial shape functions and the corresponding integrals so that *a fully symbolic computation of the equations of motion* is performed. The resulting output file can thus be considered as a black box, directly usable by a numerical simulation program without any intervention from the user, to represent the beam deformations. An example of a symbolic output file is given in Appendix A for a planar rotating beam using, for brevity, only one shape function for u_1, u_2 and ϑ_3.

The scheme illustrated in figure 9.14 represents the main steps of the ROBOTRAN symbolic algorithm, in relation with the developments of the previous sections.

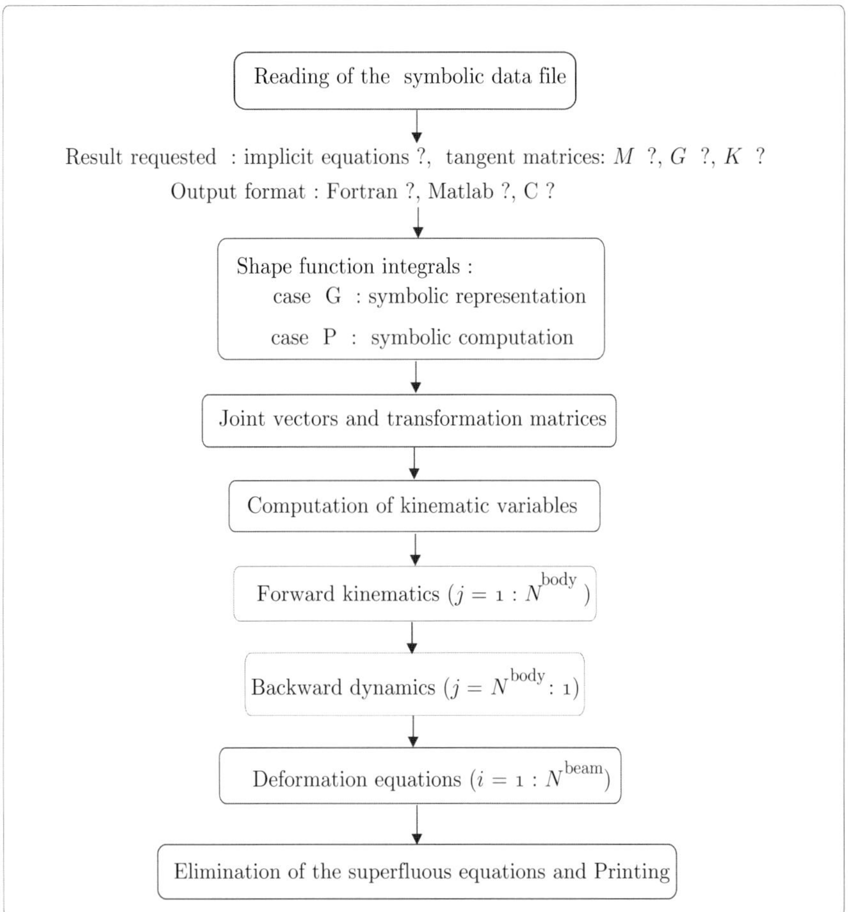

Figure 9.14: The ROBOTRAN algorithm

9.4 Numerical examples

Three typical examples illustrate the formalism. The models were generated by ROBOTRAN using monomials as shape functions to represent the beam deformations.

Rotating beam n° 1

The first example has been investigated in various publications (e.g. [47], [42]) and consists of a uniform flexible aluminium beam attached to a rotating cylinder of radius R_{cyl} as depicted in figure 9.15.

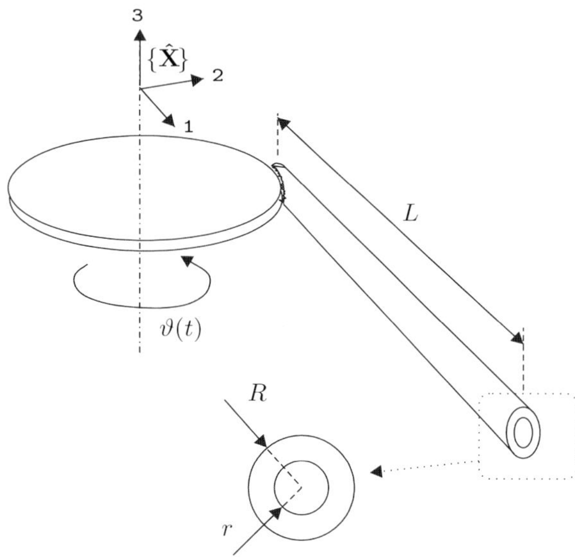

Figure 9.15: A rotating flexible beam

For this first example, the numerical data are the following: $R_{cyl} = 0$, $\rho = 3000\,\frac{kg}{m^3}$, $L = 10\,m$, $E^Y = 6.9e10\,\frac{N}{m^2}$, $G^{sh} = 2.6e10\,\frac{N}{m^2}$, $R = 0.0326\,m$ and $r = 0.0306\,m$. The hub starts to rotate from rest up to an angular speed of $6\,\frac{rad}{sec}$ according to the following law:

$$\vartheta(t) = \begin{array}{ll} \frac{2}{5}\left(\frac{t^2}{2} + (\frac{7.5}{\pi})^2\cos(\frac{\pi t}{7.5})\right) - \frac{2}{5}\,(\frac{7.5}{\pi})^2 & : \text{ if } \; t \leq 15sec \\ 6\,t - 45 & : \text{ if } \; t > 15sec \end{array} \tag{9.95}$$

The simulation is performed over 20 seconds.
The beam is modeled using series of 5, 4 and 4 monomials for τ_1, τ_2 and ρ_3 respectively. Figure 9.16 represents the lateral and longitudinal tip displacements respectively. Note that the ROBOTRAN results are identical to those produced by FEM using 10 beam elements [78]. As regards the CPU time performance

– using an implicit Runge-Kutta scheme [36] – a time reduction factor of 14 is obtained in favor of the monomial discretization for which the simulation required 16.4 seconds on a 31.9 Mflops workstation, i.e. faster than a real-time simulation.

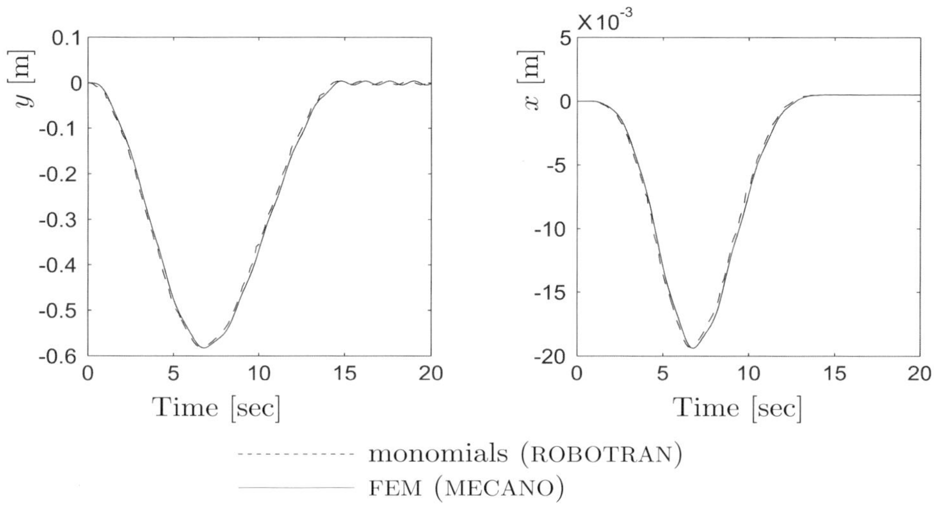

Figure 9.16: Simulation of a rotating beam

Rotating beam n° 2

The aim of the following experiment is to check how the model behaves when the beam is subjected to full 3D loading; in particular, torsion and space-flexion are combined in the proposed simulation. The system consists of a $5\,m$-long aluminium beam which has the same geometrical and physical characteristics as the previous one. A $5\,kg$ point mass is connected to the tip of the beam, through a perpendicular massless arm ($L_{arm} = 1\,m$), as depicted in figure 9.17. The system is placed in a gravity field, which acts along the negative z direction. Once static equilibrium is reached, the hub (which was at rest at $t = 0$) starts to rotate according to the law (9.95).
A monomial discretization was implemented with four functions for τ_1, τ_2 and τ_3, and three for ρ_1, ρ_2 and ρ_3 respectively. Figure 9.18 shows the evolution of the x, y, z coordinates of the beam tip and of the point mass, expressed in a frame rigidly attached to the rotating hub. The ROBOTRAN results compare favorably with those obtained via FEM using 10 beam elements. A CPU time reduction factor of 38 was observed in this case.

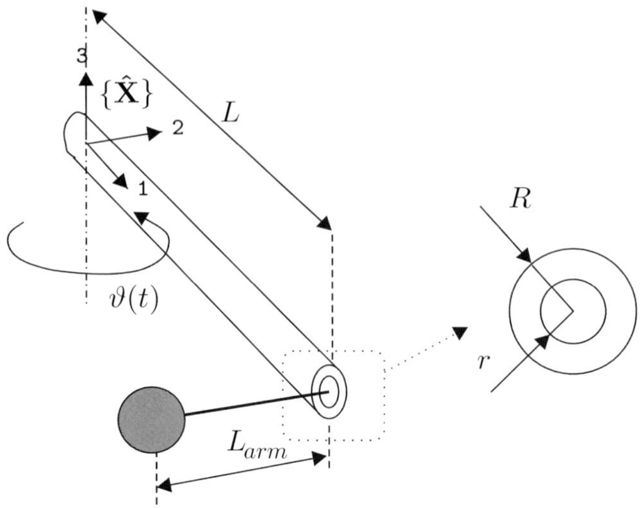

Figure 9.17: A rotating flexible beam carrying an off-centered mass

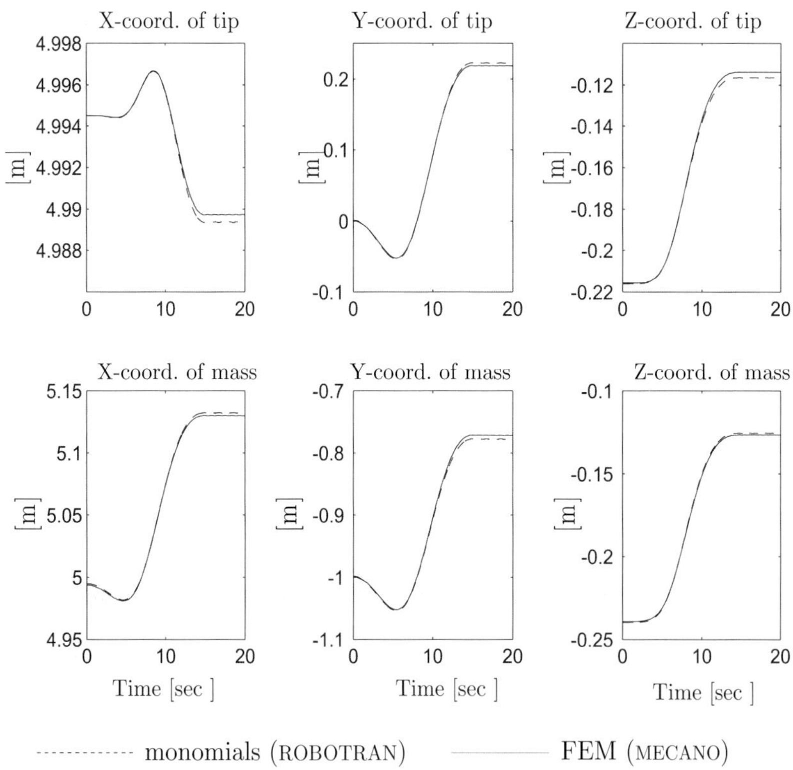

Figure 9.18: Simulation of a rotating beam carrying an off-centered mass

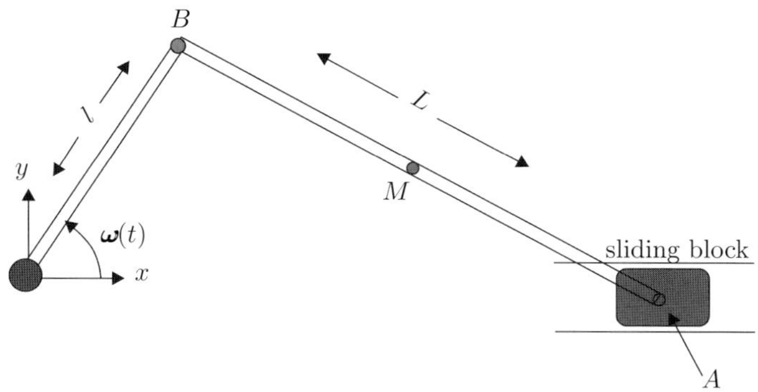

Figure 9.19: A flexible slider-crank mechanism

Flexible slider-crank mechanism

The third example is a slider-crank mechanism, investigated in [54]. Depicted in figure 9.19, the system consists of a rigid crank with length $l = 0.15\,m$, a deformable connecting rod and a rigid sliding block with half the mass of the connecting rod. The rod is $L = 0.3\,m$ long and has a circular cross-section with diameter $d = 0.006\,m$. Young's modulus is $E^Y = 0.2e12\,\frac{N}{m^2}$ and mass density is $7.87e3\,\frac{kg}{m^3}$. The crank is driven at a constant angular velocity $\omega = 150\frac{rad}{sec}$. The system is friction-free and gravity is neglected. This system contains a kinematic loop which has been taken into account by expressing the corresponding constraints at position, velocity and acceleration levels, and by using the classical Lagrange multipliers technique to compute the constraint forces. From the simulation point of view, a coordinate partitioning reduction (see chapter 4, section 4.6.1) was implemented to reduce the DAE system to a minimal ODE system before time integration. Figures 9.20 (ROBOTRAN results) and 9.21 (results from [54] using a FEM approach) compare the transverse deflection $Y(M)$ of the midpoint M and half the elongation Δl of the rod between joints A and B. In the present example, two models were envisaged in ROBOTRAN (figure 9.20): the first uses the presented approach with monomial shape functions to model the single-segment beam (continuous curve in figure 9.20). The second uses 20 finite rigid segments to model the beam (dashed curve in figure 9.20). The difference in results is mainly due to the fact that the finite segment model has no extensional d.o.f. and that the latter is excited in this simulation, as a result of inertia forces induced by the sliding block motion.
The results clearly illustrate the previous discussion relating to the fact that any transverse displacement of a point (on the centroidal axis) of a bending beam gives rise to a non negligible axial displacement.

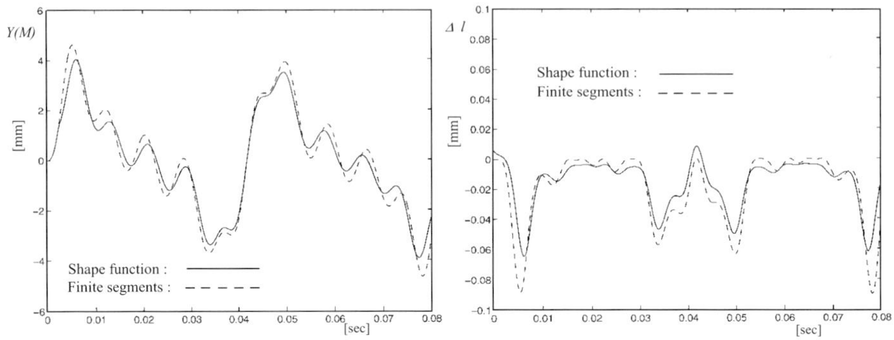

Figure 9.20: Flexible slider-crank simulation: ROBOTRAN results

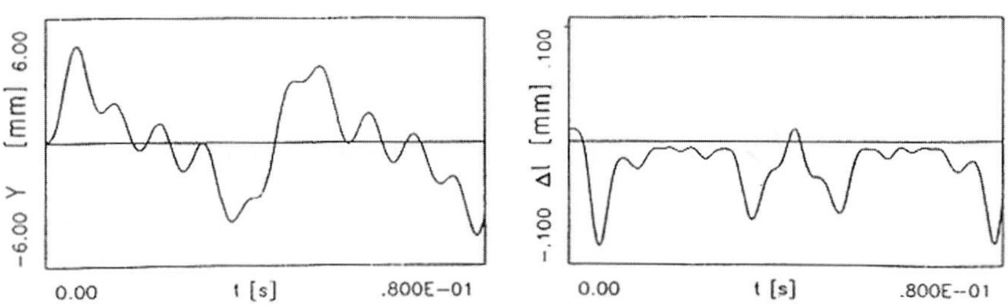

Figure 9.21: Flexible slider-crank simulation: results from [54]

Appendix A: ROBOTRAN symbolic output

```
% ROBOTRAN - Version 6.3 (build: 8 mai 2002)
function [f] = dirdynaF(q,qd,qdd,d,l,m,In,frc,trq,g,XE,L1,L2,RO,K1,K2,K3,K4,K5,K6,IY,IZ)

% Trigonometric Variables
S1 = sin(q(1));
C1 = cos(q(1));

% Auxiliary Variables (flexible beams)
XCG11 = -XE(1)+l(1,1);
IRE122 = In(5,1)+m(1)*XCG11*XCG11;
IRE133 = In(9,1)+m(1)*XCG11*XCG11;
MU1 = (m(1))/(L2(1));
ID11 = RO(1)*(IY(1)+IZ(1));
ID12 = IY(1)*RO(1);
ID13 = IZ(1)*RO(1);
L2I1 = 1.000/L2(1);
si11 = -L2I1*XE(1);
L21 = 0.500*L2(1);
L31 = 0.333*L2(1);
LS31 = 0.333*L2(1)*L2(1);

% Prelimiar Kinematic and Dynamical Terms
TI31 = q(4)*si11;
TIP31 = qd(4)*si11;
TIPP31 = qdd(4)*si11;
XI11 = -XE(1)+q(2)*si11;
XI21 = q(3)*si11;
XIP11 = qd(2)*si11;
XIP21 = qd(3)*si11;
XIPP11 = qdd(2)*si11;
XIPP21 = qdd(3)*si11;
MG11 = q(2)*L21*MU1+m(1)*XCG11;
MG21 = q(3)*L21*MU1;
GX1_1 = L21*MU1-m(1)*si11;
GY2_1 = L21*MU1-m(1)*si11;
LI11 = q(2)*GX1_1+m(1)*l(1,1);
LI21 = q(3)*GY2_1;
LIP11 = qd(2)*GX1_1;
LIP21 = qd(3)*GY2_1;
LIPP11 = qdd(2)*GX1_1;
LIPP21 = qdd(3)*GY2_1;
IE91 = IRE133+MU1*(2.000*q(2)*LS31+q(2)*q(2)*L31+q(3)*q(3)*L31);
IEP91 = MU1*(2.000*qd(2)*LS31+2.000*(q(2)*qd(2)*L31+q(3)*qd(3)*L31));
IXPP31 = qdd(4)*L21*IZ(1)*RO(1)+MU1*(q(2)*qdd(3)*L31-q(3)*qdd(2)*L31+qdd(3)*LS31);
```

```
% Forward Kinematics
OE31 = qd(1)-TIP31;
OEP31 = qdd(1)-TIPP31;
BSE11 = -OE31*OE31;
BSE51 = -OE31*OE31;

% Backward Dynamics
FO11 = LIPP11+BSE11*LI11-LI21*OEP31-2.000*LIP21*OE31...
+TI31*(LIPP21+BSE51*LI21+LI11*OEP31+2.000*LIP11*OE31);
FO21 = LIPP21+BSE51*LI21+LI11*OEP31+2.000*LIP11*OE31...
-TI31*(LIPP11+BSE11*LI11-LI21*OEP31-2.000*LIP21*OE31);
CO31 = IXPP31+IE91*OEP31+IEP91*OE31+...
MG11*(-XIPP21-BSE51*XI21-2.000*OE31*XIP11-OEP31*XI11)...
 -MG21*(-XIPP11-BSE11*XI11+2.000*OE31*XIP21+OEP31*XI21)...
 -XI11*(FO21+FO11*TI31)+XI21*(FO11-FO21*TI31);

% Flexible Beams Dynamics
FIR11 = FO11-FO21*TI31;
FIR21 = FO21+FO11*TI31;
T11_1 = -FIR11*si11;
T12_1 = -FIR21*si11;
XAR11 = XIPP11+BSE11*XI11-2.000*OE31*XIP21-OEP31*XI21;
XAR21 = XIPP21+BSE51*XI21+2.000*OE31*XIP11+OEP31*XI11;
T31_1 = T11_1-MU1*(-BSE11*LS31+L21*XAR11);
T32_1 = T12_1-MU1*(L21*XAR21-LS31*OEP31);
T41_1 = T31_1+qdd(2)*L31*MU1;
T42_1 = T32_1+qdd(3)*L31*MU1;
T51_1 = T41_1-2.000*qd(3)*L31*MU1*OE31;
T52_1 = T42_1+2.000*qd(2)*L31*MU1*OE31;
T61_1 = T51_1+q(2)*L2I1*K1(1)+MU1*(q(2)*BSE11*L31-q(3)*L31*OEP31);
T62_1 = T52_1+q(3)*L2I1*K2(1)+MU1*(q(2)*L31*OEP31+q(3)*BSE51*L31);
T72_1 = T62_1-0.500*q(4)*K2(1);
R11_1 = -CO31*si11;
OMI31 = OEP31*ID13;
R31_1 = R11_1+L21*OMI31;
R41_1 = R31_1+qdd(4)*L31*ID13;
R61_1 = R41_1+q(4)*L31*K2(1);
R71_1 = R61_1+q(4)*L2I1*K6(1);
R81_1 = R71_1-0.500*q(3)*K2(1);

% Symbolic Outputs
f(1) = CO31;
f(2) = T61_1;
f(3) = T72_1;
f(4) = R81_1;
```

Appendix B: Shape function integrals

The table summarizes the shape function integrals required by the symbolic formalism. They are formulated in the first column where $\tau_{\alpha=1,2,3}$ and $\rho_{\beta=1,2,3}$ represent a "set" of shape functions as defined in 9.11 and 9.14 respectively. The second column gives their analytical expressions when using power series:

$$\tau_\alpha(k) = \rho_\beta(k) = \left(\frac{s}{L}\right)^k \quad \text{with} \quad k = 1 : n_{t_\alpha} \text{ and } k = 1 : n_{r_\beta} \text{ respectively.}$$

The third column gives the equations in which the corresponding integrals appear, when expanded.

$\int_0^L \tau_\alpha(k)\, ds$	$\frac{L}{k+1}$	(9.57) (9.60) (9.61) (9.62) (9.81)
$\int_0^L \rho_\beta(k)\, ds$	$\frac{L}{k+1}$	(9.61) (9.63) (9.82)
$\int_0^L s\tau_\alpha(k)\, ds$	$\frac{L^2}{k+2}$	(9.61) (9.63)
$\int_0^L \tau_\alpha(k)\, \tau_\beta(l)\, ds$	$\frac{L}{k+l+1}$	(9.61) (9.63) (9.81)
$\int_0^L \rho_\alpha(k)\, \rho_\beta(l)\, ds$	$\frac{L}{k+l+1}$	(9.61) (9.63) (9.82) (9.86)
$\int_0^L \rho_\alpha(k)\, \tau'_\beta(l)\, ds$	$\frac{l}{k+l}$	(9.78) (9.86)
$\int_0^L \tau'_\alpha(k)\, \tau'_\beta(l)\, ds$	$\frac{kl}{k+l-1}\frac{1}{L}$	(9.78) (9.86)
$\int_0^L \rho'_\alpha(k)\, \rho'_\beta(l)\, ds$	$\frac{kl}{k+l-1}\frac{1}{L}$	(9.78) (9.92)
$\int_0^L \rho_\alpha(k)\, \rho_\beta(l)\, \rho_\gamma(m)\, ds$	$\frac{L}{k+l+m+1}$	(9.82)
$\int_0^L \tau'_\alpha(k)\, \rho_\beta(l)\, \rho_\gamma(m)\, ds$	$\frac{k}{k+l+m}$	(9.78) (9.86)
$\int_0^L \rho_\alpha(k)\, \tau'_\beta(l)\, \tau'_\gamma(m)\, ds$	$\frac{lm}{k+l+m-1}\frac{1}{L}$	(9.78) (9.86)
$\int_0^L \rho_\alpha(k)\, \rho'_\beta(l)\, \rho'_\gamma(m)\, ds$	$\frac{lm}{k+l+m-1}\frac{1}{L}$	(9.78) (9.92)
$\int_0^L \rho_\alpha(k)\, \rho_\beta(l)\, \rho_\gamma(m)\, \tau'_\delta(n)\, ds$	$\frac{n}{k+l+m+n}$	(9.78) (9.86)
$\int_0^L \rho_\alpha(k)\, \rho_\beta(l)\, \rho_\gamma(m)\, \rho_\delta(n)\, ds$	$\frac{L}{k+l+m+n+1}$	(9.78) (9.86)

Shape function integrals - Analytical values (monomials) - Equations involved

Chapter 10

Time integration of flexible MBS

10.1 Introduction

From an examination of system {4.1, 4.2, 4.3}, it readily appears that the equations of motion of constrained multibody systems are not only differential but differential-algebraic equations, abbreviated as DAEs [65], [40]. Time integration of DAEs has already been approached within the field of multibody dynamics using constraint stabilization [2], coordinate partitioning [92] or direct methods [66], [26].

It is worthwhile to emphasize the main advantages and drawbacks of these three techniques.

- Constraint stabilization [2] is probably one of the first approaches used in solving a DAE system. It consists in transforming the DAE into an ordinary differential system (abbreviated ODE), by differentiating the constraints and introducing *stabilization* terms [2]. The constraints are thus not satisfied exactly but oscillate with a given stabilization period and damping constants around the exact solution. While simple in concept and computationally efficient, this technique — still currently in use — lacks a theoretical background to set the values of the stabilization parameters, sometimes leading to solutions in which constraints are not satisfied exactly.

- The generalized coordinate partitioning method [92] also transforms the original DAE system into an ODE system but, as shown in section 4.6.1, does so in a quite different way, by a system *reduction*, including an accurate algebraic solution of the constraints at position, velocity and acceleration levels. As clearly indicated in [40], the method is theoretically rigorous and yields accurate results for a broad class of applications, including those to which highly nonlinear constraints are applied (see chapters 7, 8 for instance). Its drawbacks are a lack of computational efficiency induced by the associated reduction procedure (matrix operations), and also the difficulty in dealing with stiff systems since the equations are provided in a form which is not directly compatible with an *implicit* integration scheme.

- The direct methods [66], [26] exhibit quite an impressive efficiency in terms of computational speed [10]. Their basic idea is to attempt to attack the complete set of DAEs *directly* by applying, for instance, a backward differentiation formula to the variables and by solving the resulting system at each step using a Newton/Raphson type method (e.g. [26]). Since the underlying technique only requires the equations of motion in their *implicit* form, the computational effort required to generate[1] the MBS model is of order N, N being the number of degrees of freedom (see for example [10]). The implicit nature of the direct methods makes them naturally suitable for the time integration of systems whose high frequencies do not contribute significantly to the physical behavior and can thus be filtered out. This is the case of MBS containing flexible bodies whose describing equations are said to be "stiff" (see [6] for instance) since the system eigenfrequencies are distributed over a broad frequency range.
  The major drawback of the direct methods arises from the fact that, despite the development of high level mathematical theories to increase their stability and accuracy in the presence of algebraic constraints, there is a certain lack of broadly applicable and robust integrator schemes for DAE systems.

All these considerations justify the purpose of the proposed method for time integrating MBS with flexible beams (see chapter 9). Indeed, the robustness of an integration method with respect to solving kinematic constraints (e.g. *3D* body loops, wheel/rail contact, cam/follower contact, ...) is one of the most fundamental priorities. Indeed, any drift in the position constraints is unacceptable in engineering problems because of the strong physical meaning of the position constraints (they often hold components together) and because of their sometimes severe nonlinearity.

This remark justifies keeping the coordinate partitioning method (see section 4.6.1) to reduce the DAE system to an ODE, as a robust method for all multibody applications. However, the original method which is intrinsically designed for *explicit* integrator schemes, is CPU time consuming for multibody systems including flexible bodies, because of the small time integration steps required to solve these systems. Thus, it is judicious[2] to adapt the coordinate partitioning method in order to match the *implicit* integrator scheme requirements.
Rather than various implicit schemes, in this chapter we deliberately concentrate on a single one, i.e. the Newmark method [59], which offers a good compromise between efficiency and reliability over a large range of flexible multibody applications [6], [89].

[1] but not to solve them.
[2] as already suggested in [40], but in a quite different way.

10.2 Implicit integration method

The intrinsic assets of the Newmark method have been rigorously established for linear systems [6], [27]: although it is a one-step method[3] for second order systems, the Newmark scheme has second-order accuracy and is unconditionally stable for a correct choice of the parameters [6]. But it has been clearly shown in [6] that the Newmark scheme falls down from a stability point of view when it is directly applied to a DAE system. Indeed, it presents a weak instability due to the presence of the algebraic constraints and which is excited for any values of the time step. In fact, kinematic constraints introduce theoretical infinite frequencies that the Newmark scheme cannot dissipate. To circumvent this limitation, some authors introduce a numerical dissipation into the scheme to ensure at least a small damping of these frequencies[4]

Basically, it is not the *existence* of the algebraic constraints but *the fact that they are not eliminated* that requires the use of sophisticated numerical techniques to stabilize (as in [26], [6]) the implicit integration schemes for DAE system. Thus, by opting for a prior reduction of the DAE to an ODE, we still can use a classical scheme (ex. of Newmark type) with the advantages that:

1. the constraints are rigorously solved without any drift during the simulation,

2. the time step can be considerably larger than that required by explicit schemes, so that undesirable or meaningless high frequencies can be filtered out,

3. the proper elimination of the constraints and of the associated Lagrange multipliers (see section 4.6.1) circumvents the stability problems.

The major drawback of the proposed method is the numerical cost of the reduction, as pointed out in [40]. Thus attention must be paid by reducing the underlying matrix operations thanks to the symbolic computation capabilities.

10.2.1 Residual formulation of the MBS equations in a Newmark scheme

The reduced system of equations of motion 4.48 obtained by the coordinate partitioning method (see section 4.6.1):

$$M_{red}(u, v(u), t)\ddot{u} + c_{red}(\dot{u}, u, v(u), t) = \mathcal{Q}_{red} \tag{10.1}$$

represent a system of size $(n-m)$, linear in the independent accelerations $\ddot{u}$ but nonlinear in the independent positions u and velocities $\dot{u}$. In order to integrate

[3] which enables it to be self-starting.

[4] see for instance the Hilber-Hughes-Taylor algorithm proposed in [6] which retains the second order accuracy of the original Newmark method.

these equations with respect to time, let us introduce the Newmark's formulae for velocity and displacement:

$$\dot{u}_{n+1} = \dot{u}_n + (1-\gamma)\,\Delta t\,\ddot{u}_n + \gamma\,\Delta t\,\ddot{u}_{n+1} \tag{10.2}$$

$$u_{n+1} = u_n + \Delta t\,\dot{u}_n + (\frac{1}{2}-\beta)\,\Delta t^2\,\ddot{u}_n + \beta\,\Delta t^2\,\ddot{u}_{n+1} \tag{10.3}$$

where suffices $n, n+1$ characterize quantities computed at times t_n and t_{n+1} respectively, $\Delta t = t_{n+1} - t_n$ is the time step, and β and γ are free parameters. It is well-known that the most appropriate choice of these parameters corresponds to the average constant acceleration formula obtained with the particular values $\gamma = \frac{1}{2}$ and $\beta = \frac{1}{4}$ since it gives maximum accuracy (order 2) with unconditional stability – for a linear system – over the entire frequency range (see for instance [27]).

In order to obtain a solution for displacement, velocity and acceleration at time t_{n+1}, we need to write the equations of motion 10.1 at t_{n+1} to form the so-called *residual vector* R_{n+1} at time t_{n+1}. Let us first write the implicit form 2.80 of the unreduced DAE system of the MBS:

$$\Phi(\ddot{q}, \dot{q}, q, t) = \mathcal{Q} + J^T(q,t)\,\lambda \tag{10.4}$$

$$h(q,t) = 0 \tag{10.5}$$

in which $\Phi \triangleq M(q,t)\ddot{q} + c(\dot{q}, q, t, F_{ext}, L_{ext}, g)$.

According to the coordinate partitioning (relation 4.44), the following partition holds for Φ and $\mathcal{Q}$:

$$\Phi = \begin{pmatrix} \Phi_u \\ \Phi_v \end{pmatrix} \text{ and } \mathcal{Q} = \begin{pmatrix} \mathcal{Q}_u \\ \mathcal{Q}_v \end{pmatrix}$$

The residual vector computed from 10.1 can be written (see equation 2.84):

$$R_{n+1} \triangleq \Phi_u(\ddot{q}_{n+1}, \dot{q}_{n+1}, q_{n+1}, t_{n+1}) - \mathcal{Q}_u + B_{vu}^T\,(\Phi_v(\ddot{q}_{n+1}, \dot{q}_{n+1}, q_{n+1}, t_{n+1}) - \mathcal{Q}_v) = 0 \tag{10.6}$$

with B_{vu} defined, as in 4.47, by:

$$B_{vu} \triangleq -\left(J_v(q_{n+1}, t_{n+1})\right)^{-1} J_u(q_{n+1}, t_{n+1}) \tag{10.7}$$

In expression 10.6 of the residual vector, the dependent coordinates v (and their derivatives) can be expressed in terms of the independent ones (see section 4.6.1):

$$\begin{aligned} v_{n+1} &= v(u_{n+1},\, t_{n+1}), \\ \dot{v}_{n+1} &= \dot{v}(q_{n+1},\, \dot{u}_{n+1},\, t_{n+1}), \\ \ddot{v}_{n+1} &= \ddot{v}(q_{n+1},\, \dot{q}_{n+1},\, \ddot{u}_{n+1},\, t_{n+1}) \end{aligned}$$

$\dot{u}_{n+1}$ and u_{n+1} can then be expressed in terms of $\ddot{u}_{n+1}$ on the basis of the Newmark difference formulae 10.2 and 10.3 respectively. In a more synthesized form, we can thus rewrite 10.6 as a set of $(n-m)$ nonlinear algebraic equations with respect to the $(n-m)$ unknowns $\ddot{u}_{n+1}$:

$$R_{n+1} = R_{n+1}(\ddot{u}_{n+1}, (\cdots)_n) = 0 \tag{10.8}$$

10.2.2 Iterative solution of the reduced form

The nonlinear system 10.8 will be solved directly for $\ddot{u}_{n+1}$ by using the Newton/Raphson iterative method[5]. This can be achieved by considering the equations of motion in their unreduced form (10.4 - 10.5) and by supposing, as usual, the availability of an approximate solution at iteration i of time t_{n+1}, denoted[6]

$$\ddot{q}^i, \dot{q}^i, q^i, \lambda^i \tag{10.9}$$

which however exactly satisfies the constraints and their derivatives, i.e.:

$$h(q^i, t) = \dot{h}(\dot{q}^i, q^i, t) = \ddot{h}(\ddot{q}^i, \dot{q}^i, q^i, t) = 0 \tag{10.10}$$

Let now express the *corrected solution* at time t_{n+1} in the following incremental form:

$$q = q^i + \Delta q\,, \quad \dot{q} = \dot{q}^i + \Delta\dot{q}\,, \quad \ddot{q} = \ddot{q}^i + \Delta\ddot{q}\,, \quad \lambda = \lambda^i + \Delta\lambda \tag{10.11}$$

Introducing these forms into the equations of motion 10.4, we obtain

$$\Phi(\ddot{q}^i + \Delta\ddot{q}, \dot{q}^i + \Delta\dot{q}, q^i + \Delta q, t) = \\ \mathcal{Q}(\dot{q}^i + \Delta\dot{q}, q^i + \Delta q, t) + J^T(q^i + \Delta q, t)\left(\lambda^i + \Delta\lambda\right)$$

and performing a first order expansion, we obtain the following linearized equations around the current configuration (which is the configuration at iteration i at time t_{n+1}):

$$M\,\Delta\ddot{q} + G\,\Delta\dot{q} + K\,\Delta q - J^T\,\Delta\lambda = -\left\{\Phi - \mathcal{Q} - J^T\,\lambda^i\right\} + \mathcal{O}(\Delta^2) \tag{10.12}$$

where:

$$M \triangleq \left(\frac{\partial\Phi}{\partial\ddot{q}^T}\right)_i \quad \text{is the generalized mass matrix } M(q^i, t)$$

$$G \triangleq \left(\frac{\partial}{\partial\dot{q}^T}\,(\Phi - \mathcal{Q})\right)_i$$

$$K \triangleq \left(\frac{\partial}{\partial q^T}\,((\Phi - \mathcal{Q}) - J^T\,\lambda)\right)_i$$

$$\Phi = M(q^i, t)\ddot{q}^i + c(\dot{q}^i, q^i, t, F_{ext}, L_{ext}, g), \;\; \mathcal{Q} = \mathcal{Q}(\dot{q}^i, q^i, t), \;\text{and}\; J = J(q^i, t)$$

[5] contrary to the recommendations of [40]

[6] we omit the suffix $n+1$ for clarity.

As previously done in section 4.6.1, we can partition and reorganize 10.12 on the basis of 4.44 and 4.45:

$$\begin{pmatrix} M_{uu} & M_{uv} \\ M_{vu} & M_{vv} \end{pmatrix} \begin{pmatrix} \Delta\ddot{u} \\ \Delta\ddot{v} \end{pmatrix} + \begin{pmatrix} G_{uu} & G_{uv} \\ G_{vu} & G_{vv} \end{pmatrix} \begin{pmatrix} \Delta\dot{u} \\ \Delta\dot{v} \end{pmatrix} + \begin{pmatrix} K_{uu} & K_{uv} \\ K_{vu} & K_{vv} \end{pmatrix} \begin{pmatrix} \Delta u \\ \Delta v \end{pmatrix} - \begin{pmatrix} {J_u}^T \\ {J_v}^T \end{pmatrix} \Delta\lambda = -\left\{ \begin{pmatrix} \Phi_u - \mathcal{Q}_u \\ \Phi_v - \mathcal{Q}_v \end{pmatrix} - \begin{pmatrix} {J_u}^T \\ {J_v}^T \end{pmatrix} \lambda^i \right\} \tag{10.13}$$

The elimination of the Lagrange multipliers from 10.13 is obtained by performing: $(10.13)_a + {B_{vu}}^T (10.13)_b$ with $B_{vu} \triangleq -\left((J_v)^{-1} J_u \right)_i$. We then obtain[7]:

$$\begin{aligned} &\left(\begin{pmatrix} M_{uu} & M_{uv} \end{pmatrix} + {B_{vu}}^T \begin{pmatrix} M_{vu} & M_{vv} \end{pmatrix} \right) \begin{pmatrix} \Delta\ddot{u} \\ \Delta\ddot{v} \end{pmatrix} \\ &+ \left(\begin{pmatrix} G_{uu} & G_{uv} \end{pmatrix} + {B_{vu}}^T \begin{pmatrix} G_{vu} & G_{vv} \end{pmatrix} \right) \begin{pmatrix} \Delta\dot{u} \\ \Delta\dot{v} \end{pmatrix} \\ &+ \left(\begin{pmatrix} K_{uu} & K_{uv} \end{pmatrix} + {B_{vu}}^T \begin{pmatrix} K_{vu} & K_{vv} \end{pmatrix} \right) \begin{pmatrix} \Delta u \\ \Delta v \end{pmatrix} \\ &= -\left\{ \Phi_u - \mathcal{Q}_u + {B_{vu}}^T \left(\Phi_v - \mathcal{Q}_v \right) \right\} \end{aligned} \tag{10.14}$$

As regards the algebraic constraints 10.5, a first order expansion can also be considered:

$$h(q^i + \Delta q, t) = h(q^i, t) + J(q^i, t)\Delta q = 0 \tag{10.15}$$

Since we assumed that $h(q^i, t) = 0$, we can use the partition 4.44 and the definition of the matrix $B_{vu}(q^i)$ to write

$$\Delta v = B_{vu}\, \Delta u \tag{10.16}$$

and, by differentiating once and twice, we obtain

$$\Delta\dot{v} = B_{vu}\, \Delta\dot{u} \quad \text{and} \quad \Delta\ddot{v} = B_{vu}\, \Delta\ddot{u} \tag{10.17}$$

As for the velocity and displacement increments $\Delta\dot{u}$ and Δu, they can be correlated to $\Delta\ddot{u}$ via the Newmark difference formulae 10.2 and 10.3 as follows

$$\Delta\dot{u} = (\gamma\, \Delta t)\, \Delta\ddot{u}\,, \quad \Delta u = \left(\beta\, \Delta t^2\right) \Delta\ddot{u} \tag{10.18}$$

Introducing (10.16 - 10.18) into 10.14, we finally obtain the Newton/Raphson iterative formula:

$$\Im^i\, \Delta\ddot{u} = -R^i \tag{10.19}$$

[7] The λ's and $\Delta\lambda$'s are simultaneously eliminated through this operation.

where $R^i \left(= R^i_{n+1}\right)$ is the residual vector defined in 10.6 and

$$\begin{aligned}\Im^i \triangleq\ & M_{uu} + \gamma\, \Delta t\, G_{uu} + \beta\, \Delta t^2\, K_{uu} \\ & + \left(M_{uv} + \gamma\, \Delta t\, G_{uv} + \beta\, \Delta t^2\, K_{uv}\right) B_{vu} \\ & + {B_{vu}}^T \left(M_{vu} + \gamma\, \Delta t\, G_{vu} + \beta\, \Delta t^2\, K_{vu}\right) \\ & + {B_{vu}}^T \left(M_{vv} + \gamma\, \Delta t\, G_{vv} + \beta\, \Delta t^2\, K_{vv}\right) B_{vu}\end{aligned} \tag{10.20}$$

is the iteration *tangent matrix* which allows the solution of equation 10.8 to progress.

Let us summarize the Newton/Raphson implementation at time step t_{n+1} as follows:

1. To start the iterative process, we use as initial values

$$\ddot{u}^{i=0}_{n+1} = \ddot{u}_n \text{ , the result of the previous time step}$$

2. The Newton/Raphson iteration i is based on 10.19 and leads to a corrected value for the accelerations:

$$\ddot{u}^{i+1}_{n+1} = \ddot{u}^i_{n+1} + \Delta \ddot{u}^i_{n+1} \quad \text{with} \quad \Delta \ddot{u}^i_{n+1} = -\left(\Im^i_{n+1}\right)^{-1} R^i_{n+1}$$

3. According to 10.2 and 10.3, velocities and positions are then corrected as follows:

$$\begin{aligned}\dot{u}^{i+1}_{n+1} &= \dot{u}^i_{n+1} + \Delta \dot{u}^i_{n+1} \quad \text{with} \quad \Delta \dot{u}^i_{n+1} = \gamma\, \Delta t\, \Delta \ddot{u}^i_{n+1} \\ u^{i+1}_{n+1} &= u^i_{n+1} + \Delta u^i_{n+1} \quad \text{with} \quad \Delta u^i_{n+1} = \beta\, \Delta t^2\, \Delta \ddot{u}^i_{n+1}\end{aligned}$$

4. On the basis of the constraints and their derivatives 10.10 (see section 4.6.1), the dependent coordinates become

$$v^{i+1}_{n+1} = v^{i+1}_{n+1}\left(u^{i+1}_{n+1}, t_{n+1}\right) \tag{10.21}$$

$$\dot{v}^{i+1}_{n+1} = \left(B_{vu}\right)^{i+1}_{n+1} \dot{u}^{i+1}_{n+1} - \left(\left(J_v\right)^{-1} \frac{\partial h}{\partial t}\right)^{i+1}_{n+1} \tag{10.22}$$

$$\ddot{v}^{i+1}_{n+1} = \left(B_{vu}\right)^{i+1}_{n+1} \ddot{u}^{i+1}_{n+1} - \left(J_v\right)^{-1} \left\{\left(\dot{J} + \frac{\partial^2 h}{\partial t\, \partial q}\right) \dot{q} + \frac{\partial^2 h}{\partial t^2}\right\}^{i+1}_{n+1} \tag{10.23}$$

Steps 2, 3 and 4 provide a new estimation for the next iteration $i+1$. The process is stopped as soon as convergence is satisfactory, i.e. when

$$\left\|R^i_{n+1}\right\| < \varepsilon_{NR} \tag{10.24}$$

where R^i_{n+1} is the residual vector defined in 10.6 and ε_{NR} is a predefined Newton/Raphson threshold whose value is related to the order of magnitude of the forces acting on the system.

10.2.3 Local truncation error estimation

The convergence test 10.24 ensures that the iterative Newton/Raphson process converges successfully at time t_{n+1}, but it does not ensure that the quantities u_{n+1}, $\ddot{u}_{n+1}$ have been computed accurately over the time step $\Delta t = t_{n+1} - t_n$. Indeed, a local error can result from the truncation underlying the Newmark difference formulae 10.2 and 10.3. We thus need a second test to accept or reject a given time step Δt. This test is based on an estimation of the *local truncation error*, i.e. the error during an integration step.

A detailed analysis was carried out in [6] for the Hilber-Hughes-Taylor algorithm. Applied to the Newmark formulae and to the present formulation of the equations of motion, the error criterion is

$$\frac{\left(\beta - \frac{1}{6}\right)\Delta t^2}{K^{\Omega}\left(u^t\, M_{red}\, u\right)^{0.5}}\left(\Delta\ddot{u}^t\, M_{red}\, \Delta\ddot{u}\right)^{0.5} < \varepsilon_{NK} \tag{10.25}$$

where: $\Delta t = t_{n+1} - t_n$ is the current time step, the constant $K^{\Omega} = 0.012$ results from a non-dimensional cut-off frequency, M_{red} is the mass matrix of the reduced system of equations 10.1, $(u^t M_{red}\, u)^{0.5}$ is the mass norm[8] of the current displacement u, $\Delta\ddot{u}$ is the acceleration increment such that 10.24 is satisfied, and ε_{NK} is a tolerance threshold which is problem-independent. Computer experiments have shown that a value of $\varepsilon_{NK} = ...10^{-3}...$ provides adequate results in engineering problems, with a good compromise between accuracy and computational cost.

In particular, it is shown in [6] that adapting the time step according to 10.25 ensures that there will be, on average, 10 time steps per dominant period of the simulated system. Moreover, if the time step is controlled on the basis of 10.25, the sum of the amplitudes of the highest frequency components are bounded by ε_{NK}: this is a typical characteristic of implicit integration schemes which operate as a low pass filter, suitable for flexible multibody systems.

10.2.4 Contribution of symbolic generation

In spite of the apparent simplicity of the Newton/Raphson method, we should note that its implementation relies on several tricky tasks, arising not only from the implicit integration scheme - the computation $(\mathcal{O}((n-m)^2))$ and the inversion $(\mathcal{O}((n-m)^3))$ of the iteration matrix 10.20 - but also from the coordinate partitioning reduction. Indeed, this reduction leads to an expression of the iteration matrix which is costly in view of its expression 10.20. Moreover, the exact computation of the residual vector in 10.24 requires the complete evaluation of the dependent quantities v^{i+1}_{n+1}, $\dot{v}^{i+1}_{n+1}$ and $\ddot{v}^{i+1}_{n+1}$ at each Newton-Raphson iteration.

We can however temper these facts since there are ways to decrease the CPU time, as shown in the example of section 10.4.

[8] according to [6], it is more convenient to replace this norm by a reference norm $(u^t_R M_{red}\, u_R)^{0.5}$ in which u_R represents characteristic values of the displacements for the considered system. This norm is then computed once at the beginning of the simulation.

Computation of the iteration matrix

As regards the computations which result directly from the coordinate partitioning method, let us first focus on the iteration matrix $\Im^i$ whose computation is theoretically required at each Newton/Raphson iteration of each time step[9]. In view of the expression 10.20 of $\Im^i$, we note three kinds of contributions, through the following matrices:

- the mass matrix $$M = \begin{pmatrix} M_{uu} & M_{uv} \\ M_{vu} & M_{vv} \end{pmatrix}_i = \left(\frac{\partial \Phi}{\partial \ddot{q}^T}\right)_i$$

- the damping matrix $$G = \begin{pmatrix} G_{uu} & G_{uv} \\ G_{vu} & G_{vv} \end{pmatrix}_i = \left(\frac{\partial}{\partial \dot{q}^T}(\Phi - Q)\right)_i$$

- the stiffness matrix $$K = \begin{pmatrix} K_{uu} & K_{uv} \\ K_{vu} & K_{vv} \end{pmatrix}_i = \left(\frac{\partial}{\partial q^T}(\Phi - Q - J^T\lambda)\right)_i$$

The recursive extraction of the mass matrix from the inverse dynamics is straightforward (see section 3.3) and is fully generated in symbolic form by ROBOTRAN. The contributions of the damping and stiffness matrices in the iteration matrix $\Im^i$ are application dependent and they can sometimes be disregarded without adversely affecting the convergence of the Newton/Raphson iteration. Nevertheless, their recursive generation is possible using ROBOTRAN, as explained in section 5.6.
As concerns the stiffness matrix K, one can verify that the term $\frac{\partial}{\partial q^t}(J^t\lambda)$ vanishes in the final form of the iteration matrix 10.20, as a consequence of the linear combinations 10.14 and 10.16.

The last matrix involved in the expression of $\Im^i$ is the coupling matrix B_{vu}. Since ROBOTRAN recursively computes the constraint Jacobian matrix $J = \begin{pmatrix} J_u & J_v \end{pmatrix}$ (see section 4.5), the computation of B_{vu} is simplified by the symbolic approach. Moreover, on the basis of the multibody topology (i.e. the filiation of bodies and the definition of the kinematic loops), ROBOTRAN can detect the sparsity of the coupling matrix B_{vu}. Given a kinematic loop of bodies for instance, the corresponding constraints between variables involve the joint coordinates included in the loop only: this sparsity is exploited during the symbolic computation of the iteration matrix without any difficulty.

These considerations allow us to conclude the following:

- Since the elements of the iteration matrix are generated symbolically by ROBOTRAN, their computation is more efficient and certainly more reliable than the purely numerical derivation generally adopted.

- Since the iteration matrix is of size $(n - m)$, it is much smaller than the one used by direct integration methods which deal with the unreduced DAE system whose size is $(n + m)$.

[9]We shall see that a given iteration matrix can be "frozen" for several iterations, or even several time steps.

Computation of the residual vector

While the iteration matrix $\Im^i$ (eq. 10.20) may be approximated, the residual vector R^i_{n+1} (eq. 10.6) may not; since it is at the root of the convergence test 10.24. Its exact computation is thus required at each Newton/Raphson iteration of each time step, including the determination of the dependent quantities $v^{i+1}_{n+1}, \dot{v}^{i+1}_{n+1}$ and $\ddot{v}^{i+1}_{n+1}$.
For these dependent variables, all ingredients needed to compute the constraints 10.5 are generated by ROBOTRAN in a recursive manner for several kinds of kinematic loops (see section 4.1): the program provides the symbolic form of the algebraic constraints themselves $h(q)$, the Jacobian matrix $J(q)$ and the term $\dot{J}\dot{q}$. The possible sparsity of the matrix B_{vu} (see above) is also exploitable within step 4 of the iteration process to avoid useless matrix operations.
For the computation of the residual vector itself, we observe that it is obtained from the *inverse* dynamics of the multibody system in its implicit form (ex.: equations 9.94 for MBS with flexible beams). Inverse dynamics complexity being $\mathcal{O}(n)$, its recursive computation is straightforward (equation 3.42 of section 3.3).

10.3 General algorithm - optimization strategy

10.3.1 General algorithm

On the basis of the previous developments, this integration scheme was implemented in MBSOFT [23] according to figure 10.1 as a numerical extension of the symbolic program ROBOTRAN. The figure presents the general algorithm in its basic version. The time step control strategy is based on test 10.24 for the Newton-Raphson convergence and 10.25 for the local truncation error.

10.3.2 Numerical optimization

As previously suggested (see section 10.2.4), we should consider the possibility of keeping a given iteration matrix $\mathcal{J}$ (equation 10.20) for more than one Newton/Raphson iteration or even one time step, especially when, as in the present case, the estimation of this matrix is costly.

Experiences have shown that *freezing* the iteration matrix $\mathcal{J}$ for all the Newton/Raphson iterations i of a given time step $t_n \rightarrow t_{n+1}$ is often successful (see the example of section 10.4): this led us to extend it to several time steps as proposed in figure 10.2:

1. a counter i_f is introduced in order to limit the use of a frozen iteration matrix which would no longer be suitable, i.e. leading to slow convergence ($i > Iter_{jac}$) or even divergence. This counter i_f works independently of counter i which gives the number of Newton-Raphson iterations performed with the correct matrix.

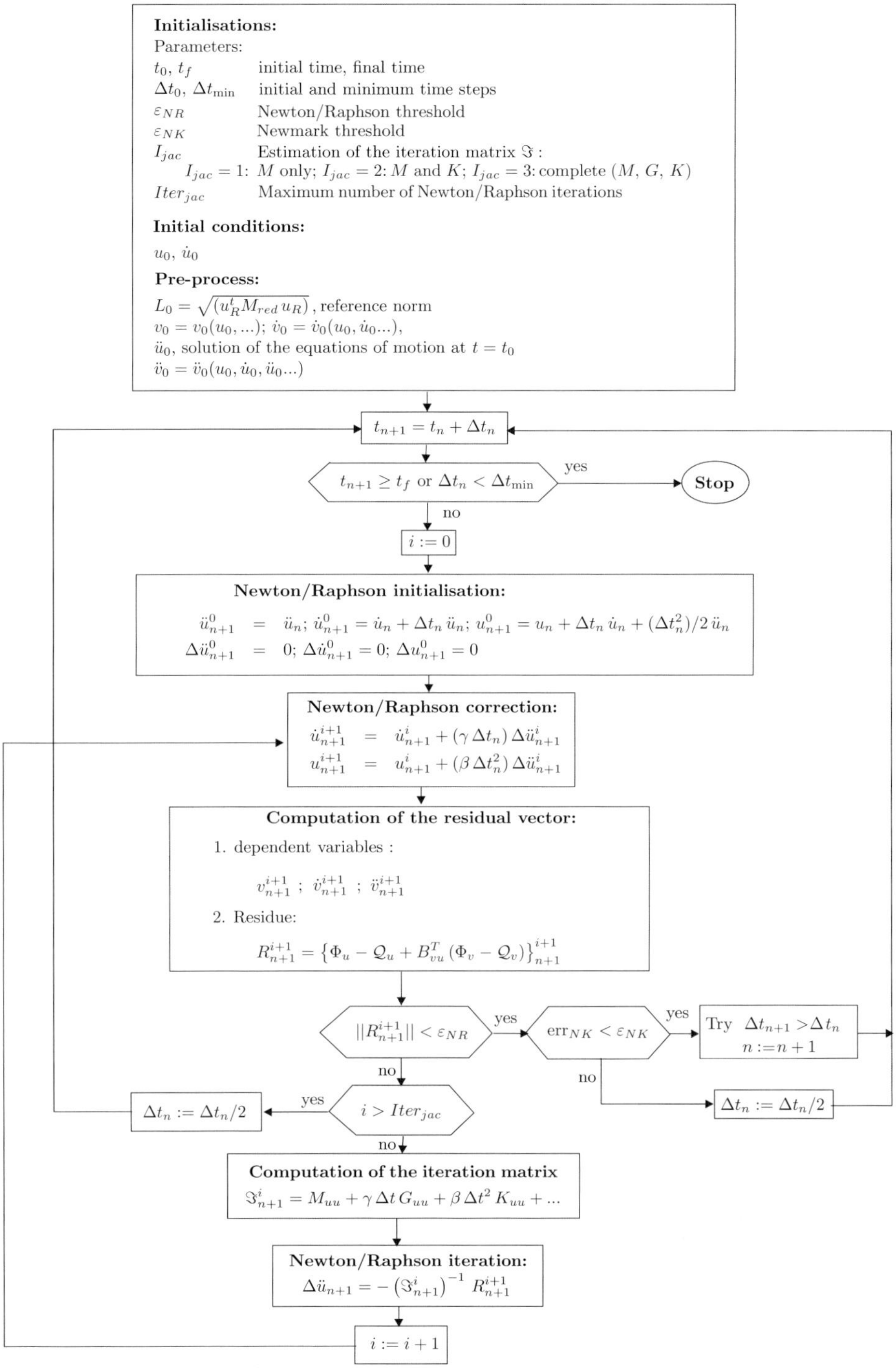

Figure 10.1: The general algorithm

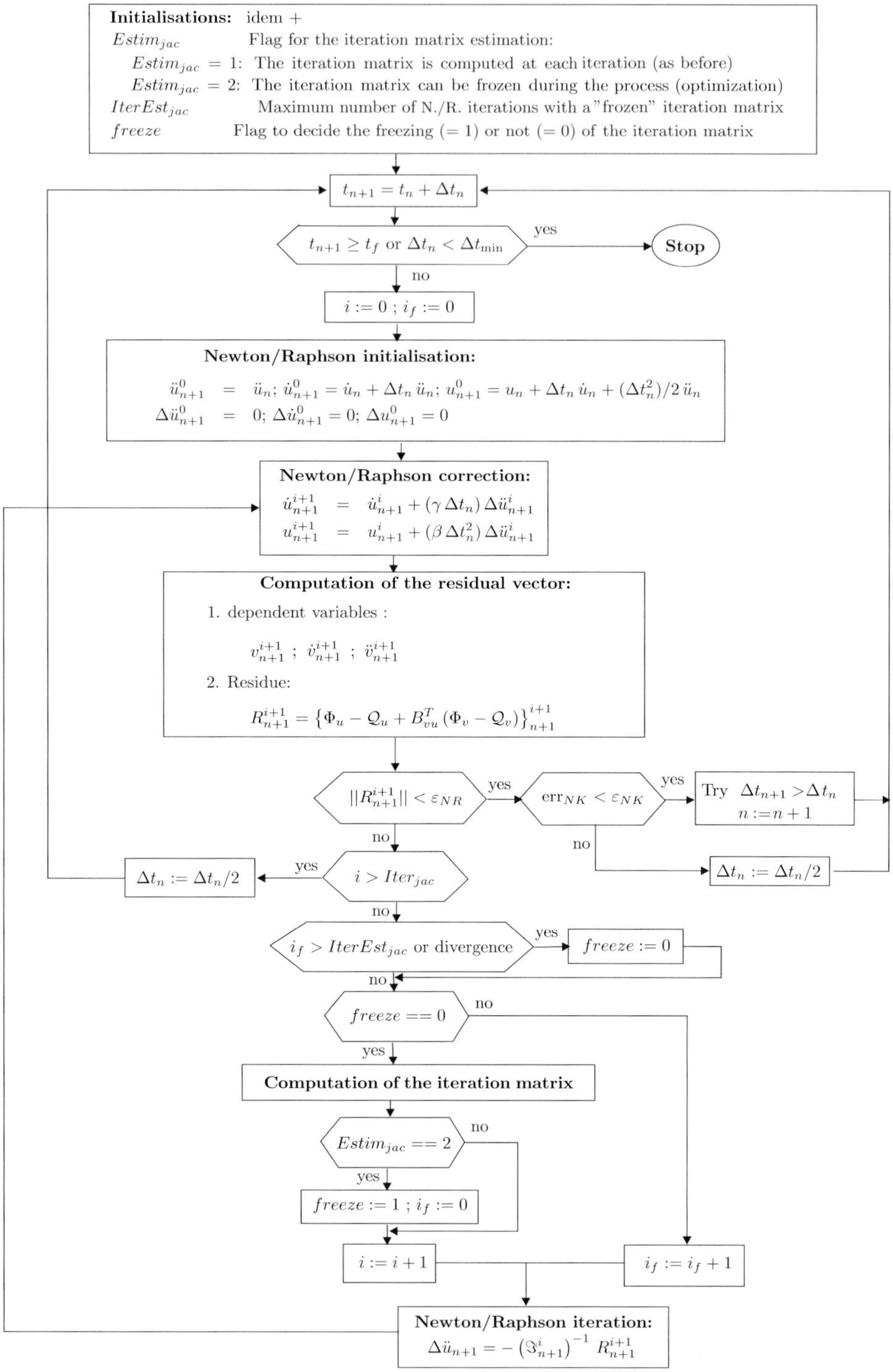

Figure 10.2: The optimized algorithm

2. if the tests 10.24, 10.25 are satisfied and if $\mathcal{J}$ is in a "frozen" state ($freeze = 1$), the iteration matrix $\mathcal{J}$ stays in this state for the next time step ($freeze = 1$). The algorithm then attempts to use it as long as possible, until the counter i_f reaches the maximum value fixed by the user ($IterEst_{jac}$).

10.4 Numerical Example

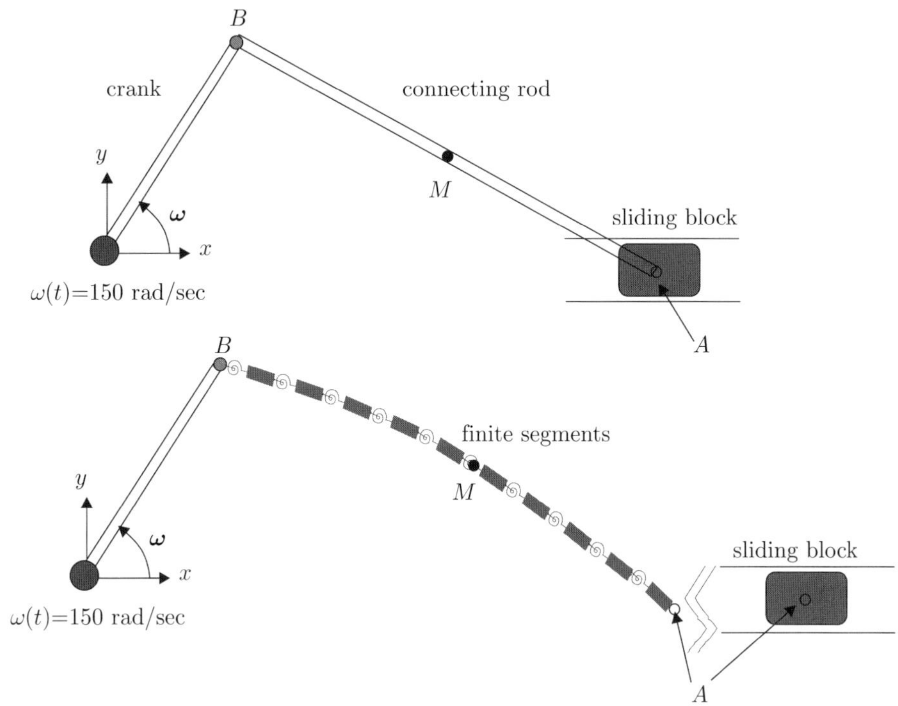

Figure 10.3: A planar flexible slider-crank mechanism

The following slider-crank mechanism has been investigated in several contributions (see e.g. [54]): it will allow us to validate our results, although the main goal of this section is to illustrate the appeal of the proposed method with respect to classical integration schemes.

As depicted in figure 10.3, the system consists of a rigid crank with length $0.15\,m$, a deformable connecting rod and a rigid sliding block with half the mass of the connecting rod. The rod is $0.3\,m$ long and has a circular cross section with diameter $d = 6\,mm$. Young's modulus is $E = 0.2 * 10^{12}\ N/m^2$ and the mass density is $\rho = 7.87 * 10^3\ kg/m^3$. The crank is driven with a constant angular velocity $\boldsymbol{\omega} = 150\,rad/\sec$. The system is frictionless, and gravity is neglected.

Since this section is not directly related to the modeling of flexible bodies, the flexibility of the connecting rod is simply modeled by 10 rigid finite segments interconnected by equivalent flexural springs (see figure 10.3) according to [42].

In order to express the loop constraint, the mechanism has been cut at joint A which links the sliding block to the tip of the connecting rod (see figure 10.3) and the loop closure is imposed at this joint: since the mechanism is planar, this leads to two kinematic constraints along the x and y directions.

10.4.1 Validation

The following table presents the modal analysis of the connecting rod (the crank being locked: $\vartheta = 0$); there is good agreement at least for the first three eigenfrequencies.

Mode	ROBOTRAN model	Analytical solution
1	131.96 Hz	131.98 Hz
2	527.57 Hz	527.91 Hz
3	1185.57 Hz	1187.79 Hz
4	2101.69 Hz	2114.45 Hz

Let us first compare the numerical results with those proposed in [54] in which the FEM technique is used to model the rod.

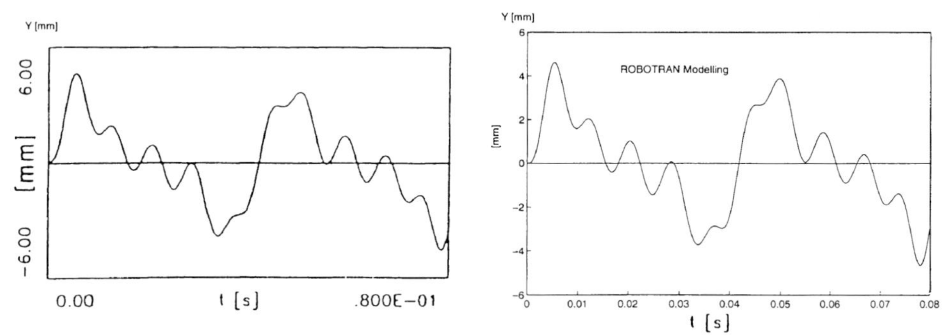

Figure 10.4: Transverse deflection of the midpoint M (left: from [54]; right: from ROBOTRAN results)

The time simulation is performed over a period of 0.08 *sec*. The initial conditions are such that $\vartheta = 0$, the three bodies being aligned with the x direction. Figure 10.4 shows the time history of the transverse deflection at the midpoint M of the connecting rod. Figure 10.5 shows the shortening of the rod, i.e. the time history of the distance between joints A and B. Both results highlight the geometrical nonlinear effects (see [54] for more details). As regards the rod shortening, the differences are due mainly to the longitudinal elasticity of the rod (considered in [54]) which was not taken into account in the finite segment model.

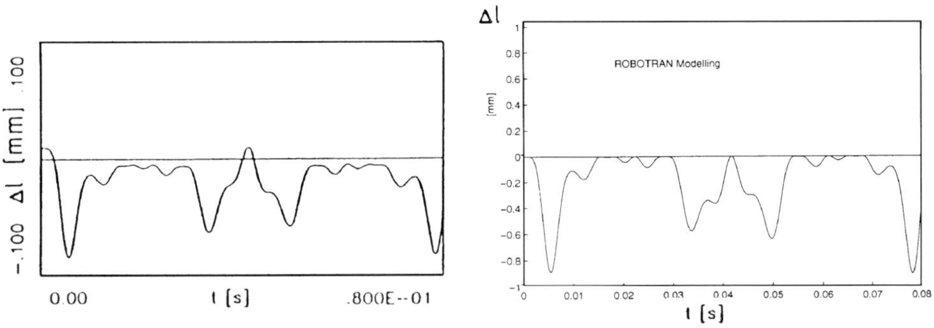

Figure 10.5: Shortening of the rod (left: from [54]; right: from ROBOTRAN results)

10.4.2 Evaluation of the proposed method

The previous slider-crank model is now used to compare the Newmark implicit method (denoted as "NKI" in figure 10.6) with an explicit integration scheme ("RK4") that uses a 4th order Runge-Kutta formula.

Let us first summarize the parameters used (or even requested by) in these two methods.

The "NKI" method:

$$\begin{array}{rr} \Delta t_{min} = 1e^{-6} & I_{jac} = 2 \\ \Delta t_{\max} = 1e^{-1} & Iter_{jac} = 4 \\ \varepsilon_{NK} = 1e^{-3} & Estim_{jac} = 2 \\ \varepsilon_{NR} = 1e^{-4} & IterEst_{jac} = 7 \end{array}$$

The "RK4" method:

$$\begin{array}{rr} \Delta t_{min} = 1e^{-8} & \varepsilon_{relative} = 1e^{-3} \\ \Delta t_{\max} = 1e^{-3} & \varepsilon_{absolute} = 1e^{-5} \end{array}$$

As shown on figure 10.6_{left}, the numerical results are equivalent. However, this is not the case for the CPU time requirements as shown in the next table:

"NKI"	"RK4"
$1.9e+4\ sec$	$1.7e+5\ sec$

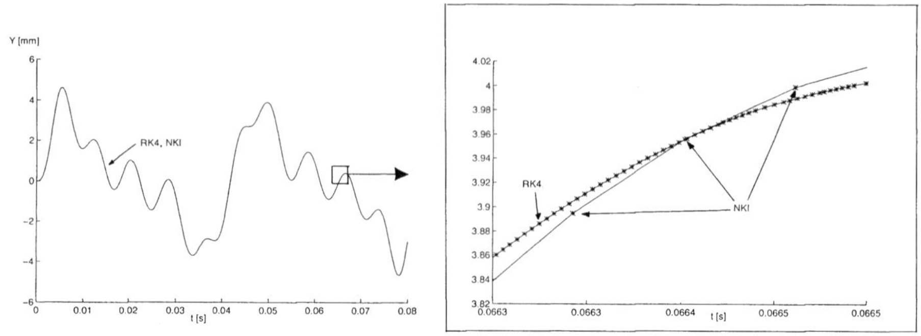

Figure 10.6: Comparison of the time step sizes

This can be easily explained by zooming in on a portion of the simulation (e.g. from 0.0663 sec. to 0.0665 sec. - see figure 10.6_{right}): the "NKI" method operates correctly as a low pass filter, allowing larger time steps than the other method, for equivalent error tolerance.

Figure 10.7 shows the evolution of the norm of the constraints (at position level) for the "NKI" method; note that the algebraic constraints are rigorously satisfied at each time step thanks to the coordinate partitioning method (equation 10.21).

Figure 10.8 represents the time history of the transverse acceleration of mid-point M: despite high fluctuations, no instability appears such as those that arise when using the Newmark scheme directly on the DAE system.

Figure 10.9 gives the number of Newton/Raphson iterations required by the "NKI" method, and figure 10.10 the number of time steps covered by a frozen iteration matrix: this clearly shows the benefit of the optimized algorithm depicted in figure 10.2.

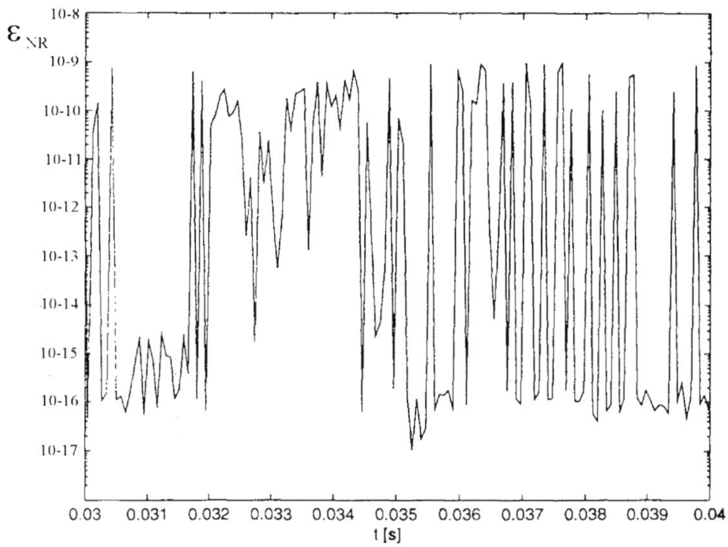

Figure 10.7: Constraints solution

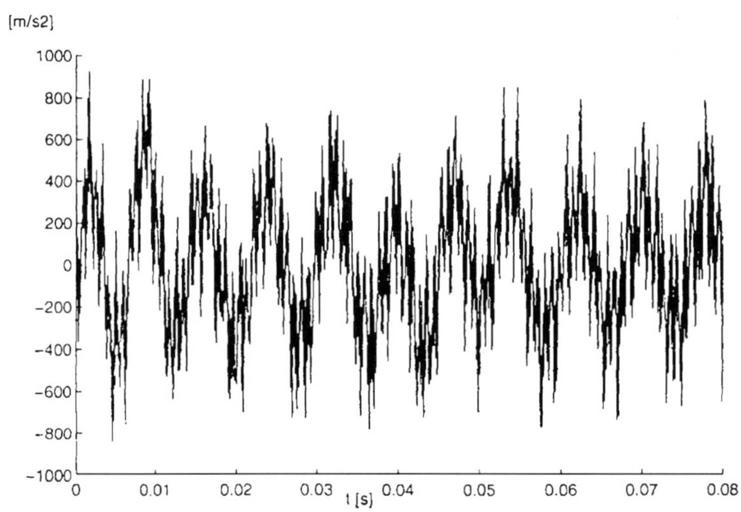

Figure 10.8: Midpoint transverse acceleration

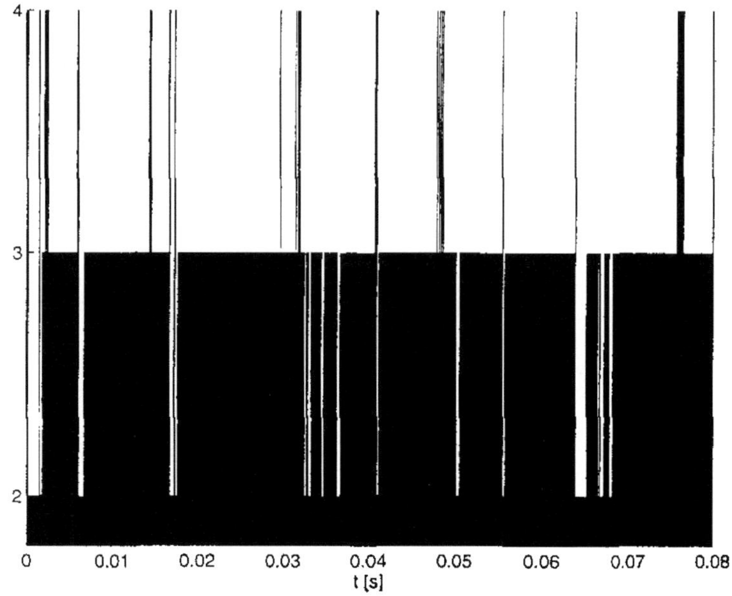

Figure 10.9: Number of Newton-Raphson iterations

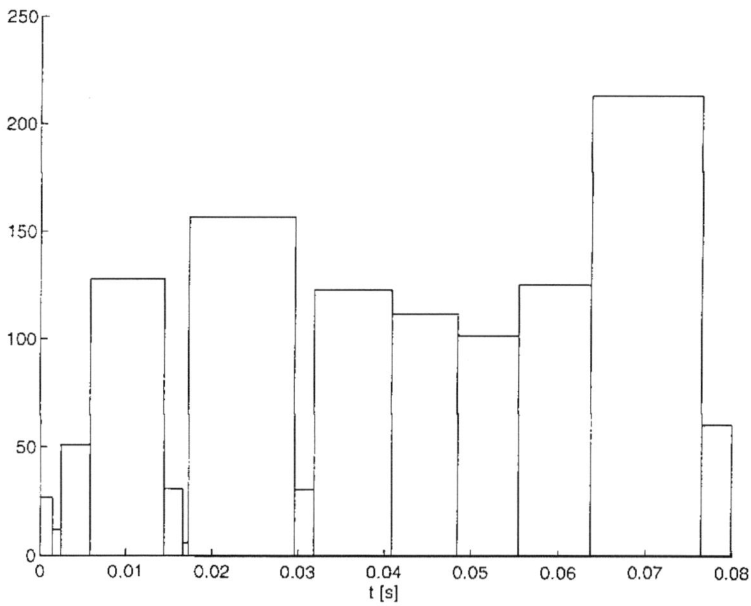

Figure 10.10: Number of time steps with a "frozen" matrix $\mathcal{J}$

Part III

Tutorial

Chapter 11

Introduction

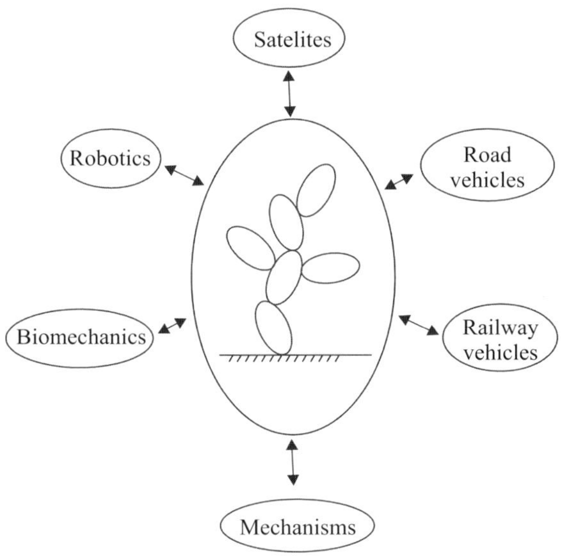

Figure 11.1: Families of multibody systems (examples)

The present tutorial mainly focuses on the modeling process (from system hypotheses to simulation results), rather than on program capabilities, which is more the subject of a user's manual. MBSOFT [23] to which this tutorial refers is the numerical complement of the ROBOTRAN program (see chapter 5) in the MATLAB environment. The objective of these softwares is the modeling (ROBOTRAN) and analysis (MBSOFT) of multibody systems, defined as being any mechanical system made up of articulated bodies. As can be seen from figure 11.1, the fields of application are numerous and varied: they range from robotics to biomechanics, and encompass the analysis of road and track vehicles.

One should keep in mind that one of the main advantages of the approach is its capacity to generate equations of motion in *symbolic* form. This symbolic form is both compact and user-friendly, thereby allowing the user to easily understand and set up his model. The symbolic equations are then implemented in the numerical study which makes up the second processing stage, viz. the MBsim module (see figure 11.2).

The current tutorial is intended to familiarize the reader with new concepts required to carry out a specific study. It is made up of a series of examples of gradually increasing complexity, each example introducing one or more new concepts.

The best way for a first-time user to take full advantage of the tutorial is to read and solve the problems in the suggested order. A user who is already familiar with the modeling of multibody systems can proceed directly to the specific example containing the required new concept.

> rem.: The various theoretical notions, as well as all definitions, chosen terminology etc. are given in detail in the previous Parts of the book and will therefore not be systematically expanded here, except when considered necessary for the understanding of the problem.

11.1 Methodology

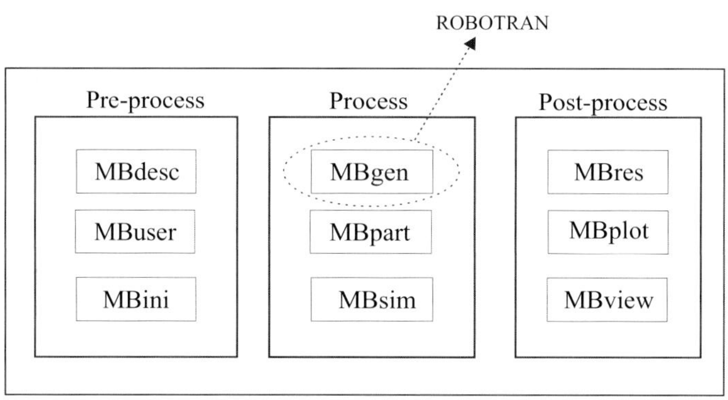

Figure 11.2: MBSOFT modules: 1. Pre-process ; 2. Symbolic/Numerical processes ; 3. Post-process

Throughout the current tutorial, we shall be attempting to inculcate a methodology to efficiently approach a new problem and express it using our formalism so as to make it amenable to treatment via a numerical program like MBSOFT, whose various modules are illustrated in figure 11.2.

In order that all aspects of a new problem be properly and rigorously covered, we propose the following methodology, at all times:

We state the problem within a box, with a clear definition of the expected results.

Originality of the problem: this heading announces the new features which is introduced in the current problem. This is how the user can choose the most relevant problems in preparing his own application.

11.1.1 Analysis

Description of the system

In this first stage, a maximum amount of information on the system must be gathered, as to:

- The number of bodies involved
- The physical characteristics of these bodies (masses, inertias, ...)
- Types of joints (e.g.: Cardan joint, ball joint, ...), number of allowed degrees of freedom (noted as "d.o.f."), preliminary analysis of possible motions and dynamical characteristics
- Kinematic constraints (2-D system, closed-loop structure, ...)
- Total number of degrees of freedom of the multibody system
- List of prevailing forces and torques (both internal and external, gravity)
- Analysis of the environment (road profile, contact interfaces, ...)

Modeling hypotheses

- Determination of all parameters which have little or no influence in the intended simulations (e.g.: spurious d.o.f., body having negligible mass or inertia, frictionless joint, ...)
- Modeling of prevailing forces and torques: setting up of their constitutive equations
- Detailed list of expected results so as to complete the model (e.g.: choice of a sensor, determination of a specific force, parameterization)

Multibody model

From here on, we shall be entering the ROBOTRAN multibody formalism: we choose a "potato" graphical representation as typified by that of figure 11.3 (which is drawn from one of the example problems). This gives an idea of how the bodies relate to one another.

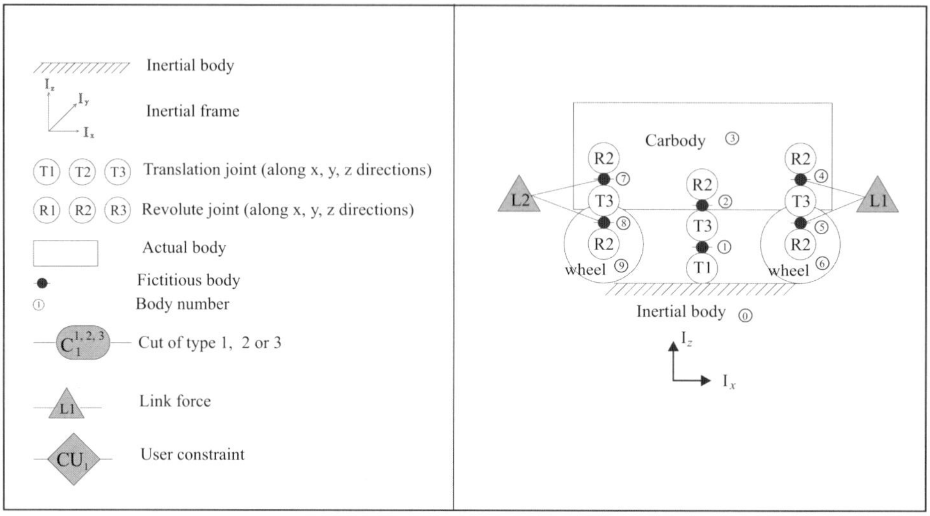

Figure 11.3: Key words and "potato" graph

- An inertial reference frame is chosen, which conforms to the underlying assumptions of some of the specific multibody models (in the case of wheeled mechanisms or cam systems for instance).

- A local reference frame is associated with each body (with axes generally aligned along the principal axes of inertia).

- All complex joints are broken down into a succession of 1 d.o.f. joints linking fictitious bodies (see Part I, chapter 5).

- Closed-loop analysis: choice of cuts (type and location) needed to recover a tree-like structure (see Part I, section 4.1):

  - cut of a body: denoted as a cut of type 1 (C^1 cut)
  - cut in a ball joint: denoted as a cut of type 2 (C^2 cut)
  - cut of a connecting rod: denoted as a cut of type 3 (C^3 cut)

- Bodies are numbered.

- A reference configuration is chosen. The "potato" graph is drawn in this configuration.

- Bodies carrying non-articular forces/torques are identified.

11.1.2 Program run

The nine modules of the MBSOFT program illustrated here are briefly described in this section because they represent the main steps to model and analyze multibody systems. We will systematically refer to these steps in all problems of the tutorial.

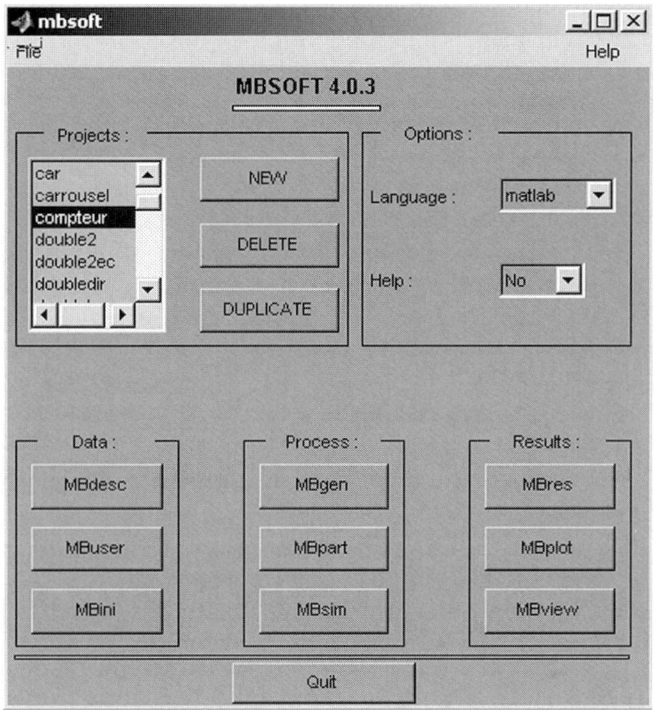

Figure 11.4: MBSOFT: the main window

Preprocessor: *MBdesc*, *MBuser*, *MBini*

The three modules *MBdesc*, *MBuser* et *MBini* were developed in order to input data and assist the user in customizing his model to the targeted application.

***MBdesc*: logical set of dialogue boxes allowing the introduction of data pertaining to:**

- Bodies and joints (in the reference configuration)
- Kinematic constraints (loop-, "user"-, inside or inter sub-system type)
- Constraint forces between bodies (other than joint forces)

- Kinematics of a sensor or of any body
- The wheels of a vehicle-type multibody system
- Cam/follower type contacts

Both the symbolic data (for MBgen) and the numerical data (for MBsim) are generated at this stage.

***MBuser*: the user creates and fills problem-specific files pertaining to:**

- Kinematically driven variables (see chapter 4, section 4.2.2)
- Joint forces
- "User" forces
- "User" constraints (see chapter 4, section 4.2)
- Wheel/ground contact models (see chapter 6)
- Road profiles
- Cam and follower profiles (see chapter 8)

***MBini*: Introduction of numerical initial conditions pertaining to**

- Joint variables
- "Cam/follower"-type contacts

Processor: *MBgen*, *MBpart*, *MBsim*

***MBgen:* Symbolic treatment:** After checking the symbolic data generated by MBdesc (number of sub-systems, number of bodies, of links, ...), the symbolic equation generator MBgen[1], takes a few seconds to turn out the symbolic model required by the specific application, viz. equations of motion, constraint equations, wheel kinematics, etc. It should be emphasized that MBgen is run only once for any given application, unless, in the course of a study, the user wishes to add bodies or input quantities (e.g. a vector component) which did not previously exist and were therefore equal to zero in the original model.

[1] which is in fact a module of ROBOTRAN

***MBpart*: Coordinate Partitioning** According to the coordinate partitioning method (see chapter 4, section 4.6.1), the joint variables q of any constrained multibody system must be partitioned into independent (u) and dependent (v) variables, on the basis of the full set of kinematical constraints. In order to perform this partitioning, the user has two possibilities: either by selecting himself the sub-set of independent variables, or by asking the program to do it on the basis of the constraint Jacobian numerical conditioning.

Beside this partitioning, the user must also indicate in MBpart the indexes of the kinematically driven joint variables (which obviously cannot be dependent), whose value ($f(t)$) is introduced via the MBuser module, in a specific user function.

***MBsim*: Numerical treatment** The numerical data (MBdesc), the application specificities (MBuser), the initial conditions (MBini), the symbolic model (MBgen) and the coordinate partitioning (MBpart) are fed into the numerical model MBsim which then performs the numerical analysis requested by the user. After checking the coherence of the choice of variables (independent, dependent and kinematically driven), the number of degrees of freedom and the possible redundancy of certain constraints, MBsim then offers the user a choice of various numerical treatment menus, such as:

- Kinematics
- Direct or inverse equilibrium
- Modal analysis
- Time simulation

Post-processor: *MBres, MBplot, MBview*

Three modules are available to the user:

1. *MBres* is an interface which allows the user to visualize result files provided by MBsim. These files display the results as documented columns and tables.
2. *MBplot*: provides (x, y) graphs of the various results (or combinations thereof) provided by MBsim.
3. *MBview*: generates a graphical animation of the multibody system from:
   - a 3-D graphical representation of the non fictitious bodies of the multibody system
   - the topology of the multibody system (filiation, location of the joints),
   - numerical results provided by MBsim (sequence of different configurations (as a result of a modal analysis or a time simulation).

11.2 Problem statements

We propose eleven exercise problems ordered so as to progressively and gradually familiarize the user with more and more complex features. With this in mind and so as to allow the user to directly start with problems corresponding to his level of expertise, we shall now briefly summarize their contents.

11.2.1 Problem 1: a double spring-mass system

The first-time user is introduced to the multibody approach and vocabulary via a simple example. He also learns how to input joint forces.

11.2.2 Problem 2: a merry-go-round

The novelty in this case is the modeling of a multi-d.o.f. joint and the time control of a joint variable. The user therefore needs to time integrate the equations of motion. The concept of a sensor is also introduced.

11.2.3 Problem 3: a small cart

Here we learn how to introduce a tired wheel into a model, including the representation of the road profile. We also learn how to model a simple joint whose axis is not aligned with one of the axes of the reference frames associated with the connected bodies.

11.2.4 Problem 4: a slider-crank mechanism

This is the first problem requiring the introduction of kinematic constraints. It is also the first example of what we call a "kinematically driven system" (i.e. a system which has no independent variable, thus no degree of freedom).

11.2.5 Problem 5: small cart 2

The objective of this problem is threefold; first, internal links are introduced between two bodies of the system, then a sensitivity analysis of a parameter is performed as to its effect on the equilibrium configuration and the vibrational modes. Finally, an animation of some of the cart modes is attempted.

11.2.6 Problem 6: a five-point suspension

This is the first 3-D problem of the series. Among the new features introduced: a static analysis, the inverse equilibrium and the kinematics of a given body.

11.2.7 Problem 7: a jeep suspension

As was the case of the previous problem, this one corresponds to an actual system. It finds its place here since it is quite comprehensive, as it features links, kinematic constraints and the possibility of being broken down into sub-systems. This problem features in an internationally recognized benchmark for multibody system softwares, which makes the exercise of solving it all the more interesting.

11.2.8 Problem 8: a jeep

This problem is in fact the above-mentioned benchmark and is intended for users studying systems consisting of a large number of bodies. It allows us to see how a system can be split into several sub-systems. The jeep is modeled as 4 x [a quarter chassis + suspension] to make up the complete vehicle: the four quarter-vehicles are assembled via inter-sub-system constraints. The problem also features a more detailed description of a road profile.

11.2.9 Problem 9: a flexible slider-crank

The main feature of this problem is the artificial introduction of a flexible body. We show how to replace a beam by rigid elements connected through springs which amount to an "equivalent flexibility". The problem also allows a discussion as to the consequences of "stiff" problems (i.e. featuring eigenfrequencies of largely different orders of magnitude) on the numerical treatment.

11.2.10 Problem 10: a radiation counter

The radiation counter allows us to show how a model can be extended to meet the particular needs of the user. A "user constraint" is introduced but is not processed automatically by the program: the user must therefore include it in his model.

11.2.11 Problem 11: a "cam/follower" device

To model systems featuring contacts between shaped bodies, the "cam/follower" model of chapter 8 has been developed within MBSOFT: this problems highlights its performance in an elementary application. Two shapes are compared, both from a kinematic and a dynamical point of view.

Chapter 12

Problems

12.1 A double spring-mass system

The objective is to determine the equilibrium configuration as well as the eigenfrequencies of a double spring-mass system where both masses are constrained to move in a vertical plane.

Originality of the problem: The main objective of this elementary exercise is to familiarize the user with the methodology presented in the introduction. Generalized joint forces $\mathcal{Q}(q)$ will also be introduced.

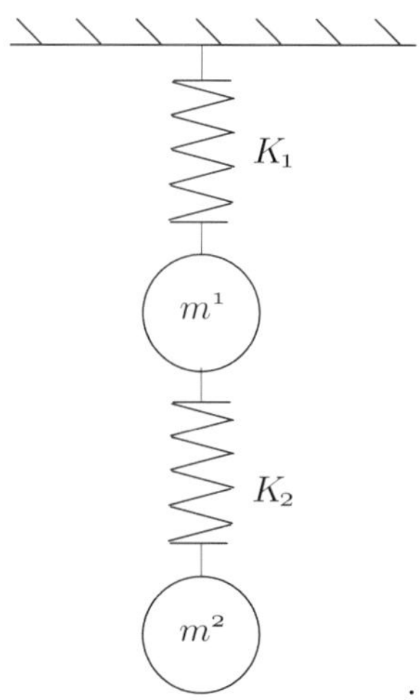

12.1.1 Analysis

Description of the system

- Number of bodies: 2, both with 1 kg mass

- Joints: two vertically directed prismatic joints.
- Degrees of freedom: 2.
- Forces and torques: elastic spring forces
- Environment: the system is placed in a gravity field

Modeling hypotheses

- The bodies are assumed to be point masses.
- Spring constitutive equations: assumed to be linear, with stiffness $K = 100\,N/m$, and zero unstrained length.

12.1.2 Multibody model

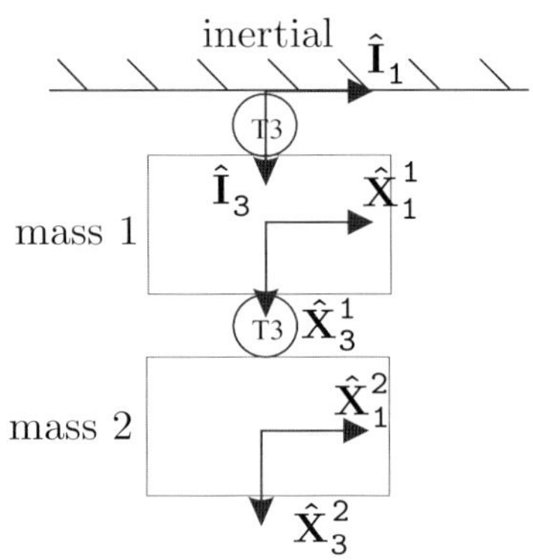

- Choice of inertial reference frame: the origin is chosen at the anchorage point of the first spring on the inertial body with a downwards $\hat{\mathbf{I}}_3$ direction.
- Body-fixed frames: in accordance with the multibody formalism, a reference frame is associated with each body.
- The depicted configuration corresponds to the reference configuration: in this configuration, all body frames must be aligned with the inertial reference frame.
- Numbering of joints:
  - Joint 1: $T3$ prismatic joint relating to strain of first spring
  - Joint 2: $T3$ prismatic joint relating to strain of second spring
- Numbering of bodies: a body following a joint conventionally carries the same number as that joint:
  - Mass 1 = body 1
  - Mass 2 = body 2

12.1.3 Computer pre-process

System data (MBdesc)

As our formalism uses only single-degree of freedom joints (later we shall see how to model multi-degree of freedom joints), there are always as many bodies as joints. The first item of data to input for any body is its parent, viz. the body located upstream in the tree-structure of the system. After having input the names of the body and joint, as well as the latter's degree of freedom, vector $\mathbf{d}^{ii}$ $(d_1^{ii}, d_2^{ii}, d_3^{ii})$ then specifies the position of the body center of mass relative to the associated joint. Vector $\mathbf{d}^{ij}$ $(d_1^{ij}, d_2^{ij}, d_3^{ij})$ specifies the position of the body joint relative to the preceding joint (on the parent body).

Advantage can be taken of the fact that some quantities are equal to zero or approximated as such, by setting their symbolic value[1] to zero (in the present exercise, since the masses are considered to be dimensionless, the corresponding $\mathbf{d}^{ii}$ and $\mathbf{d}^{ij}$ vectors are zero, along with the inertia matrices). Therefore, ROBOTRAN (MBgen module) will generate the equations with many symbolic simplifications, notably by bypassing all zero terms or factors.

User's model (MBuser)

To introduce specific constitutive equations of forces and/or prescribed laws for driven joint variables, the user must fill "user's functions" for which input/output arguments are clearly defined. For a MBSOFT application, the corresponding files are located in the project user directory. In order to input the joint forces Q^i caused by the springs, the "user's function" must contain the following statements (MATLAB syntax using arrays):

```
Q(1) = - 100 * q(1);
Q(2) = - 100 * q(2);
```

Considering that q^i characterizes the position (orientation) of body i relative to the preceding body, that Q^i is the joint force (torque) exerted along the joint axis on body i by its parent, it is easy to understand that the above two lines relate to the restoring force of a spring with stiffness 100 N/m and zero neutral length.

In the current case, variable q^1 represents the displacement of mass 1 relative to the inertial body, whereas q^2 represents the displacement of mass 2 relative to mass 1. This serves as a reminder that we use relative coordinates: the position/orientation of a body is always taken relative to its parent.

Initial conditions (MBini)

Whatever the numerical treatment, we must input initial conditions for the joint positions q^i and possibly for the joint velocities $\dot{q}^i$: in the present case, we simply set all q^i and $\dot{q}^i$ to zero. This amounts to choosing a starting configuration where the two masses coincide with the origin of the inertial reference frame.

[1] and their numerical value, automatically

12.1.4 Computer process

Symbolic treatment (ROBOTRAN - MBgen)

After having generated the data files and the user's models, we can then activate the ROBOTRAN program (MBgen module) so as to generate the equations of motion in symbolic form, namely the mass matrix $M(q)$ and the vector $c(q, \dot{q}, ...)$ of equation 3.43 (menu: Direct models - Tree-like - Direct dynamics - NER; see figure 5.17). This entails the creation of files (in MATLAB syntax) which appear in a sub-directory of the project. The user should never have to modify these symbolic files.

Coordinate partitioning (MBpart)

Before starting the numerical process, we must partition the joint variables q into independent (u) and dependent (v) ones, on the basis of the set of kinematic constraints. Since only implicit constraints arise here (resulting from the choice of relative coordinates), there are no – explicit – constraints on q_1 and q_2 which are thus both independent variables.

At this level also, possibly driven joint variables ($q = f(t)$) must be indicated (none in this problem).

Numerical treatment (MBsim)

The numerical treatment, which will produce the requested results, is now ready to be run[2]. We first seek to determine the equilibrium configuration of our double spring-mass system subjected to gravity. Equilibrium can be obtained using the Newton-Raphson iterative method (in the same way as for the kinematic loop closure, see equation 4.49) applied to equation 3.43 in which $\dot{q} = \ddot{q} = 0$:

$$\text{find } q_{equil} \text{ such that } c(q_{equil}, ...) - \mathcal{Q}(q_{equil}, ...) = 0.$$

Since the Jacobian $\frac{\partial c}{\partial q^T}$ is not available analytically, it is approximated numerically using a finite difference technique.

Around the system equilibrium, a modal analysis then allows us to determine the eigenvalues of the system: this process requires two consecutive steps:

1. linearization of the equations around the previously found equilibrium position (by numerical means)

2. eigenvectors and eigenmodes evaluation.

[2] It is obvious that the numerical methods mentioned hereunder are not indispendable to solve this elementary problem for which analytical solutions exist for equilibrium and modal analysis. This first exercise is mainly proposed to illustrate the methodology.

12.1.5 Computer post-process

The boxes below display the requested numerical results. Further details (e.g. eigenvectors corresponding to the various eigenvalues, nature of the mode) can be obtained by sorting them properly and by observing the eigenvector components expressed in polar form for instance (amplitude, phase).

Equilibrium configuration

Mass 1 relative displacement: $q^1 = 0.1962\,m$
Mass 2 relative displacement: $q^2 = 0.0981\,m$

Modal analysis

$\nu_1 = 0.98\,Hz$ (in-phase mass motion)
$\nu_2 = 2.58\,Hz$ (anti-phase mass motion)

12.2 A merry-go-round

> One objective is to determine the slant angles of the arms of a merry-go-round which is rotating at an angular speed of a quarter revolution per second.
> A further question is the determination of the maximum norm of the acceleration that a passenger will undergo in a nacelle of the merry-go-round when the speed ramp of the device is given.

Originality of the problem: This problem allows us to introduce the concept of driven variables and the modeling of multi-degree of freedom joints. The concept of a kinematic sensor is a further novel feature. Moreover, this problem allows us to stress the important influence the choice of initial configuration can have on the determination of the equilibrium configuration.

12.2.1 Analysis

Description of the system

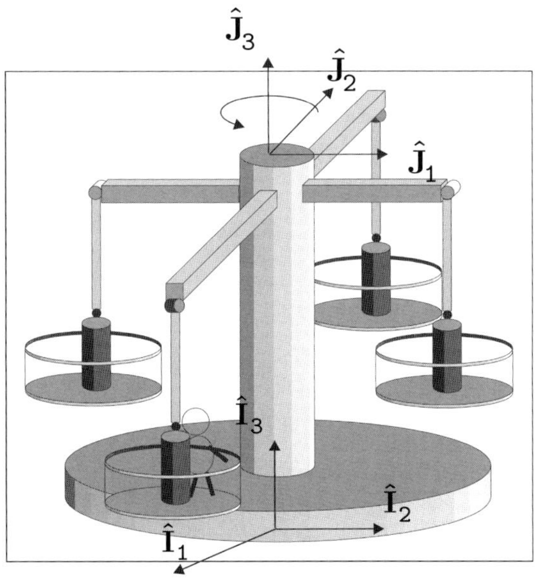

- Number of bodies: $9 + 1$ inertial:
  - the merry-go-round base (inertial),
  - the central pole, including the horizontal beams (fixed relative to the central pole),
  - the four arms,
  - the four nacelles.

- Physical[3] joints: 9
  - vertical rotation between the central pole and the base,
  - simple R-joint between the beams fixed to the central pole and the four arms,
  - ball joint (rotations being possible in all three directions) between arms and nacelles.
- Degrees of freedom: 17. One degree of freedom follows from each beam/arm joint, three from each arm/nacelle joint, and one is related to the central pole rotation.
- Forces and torques: viscous forces in the joints between the central pole and the four arms as well as between arms and nacelles.
- Environment: the system is in a gravity field.

Modeling hypotheses

- It is assumed that viscous forces in the joints can be modeled as velocity-linear damping.
- It is assumed that a nacelle cannot rotate around its axis of symmetry.
- The passenger's mass is neglected.
- The speed ramp is given by a mathematical law (see below) which provides a good approximation of the performance of the drive motor.

[3]means that they denote the actual joints which can have up to 6 d.o.f. depending on the involved mechanical device.

12.2.2 Multibody model

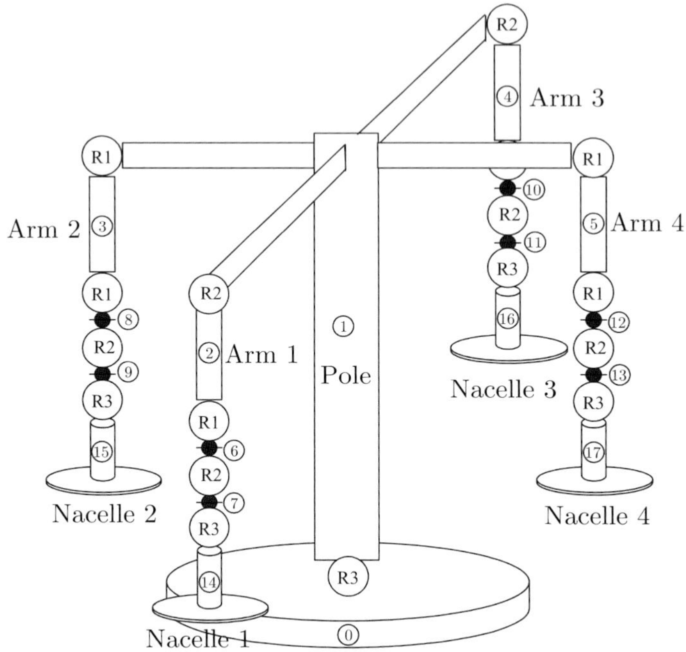

- Choice of inertial frame: this frame is centered on the central pole, where it is joined to the base (which itself belongs to the inertial body).
- Reference configuration: the reference is that depicted on the above figure. It corresponds to a situation where the arms are aligned with the central pole (i.e. vertical) and the axis of symmetry of the nacelles.
- Body frames: as usual, a reference frame is chosen with its origin at the center of mass of each body. In the reference configuration, all frames are aligned.
- Numbering of bodies and joints: the new feature here is the modeling of the joint between arm and nacelle. This three-degree of freedom joint is replaced by three sequential single-degree of freedom joints (see section 5.8.1). As a body needs to be associated with each joint, we insert so-called *fictitious* bodies, viz. dimensionless bodies having neither mass nor inertia (which means that the successive degrees of freedom are all concentrated at a same point). The single d.o.f. joints are the following:
  - Joint 1: R(revolute)-joint between central pole and base (inertial).
  - Joints 2 to 5: R-joints between each arm and the central pole. As there are four of them and that they are at right angles in the reference configuration, there are two $R1$ type joints and two $R2$ type joints.

- Joints 6 and 7, 8 and 9, 10 and 11, 12 and 13: these joints account for the first two degrees of freedom of each nacelle.
- Joints 14 to 17: these joints allow each nacelle to enjoy a rotational degree of freedom around its axis ($R3$ type).
- Note that any joint is always associated with the body that follows it. The numbers on the figure correspond to those of the bodies and consequently to those of the joints which are directly upstream of each of these bodies (whether they be actual or fictitious bodies).

A *kinematic sensor* is placed on nacelle 14, in the passenger's seat. This simply tells ROBOTRAN to generate the absolute position, velocity and acceleration of that specific point (according to section 4.5), just as if an actual sensor had been placed at that location.

12.2.3 Computer pre-process

System data (MBdesc)

The multibody data can be summarized as follows.

- central pole:
  - height $= 4.5\,m$, length of horizontal beams $= 1\ m$
  - height of center of mass $= 2\ m$, mass $= 600\ kg$
  - inertia $= 16\ kg\ m^2$ (around vertical axis), $454.8\ kg\,m^2$ (around transverse axis)
- Arms:
  - length $= 3\,m$, center of mass $=$ geometrical center
  - mass $= 44\,kg$, inertia $= 0.091\,kg\,m^2$ (around axis)
  - $15.9\,kg\,m^2$ (around transverse axis)
- Nacelle:
  - center of mass: $1\,m$ below the ball-joint, mass $= 123.3\,kg$
  - inertia $= 61.7\,kg\,m^2$ (around axis of symmetry) and $15.4\,kg\,m^2$ (around transverse axis)

We then need to locate the sensor (see section 4.5, figure 4.21 and example 5.9). This requires the number of the body which is supposed to carry the sensor (body 14) as well as its position vector relative to the joint which precedes that body:

$$\mathbf{d}_c = [\hat{\mathbf{X}}^{14}]^T \begin{pmatrix} -0.375 \\ 0 \\ -0.5 \end{pmatrix}$$

and the index of the reference frame, i.e. the frame relative to which we wish to express and compute the kinematics. In the present case, the inertial frame (absolute kinematics) is chosen.

User's model (MBuser)

The joint damping needs to be input into a user function which will contain the following statements (in which qd(i) stands for $\dot{q}_i$).

```
Q(2)=-1000*qd(2);Q(3)=-1000*qd(3);
Q(4)=-1000*qd(4);Q(5)=-1000*qd(5);
Q(6)=-300*qd(6);Q(7)=-300*qd(7);Q(14)=-300*qd(14);
Q(8)=-300*qd(8);Q(9)=-300*qd(9);Q(15)=-300*qd(15);
Q(10)=-300*qd(10);Q(11)=-300*qd(11);Q(16)=-300*qd(16);
Q(12)=-300*qd(12);Q(13)=-300*qd(13);Q(17)=-300*qd(17);
```

We still have to complete the function file that will allow us to control the central pole rotation. In a first phase of the problem, the central pole will be rotated at a steady-state angular speed of a quarter revolution per second. The function describing the driven variables needs to have the following statements:

```
q(1)  = pi*t/2;
qd(1)  = pi/2;
qdd(1)  = 0;
```

When simulating the merry-go-round as a fairground device (second phase), this file will have to feature the control law expressing the imposed speed ramp.

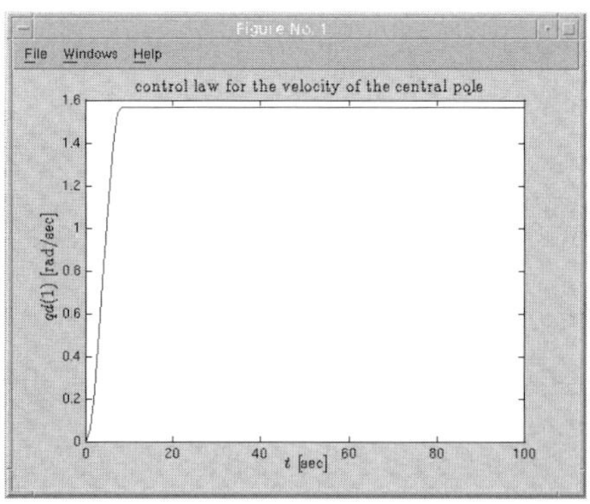

This control law is depicted in the adjoining figure and allows us to spin up the central pole in order to reach its steady-state value ($\frac{\pi}{2}\ \frac{rad}{sec}$) via continuous speed and acceleration laws.

```
if t <= 8
   q(1)=pi/4*t-2*sin(-pi/8*(t-8));
   qd(1)=pi/4*(1+cos(-pi/8*(t-8)));
   qdd(1)=pi*pi/32*sin(-pi/8*(t-8));
elseif t > 8
   q(1)=pi*(t-2)/2;
   qd(1)=pi/2;
   qdd(1)=0;
end
```

Furthermore, in order to comply with the stated hypotheses, we must prevent the nacelles from spinning around their axes of symmetry; this is achieved by setting in this user's function the rotation variables `q(14),q(15),q(16),q(17)` to zero for the numerical integration.

> Note: the latter settings are not necessary when determining the equilibrium configuration. The Newton/Raphson method, when computing the iterative Jacobian matrix, can sense[4] that these so-called *ignorable* variables have no influence on the equilibrium configuration and will automatically set them to zero.

Initial conditions (MBini)

For this application, an a priori initial configuration is chosen such that all variables (q, $\dot{q}$) are equal to zero (= reference configuration).

12.2.4 Computer process

Symbolic treatment (ROBOTRAN - MBgen)

According to the requested results, we must generate the equations of motion (direct dynamics) and the sensor's kinematics (menu: Toolbox - Sensor kinematics).

Coordinate partitioning (MBpart)

As detailed previously, there are five driven variables: q^4 $(= f(t))$ for the central pole and q^{14} through q^{17} $(= 0)$ for the axial rotation of the nacelles. The other variables are independent because there are no kinematic constraints in the system.

Numerical treatment (MBsim)

We will first investigate the importance of the choice of initial configuration in determining the equilibrium configuration, which we shall denote "quasi-static" since it relates to a system which is rotating at constant speed.

[4] in MBsim

For the equilibrium of the system undergoing a constant rotation at a quarter revolution per second, we choose the a priori initial configuration (where all variables, except q^4, are set to zero). After running the equilibrium procedure (Newton/Raphson method with numerical evaluation of the Jacobian), the resulting equilibrium configuration is the following:

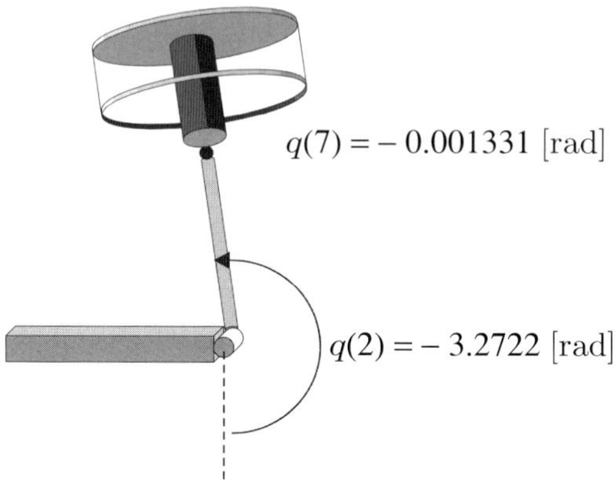

This equilibrium configuration is obviously not the expected one ! This one is unstable. To make sure of this, we can simply perform a modal analysis around the configuration: eigenvalues with positive real parts ensue, revealing the instability of the configuration.

There can indeed be many possible equilibrium configurations. The user should take care to start with initial conditions which are close to the desired equilibrium configuration: this is of course closely related to the iterative nature of the Newton/Raphson method. In the present case, an appropriate initial configuration consists in placing the arms at a 45° angle ($q^2 = q^3 = -\frac{\pi}{4}$; $q^4 = q^5 = \frac{\pi}{4}$). This allows us to determine the values of the joint coordinates q at the requested equilibrium and, in particular, the values of variables 2, 3, 4, and 5 which we are most interested in (see the final solution).
A different approach would have been to simulate (via a numerical integration) the merry-go-round spinning at a quarter revolution per second over a "quite long" period in order to ensure reaching steady state, but this would be much more costly in terms of processing time than the equilibrium procedure (this comment becomes really relevant when dealing with large systems like a full car, a tramway, ...).

With the second set of driven variables (speed ramp), we can now simulate the time motion[5] of the merry-go-round (total time: 100 seconds with a request to save the configuration, e.g. every second). We take the set of initial conditions where all joint variables (including q^4) are equal to zero, which in this case relate well to the merry-go-round at a standstill (allowing passengers to embark !).

The simulation results provide the positions, velocities and accelerations of the joint variables ($q(t)$, $\dot{q}(t)$, $\ddot{q}(t)$). Using these values, from each configuration,

[5] A 4th order Runge Kutta method with controlled time step is proved to be reliable for all the proposed exercises.

we can easily work out the absolute position, velocity and acceleration of the sensor (placed on body 14) whose kinematics was generated symbolically by ROBOTRAN according to the recursive computation of section 4.5.

12.2.5 Computer post-process

As concerns the sensor absolute acceleration, $\ddot{\mathbf{x}}^{sen} = [\hat{\mathbf{I}}]^T \ddot{x}^{sen}$, we can set up the graph of the acceleration norm versus time, taking care to compute the square root of the sum of squares of its three components:

$$||\ddot{\mathbf{x}}^{sen}||\,(t) = \sqrt{(\ddot{x}_1^{sen})^2 + (\ddot{x}_2^{sen})^2 + (\ddot{x}_3^{sen})^2}$$

Equilibrium configuration

Arm angle at constant rotational speed: $0.7014\, rad$

Acceleration norm

Norm of maximum acceleration: $9.75\, m/sec^2$ i.e. approx.1 g

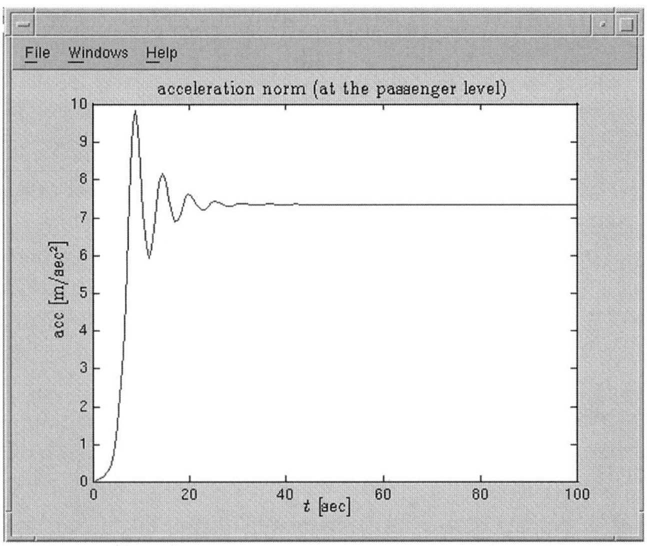

12.3 A small cart

The objective is to determine the equilibrium configuration as well as the eigenfrequencies of a cart at rest.
A further question is the determination of the forces exerted by the ground on the wheels in this configuration.

Originality of the problem: Via this exercise, we shall introduce wheel-type bodies. Moreover, we shall illustrate how to model a joint whose axis is in any given direction (see section 5.8.1). Indeed in that case, it is not possible, without resorting to some kind of trick, to find a reference configuration where all joint axes correspond to a principal direction of the inertial frame.

12.3.1 Analysis

Description of the system

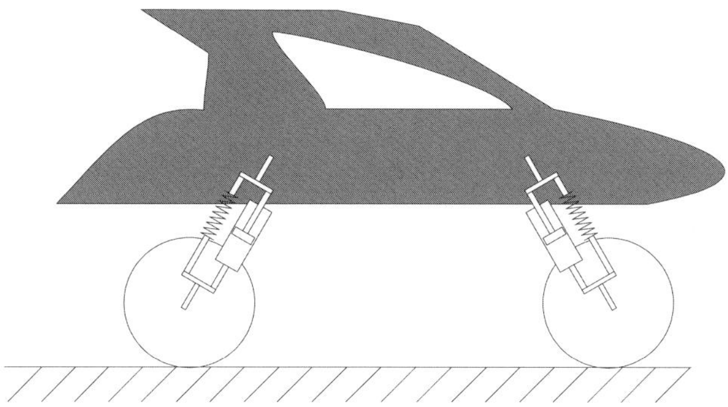

- Number of bodies: 5 (the chassis and the four wheels)
- Joints: The chassis is connected to the ground (inertial body) through a free 6 degree-of-freedom joint and each wheel has two degrees of freedom relative to the chassis (translation in the direction of the damper and rotation around its axis).
- Degrees of freedom: $14 = 6 + 4 \times 2$.
- Forces and torques requiring description: forces in the springs and dampers, wheel-ground forces.
- Environment: the system is in a gravity field.

Modeling hypotheses

- Assuming perfect right to left symmetry of the vehicle, we may carry out our simulations on a half-vehicle and therefore restrict the analysis to a planar system. Three degrees of freedom are only taken into account for the chassis: horizontal, bounce and pitch motions. Our model therefore has seven ($= 3 + 2 \times 2$) degrees of freedom.

- The suspension spring and damper exhibit a linear behavior (as regards both displacement and velocity).

- The wheel-ground contact force $\mathbf{F}_w$ is assumed to be purely vertical. This assumption is quite acceptable considering the required results (vehicle at rest). The constitutive equation of the vertical force is (see chapter 6):

$$F_{vert} = K_{rad}\ p_w \tag{12.1}$$

p_w being the tire penetration computed from equation 6.18.

12.3.2 Multibody model

- Choice of inertial reference frame: in the reference configuration, we place its origin at the chassis center of mass, with the x-axis pointing forwards and the z-axis upwards. This choice of axes is in the present case imposed by the presence of wheels (see chapter 6).

- The wheel-fixed frames are such that, in the reference configuration:

  - The x-axis points forwards,
  - The y-axis is aligned with the wheel's axle, and points to the left,
  - The z-axis points upwards.

This choice is imposed by the wheel/ground model convention.

- The translational motion of the wheels (relative to the chassis) occurs around an axis which is not aligned with any of the chassis or wheel frame axes: in our case, there is an angle of $\pm\frac{\pi}{6}$ *rad* between the chassis's vertical axis and the spring axis. We need therefore to introduce an additional $R2$-type revolute joint associated with a fictitious body (see section 5.8.1). If we note the corresponding joint variable q^i, its value will be 0 in the reference configuration. Thereafter, during the numerical treatment, the angle will be taken into account by imposing a *constant* value $q^i = \frac{\pi}{6}$ for the front wheel and $-\frac{\pi}{6}$ for the rear wheel at all times. We therefore consider it to be a *driven* variable.

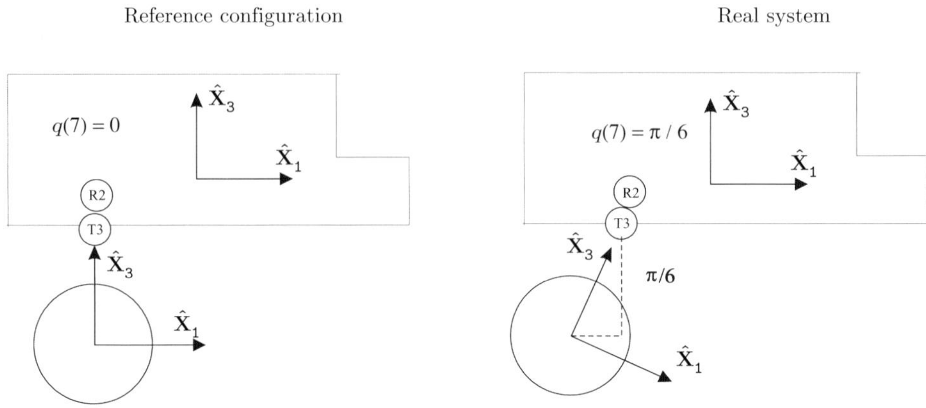

- The force produced by the spring and damper is modeled as a joint force, as in problem 1. The wheel/ground contact force is considered as external and is modeled via a constitutive equation to be provided in a user function (see further on).
- Numbering of bodies and joints:
  - Joint 1: vehicle translation in its forward direction of motion
  - Joint 2: vertical bouncing translation of the vehicle
  - Joint 3: rotation around an axis perpendicular to the vehicle plane, accounting for pitch
  - Joints 4 and 7: fictitious joints (rotation fixed at $\pm\ \pi/6$) allowing translation in a direction which is not aligned with one of the preceding body's reference axes (i.e. the chassis)
  - Joints 5 and 8: suspension translation joints
  - Joints 6 and 9: wheel spin R-joints

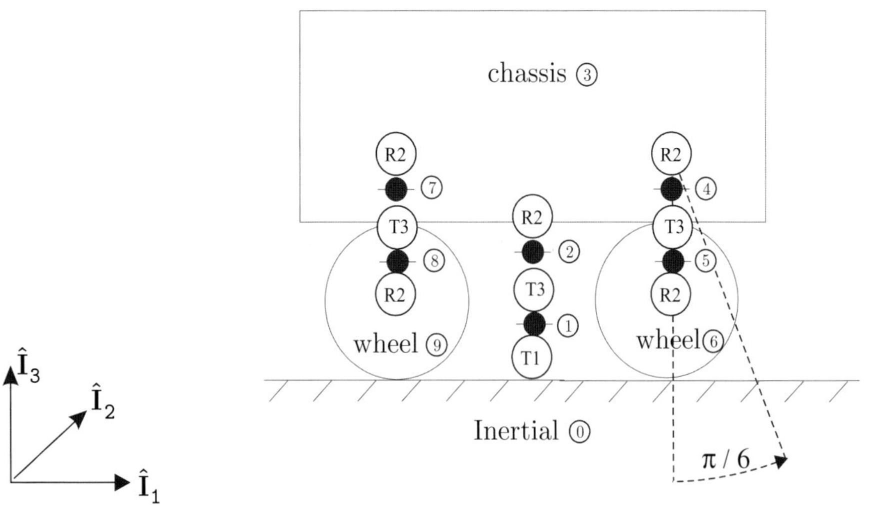

- Reference configuration: the reference configuration is such that the chassis center of mass is at zero elevation from the ground (variable q^2 will therefore directly provide the height of the chassis), and the wheel turning axes are vertical. The displacements in the suspensions are equal to zero (the wheels are "stuck" to the chassis and thus $q^5 = 0$ and $q^8 = 0$).

12.3.3 Computer pre-process

System data (MBdesc)

Presented hereafter the multibody data required for this application:

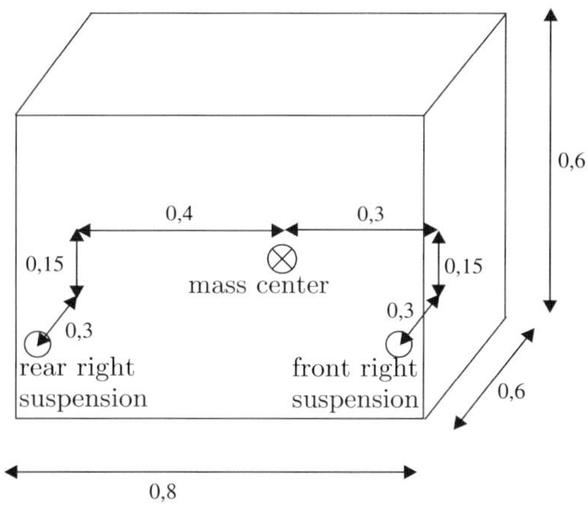

Chassis:

- Geometrical dimensions:
  - Position vector components[6] in the chassis reference frame for the forward suspension attachment point: $(0.3, -0.3, -0.15)$.
  - Position vector components in the chassis reference frame for the rear attachment point: $(-0.4, -0.3, -0.15)$.
  - The center of mass coincides with the attachment point of joint 3 and thus: $d_1^{33} = d_2^{33} = d_3^{33} = 0$).
- Mass and inertia of the chassis (considered to be a rectangular parallelepiped):
  - Mass: $200\,kg$. To model a half-chassis, $100\,kg$ will be taken.
  - Inertia: since the model is planar, the only component needed for the inertia matrix is: $I_{22} = \frac{200(0.8^2+0.6^2)}{12} = 16.667\,kg\,m^2$. For a half-chassis, only half of that is needed (for I_{22})$=>$ $I_{22} = 8.333\,kg\,m^2$.

[6] (x, y, z), units: $[m]$

Wheel:

- Nominal radius $= 0.3\ m$, mass$= 10\ kg$, inertia: $I_{22} = 0.45\ kg\, m^2$
  The center of mass coincides with the wheel geometrical center.

Since the wheel-ground contact force is considered to be external to the system, the external forces F^i_{ext} and torques L^i_{ext} symbolic components must appear in the main ROBOTRAN data file for each wheel i.

User's model (MBuser)

For this vehicle application, the following user function must be created.

1. For the road profile: In order to define the road profile, the user must record three analytical functions defined in the inertial reference frame: the road profile $\mu(x, y)$, and the derivative in the x and y-direction: $\frac{\partial \mu}{\partial x}$ and $\frac{\partial \mu}{\partial y}$ respectively (see chapter 6).

In the present case, the profile description is trivial (flat ground) and is given below:

```
mu = 0.0;
dmudx = 0.0;
dmudy = 0.0;
```

A more complex profile will appear in problem 5.

2. For the tire model: According to equation 12.1, the model is simply described in this case by a radial elastic force whose stiffness is $K_{rad} = 20.000\ \frac{N}{m}$ which only appears if the wheel penetration p_w is positive.

```
Fvert = 20000 * (rnom - rw);
Flat = 0;
Flong = 0;
Myaw = 0;
```

Let us point out that ROBOTRAN (menu "sensor kinematics", see chapter 5) is used here to provide the following useful symbolic ingredients:

- the wheel center absolute position $\mathbf{x}(q)$, whose z-component x_3 is necessary to compute p_w for any wheel configuration;

- the rotation matrix $R^{i,0}$ between the wheel i and the inertial frame. The latter is indeed necessary to project the above force in the wheel reference frame.

3. For the suspensions: The spring (stiffness $k = 6000\ \frac{N}{m}$ and unstrained displacement $l_0 = -0.25\ m$) and the damper (damping coefficient $c = 300\ \frac{Ns}{m}$) of each suspension are input as joint generalized forces $\mathcal{Q}$. We therefore need to input the following statements in a user's function:

```
Q(5) = -6000*(q(5)+0.25)-300*qd(5);
Q(8) = -6000*(q(8)+0.25)-300*qd(8);
```

4. For the driven variables:

- Set to a fixed value of $\pm\ \frac{\pi}{6}$ the two R-joint variables which had to be added to take into account the slant of wheel turning axes.

The forward motion of the vehicle and the wheel spin have no influence on the equilibrium configuration and the modal analysis. The corresponding variables are said *ignorable*. Such variables can therefore also be set to zero (although this is not mandatory in MBSOFT which would automatically detect that they have no influence on the system equilibrium).

We therefore add the following statements in the user's model:

```
% slant angles
  q(4) = -pi/6;
  q(7) = pi/6;
% Longitudinal position (ignorable)
  q(1) = 0;
% wheel rotational position (ignorable)
  q(6) = 0;
  q(9) = 0;
```

Initial conditions (MBini)

We choose as initial conditions, a $50\,cm$ chassis height ($q^2 = 0.5$), zero pitch ($q^3 = 0$) and zero suspension strain ($q^5 = q^8 = 0$).

12.3.4 Computer process

Symbolic treatment (ROBOTRAN - MBgen)

According to the requested results, we must generate the equations of motion (direct dynamics) and the sensor's kinematics, each of them being placed on a wheel center (menu: Toolbox - Sensor kinematics).

Coordinate partitioning (MBpart)

As detailed above, there are six driven variables: q^1, q^4, q^6, q^7, q^9. The other variables are independent because there are no kinematic constraints in the system.

Numerical treatment (MBsim)

Further to the requested results, static equilibrium (using a Newton/Raphson type method) followed by a modal analysis (with a prior numerical linearization of the equations around the previously found equilibrium) must be performed.

Let us point out that to ensure a good convergence of the Newton/Raphson method, we recommend to artificially remove the nonlinearity underlying equation 12.1 induced by the computation of p_w (see equation 6.18): in other words, during the equilibrium iterative process, we accept wheel/ground traction forces. We must of course check that in the Newton/Raphson final solution, a compression force is found !

12.3.5 Computer post-process

The results are the following.

Equilibrium configuration

Chassis height: 0.5737 m
Chassis pitch: 0.0317 rad
Front suspension strain: 0.1672 m (spring unstrained length: 0.25 m)
Rear suspension strain: 0.1909 m (spring unstrained length: 0.25 m)
Front wheel/ground radial contact force: 661.7 N
Rear wheel/ground radial contact force: 515.5 N

Modal analysis

Chassis pitch mode: 1.53 Hz
Chassis bounce mode: 1.82 Hz
Rear suspension motion: 6.66 Hz
Front suspension motion: 6.77 Hz

12.4 A slider-crank mechanism

> The objective is to analyze a slider-crank system and to set up a graph displaying the time evolution of the displacement of the piston along the slider, when the angular velocity of the crank is set to a fixed value $\omega(t) = 150\, rad/sec$. A time-graph of the force exerted by the connecting rod on the piston is also requested, as well as the driving torque applied to the crank.

Originality of the problem: This seemingly simple system introduces two new concepts, viz. closed-loop structures (constraints and Lagrange multipliers) and control laws.

The examples in the figure below illustrate what we mean by closed-loop structure. The classical example of the four-bar mechanism shows that its motion implies constraints between the joint variables q. These constraints are algebraic relations between the variables and their first and second derivatives.

The same holds for a suspension; this system will be studied in more detail further on in the tutorial.

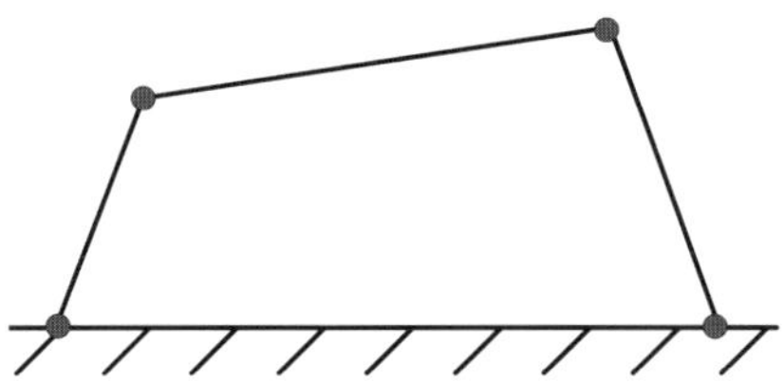

Four-bar mechanism

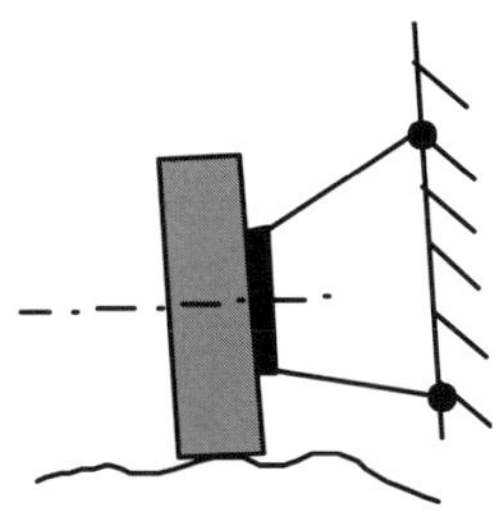

Suspension

The rigid body motion of a slider-crank system is completely determined when the crank rotation is imposed. Since the relevant variable is driven, there is no remaining independent variable and the number of degrees of freedom is equal to zero. We are therefore left with a kinematic study, which can be followed – if so required – by a computation of the inverse dynamics in order to determine the efforts which are needed to produce the imposed motion.

12.4.1 Analysis

Description of the system

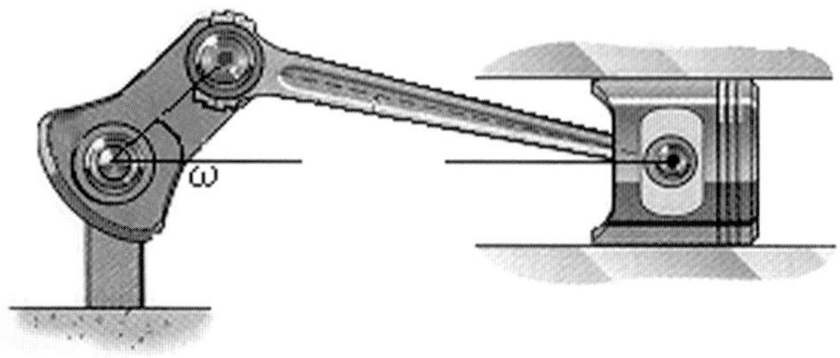

- Number of bodies: 3
  - Crank
  - Connecting rod
  - Piston
- Joints: 4
  - A revolute joint (1 degree of freedom) around a fixed point, for the crank
  - A revolute joint between the connecting rod and the crank
  - A revolute joint between the connecting rod and the piston
  - A prismatic joint between the piston and the cylinder
- Kinematic loop: 1, imposing closure of the system
- Degrees of freedom: 4 (the joints) - 3 (planar loop) = 1
- External forces and torques requiring description: none, since the driving torque required to ensure rotation of the crank follows from the latter's control law (the torque will be equal to the associated Lagrange multiplier).
- Environment: the system is not in a gravity field.

Modeling hypotheses

- The crank and connecting rod are modeled as homogeneous rigid cylinders and the piston as a homogeneous rectangular parallelepiped.
- The joints are assumed to be frictionless, which implies that we are able to introduce a cut at one of the joints.
  rem.: in presence of joint friction, this type of cut would prevent us from easily introducing the corresponding torque, since the cut joint is no longer part of the model (no joint variable or force).

12.4.2 Multibody model

- Choice of inertial reference frame: this frame is located on the rotation axis of the crank, with its x and y axes in the plane of motion.

- Loop cut: as required by the methodology, we must open the loop. Since the joints are frictionless, we may choose to cut the system at the joint between connecting rod and piston as depicted in the figure below. This results in two parts which are assembled by writing the 3 kinematic constraints associated with a "C^2 cut" according to section 4.1.2. For a planar system, the third constraint is trivial: $(0 = 0)$. It will be detected as such and considered as redundant by MBSOFT.
  For this type of cut, the Lagrange multipliers provide the force components in the joint, as exerted by the secondary body (the piston) on the primary body (the connecting rod), in a frame linked to the ascendant which is common to the two kinematic chains leading to the cut (in our case, this common ascendant is simply the inertial body). These components will allow us to answer the second question.

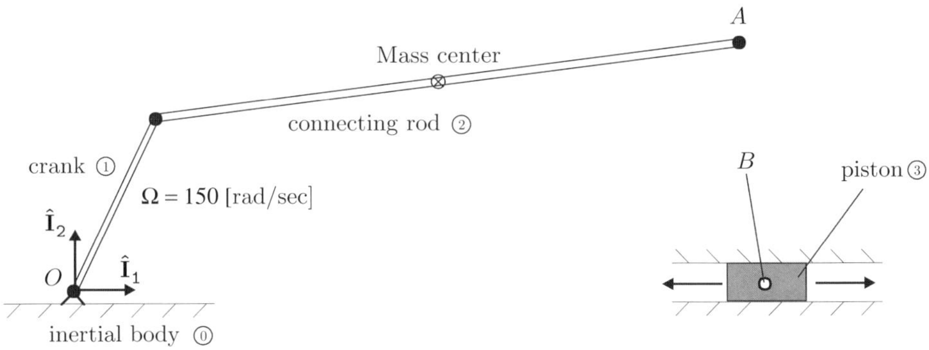

- A reference frame is associated with each body.

- Numbering of bodies and joints:

  – Joint 1: rotation of the crank (body 1) relative to the inertial body
  – Joint 2: rotation of the connecting rod (body 2) relative to the crank
  – Joint 3: translation of the piston (body 3) relative to the inertial body

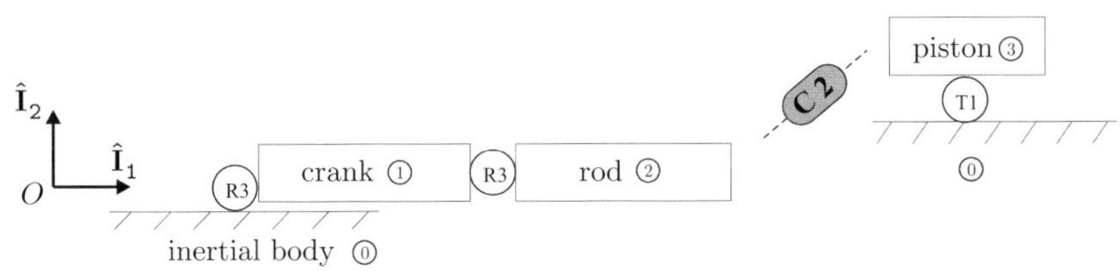

- Reference configuration: as a consequence of the way in which the bodies are defined, the reference configuration is such that the crank and connecting rod are both horizontal, with the piston at a distance along $\hat{\mathbf{I}}_1$ equal to the sum of the lengths of the crank and connecting rod.

Note: the definition of the reference configuration of a closed-loop system requires that the loops first be opened and a tree-like structure be established, so that the body reference frames can be aligned as imposed by the ROBOTRAN conventions.

12.4.3 Computer pre-process

System data (MBdesc)

As regard the joints and bodies, this system is easy to define. The numerical values required are as follows:

- Crank: length $= 0.15\,m$; the mass and inertia matrix are not required since the corresponding joint variable is given a constant angular velocity.

- Connecting rod: length $L = 0.3\,m$, radius $R = 3\,mm$, mass $= 0.0667557\,kg$, inertia:
  $I_{33} = m\left(\frac{L^2}{12} + \frac{R^2}{4}\right) = 5.00817968e^{-4}\,kg\,m^2$.
  Center of mass = geometrical center

- Piston: position in the reference configuration: $\mathbf{d}^{03} = [\hat{\mathbf{I}}]^T\,(0.45 \;\; 0 \;\; 0)^T$;
  mass $= 0.033377851\,kg$.
  Center of mass = geometrical center.

The following step is to define the positions of the attachment points A and B of the C^2 cut joint. According to our modeling conventions (see chapters 4 and 5), vector $\mathbf{d}_{lp}$ is the position vector of the cutting point on the primary body (body 2) relative to the preceding joint, i.e. the vector joining the crank-connecting rod joint to point A, expressed in body 2's reference frame, and vector $\mathbf{d}_{ls}$ is defined in a similar manner, giving the position of attachment point B relative to joint 3 on the secondary body (body 3) in the reference frame associated with the latter (here in fact: $\mathbf{d}_{ls} = 0$).
During the numerical process, the loop must be numerically closed (ex. using the Newton/Raphson method), as long as the initial open configuration was well chosen.

User's model (MBuser)

In the same way as for the merry-go-round, we need to control the rotation angle variable of the crank. To this end, we add the following statements to the model, as a user's function:

```
q(1) = 150*t;
qd(1) = 150;
qdd(1) = 0;
```

Initial conditions (MBini)

The initial configuration is identical to the reference configuration. All joint coordinates are therefore set to zero.

12.4.4 Computer process

Symbolic treatment (ROBOTRAN - MBgen)

For the problem at hand, we need to symbolically generate:

- the equations of motion of the "open" system (menu: Direct models - Tree-like - Direct dynamics, see figure 5.17),
- the algebraic loop constraints (and their time derivatives) between generalized coordinates, which impose the closure of the system (menu: Toolbox - Constraints, see figure 5.17).

Coordinate partitioning (MBpart)

Variable q^1 is driven and, due to the kinematic loop, there is no more degree of freedom as already mentioned. Thus variables q^2, q^3 and q^4 are all dependent and the coordinate partitioning is unique in this case.

Numerical treatment (MBsim)

This problem is of particular interest since, if we control the crank's rotation variable, there is no remaining independent variable, hence no degree of freedom. The first part of the problem is therefore an exercise in pure kinematics. Since the crank must rotate at 150 rad/sec, each revolution takes 0.0419 sec (which will be the total simulation time). For this type of mechanism, it should also be noted that computations relating to the kinematics and to the dynamics are completely uncoupled. Modifying a dynamical characteristic (mass, center of mass, ...) obviously has no influence on the kinematics of the slider-crank mechanism.

In order to solve the second part of the problem, we must compute the inverse dynamics problem from the previous kinematical results, i.e. a set of consecutive configurations α: $\{t_\alpha, q_\alpha, \dot{q}_\alpha, \ddot{q}_a\}$. Then, using equation 4.50 we can compute the Lagrange multipliers as time functions:

- The Lagrange multiplier λ_1 associated with the driven variable $q^1(t)$ corresponds to the crank input torque (in $[Nm]$).

- The two Lagrange multipliers λ_2 and λ_3 associated with the loop constraints provide the force components (in the x and y direction of the $\{\hat{\mathbf{I}}\}$ frame) in the cut joint, as exerted by the piston on the connecting rod.

12.4.5 Computer post-process

The piston motion is directly provided by the time history of variable q^3.

As regards the constraint forces:
$F_x^{con.rod \rightarrow piston} = -\lambda_2$
$F_y^{con.rod \rightarrow piston} = -\lambda_3$
$F_z^{con.rod \rightarrow piston} = 0$ (the third component $= 0$ for a planar system)
$M_z^{crank} = \lambda_1$
The following graphs answer the terms of the problem.

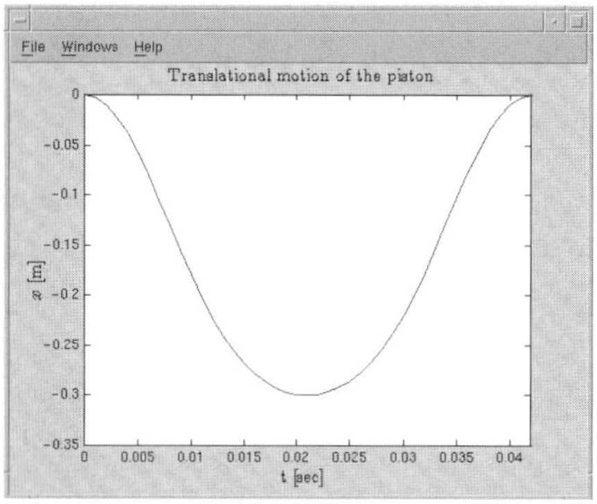

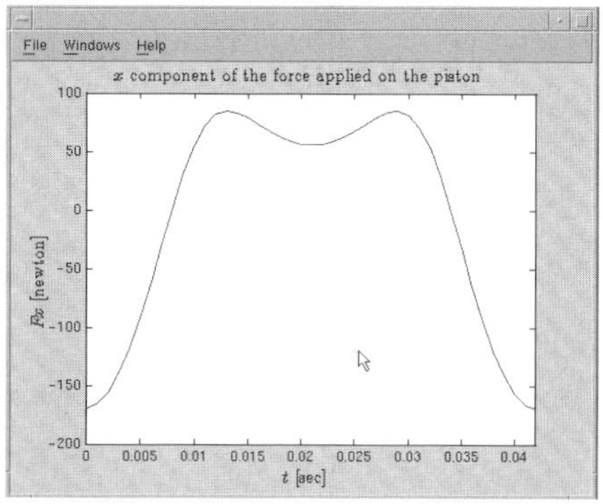

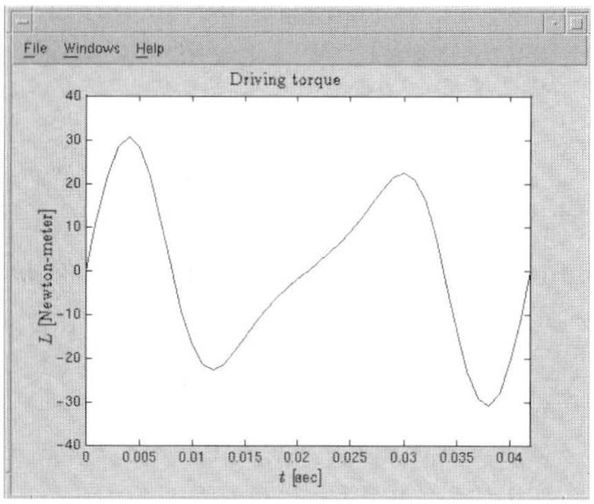

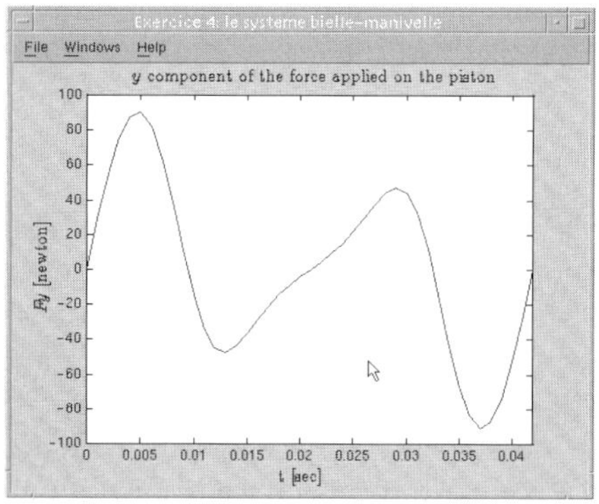

12.5 Small cart 2

Starting from the cart of problem 3, the objective here is to perform a parameter sensitivity analysis to find a stiffness coefficient k of the suspension spring so that the chassis bounce frequency has a value around 1.3 Hz, close to that of a human heartbeat.

Originality of the problem: We shall be touching on a new modeling specificity, viz. point-to-point links which are internal to the multibody system (see section 4.3) and a new numerical analysis namely, the parameter sensitivity study allowing us to optimize the modal behavior of the vehicle.

12.5.1 Analysis

Description of the system

The reader is referred back to problem 3.

Modeling hypotheses

Identical

12.5.2 Multibody model

- Choice of inertial reference frame and frames associated with the various bodies: as in problem 3.
- Contrary to what we did in problem 3, we choose here to model the spring and damper forces as *point-to-point links.*
  - Link 1: between bodies 4 (following the fictitious joint $q^4 = -\frac{\pi}{6}$) and 5
  - Link 2: between bodies 7 (following the fictitious joint $q^7 = \frac{\pi}{6}$) and 8
  - The wheel-ground contact forces are considered as external to the system (see problem 3)
- Numbering of bodies and joints:
  - Joint 1: rigid translation of the chassis in its forward direction of motion
  - Joint 2: rigid chassis bounce
  - Joint 3: rigid chassis pitch
  - Joints 4 and 7: fictitious joints (rotation set to $\pm\pi/6$) allowing translation in a direction which is not aligned with one of the preceding body's reference axes (i.e. the chassis)

- Joints 5 and 8: wheel translation joints
- Joint 6 and 9: wheel spin R-joints

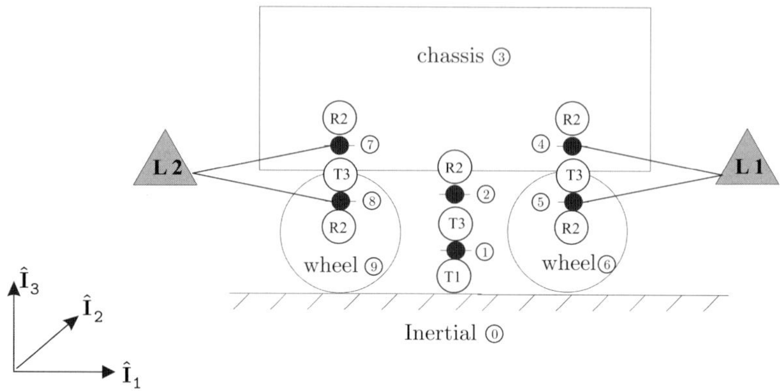

- Reference configuration: the reference configuration is identical to that of problem 3.

12.5.3 Computer pre-process

System data (MBdesc)

As a complement to problem 3, we still need to define the two internal links of the multibody system (see section 4.3).

In order to model these links, we must first define their attachment points, viz. on the one hand the bodies to which they are attached: primary body (body 4 for link 1, body 7 for link 2) and secondary body (body 5 for link 1, body 8 for link 2), and on the other hand, the position vectors $\mathbf{d}_{bp}$, $\mathbf{d}_{bs}$ of the attachment points on the primary and secondary bodies respectively. In the present model, these two vectors are simply equal to zero since the springs and dampers are directly attached to the joint attachment points.

User's model (MBuser)

The user is referred to problem 3 as regards driven variables, road profile and tire model.

We need to introduce the constitutive equation of the links: in the present case, it will be a linear spring with stiffness $k = 6000\,\frac{N}{m}$, neutral length $l_0 = 0.25\,m$, combined with a linear damper whose damping coefficient is $c = 300\,\frac{Ns}{m}$.

Initial conditions (MBini)

We choose as initial conditions those corresponding to the equilibrium configuration found in problem 3.

12.5.4 Computer process

Symbolic treatment (ROBOTRAN - MBgen)

The sequence of operations is similar to that of problem 3:

- Generation of the dynamical equations of the system (Direct dynamics)
- Generation of the wheel kinematics (menu: Toolbox - Sensor kinematics)
- Generation of the point-to-point links (menu: Toolbox - Link forces, see figure 5.17).

Coordinate partitioning (MBpart)

see problem 3

Numerical treatment (MBsim)

The first stage consists in checking that our new model is in fact equivalent to that of problem 3, where the suspension loads where modeled as joint forces. To do this, we compute the system equilibrium and perform a modal analysis, and note that the results are absolutely identical.

The intended sensitivity analysis can be carried out by performing successive equilibrium/modal analyses of the system for different values of the parameter (k). In MBSOFT, this can be done automatically using the Equilibrium - Modal analysis/Sensitivity study menu:

- Request a modal analysis for each configuration
- Choose 2 parameters to be changed: k for link 1 and k for link 2
- Ask for e.g. 30 configurations
- Change the spring stiffness from $6000 N/m$ to $3000 N/m$

  The program then works out the 30 equilibrium configurations and provides a modal analysis for each one.

12.5.5 Computer post-process

By simply scanning the results, we can check that for a stiffness equal to $3200 N/m$, the bounce frequency (which for the present system is the second frequency, the first one corresponding to chassis pitch, see problem 3) is $1.3 Hz$, which will be our optimum.

The boxed field at the end of this section describes the "performance" of our optimized vehicle.

Eigenvectors are helpful to interpret eigenmodes. This interpretation is however usually quite difficult for large multibody systems, which is why it is really interesting to animate one or more modes. In MBSOFT, it is the purpose of the MBview module: the choice of the mode to be displayed can be made in two different ways: either via the number attached to any eigenmode or via a variable which one wishes to single out. For example, let us pick variable q^2 (bounce): MBSOFT will set up the files required by the animation, each time the bounce motion has at least a 30% involvement compared to the dominant variable.

Let us look at the first bounce mode (mode 2: eigenvalue $= -4.02222263 \pm 8.1329852\,i$, $\nu_2 = 1.29 Hz$). Here is the eigenvector:

variable q	motion	amplitude r	phase ϕ [°]
q^2	carbody bounce	0.442458	−56
q^3	carbody pitch	1.000000	0
q^5	front suspension	0.372295	69
q^8	rear suspension	0.902191	146

The dominant variable for this mode is pitch (variable q^3). This pitch is due mainly to the excursions of the rear wheel, with a small involvement of the front wheel, which is 77° out of phase from the rear wheel (146 − 69). An animation of this mode provides a better understanding of the motion.

The final solution is summarized hereunder.

Equilibrium configuration

Chassis height: 0.51422 m
Chassis pitch: 0.0892 rad
Front suspension strain: 0.08037 m (spring unstrained length: 0.25 m)
Rear suspension strain: 0.15219 m (spring unstrained length: 0.25 m)
Front wheel/ground radial contact force: 696.5 N
Rear wheel/ground radial contact force: 480.7 N

Modal analysis

Chassis pitch mode: 0.9 Hz
Chassis bounce mode: 1.29 Hz
In-phase suspension excursions: 5.9 Hz
Opposed-phase suspension excursions: 6.41 Hz

12.6 A five-point suspension

The objective is to carry out a kinematic and dynamical investigation of a 5-point suspension, mounted on a chassis fixed in a quasi-static equilibrium configuration (forward velocity: $v = 30m/s$).
The expected results are:
- *Kinematics*: time history of the camber and steer angles, when the height of the wheel center is altered within a range of plus or minus $20\,cm$ around the equilibrium configuration.
- *Dynamics*: system behavior (height of wheel center) when the ground undergoes a downwards step of 10 cm.

Originality of the problem: This is the first complex $3D$ problem, which involves quite a number of previously encountered notions such as: multi-degree of freedom joints, algebraic constraints between joint variables, links between bodies, wheel model and kinematic sensors. A new numerical analysis will also be used, viz. inverse equilibrium. This problem has been proposed as an international vehicle benchmark in 1991 by the IAVSD[7]. The complete description of the benchmark has been published in [49].

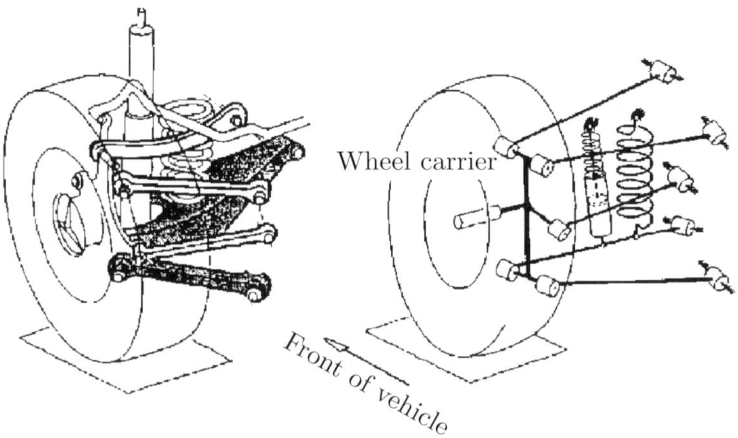

12.6.1 Analysis

Description of the system

- Number of bodies: 8
  - Chassis
  - Five suspension bars
  - Wheel carrier

[7]International association for vehicle system dynamics

  - Wheel

- Internal joints of the suspension:
  - 5 ball joints connecting the bars to the chassis (ball joint = 3 degrees of freedom)
  - 5 ball joints connecting the bars to the wheel carrier
  - 1 simple revolute joint for wheel spin

- Degrees of freedom: 7
  - 1 d.o.f. for suspension deformation
  - 5 $\times$ 1 d.o.f. for the rotations of the 5 bars around their main axes
  - 1 d.o.f. for wheel spin
  - Forces and torques:
  - Force developed by the spring-shock absorber pair
  - Ground to tire contact force

- Environment:
  - The system is in a gravity field
  - The ground is "flat"

Modeling hypotheses

- The inertia effects of the bars are neglected.
- Since the chassis is assumed to be fixed, it is chosen as inertial body (body 0).
- We assume that the 5 bars cannot rotate freely around their main axis. We model the 5 chassis-side ball joints as 2 d.o.f. universal joints, which therefore leaves us with 2 degrees of freedom for the full system, i.e. suspension excursion and wheel spin.
- No friction in the joints.
- Spring and shock absorber constitutive equations: linear spring and damper (stiffness $k = 14000\,N/m$ and damping coefficient $c = 1400\,Ns/m$). The neutral length l_0 of the spring is not known at the outset and remains to be determined given that the initial configuration is the equilibrium configuration.
- The force arising from wheel-tire interaction is modeled via a simple vertical linear spring (stiffness $K_{rad} = 200000\ N/m$). The nominal radius of the wheel is 30 cm (unloaded wheel). In the equilibrium configuration, this radius reduces to 27.797933 cm (benchmark datum).

Analysis of expected results

Since we have to determine the vertical position, velocity and acceleration of the wheel center, we need to add a position, velocity and acceleration sensor.

The camber angle φ_w has been defined in equation 6.11. In accordance with the standards prevailing in vehicle dynamics, the steer angle δ_w is defined as:

$$\delta_w = arctan\left(\frac{\hat{\mathbf{R}}_1 \, . \, \hat{\mathbf{I}}_2}{\hat{\mathbf{R}}_1 \, . \, \hat{\mathbf{I}}_1}\right) \tag{12.2}$$

in which according to section 6.4, the unit vector $\hat{\mathbf{R}}_1$– which belongs to the wheel plane – is tangent to the wheel circumference and to the ground at the contact point Q. The inertial reference frame $\{\hat{\mathbf{I}}\}$ is defined as follows:

- $\hat{\mathbf{I}}_1$: vehicle longitudinal axis (pointing forwards);
- $\hat{\mathbf{I}}_2$: vehicle lateral axis (pointing left);
- $\hat{\mathbf{I}}_3$: vertical axis (pointing upwards).

12.6.2 Multibody model

- Choice of the inertial reference frame origin: $\{\hat{\mathbf{I}}\}$ is fixed to the chassis. Its origin lies 0.27797933 m above the ground (before the latter's downward step). This value corresponds to the wheel radius in the given equilibrium configuration and therefore to the height of its center: this very precise value results from the benchmark data.
- The force produced by the spring and shock absorber is modeled as a link between the chassis (body 0) and the center point of bar 5 (see figure).
- Cutting of loops: four "C^2 cut" are made at the connections between bars 1, 2, 3, 4 and the wheel carrier: this induces 12 loop constraints (see figure).
- Numbering of bodies and joints:
  - joints 1 and 2: between bar 1 and chassis (body 0)
  - joints 3 and 4: between bar 2 and chassis (body 0)
  - joints 5 and 6: between bar 3 and chassis (body 0)
  - joints 7 and 8: between bar 4 and chassis (body 0)
  - joints 9 and 10: between bar 5 and chassis (body 0)
  - joints 11, 12 and 13: universal joints between bar 5 and wheel carrier (body 13)
  - joint 14: wheel spin (body 14)

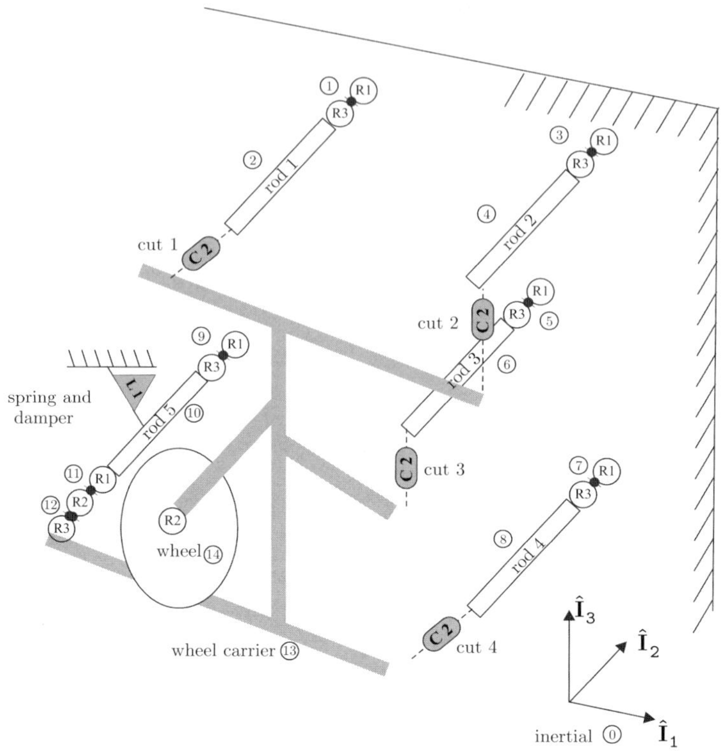

Figure 12.1: 5-point suspension reference configuration

- In the reference configuration, all the suspension bars are horizontal, as well as the spindle axis of the wheel carrier and the wheel spin axis. (As a reminder, this configuration does not conform to the closure of kinematic loops).

12.6.3 Computer pre-process

System data (MBdesc)

With the original data of the problem (see [49]), the suspension is defined using the absolute coordinates of the various attachment points of the joints, the spring and the shock absorber (this is not the natural way of doing things, since it is easier to measure e.g. the length of a bar rather than the absolute coordinates of its end-points). In the same way, the inertia matrix of the wheel carrier is defined in the inertial reference frame instead of that associated to the carrier. These data have been recalculated (this is a tedious task which we see no point in recording here) so as to be compatible with our formalism which uses relative joint coordinates.

- Bars:

Bar/joint	Mass	Length	Joint location
	$[kg]$	$[m]$	$[m]$
1 / 1	2.0	0.223725	$d_1^{01} = -0.064$; $d_2^{01} = 0.413$; $d_3^{01} = 0.327$
2 / 3	2.0	0.279394	$d_1^{03} = -0.303$; $d_2^{03} = 0.432$; $d_3^{03} = 0.295$
3 / 5	4.0	0.297851	$d_1^{05} = -0.093$; $d_2^{05} = 0.366$; $d_3^{05} = 0.004$
4 / 7	1.0	0.41905	$d_1^{07} = -0.236$; $d_2^{07} = 0.388$; $d_3^{07} = -0.109$
5 / 9	3.0	0.411638191	$d_1^{09} = 0.2115$; $d_2^{09} = 0.384$; $d_3^{09} = -0.1$

rem.:

1. The inertia of the bars being negligible are set to 0.
2. The lengths (third column) are used for setting the y component (see the reference configuration on the previous figure):
   - of the loop cut position vectors $\mathbf{d}_{lp}$ (for bars 1, 2, 3 and 4)
   - of the joint position vector $\mathbf{d}^{5,11}$ (locating joint 11 on bar 5).

- Wheel carrier ($i = 13$)
  - Mass: $15\,Kg$
  - Inertia matrix:
    * $I_{11}^i = 0.3\,kg\,m^2$; $I_{12}^i = 0.0101822\,kg\,m^2$; $I_{13}^i = 0.0148769\,kg\,m^2$
      $I_{22}^i = 0.20121\,kg\,m^2$; $I_{23}^i = 0.0499851301\,kg\,m^2$
      $I_{33}^i = 0.19870669145\,kg\,m^2$
  - Position vector of mass center:
    * $d_1^{ii} = -0.1055\,m$; $d_2^{ii} = -0.2712847\,m$; $d_3^{ii} = 0.224325\,m$
  - Location of cuts on wheel carrier:

C^2 cut	$d_{ls_1}\,[m]$	$d_{ls_2}\,[m]$	$d_{ls_3}\,[m]$
1	-0.0665	-0.0951515	0.480196
2	-0.1195	-0.0952368	0.4732
3	-0.1905	-0.08864	0.112089
4	-0.0075	4.877686e-05	3.9997e-03

- Wheel ($i = 14$)
  - unconstrained radius: $r_{nom} = 0.30\,m$
  - Mass: $12\,Kg$
  - Inertia: $I_{22}^i = 1.46\,kg\,m^2$; $I_{11}^i = I_{33}^i = 0.74\,kg\,m^2$
  - Position vector of (wheel) joint on the carrier with respect to joint 13:

  * $d_1^{13,14} = -0.0025\,m$; $d_2^{13,14} = 0.03263172\,m$; $d_3^{13,14} = 0.133612\,m$

- Characteristics of the suspension (point-to-point link):
  - Position vector $\mathbf{d}_{bp}$ of attachment point on the chassis:
    * $d_{bp_1} = 0.09\,m$; $d_{bp_2} = 0.525\,m$; $d_{bp_3} = 0.17\,m$
  - Position vector $\mathbf{d}_{bs}$ of attachment point on bar 5 (body 10):
    * $d_{bs_1} = 0$; $d_{bs_2} = 0.20582\,m$; $d_{bs_3} = 0$
  - Constitutive equation parameters:
    * $k = 14000\,N/m$
      $c = 1400\,Ns/m$
      l_0: to be determined (see below)

Finally, to provide the required results, we need to place a kinematic sensor at the wheel center.

User's model (MBuser)

- Set the wheel spin to a constant $\frac{30}{0.3}\;rad/s$ value
- Introduce the vertical ground/tire force as in problem 3.
- Define the road profile $\mu(x, y)$:

```
mu = -0.27797933;
dmudx = 0.0;
dmudy = 0.0;
```

Initial conditions (MBini)

As the suspension has only one degree of freedom (since the wheel spin is driven), there is only one independent joint variable: we choose q^3, which is the rotation around a longitudinal axis (joint "$R1$") between the chassis and bar 2.

For the imposed configuration (given by the benchmark), a simple geometrical evaluation gives $q^3 = 0.20774326\,rad$.

As initial configuration, we choose $q^3 = 0.20774326\,rad$ from which we then determine a good initial guess for the dependent variables in the proposed configuration. For example, the following values are quite satisfactory:

$q^1 = 5^\circ\ (\times\pi/180)$	$q^9 = -5^\circ\ (\times\pi/180)$
$q^2 = 0^\circ\ (\times\pi/180)$	$q^{10} = 30^\circ\ (\times\pi/180)$
$q^4 = -45^\circ\ (\times\pi/180)$	$q^{11} = 5^\circ\ (\times\pi/180)$
$q^5 = 5^\circ\ (\times\pi/180)$	$q^{12} = 0^\circ\ (\times\pi/180)$
$q^6 = 20^\circ\ (\times\pi/180)$	$q^{13} = -30^\circ\ (\times\pi/180)$
$q^7 = -5^\circ\ (\times\pi/180)$	$q^{14} = 0^\circ\ (\times\pi/180)$
$q^8 = -35^\circ\ (\times\pi/180)$	

12.6.4 Computer process

Symbolic treatment (ROBOTRAN - MBgen)

The following operations need to be performed:

- Generation of the dynamical equations of the system
- Generation of the algebraic constraint equations
- Generation of the constitutive equations of the link
- Generation of the wheel kinematics
- Generation of sensor kinematics (for the wheel center)

Coordinate partitioning (MBpart)

Variable q^3 will be independent, corresponding to the suspension vertical d.o.f. ("wheel hop"). The rotation of the wheel around its axis, q^{14}, being "ignorable" for the present problem will be locked ($q^{14} = 0$) . The other 12 variables (q^1, $q^2, q^4 \ldots q^{13}$) are dependent and will be computed from the 12 kinematic loop constraints.

Numerical treatment (MBsim)

Kinematics Starting from the initial conditions, we perform a first closure of the system with the Newton-Raphson method. We then check the absolute position of the wheel center to which corresponds our kinematic sensor.

The components of the sensor position vector are $(0 \quad 0.768 \quad 0)^T$ in the inertial reference frame, and these correspond exactly to the absolute coordinates of the wheel center, as given by the benchmark[8]. Our system is thus correctly positioned in the imposed configuration.

We must now look at all the possible excursions of the suspension between two extreme values of the wheel center height z_w (between $z_w = -0.2$ and $0.2\ m$). The first step is to determine the value of variable q^3 so as to obtain $z_w = -0.2$ and 0.2. To this end, we perform a kinematic closure of the system by updating q^3 within the current configuration. Note that the Newton/Raphson method (loop closure) might encounter convergence problems if the user requests a loop closure with initial conditions too far removed from the actual solution. He then has to proceed "stepwise" in incrementing q^3. Final values are $z_w = -0.206\,m$ with $q^3 = -1.1\ rad$ and $z_w = 0.201\,m$ with $q^3 = 1.2\ rad$, and thus: $q^3 \in [-1.1, 1.2]$.

After (at least) one "step" at $q^3 = 0$ (a direct jump from the equilibrium configuration $q^3 = 0.2077...$ to $q^3 = -1.1$ is not possible due to lack of convergence of the Newton/Raphson procedure), we are now in a position to carry out the kinematic analysis and determine e.g. 50 configurations from -1.1 to 1.2.

[8] The zero value of the z-component is no coincidence. It follows from the choice (see benchmark) of origin for the inertial reference frame (see above).

For each configuration, we have access to the absolute position of the wheel center via the kinematic sensor.

In order to determine the camber and steer angles as functions of the wheel center height, we need to write a short (e.g. MATLAB) program to compute equations 6.11 and 12.2 respectively. The latter can be easily computed from the elements of the rotation matrix of the wheel $R^{14,0}$.

Inverse equilibrium Before running a dynamical simulation, we need to determine the neutral length l_0 of the spring in parallel with the shock absorber. Indeed, according to the benchmark (see [49]) the equilibrium configuration is *imposed* to be the initial configuration.
This neutral length can be computed via an *inverse* equilibrium procedure for which the Newton-Raphson method does not iterate on the independent generalized coordinate q^3 (=direct equilibrium) but on parameter l_0, q^3 being fixed (=inverse equilibrium).
In other words, find iteratively l_0 such that:

$$c_{red}(q, \dot{q} = 0, ..., F^i_{ext}(l_0), L^i_{ext}(l_0), ...) - \mathcal{Q}_{red}(q) = 0$$

After a few iteration steps, the equilibrium procedure comes up with $l_0 = 0.8923\,m$. At this stage, the suspension is in the equilibrium configuration specified by the benchmark.

Dynamics We can now launch the requested simulation. In the user file describing the ground, reduce the ground height $\mu(x, y)$ by $10\,cm$. Then run a time integration for a $0.6\,sec$ period with a maximum time step of $10^{-2}\ sec$ so as to ensure a minimum of 60 points (a fourth-order Runge-Kutta scheme is perfectly suited).

After completing the simulation, retrieve the time history of absolute vertical position, velocity and acceleration of the wheel center, using the kinematic sensor symbolic equations.

12.6.5 Computer post-process

Kinematics

For the requested kinematics, the numerical analysis produces the following graphs: the wheel camber angle φ_w, the wheel steer angle δ_w and the wheel center longitudinal position x_w, versus the wheel center vertical position z_w.

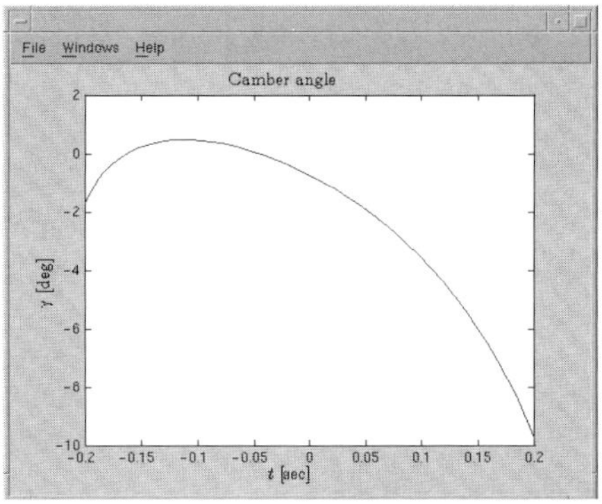

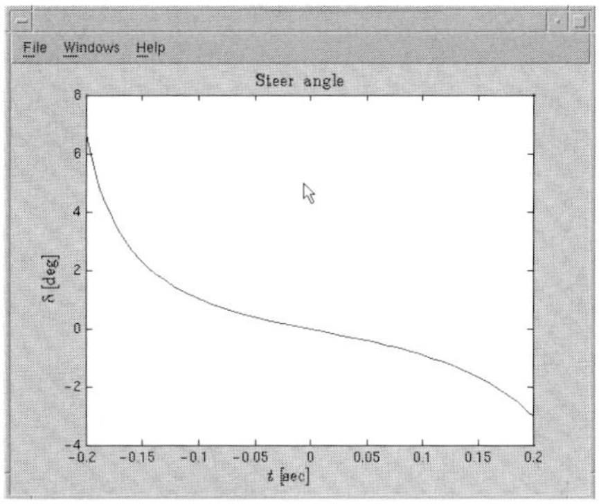

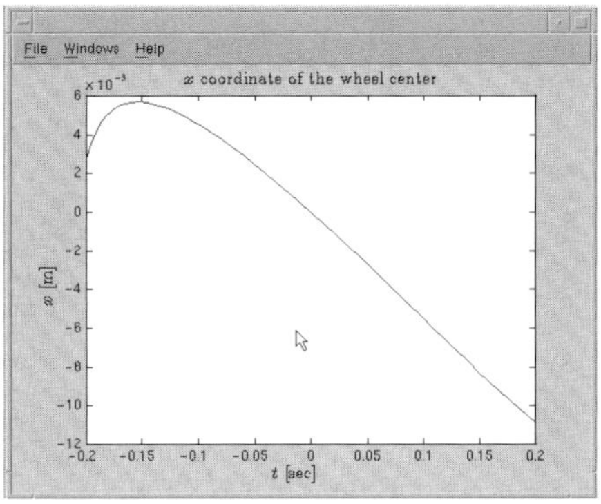

Dynamics

For the simulation results of the part 2 of the problem, the following graphs are produced (note the shift on the x-axis since the ground drops away only after 0.2 seconds):

- Vertical position of the wheel center versus time

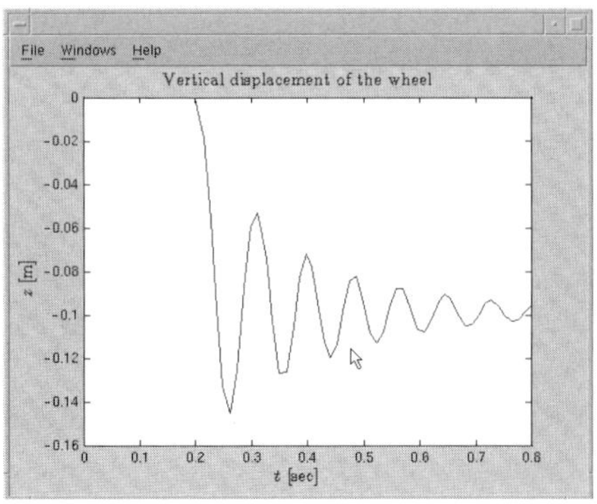

- Vertical velocity of the wheel center versus time

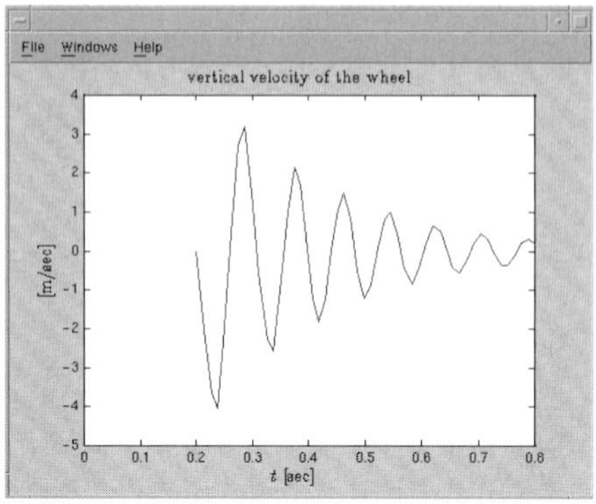

- Vertical acceleration of the wheel center versus time

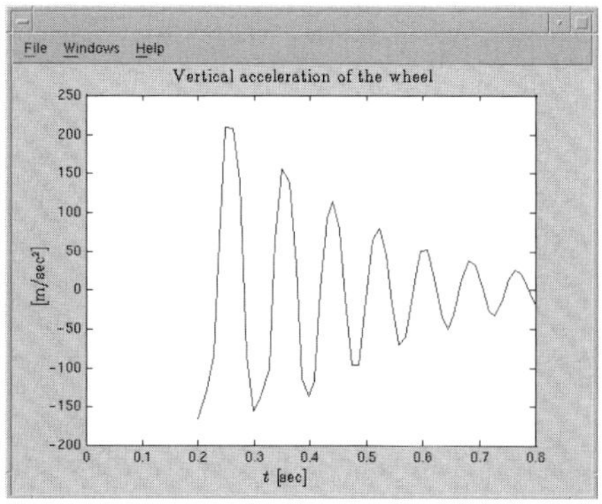

12.7 A jeep suspension

> Determine the time history of the wheel steer angle as a function of the excursion of the front right suspension for the Bombardier jeep depicted below[9]. A full set of data is also available in reference [49].

Originality of the problem: The final objective is to carry out a series of simulations (kinematics and dynamics) of the complete vehicle (see problem 12.8). The present problem looks at a single suspension and focuses on the kinematics only.

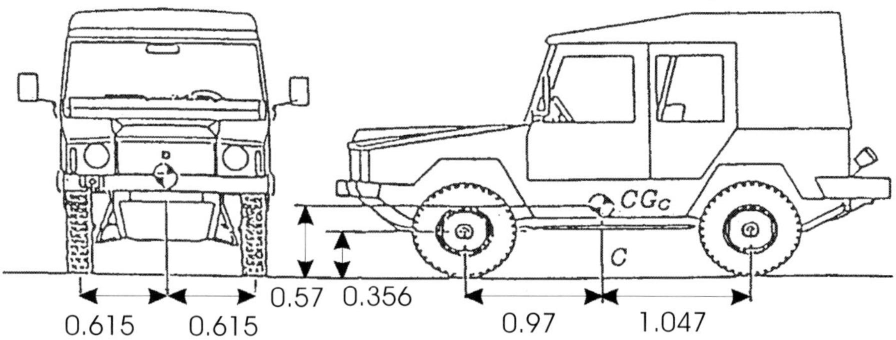

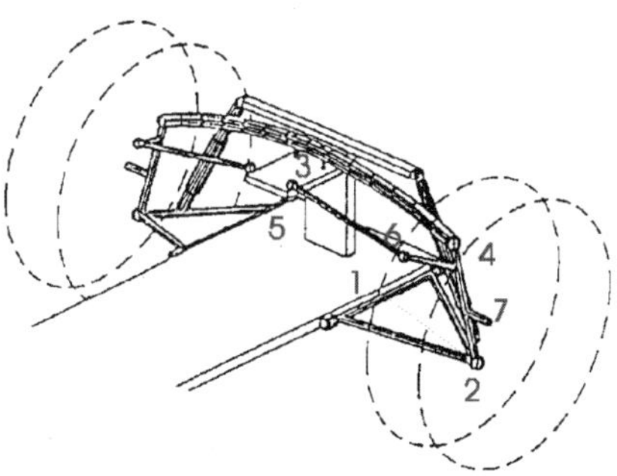

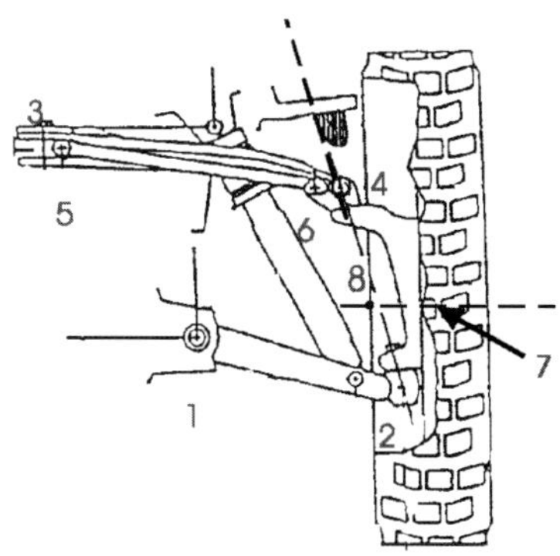

[9] This problem was also proposed as a 'Benchmark' by the IAVSD.

12.7.1 Analysis

Description of the system

- Number of bodies in the analysis of one suspension: 7
  - the chassis
  - the A-arm (between points 1 and 2)
  - the leaf spring (between points 3 and 4)
  - the rack (left from point 5)
  - the steering rod (between points 5 and 6)
  - the wheel carrier (on which points 2, 4, 6 and 7 are defined)
  - the wheel
- Internal joints of the suspension:
  - chassis - A-arm (pt 1): revolute joint
  - A-arm - rotation axis of the wheel carrier (pt 2): revolute joint
  - chassis - rack: prismatic joint
  - rack - steering rod (pt 5): universal joint
  - steering rod - wheel carrier (pt 6): spherical joint
  - bending of leaf spring
  - leaf spring - rotation axis of wheel carrier (pt 4): revolute joint
  - rotation of wheel carrier around its axis (passing through points 2 and 4): revolute joint
  - wheel carrier - wheel (pt 7): revolute joint
- Degrees of freedom: 9 + leaf spring deformation d.o.f. (theoretically an infinite number)
  - 6 degrees of freedom for the chassis
  - 1 degree of freedom for the in-plane deformation of the suspension
  - 1 degree of freedom for the steering
  - 1 degree of freedom for the wheel spin
- Forces and torques
  - force exerted by the spring-shock absorber assembly between the chassis and the $A-$arm
  - tire contact force

Note: One should be aware that any deformation of the leaf spring produces forces which require description, depending on the modeling hypotheses which are described in the next section.

- Environment:
  - The system is placed in a gravity field
  - Flat ground

Modeling hypotheses

- When evaluating the wheel steer angle δ_w, the chassis can be assumed to be fixed and thus assimilated to the inertial body, and the steering frozen (zero rack displacement). Wheel spin is of no interest. These degrees of freedom do not therefore need to be modeled for the kinematic results requested in this problem. Our model thus exhibits only one remaining degree of freedom. This suspension will however be used in problem 12.8 relating to the dynamical analysis of the full vehicle, which is why we model all the possible motions of the suspension and simply freeze (driven variables) the spurious degrees of freedom.

- In the requested kinematic analysis, no forces need to be modeled, and there is no need to introduce the dynamical parameters of the bodies (mass, inertia, position of center of mass).

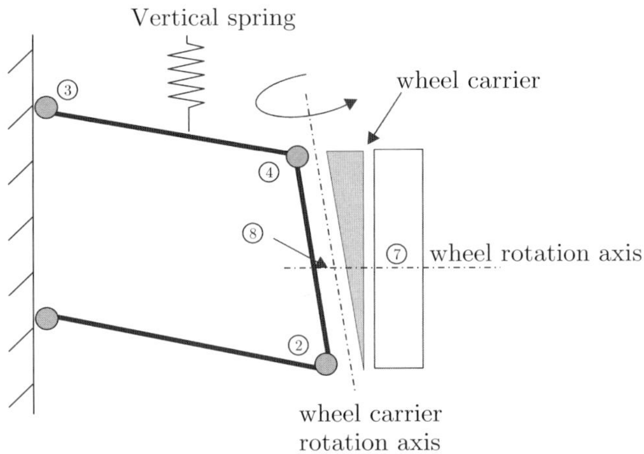

- The axis of the revolute joint which steers the wheel carrier coincides with the line passing through points 2 and 4. We may therefore locate the joint anywhere along this axis: we choose the point where the spin axis of the wheel intersects the 2-4 line, and name this point "point 8".

- The leaf spring is modeled as a rigid bar attached to the chassis (pt 3) and to the axis of the wheel carrier (pt 4) via revolute joints and a spring which is constrained to remain perpendicular to the chassis. The characteristics of the spring and its attachment point on the rigid bar will be specified in problem 12.8.

Analysis of expected results

The steer angle δ_w of the wheel is that between the horizontal track of the wheel mid-plane and the axis of symmetry of the vehicle (see equation 12.2 in problem 12.6).

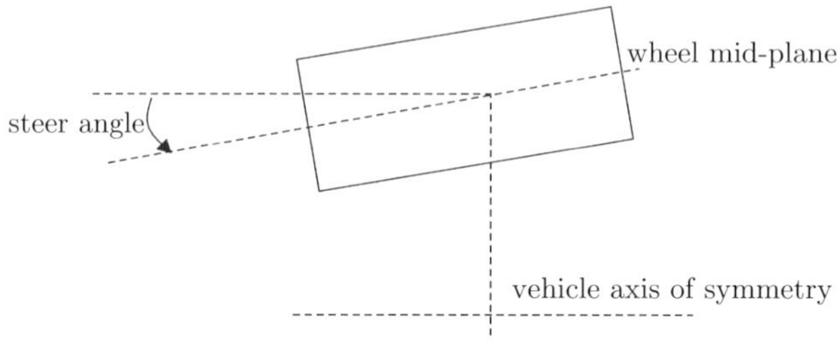

A reasonable approximation to this steer angle is given by the rotation angle of the wheel carrier around its axis (2-4 line): we shall therefore attempt to determine the latter value.

12.7.2 Multibody model

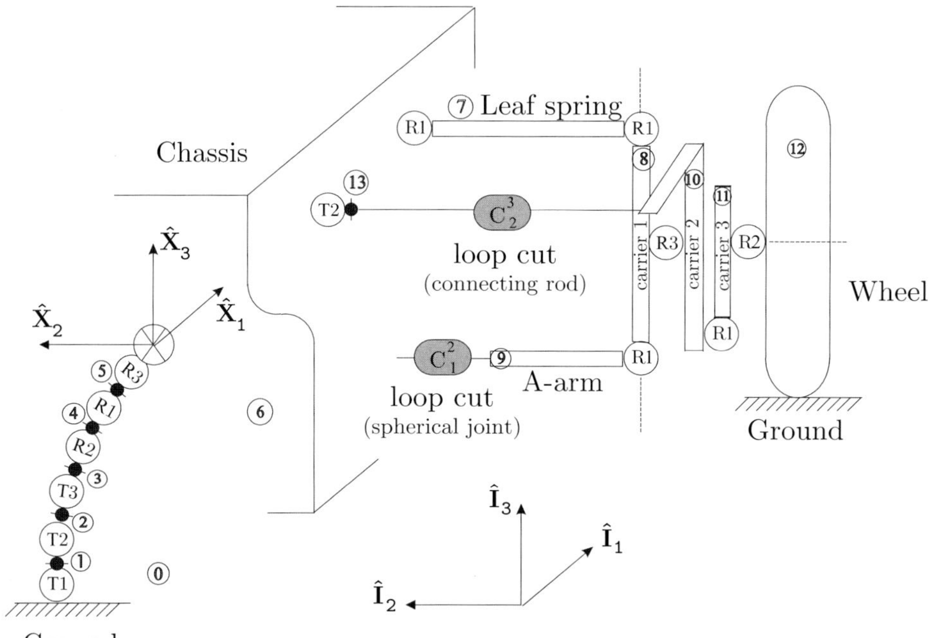

- Choice of inertial reference frame: that of conventional vehicle dynamics (x-axis: forwards, z-axis: upwards, see chapter 6).
- Since the axis of rotation of the wheel carrier and the wheel spin axis are not perpendicular, we cannot put the system in its reference configuration without introducing an additional revolute joint $R1$ (combined with

a fictitious body called "carrier 3" in the previous figure). The corresponding variable q^{11} is equal to 0 in the reference configuration and is set to $0.27128838\,rad$ for the simulations.

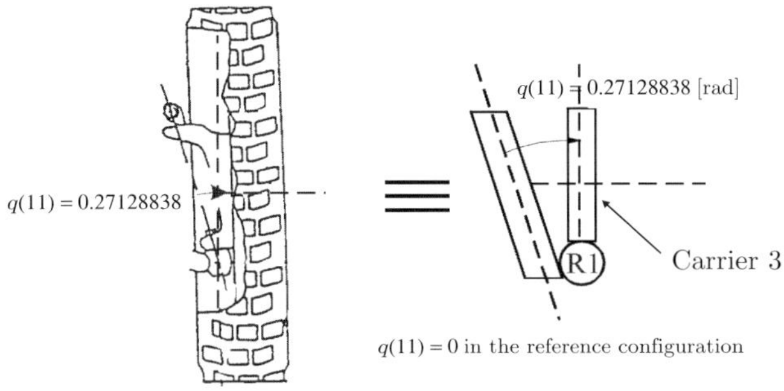

- In order to place the system in its reference configuration, we need to open up the three-bar system made up of the leaf spring, wheel carrier 1 and the A-arm. We opted for a C^2 cut on the joint between the chassis and the A-arm. This cut entails three algebraic constraints, one being trivial since the three-bar system is planar. In the same way, we need to open up the loop made up of the chassis, leaf spring, wheel carrier 1, wheel carrier 2 and steering bar. Since the mass and inertia of the steering bar are negligible compared to those of the other bodies (chassis, wheel carrier), an efficient way of opening the loop consists in discarding this bar and imposing a constant distance between its two attachment points: that is a C^3 cut.

- Numbering of bodies and joints:
  - joints 1 to 6: six degrees of freedom of the chassis
  - joint 7: revolute joint between the leaf spring (body 7) and the chassis (body 6)
  - joint 8: revolute joint between the arm (body 8) and the leaf spring (body 7)
  - joint 9: revolute joint between the A-arm (body 9) and carrier 1 (body 8)
  - joint 10: revolute joint between wheel carrier 2 (body 10) and carrier 1 (body 8)
  - joint 11: fictitious joint (frozen rotation) to take into account the angle between the wheel carrier rotation axis and the wheel plane (wheel carrier 3)
  - joint 12: wheel spin (body 12)
  - joint 13: prismatic joint between the rack (body 13) and the chassis (body 6)

- Reference configuration: the reference configuration is such that the chassis center of mass lies 57 cm above the ground (nominal configuration specified in the benchmark).

12.7.3 Computer pre-process

System data (MBdesc)

- The geometrical dimensions are as follows (in relative coordinates).

- The components of position vector $\mathbf{d}^{01}$ relative to the first joint (chassis/ground) are: $(0\ \ 0\ \ 0.57)^T$: this vector is not essential but allows a positioning of the chassis center point in a reference configuration which is "close" to the final equilibrium configuration.

- The components of the joint position vectors on the chassis in the latter's reference frame are:

Point	x	y	z	Comments
A-arm attachment	0.97	-0.259	-0.268	C^2 cut: $\mathbf{d}_{lp}$ vector
Leaf spring attachment	0.97	-0.1585	0.03	$\mathbf{d}^{67}$ vector
Rack center point	0.83	0.0	0.03	$\mathbf{d}^{6,13}$ vector

- Length of the rigid bar modeling the leaf spring: 0.33664707 m

- The components of the position vectors of the revolute joints (in the frame attached to carrier 1) are:

Point	x	y	z	Comments
A-arm attachment	0.0	0.0	-0.31346451	$\mathbf{d}^{89}$ vector
Wheel carrier 2 attachment	0.0	0.0	-0.18164334	$\mathbf{d}^{8,10}$ vector

- Length of A-arm: 0.32146 m (to define the $\mathbf{d}_{ls}$ vector of the C^2 cut)

- In the frame attached to wheel carrier 2, we still need to define the position vector of the fictitious joint $R1$ (variable q^{11}) and the attachment point of the steering bar:

Point	x	y	z	Comments
Joint 11	0.0	0.0	-0.13182117	$\mathbf{d}^{10,11}$ vector
C^3 cut	-0.14	0.038537057	0.19236225	$\mathbf{d}_{lp}$ vector

- Steering bar attachment point on the rack: $(0\ \ -0.07\ \ 0)^T$ ($\mathbf{d}_{ls}$ vector of the C^3 cut)

- Length of the steering bar: 0.38424601 m (length of the C^3 cut connecting rod)

- The components of the position vector of the wheel spin joint (joint 12) in the frame attached to wheel carrier 3 are: $(0\ \ -0.043\ \ 0.127)^T$.

- Perform the two cuts: C^2-type between the chassis (primary body 6) and the A-arm (secondary body 9) and C^3-type between wheel carrier 2 (primary body 10) and the rack (secondary body 13).

Body 12 does not have to be input as a wheel: as we know, no tire model needs to be introduced in the kinematic analysis of the present problem.

User's model (MBuser)

Variables q^1 to q^6 (degrees of freedom of the chassis) as well as the wheel carrier fictitious angle ("variable" q^{11}), the wheel spin (variable q^{12}) and the rack excursion (variable q^{13}) are locked:

$$\boxed{\begin{array}{l} q^1 \text{ to } q^6 = 0 \\ q^{11} = 0.27128838 \\ q^{12} = q^{13} = 0 \end{array}}$$

Initial conditions (MBini)

The initial configuration is prescribed in the benchmark. It corresponds to a suspension angle $q^7 = 0.206425\,rad$. The dependent variables can be set to zero. Note that this configuration is not an equilibrium configuration. The latter will be computed as part of the next problem.

12.7.4 Computer process

Symbolic treatment (ROBOTRAN - MBgen)

All that is needed is to generate the loop kinematic constraints.

Coordinate partitioning (MBpart)

- q^7: independent variable
- q^1 to q^6, q^{11} to q^{13} : driven variables (locked)
- q^8 to q^{10}: dependent variables

As foreseen, a trivial (redundant) constraint appears ($0 = 0$): indeed, the suspension is planar.

Numerical treatment (MBsim)

- Perform a first closure of the system in order to determine the values of the dependent joint variables.
- Give variable q^7 an increment from 0 rad to $0.6\,rad$ (e.g. perform 100 configurations).

12.7.5 Computer post-process

The solution is given by the graph of angle q^{10} versus the suspension angle (independent variable q^7). This graph is shown below. The angle is zero in the initial configuration. Around this configuration, when the suspension drops down (decrease in q^7), the steer angle varies from zero to a slightly negative value. Conversely, when q^7 increases, the steer angle becomes positive and increases correspondingly. It remains however that the variations of the ILTIS steer angle (frozen steering) around its nominal value are very limited.

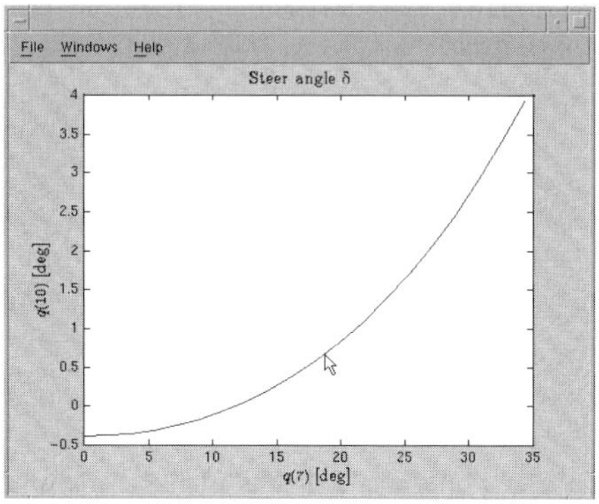

12.8 A jeep

- Determine the static equilibrium configuration of the jeep described in problem 12.7. The results expected are of geometrical nature (chassis height, pitch angle) and also of dynamical nature (forces exerted by the leaf springs, tires, spring-shock absorber assembly, as well as the loads appearing in the steering bar.
- Compute the eigenvalues relative to the vehicle operating on a straight at a constant $20\,m/s$ velocity.
- Finally, analyze the behavior of the vehicle at 5 and $10\,m/s$ over a given road profile (cosine bump).

Originality of the problem: By means of this quite comprehensive problem, various new and important features will be introduced:

- the decomposition of a complex system into a set of sub-systems, with the introduction of re-assembly constraints (see section 4.4).
- the model via a "user-type" link force associated with the loads induced by the leaf spring
- a more elaborate road profile and the use of the *Calspan* lateral wheel/ ground model (see chapter 6).

12.8.1 Analysis

Description of the system

The rear suspensions are mirror images of the front ones. The only difference lies in the absence of a rack: the steering bar is attached directly to the chassis. As regards the detailed description of a suspension, the user is referred back to the previous problem.

Modeling hypotheses

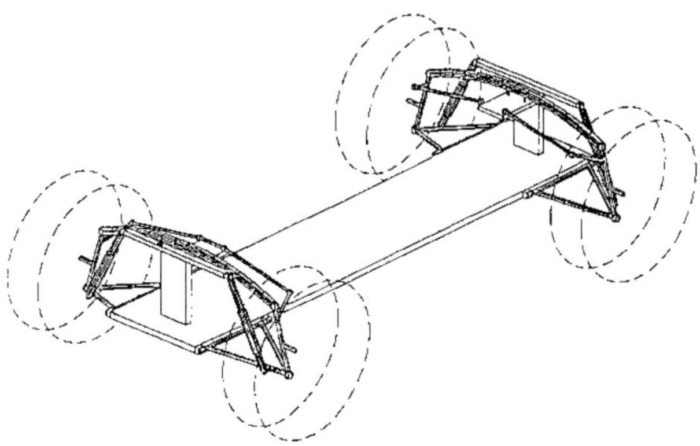

- It is assumed that all masses and inertias are redistributed onto the following bodies: chassis, A-arm, wheels. This is as imposed by the benchmark.
- Other modeling simplifications are proposed in the benchmark and relate to the wheel/tire contact force, the leaf spring and the spring-shock absorber assembly.
  - The tire model consists of a radial force (perpendicular to the ground) modeled (when contact) by a spring with stiffness $463000\,N/m$. The nominal radius of the front wheels is $0.364325\ m$ and that of the rear wheels is $0.363812\,m$. The lateral forces are computed from the "Calspan" model described in chapter 6.
  - The leaf spring is replaced by a rigid bar and a spring acting[10] along the chassis z-axis, between the end of the bar and the chassis. The spring rate is $k = 35906\,N/m$, its neutral length $l_0 = 0.145\,m$ (given data). When the excursion of the rigid bar is equal to zero (horizontal bar), it hits a stop block on the chassis. This stop block has a $10^7\ N/m$ stiffness and acts in the same direction as the spring (z-axis of the chassis).
  - For the spring-shock absorber assembly, the following formulae were set up using experimental data.

Spring:

Let us denote x the distance between the primary attachment point on the chassis and the attachment point on the A-arm. In the nominal configuration (given in the terms of the problem), $x = 0.440557\,m$. The spring force is given by:

$$F_{spr} = -4.0092.10^6 + 2.8397.10^7\,x - 6.7061.10^7\,x^2 + 5.2796.10^7\,x^3\ \ [N] \quad (12.3)$$

Damper:

The damping force depends on the relative velocity $\dot{x}$ ($[m/s]$) between these two points:

$$\begin{aligned}
&\text{if } \dot{x} < -0.2\\
&\quad F_{damp} = -416.42 + 1844.3\,\dot{x}\\
&\text{if } -0.2 \le \dot{x} < 0.21\\
&\quad F_{damp} = 9945.627\,\dot{x} + 33955.72\,\dot{x}^2 - 59832.25\,\dot{x}^3 - 395651.0\,\dot{x}^4\\
&\text{if } \dot{x} \ge 0.21\\
&\quad F_{damp} = 1919.1638 + 1634.727\,\dot{x}
\end{aligned} \quad (12.4)$$

Analysis of expected results

The force in the steering bar is computed from the Lagrange multiplier associated with the C^3 type constraint.

[10] for any configuration of the vehicle.

12.8.2 Multibody model

- Number of sub-systems: 4 (front right suspension, front left suspension, rear right suspension, rear left suspension)

If we had chosen to model the system as a single entity, this would have entailed 31 bodies (6 for the chassis, 4 × 6 bodies for a full suspension (including steering bar and wheel) + 1 rack). 16 constraints would be needed (4 × 3 for the C^2 cuts + 4 × 1 for the C^3 cuts), leaving 15 degrees of freedom (6 for the chassis + 4 × 1 for the suspensions +4 × 1 for wheel spins +1 for the steering). The size of the matrices involved in the equations of motion would become quite large for such a system, whence the idea of breaking down the vehicle into four sub-systems. Each sub-system models a quarter-chassis and a suspension. ROBOTRAN generates the equations of each sub-system independently. Although there is no rack at the rear, to make things easier each sub-system is assumed to consist of 13 bodies (see problem 12.7). The racks at the rear are frozen during the numerical simulation.

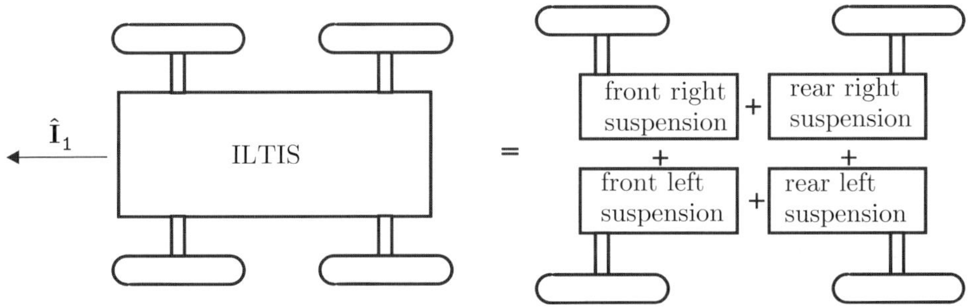

- Since the sub-systems are not independent, additional constraints are required. These constraints are quite elementary and consist simply in imposing that the 4 quarter-chassis be properly re-assembled, and also that the two parts of the rack of the two front suspensions have identical displacements. In other words, we impose identities between the 6 joint variables modeling the 6 degrees of freedom of each quarter-chassis, as well as between the two variables modeling the translational motion of the forward rack. These user-type constraints between sub-systems will lead to 19 additional Lagrange multipliers (see section 4.4).
  Compared with the model of the previous problem, we also need to add:
  - a link between the chassis and the A-arm in order to model the spring-shock absorber assembly,
  - a "user"-type force between the chassis and the rigid bar to model the leaf spring restoring force. The reason we implement a "user"-type force is that the spring must at all times remain parallel to the z-axis of the chassis frame (see terms of the problem), meaning that its attachment point on the chassis is consequently not fixed. A conventional link (see section 4.3) would therefore not be appropriate.

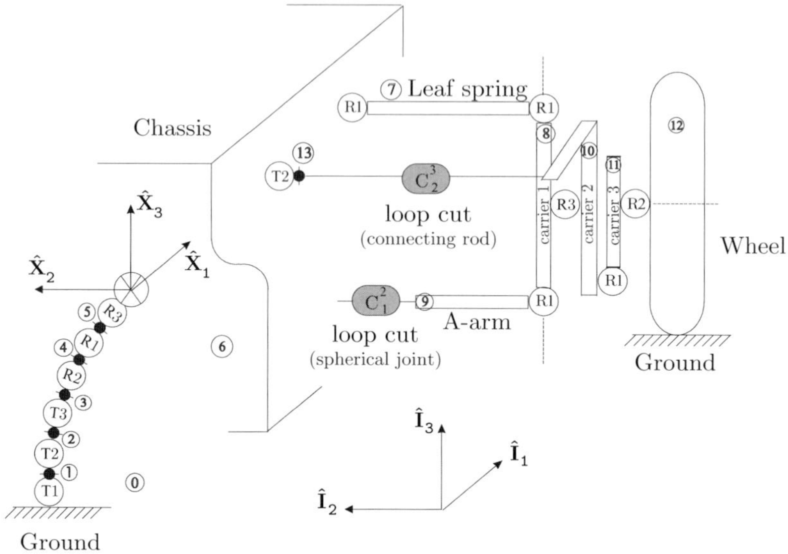

Figure 12.2:

12.8.3 Computer pre-process

System data (MBdesc)

Input the symbolic external forces and torques on bodies 6, 7, 9 and 12 of each sub-system, in the ROBOTRAN symbolic data files.

- Dynamical data:
  - Mass of the chassis: 1260 kg
  - Inertia of the chassis (in the latter's reference frame):

$$\begin{pmatrix} 130 & 0 & 0 \\ 0 & 1620 & 0 \\ 0 & 0 & 1670 \end{pmatrix} kg\, m^2$$

    These mass and inertia can either be associated with the chassis of the first sub-system, or shared between the four quarter-chassis.
  - Mass of a wheel and its braking system (body 12): 57.35 kg.
  - Inertia of a wheel in its body frame:

$$\begin{pmatrix} 1.2402 & 0 & 0 \\ 0 & 1.908 & 0 \\ 0 & 0 & 1.2402 \end{pmatrix} kg\, m^2$$

  - Mass of an A-arm (body 9): 6 kg

– Inertia of an A-arm in its body frame:

$$\begin{pmatrix} 0.052099 & 0 & 0 \\ 0 & 0.023235 & 0 \\ 0 & 0 & 0.068864 \end{pmatrix} kg\,m^2$$

– All other bodies are assumed to have negligible mass.

- Introduce a link between bodies 6 and 9 of each sub-system (for sub-system "front right suspension" - see problem 12.7)

Shock absorber attachment	x	y	z	Comments
on the chassis (body 6)	1.045	−0.297	0.062	$\mathbf{d}_{bp}$ vector
on the A-arm (body 9)	0.075	0.0728438	−0.0046671	$\mathbf{d}_{bs}$ vector

The constitutive equation of the link is introduced in a user function according to equations 12.3 and 12.4.

- Input the data for the three other sub-systems, taking into account:
  – the left to right symmetry for the signs of y-components of all position vectors, and front to rear symmetry for the signs of x-components of position vectors
  – the position of the vehicle center of mass (located forward, see previous problem), which changes the x-component of all position vectors defined on the chassis, by an amount: $1.047 - 0.97 = 0.077\,m$.
  – the attachment of the rigid bar
  – the attachment of the link
  – the attachment of the rack
  – the attachment of the C^2 cuts between the chassis and the A-arm.

User's model (MBuser)

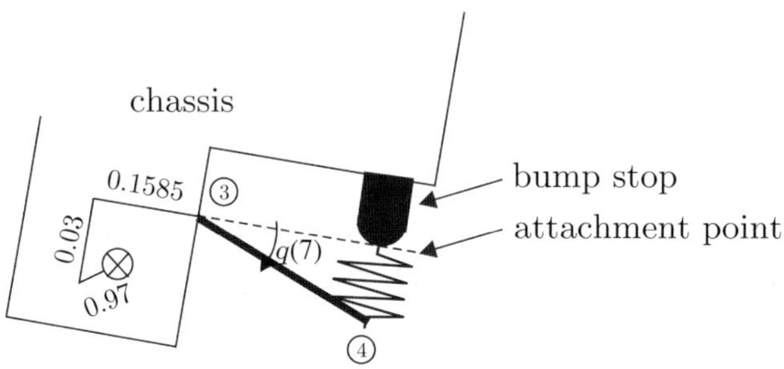

- For each sub-system, we start by introducing the force between the rigid bar and the chassis, in the form of a "user" force. This requires a short preliminary calculation (see the above figure). The spring attachment points used in modeling the leaf spring are located: one at an end of the rigid bar (point 4), the other at the intersection of a perpendicular to the chassis $\{x, y\}$-plane through point 4 and a parallel to the chassis y-axis through the other end of the rigid bar (chassis side, point 3). This intersection is not a fixed point on the chassis.

- For sub-system 1, "front right suspension", its coordinates in the chassis fixed frame are:

along x:	$0.97\ [m]$
along y:	$0.1585 + 0.33664707 * \cos q^7\ [m]$
along z	$0.03 - 0.33664707 * \sin q^7\ [m]$

- The spring length is $l = 0.3664707 * \sin q^7\ \ [m]$
- The spring hits the stop block when $q^7\ \leq\ 0$ for the right hand side suspensions and $q^7 \geq 0$ for the left hand side suspensions.

- The next step is to input the driven variables:
  - the angle of the wheel carrier (variable q^{11}) for each sub-system (paying attention to left to right symmetry),
  - the wheel spin rotations,
  - the displacement of a front rack (the vehicle is on a straight), the other one being determined via a simple constraint (see below).
  - the position of the two rear racks, which are frozen at a zero value.
- The 19 elementary constraints between sub-systems must be introduced.
- The final input relates to the road profile.
  The imposed profile is a $5\,m$-long cosine bump with a maximum height of 20 cm. It is described by its analytical equation $\mu(x, y)$, which the user must introduce in his multibody model. In the initial configuration, the front wheels lie against the rise of the bump (contact point at the transition between flat ground and cosine bump).
  The road profile function and spatial derivatives are:

$$
\begin{aligned}
&\mu = 0 \\
&\frac{\partial \mu}{\partial x} = \frac{\partial \mu}{\partial y} = 0 \\
&\text{if } \ 0.97 \leq x < 5.97 \\
&\quad \mu = 0.1\,\sigma\left\{1 - \cos(\frac{2\pi}{5}\,(x - 0.97))\right\} \\
&\quad \frac{\partial \mu}{\partial x} = 0.04\,\pi\,\sigma\sin(\frac{2\pi}{5}\,(x - 0.97)) \\
&\text{end}
\end{aligned}
$$

In the case of an in-phase bump, $\sigma = 1$ for all four sub-systems. For a contrary-phase bump, $\sigma = 1$ for the sub-systems on the right and -1 for those on the left.

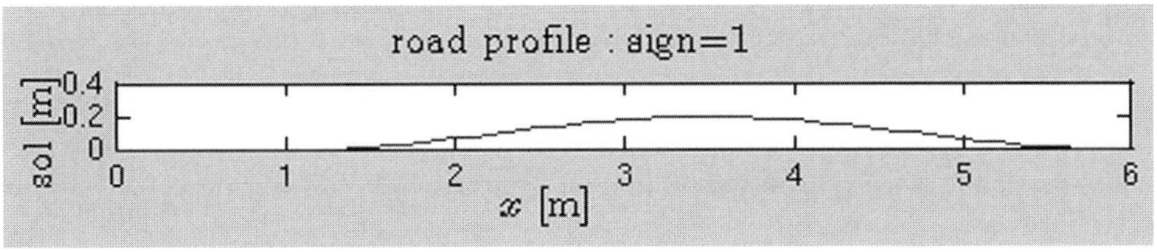

Initial conditions (MBini)

The initial configuration (resulting from equilibrium) is identical to that chosen in the previous problem (pay attention to left/right symmetry).

12.8.4 Computer process

Symbolic treatment (ROBOTRAN - MBgen)

Using ROBOTRAN (MBgen), we generate:
- the dynamics,
- the loop kinematic constraints,
- the links,
- the wheel kinematics.

Coordinate partitioning (MBpart)

- Driven variables[11]:
  - sub-system 1 "front right suspension": variables q^1, q^{11}, q^{12}, q^{13}
  - sub-system 2 "front left suspension": variables q^{11}, q^{12} but not variable q^{13} which is constrained to be equal to variable 13 of sub-system 1)
  - sub-system 3 "rear right suspension": variables q^{11}, q^{12}, q^{13}
  - sub-system 4 "rear left suspension": variables q^{11}, q^{12}, q^{13}
- Independent variables:
  - sub-system 1 "front right suspension": variables q^2, q^3, q^4, q^5, q^6 q^7
  - sub-system 2 "front left suspension": variable q^7
  - sub-system 3 "rear right suspension": variable q^7
  - sub-system 4 "rear left suspension": variable q^7

[11] local numbering, i.e. from q^1 to q^{13} for each sub-system

Numerical treatment (MBsim)

Static equilibrium In the benchmark, it is first asked to determine the static equilibrium configuration (vehicle at rest[12]). Based on this configuration, determine the forces in the links and tires and the Lagrange multipliers associated with the C^3 constraints.

Note that the multiplier associated with a C^3 type (rod) constraint corresponds to the tensile force per unit length in the rod (see section 4.1.3)

Modal analysis at 20 m/sec speed Starting from the previous static equilibrium, change the forward velocity of the vehicle (driven variable q^1 in subsystem "front right suspension": $q^1 = 20\,t$; $\dot{q}^1 = 20$; $\ddot{q}^1 = 0$) along with the spin velocities of the four wheels[13] ($\dot{q}^{12} = 20/\text{nominal radius}$). Note that this is an average angular velocity, based on the nominal radius of the wheels. This approximation is rather valid for the requested modal analysis.

First perform a direct equilibrium along with a modal analysis. In order to limit the computing time, choose the previous static equilibrium configuration as initial configuration for starting the present quasi-static equilibrium.

Displacement over an irregular road profile Two simulations are performed. The first one uses the "in-phase" road profile and $\dot{q}^1 = 5\,m/s$, the second one a "contrary-phase" profile and $\dot{q}^1 = 10\,m/s$. The vehicle's behavior must be determined over at least 15 meters.

- We must properly adapt the wheel angular velocity (the previous approximation is still valid).

- We then perform the numerical integration, where we suggest using a "stiff" integrator. A significant variable (e.g.: chassis height) can be monitored to check on the behavior of the vehicle.

[12] In this case, only the vertical ground/tire force are considered since the Calspan model is irrelevant at rest (and mathematically singular).

[13] Locking the wheel spin velocity is justified by the fact that there is no longitudinal model for the contact forces.

12.8.5 Computer post-process

Static equilibrium

Chassis height[14]:	$d_z^{01} - 7.067\,10^{-3} m$
Chassis pitch:	$0.002552\, rad$
Vertical force on front wheel:	$3831.31\, N$
Vertical force on rear wheel:	$3591.92\, N$
Force in the front steering bar[15]:	$(17.115\ \times\ 0.384246)\ N$
Force in the rear steering bar:	$(15.964\ \times\ 0.384246)\ N$
Internal load at A-arm/front carrier joint:	$3270.68\, N$
Internal load at A-arm/rear carrier joint:	$3027.36\, N$

Modal analysis

Description	Eigenvalue λ	Frequency $[Hz]$[16]	Damping [%]
Vehicle lateral (rigid)	$0.0 + 0.0\ i$	/	/
Vehicle yaw (rigid)	$0.0 + 0.0\ i$	/	/
Cabin lateral/roll	$-4.05 \pm 3.78\ i$	0.88	73.04
Cabin pitch	$-3.94 \pm 9.48\ i$	1.63	38.38
Cabin bounce	$-4.90 \pm 11.1\ i$	1.93	40.33
Cabin roll	$-23.8 \pm 13.5\ i$	4.36	86.97
Wheel hop / Cabin roll	$-27.6 \pm 71.5\ i$	12.20	36.06
Wheel hop / Cabin bounce	$-29.8 \pm 82.0\ i$	13.89	34.17
Wheel hop / Cabin pitch	$-29.7 \pm 82.9\ i$	14.01	33.71
Wheel hop / Cabin yaw	$-29.3 \pm 85.9\ i$	14.45	32.29

Time integration

1. In-phase bump ($v = 5\, m/s$)
   The graphs below depict the time history of the chassis height, its pitch angle and the radial forces on the tires (when one of these forces becomes zero, this means the corresponding wheel has left the ground).

---

[14] recall that $d_z^{01} = 0.57\ m$

[15] The force in a connecting rod associated with a C^3 cut is the corresponding Lagrange multiplier (units: $[N/m]$) $\times$ the rod length $[m]$.

[16] frequency of the undamped mode= $\sqrt{re^2(\lambda) + im^2(\lambda)}$

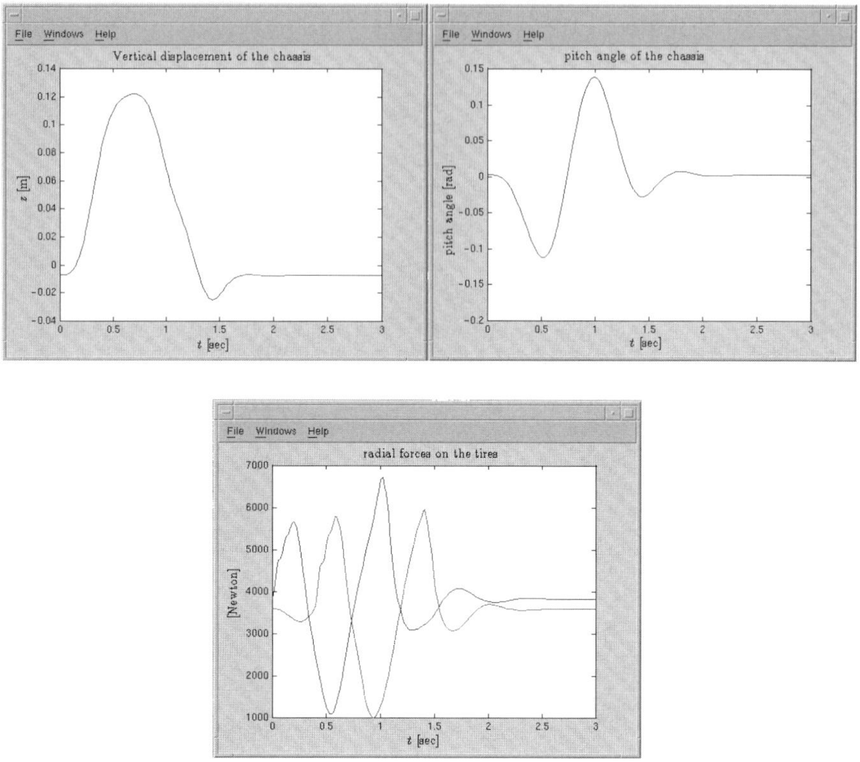

2. Contrary-phase bump ($v = 10\, m/s$)

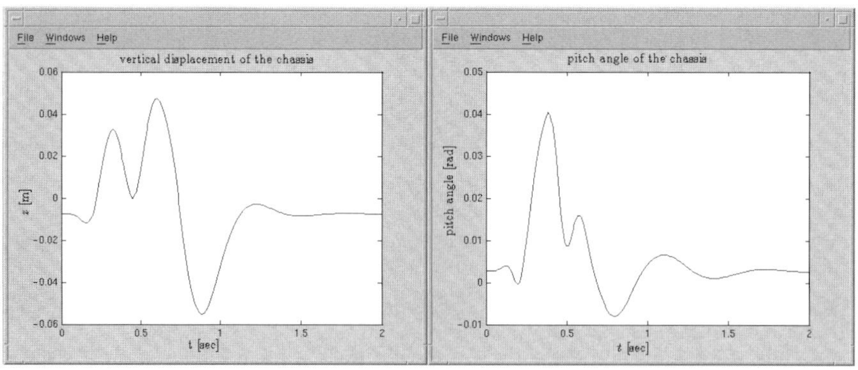

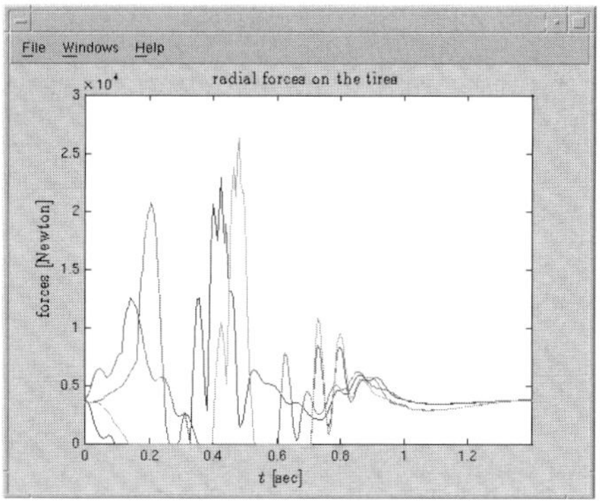
File Windows Help
radial forces on the tires
forces [Newton]
t [sec]

12.9 A flexible slider-crank

Starting from the slider-crank mechanism of problem 12.4, we now include the flexibility of the connecting rod. Carry out a modal analysis of the system (at rest) and determine the transverse deflection of the connecting rod mid-point as well as its shortening when the crank is given a constant 150 $\frac{rad}{sec}$ rotational speed

Originality of the problem: Showing that it is feasible to model a flexible beam in a quite simple manner by using a set of rigid bodies which are directly compatible with a rigid multibody formalism.

12.9.1 Analysis

Description of the system

- Number of bodies: 3
  - Crank (rigid)
  - Connecting rod (flexible, homogeneous and isotropic material, circular cross section, Young's modulus $E = 0.2\ 10^{12}\ \frac{N}{m^2}$)
  - Piston
- Joints: 4 (see problem 12.4)
- Kinematic loop: 1. This loop reduces the number of degrees of freedom by 2 (planar system)
- Degrees of freedom: 1 rigid + an infinity resulting from the flexibility of the connecting rod
- Forces and torques: a torque on the crank ensuring a constant rotational speed (see problem 12.4)
- Environment: The system is not in a gravity field.

Modeling hypotheses

- The system is assumed to be frictionless (also as regards structural damping in the flexible connecting rod)
- We assimilate the flexible connecting rod to a beam undergoing pure bending (zero extension and shear) and it is modeled using the finite segment technique suggested in [42]: the latter consists in replacing the beam by a series of rigid bodies (10 of these are sufficient for the requested simulations) linked together by springs modeling the flexural stiffness of the beam.

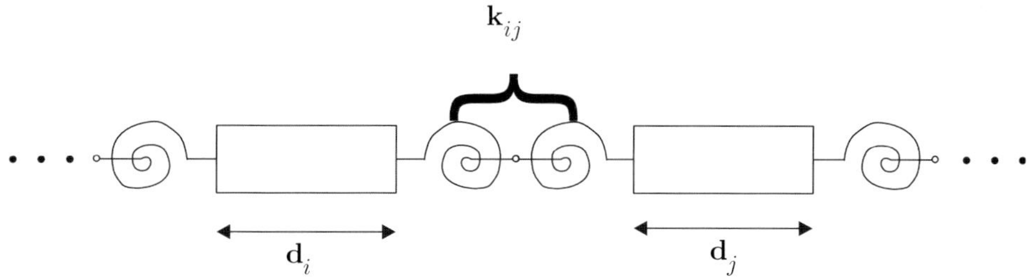

Each spring has a stiffness:

$$k_{ij} = \frac{2E_i^Y E_j^Y I_i I_j}{(E_i^Y I_j d_j + E_j^Y I_i d_i)} \tag{12.5}$$

In our case, segments i and j are identical (1/10th of the connecting rod length), whence:

$$\text{if } i = j \Longrightarrow K_{ij} = \frac{E^Y I}{d}$$

where:

- E is the Young's modulus of the connecting rod $= 0.2\ 10^{12}\ N/m^2$
- I is the geometrical moment of inertia of a segment cross section $[m^4]$, relative to its geometrical center, around the axis perpendicular to the plane formed by the connecting rod and the crank, i.e. $I = \frac{\pi R^4}{4}$ where R is the segment radius $= 3\,mm$.
- d is the length of a segment (and thus d = connecting rod length/10)

Overall system representation:

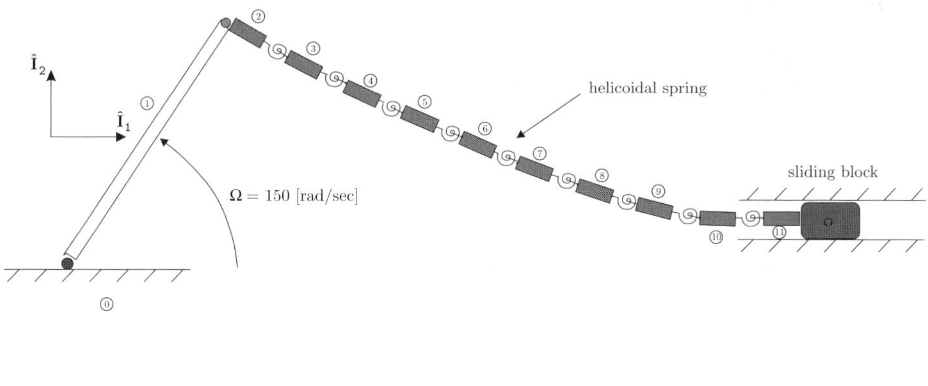

Analysis of expected results

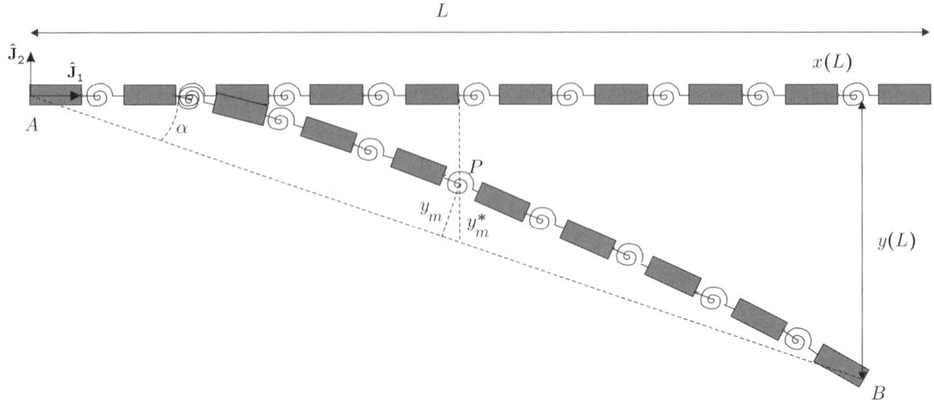

The transverse deflection y_m (≥ 0 in the figure) of the mid-point P and the beam shortening ($||\overrightarrow{AB}|| - L$) can be computed on the basis of a simple geometrical reasoning:

$$y_m = -\cos(\alpha) * y_m^\star \quad \text{and} \quad ||\overrightarrow{AB}|| - L = \sqrt{x(L)^2 + y(L)^2} - L \tag{12.6}$$

where, using similar triangles:

$$y_m^\star = -y\left(\frac{L}{2}\right) + \frac{x(\frac{L}{2})}{x(L)} * y(L) \quad \text{and} \quad \alpha = \arctan\frac{y(L)}{x(L)}$$

$x(L)$, $y(L)$ are the coordinates of the end-point of beam AB, in a reference frame $\{\hat{\mathbf{J}}\}$ attached to the first segment A. They can be easily obtained via a position sensor (attached to the end of the 10th segment (point B on body 11)) and expressed in the frame attached to the first segment (body 2). Thanks to a second sensor, it is also possible to determine the $x\left(\frac{L}{2}\right)$,$y\left(\frac{L}{2}\right)$ coordinates of the mid-point (end-point of segment 5, i.e. body 6) expressed in the same frame.

12.9.2 Multibody model

- Choice of inertial frame: it is placed along the crank rotation axis, with its axes $\hat{\mathbf{I}}_x$ and $\hat{\mathbf{I}}_y$ in the plane of motion (see problem 12.4).
- Loop cut: in the same way as in problem 12.4, the C^2 cut is made at the joint between connecting rod and piston, and results in three algebraic constraints (one being trivial (planar system)).
- Numbering of bodies and joints:
  - Joint 1 : rotation of the crank relative to the inertial reference frame,
  - Joint 2 : rotation of the first connecting rod segment relative to the crank,

  - Joints $3, 4, 5, 6, 7, 8, 9, 10$ and 11 : internal connecting rod rotations,
  - Joint 12 : translation of the piston relative to the inertial reference frame.

- Reference configuration: The reference configuration is such that the crank and all connecting rod segments lie horizontally, the piston lying along axis $\hat{\mathbf{I}}_x$ at a distance equal to crank + connecting rod lengths.

12.9.3 Computer pre-process

System data (MBdesc)

- The numerical data are copied from problem 12.4. As concerns the flexible connecting rod, the data appear above (flexural stiffness, number of segments, ...).
- Carry out the cut (C^2-type) between the end of the connecting rod and the piston.
- Place two position sensors (the first one at the end of body 6, the other at the end of body 11), for which the kinematics will be computed symbolically with respect to body 2.

User's model (MBuser)

- Control the rotation angle of the crank. At first, for the modal analysis, the crank is at rest ($q^1 = \dot{q}^1 = 0$).
- Add stiffness to joints 3 to 11, in MATLAB syntax for instance:

```
k = 0.2e12 * pi * 0.003^4 / 4 / 0.03;
for  i = 2 :  11
   Q(i) = -k * q(i);
end
```

Initial conditions (MBini)

In the initial configuration, the connecting rod-crank system lies horizontally. All the joint coordinates q are thus equal to zero.

12.9.4 Computer process

Symbolic treatment (ROBOTRAN - MBgen)

- Generate the dynamical equations of the system,
- Generate the constraint equations associated with the loop,
- Generate the kinematic equations relating to the two sensors.

Coordinate partitioning (MBpart)

As regards partitioning of the variables, an acceptable choice of dependent variables is q^2 and q^{12}. Variable q^1 is driven like in problem 12.4.

Numerical treatment (MBsim)

The first task is a modal analysis of the system at rest (with the crank lying horizontally). Note that the reference configuration is also an equilibrium configuration as there are no gravity effects. The modal analysis results can compared with the analytical values (flexible simply-supported beam).

Mode	Numerical results	Analytical results
1	$131.96\,Hz$	$131.98\,Hz$
2	$527.57\,Hz$	$527.91\,Hz$
3	$1185.57\,Hz$	$1187.79\,Hz$
4	$2101.69\,Hz$	$2114.45\,Hz$

The first three modes are quite satisfactory[17]; our 10-segment beam model is therefore valid for the requested simulation. Better results could be obtained by increasing the number of segments, but this would uselessly burden the model.

The second task is a numerical time integration of the system.

In order to produce this result, the best approach is to impose a continuously increasing rotation of the beam from $0\,rad/sec$ to a constant $150\,rad/sec$ by appropriately controlling the driven crank rotation $\dot{q}^1$. Indeed, the task consists in simulating the mechanism at a constant $150\,rad/sec$ speed. However, most integrators will not stand a sharp discontinuity (instantaneous shift from 0 to $150\,rad/sec$). Moreover, the choice of integrator is of utmost importance as regards processing speed. The eigenfrequencies of the system are of quite different orders (from $132\,Hz$ to $2102\,Hz$ for the 4th mode). This is called a *stiff* problem and the implicit Newmark scheme (see chapter 10) is well suited to the task. Note that a different choice of integration scheme would have entailed much longer CPU times, as can be seen from the following table:

Integration scheme	Newmark	Runge-Kutta method[18]	Stiff method[19]
relative CPU times	1	9	6

In order to implement the Newmark scheme, the user needs to specify (see chapter 10):

- A "cut-off" frequency (see section 10.2.3): this allows the scheme to propose an initial time increment. Choose e.g. the fourth frequency, i.e. $2115\,Hz$.

[17] Note that the computed frequencies slightly underestimate the analytical frequencies: the finite segment method implemented here is not a Rayleigh-Ritz type method.

[18] ode45 MATLAB

[19] ode15s MATLAB

- Choose as Jacobian the mass matrix + stiffness matrix and do not request a computation at each time step (see figure 10.2).

- An order of magnitude of the various generalized variables (see section 10.2.3). This order of magnitude does not have to be accurate. However, the loop closure procedure must converge for the given choice. It allows the scheme to adapt its error tolerance to the chosen application.
  A reasonable estimation consists in choosing a 1° opening ($\pi/180\,[rad]$) at each internal joint of the beam. This leads to:

$$\boxed{\begin{array}{l} q^1 = 0 \\ q^3 \text{ to } q^{11} = \frac{\pi}{180} \\ \hat{q}^2 = \hat{q}^{12} = 0 \text{ (dependent)} \end{array}}$$

The evolution of $x(L)$, $y(L)$ and $x\left(\frac{L}{2}\right)$, $y\left(\frac{L}{2}\right)$ during (or after) the simulation is obtained from the kinematics of the two position sensors. The requested results (mid-point transverse deflection and rod shortening: equations 12.6) can then be obtained via a short user program on the basis of the previous results.

12.9.5 Computer post-process

In order to visualize the requested results, viz. the time history of the mechanism (with a constant speed crank), one simply needs to "window" 0.08 seconds of the simulation results, starting at a point where the connecting rod deflection is zero. This produces the following graphs, which compare favorably with those resulting from an involved finite element computation [54].

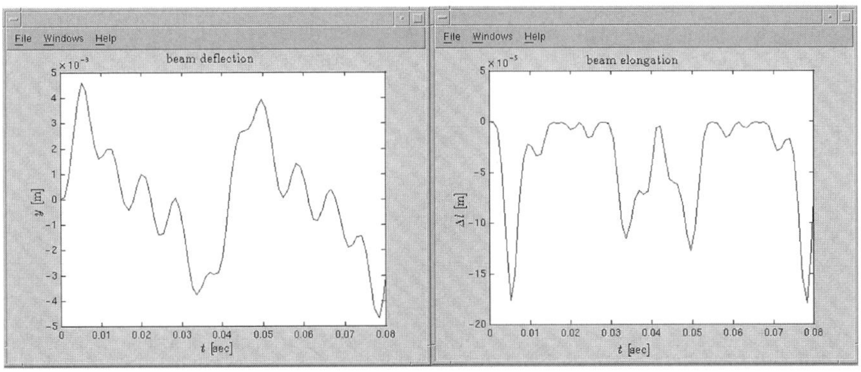

12.10 A radiation counter

The depicted radiation counter is actuated via a worm screw. Determine the value of the torque M applied to the screw at equilibrium, versus the opening angle ϑ.

Originality of the problem: There is a specific kinematic constraint between the screw and its nut. The user will therefore need to set it up using the available symbolic tools. We therefore have an example here of a user constraint (see chapter 4, section 4.2).

12.10.1 Analysis

Description of the system

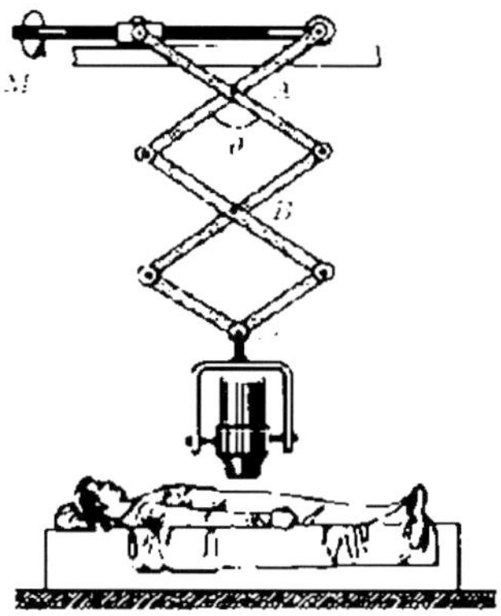

- Number of bodies: 9
  - The screw
  - The nut
  - Six bars
  - The radiation counter
- Number of joints: 12
  - Rotation of the screw
  - Translation of the nut
  - 9 revolute joints between bars
  - A revolute joint between the bars and the counter
- Number of kinematic constraints: 10

  - 1 constraint between the screw and the nut
  - 3 kinematic loops, each entailing 3 constraints

- Degrees of freedom: $2 = 12$ one-d.o.f. joints -1 screw-nut constraint -9 loop constraints

  - 1 upwards-downwards d.o.f. for the complete system (screw-nut-bars) driven by the rotation of the screw
  - 1 d.o.f. for the counter's orientation

- Forces and torques involved: a torque M applied to the screw.

- Environment: the system is in a gravity field.

Modeling hypotheses

- We shall perform a static analysis of the system. The counter orientation will be frozen and we shall drive the rotation of the screw.

- Since the bars have a negligible mass compared to that of the counter, only the latter's mass will be taken into account and assumed to be a point mass located at C ($m = 2\,kg$).

- The joints are assumed to be frictionless.

12.10.2 Multibody model

- Choice of inertial frame: located at the left end of the screw, its x-axis aligned with the latter's axis of rotation and pointing to the right, the y-axis pointing upwards. The z-axis forms a right-handed orthonormal frame.

- The loops are cut in order to recover an open tree-like structure. Since the joints are frictionless, we can cut the three middle joints (A, B and C) between bars (C^2 cut, see next figure).

- There is also a constraint between the screw rotation and the axial displacement of the nut, viz. axial displacement $= \alpha \times$ screw rotation where α is the lead of the screw.

- A frame is associated with each body.

- Numbering of bodies and joints:

  - Joint 1: $R1$-rotation between the screw and body 0.
  - Joint 2: $T1$-translation between the nut and body 0 (not body 1, otherwise the system would rotate around the x-axis along with the screw).

- Joints 3, 4, 5, 7, 8, 9: $R3$-rotations between bars.
- Joint 6: $R3$-rotation allowing to rotate the counter (frozen in the sequel, see hypotheses).

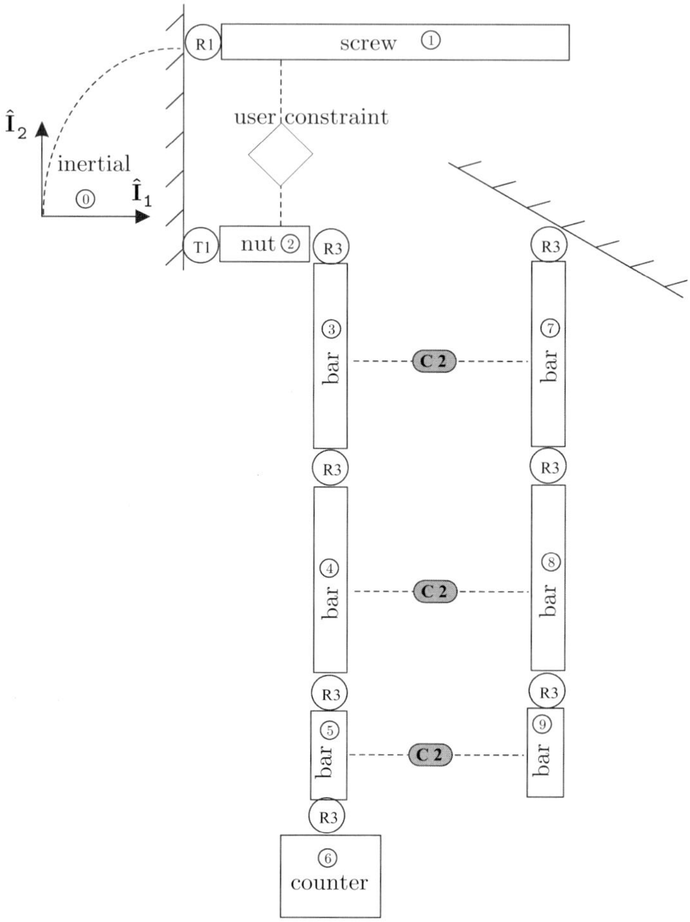

12.10.3 Computer pre-process

System data (MBdesc)

The numerical data required are:

- bar lengths: $L = 1\,m$ and $0.5\,m$ for the last two
- length of the screw: the fixed point on the right lies 1 m from the origin of the inertial frame (along $\hat{\mathbf{I}}_1$)
- lead of the screw: $\alpha = 0.01\,m/rad$
- counter mass: $m = 2\ kg$

Since we perform a static analysis, we may set the velocities $\dot{q}^i$ and accelerations $\ddot{q}^i$ to zero for all variables.

User's model (MBuser)

We need to generate a user-type constraint for the screw-nut connection. According to section 4.2, the following statements must be provided, for the constraint $h(q)$, the Jacobian $J(q)$ and the acceleration term $\dot{J}\dot{q}(q, \dot{q})$:

```
  alpha = 0.01;  % lead of the screw
% Constraint
  h(1) = q(2) - alpha * q(1);
% Jacobian
  J = zeros(1,9);
  J(1,1) = -alpha;
  J(1,2) = 1;
% J̇q̇ term
  jdqd(1) = 0;
```

As specified in the hypotheses, the angular position of the counter is frozen and the screw rotation is driven in such a way that all possible excursions of the system are reached. Since the lead of the screw is $0.01\,m/rad$, the screw must rotate from 5 to 95 rad to make the nut travelling from 5 to 95 centimeters.

The corresponding driven variable q^6 and q^1 will be computed as follows:

```
q(1) = t; % with t varying between 5 and 95
q(6) = 0; % to freeze the counter rotation
```

Initial conditions (MBini)

Configuration $q^i = 0\;\forall\, i$ amounts to aligning all the bars with the screw. This limit configuration could lead to loop closure problems, which is why we choose more realistic values. A simple solution consists in choosing $\vartheta = \pi/2\,rad$, which entails:

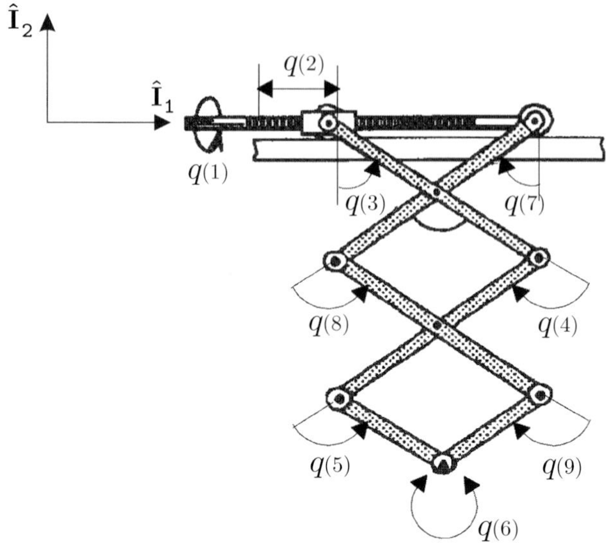

$q(1) = 29.28932188;$
$q(2) = 0.2928932188;$
$q(3) = 0.7853981634; (\pi/4)$
$q(4) = -1.5707963267; (-\pi/2)$
$q(5) = 1.5707963267; (\pi/2)$
$q(6) = 0;$
$q(7) = -0.7853981634; (-\pi/4)$
$q(8) = 1.5707963267; (\pi/2)$
$q(9) = -1.5707963267; (-\pi/2)$

12.10.4 Computer process

Symbolic treatment (ROBOTRAN - MBgen)

We are interested in obtaining:

- The dynamical equations of the system (in which the $M(q)\,\ddot{q}$ term is of no interest because $\ddot{q}^i = 0, \forall\, i$).
- Algebraic constraints between the joint variables. We have three C^2 cuts ('ball-joint' cuts) and one user constraint.

Coordinate partitioning (MBpart)

Generalized variables 1 and 6 are driven. Three constraints are redundant (i.e. trivial), since the system is planar. The number of independent variables is thus equal to zero: we are indeed in a static situation.

Numerical treatment (MBsim)

The loops can be closed analytically or by using a Newton-Raphson algorithm. Since variable q^1 is set equal to t, we take $t = 29.28932188$ for the initial configuration. In this configuration, we determine the Lagrange multipliers from equation 4.50. The results are summarized in the next table.

Note: Let us recall that the Lagrange multipliers for a C^2-type cut correspond to components in the frame of the common ancestor (of the loop) of the force exerted by the secondary body on the primary body. Taking the example of joint A, they give the force exerted by body 7 on body 3, i.e. a positive horizontal component along $\hat{\mathbf{I}}_1$ (to the right).

Constraint 1:	78.48 N	horizontal force in joint A
Constraint 2:	0.0 N	vertical force in joint A (0 due to symmetry)
Constraint 3:	0.0 N	transverse force (0, planar mechanism)
Constraint 4:	−39.24 N	horizontal force in joint B
Constraint 5:	0.0 N	vertical force in joint B (0 due to symmetry)
Constraint 6:	0.0 N	transverse force (0, planar mechanism)
Constraint 7:	9.81 N	horizontal force in joint C
Constraint 8:	9.81 N	vertical force in joint C
Constraint 9:	0.0 N	transverse force (0, planar mechanism)
Constraint 10:	−49.05 N	force required to move the nut
Driven joint 1:	−0.4905 Nm	torque required to turn the screw
Driven joint 6:	0.0 Nm	torque required to rotate the counter

The multiplier associated with the user constraint (constraint 10) gives the force on the nut in the current configuration. Indeed, the user constraint reads:

$$q^2 - \alpha\, q^1 = 0$$

and its Jacobian is

$$J(q) = (-\alpha \quad 1 \quad 0 \quad 0 \quad ...);$$

In the equations of motion associated with the first two joints, the generalized torque and force associated with this constraint become:

$$\mathcal{Q}' = J^T \lambda = \begin{pmatrix} -\alpha \\ 1 \end{pmatrix} \lambda \Longrightarrow \lambda = \mathcal{Q}'_{(2)} \; [N]$$

The torque M on the screw is simply the multiplier associated with the q^1 driven variable. In the initial configuration, we find $M = -0.4905\ Nm$ (see in the above table). We now need to calculate this multiplier for all the requested configurations; this can be achieved for instance by taking a time interval between 5 and 95 (recall that $q^1 = t$) and requesting a 1-*sec* time step to produce 91 configurations for which the Lagrange multipliers can be computed.

12.10.5 Computer post-process

Since the opening angle $\vartheta = 2\, q^3$ (see the first figure), we can plot M versus $2\, q^3$ which gives:

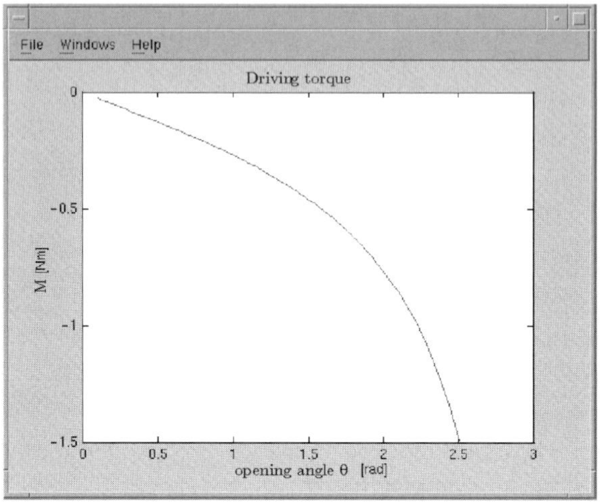

Note: the analytical solution $M = -\frac{5}{2}\, mg\ \alpha \tan\frac{\vartheta}{2}$ – which can be easily found using the virtual power principle – perfectly matches the results.

12.11 A "cam/follower" device

Determine the time history of the angle of a follower arm, for two different follower profiles, as well as the angular velocity and acceleration of the arm and the normal contact force, considering a cam rotating at constant 180 rad/sec angular velocity

Originality of the problem: This last problem will allow us to introduce a contact model between a cam and a follower and to compare the two contact models detailed in chapter 8, viz. the "combined" and the "compliant" model (spring and damper), as regards simulation results and processing time.

12.11.1 Analysis

Description of the system

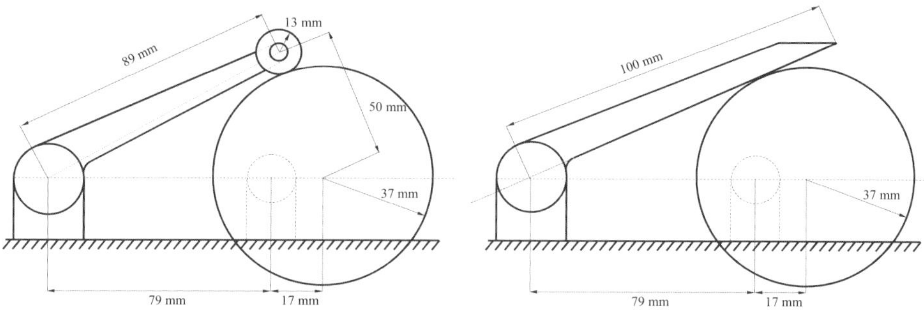

- Number of bodies:
  - 3 (the cam, the follower and the roller) for profile 1,
  - 2 (the cam and the follower) for profile 2.
- Joints: respectively 3 and 2 simple R-joints.
- Degrees of freedom: for the first profile, a contact-free configuration affords two degrees of freedom, viz. the angular displacement of the follower arm and the rotation of the roller (the cam rotates at constant velocity). If the follower and cam are in contact, the angular displacement of the arm is directly related to the cam's rotation angle, and the number of d.o.f. drops to one. For the second profile, the number of d.o.f. is one, viz. the angular displacement of the follower arm if the latter is contact-free, and zero if this is not the case.
- Forces and torques to be modeled: "cam/follower" contact forces (the contact is assumed to be frictionless: $\mu = 0$) and a torsion stiffness inside the follower/bearing joint. We shall be considering two different springs, the first stiffness being 1000 Nm/rad and the second one 100 Nm/rad.

- Environment: the system is placed in a gravity field.

Modeling hypotheses

- Rolling without slipping between the roller and the cam. The two systems then have the same number of d.o.f. (0 or 1 depending on contact condition).
- If we assume that the inertia of the roller is negligible, we can neglect it's rotation.
- The internal torsion spring has a perfectly linear characteristic curve (versus rotation).

12.11.2 Multibody model

- Choice of inertial frame: its origin is placed at the arm joint's center point, with the x-axis pointing to the right and the y-axis upwards. This choice of axes is consistent with the convention of chapter 8.
- A frame is associated with each body. Those relative to the cam and follower are such that their profiles are defined in the plane determined by the local x and y axes, to be consistent with the convention of chapter 8.
- "Combined" model: the normal contact between cam and follower can be modeled via an algebraic constraint (see section 8.3). The corresponding Lagrange multiplier allows to recover the normal force during the contact phase. When there is loss of contact, this constraint disappears, and the number of d.o.f. increases by 1. When contact is resumed, this force is temporarily modeled via a spring and parallel damper acting along the common normal to the two profiles. When contact is fully re-established, the constraint between cam and follower is active and the number of d.o.f. decreases by 1. This is called the "combined model".
- "Compliant" model: instead of imposing an algebraic constraint between the cam and the follower throughout the contact phase, we constantly model the normal contact force via a spring and parallel damper (as in [5] and see section 8.4, equation 8.48). In the event of loss of contact, this device simply disappears. This is called the "compliant" model.
  - advantage: the number of d.o.f. stays the same, whether there is contact or not.
  - drawback: due to the high stiffness of the spring modeling the normal contact, the time required for a numerical integration can be quite important (since the time step has to be very small).

Both models will be used and compared.

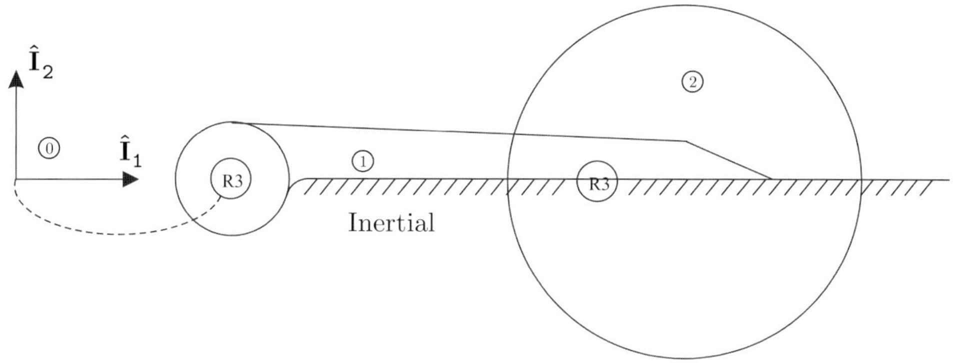

- Numbering of bodies and joints:
  - Joint 1: $R3$ revolute joint between the bed-plate (inertial body) and the follower
  - Joint 2: $R3$ revolute joint between the bed-plate and the cam

12.11.3 Computer pre-process

System data (MBdesc)

- The geometric data appear on the first figure describing the mechanism.
- Dynamical data:
  - masses of cam and follower: $1\,kg$
  - moment of inertia of cam and follower around the z-axis: $0.01\,kg\,m^2$
  - follower center of mass: 50 mm from the follower's origin (along its longitudinal axis)
  - stiffness of the torsion spring inside joint 1: 100 Nm/rad or 1000 Nm/rad (depending on simulation case - see below)
  - the stiffness of the spring used to model the normal contact force between the cam and follower is 10^7 N/m and the linear damper has a 10^5 Ns/m damping rate.

User's model (MBuser)

- Introduce the follower's internal spring:

```
Q(1) = -1000 * q(1); % simulation 1
% Q(1) = -100 * q(1); % simulation 2
```

- Drive the cam's rotational variable. To this end, we write the following statements:

```
% Cam rotation (function of time t)
omega = 180;
q(2) = omega * t; qd(2) = omega ; qdd(2) = 0;
```

- Define the cam/follower profiles according to chapter 8. The profiles implemented are quite simple:
  - As regards the cam: we have a closed profile, drawn using a single arc-type primitive whose center is offset from the center of rotation. A single center point and one offset vector are therefore all that are needed to define this profile.
  - As regards follower 1: in the same way as for the cam, the roller is drawn as an arc, whose center is offset from the follower's joint.
  - As regards follower 2: we use an open profile based on a segment-type primitive. Only two points are needed to define the latter, e.g. those with coordinates $(0,0)$ and $(0.1,0)$ $(0.1 > 0.089)$.

These data related to the cam and follower profiles must be introduced in the model according to the conventions explained in chapter 8. In particular the computation of figure 8.2 must be performed as a pre-process of the simulations[20].

Initial conditions (MBini)

As previously done for constrained systems, a good guess must be done for the initial value of the dependent joint coordinates. For the present problem, the driven variable q^2 being equal to 0 for $t = 0$, a valuable choice for the dependent variable is $q^1 = 0.52\, rad$ ($\simeq 30\, deg$).

According to section 8.3, since an iterative procedure is also required to find — at each simulation step – the geometrical contact between the cam and the follower, a good initial guess must also be provided for the auxiliary profile variables η_p and η_s. For the present problem, according to the profile descriptions, acceptable initial conditions for η_p, η_s are:

$$\eta_p = 0.058 \; ; \; \eta_s = 0.06 \text{ for } t = 0$$

12.11.4 Computer process

Symbolic treatment (ROBOTRAN - MBgen)

- Generate the dynamical equations of the open system
- Generate the – elementary – kinematics of the cam and follower.

[20]In the MBSOFT program, a specific module – MBprof – has been developed to describe cam/follower profiles and to perform the necessary pre-process upstream from the numerical simulations.

Coordinate partitioning (MBpart)

- From the problem formulation, we know that q^2 is driven (constant speed rotation).

- Depending on the chosen model, variable q^1 will be dependent ("combined" model) or independent ("compliant" model). In the first case however, we must absolutely check if the cam/follower contact is effective or not for the initial configuration of the system; for this purpose, we need to compute the initial value of the Lagrange multiplier associated with the normal contact constraint (see equation 8.42): a positive value corresponds to a compressive force and thus an actual contact. A negative force would indicate that an initial contact would be artificial (traction normal force): the contact constraints must be thus momentarily disregarded and q^1 is - momentarily - an independent variable.

Numerical treatment (MBsim)

We start by examining the first profile (profile 1). As a preliminary to the expected results, we suggest a comparison between the two types of models which can be implemented to represent the contact between cam and follower (for profile 1). In both cases, we propose to compute the time history of the follower arm's angle of rotation (joint variable q^1) versus the cam's angle of rotation (driven variable q^2).

- For the "combined" model, variable q^1 will be dependent and determined from the algebraic solution of the contact constraints. For the initial configuration, the Newton-Raphson procedure finds: $q^1 = 0.5422\,rad$. The whole simulation can be performed via a purely kinematic analysis because there is no more d.o.f[21]. The rotation period of the cam is $2 * \pi/180\,\text{sec}$ ≈ 0.035 sec. We choose a 10^{-3} sec time step.

- For the "compliant" model, variable q^1 becomes independent. Since we now have one degree of freedom, a numerical time integration needs to be performed. The first step will be to determine the static equilibrium of the system, without forgetting to set the cam rotation to zero. Once the equilibrium has been determined, we must restore the cam control law (constant speed rotation) and initiate the integration (preferably with an integrator dedicated to stiff ODE). Choose a time step no greater than 10^{-3} *sec* (minimum time step 10^{-9} *sec*).

[21] assuming a compressive normale force, which can be checked via the sign of the Lagrange multiplier associated with the normal contact constraint.

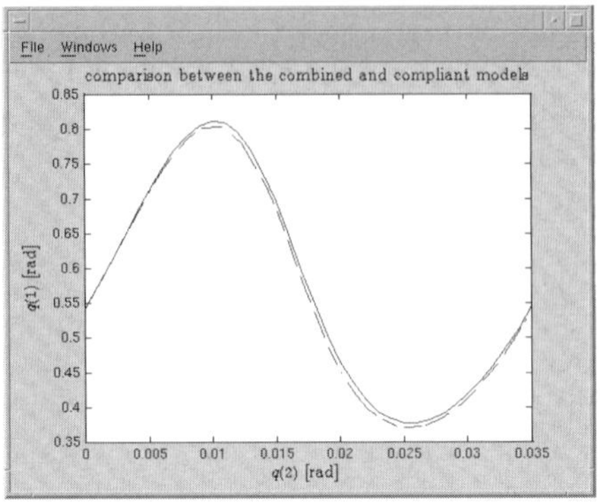

As can be seen from this graph, the results are quite similar. It can be seen that the "compliant" model (dashed line) gives slightly lower values; this is due to the contact model which assumes that the two bodies interpenetrate: the follower penetrates the cam as a result of the finite stiffness implemented ($K^m = 10^7 \ N/m$).

The processing time required by the "combined" model is much more lower than that of the "compliant" model. This difference accrues from the fact that the "combined"[22] model only requires solving a kinematic problem, viz. the algebraic contact equations, without any kind of numerical integration, the latter being mandatory with the "compliant" model.

A further advantage of the "combined" model is that, if we wish to change a dynamical parameter of the system (e.g. an inertia, the follower's internal torsion spring, the coefficient of friction of the contact force), the kinematic analysis remains unchanged. However, the numerical integration, which must be performed in the "compliant" model, needs to be rerun with any new parameter values.

12.11.5 Computer post-process

Let us now consider only the "combined" model. Still under the assumption of a permanent contact, the kinematic analysis provides the following results for the follower angle, angular velocity and acceleration respectively (continuous line: profile 1 ; dotted line: profile 2).

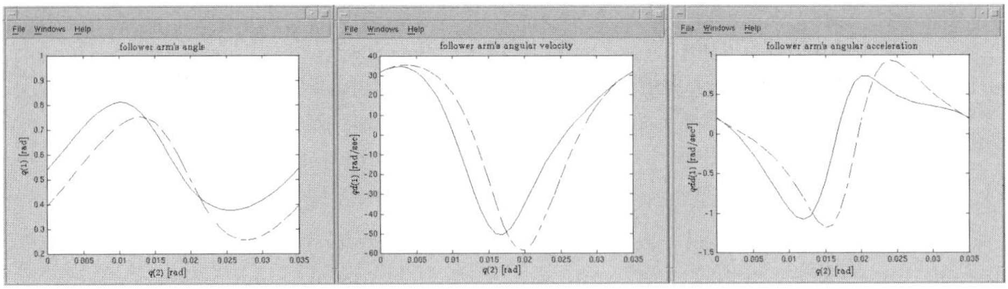

[22] where the contact is assumed to be permanent.

Profile 1 (continuous line):
- maximum angle: $0.81\,rad$ i.e. $46.5°$
- minimum angle: $0.38\,rad$ i.e. $21.6°$
- maximum speed: $3.49\,10^2\,rad/sec$
- maximum acceleration: $7.42\,10^5\,rad/sec^2$

Profile 2 (dashed line):
- maximum angle: $0.75\,rad$ i.e. $43.1°$
- minimum angle: $0.26\,rad$ i.e. $14.7°$
- maximum speed: $3.556\,10^2\,rad/sec$
- maximum acceleration: $9.29\,10^5\,rad/sec^2$

For the first profile, we will now evaluate the normal contact force between cam and follower, for both values of the torsional spring inside joint 1. We thus need to determine the Lagrange multipliers for the previous kinematic analyses.

The results are represented on the two next plots and compare profiles 1 (continuous line) and 2 (dashed line). According to well-established conventions of static analysis, a compression force is negative while a traction force is positive. This convention has been adopted here; the normal contact force $F_n^{contact}$ is thus computed by inverting the sign of the Lagrange multiplier in equation 8.40:

$$F_n^{contact} \triangleq -\,||\mathbf{n}^p||\,\lambda_3$$

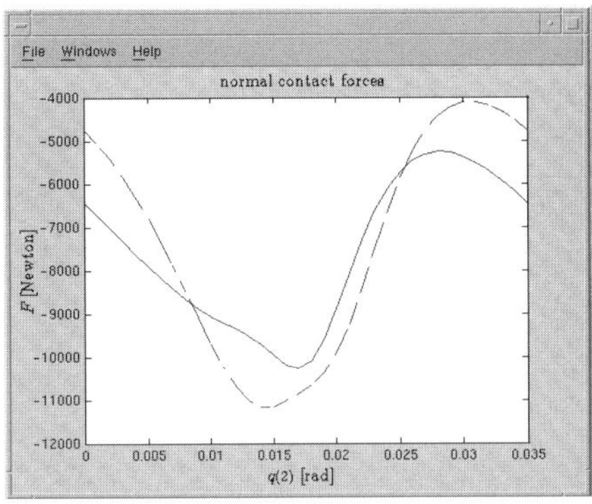

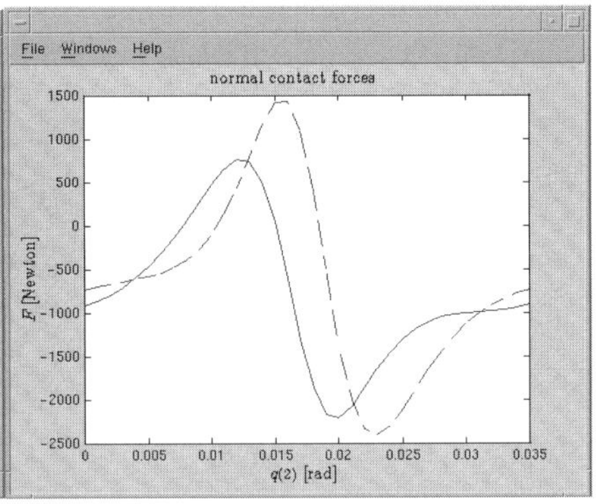

It can be observed that for the $100\,N/m$ torsional spring (right plot), the normal contact force becomes positive[23] for both profiles, which normally means that losses of contact have occurred. The dynamical calculations based on the previous kinematic analysis are not valid in this situation and it is therefore necessary to carry out a numerical integration with the "combined" model which manages the intermittent contact using the sign of the normal contact force (see section 8.4).

- If $F_n^{contact}$ is negative (i.e. positive Lagrange multiplier), the loop between the cam and follower must be closed. This loop constrains one d.o.f. (variable q^1 in our case).

[23] which in this case corresponds to an irrelevant "traction" force between cam and follower

- If $F_n^{contact}$ is (or becomes) positive (i.e. negative Lagrange multiplier), the program must switch models: there is no longer any loop closure. Variable q^1 becomes independent: the follower is no longer constrained by its contact with the cam.

The contact between bodies is re-established via a "compliant" transient model. The latter introduces a parallel spring and damper during the transient contact resumption phase, until contact stabilization is achieved (i.e. until velocity and acceleration along the contact normal fall below a given threshold).

We can now start the time integration of the system (preferably with a "stiff" integrator) with final time $= 2 * \frac{\pi}{180}$ sec. If allowed by the algorithm, it is interesting to input minimal, maximal and initial values for the time step Δt, namely: Δt_{min}, Δt_{max} and Δt_{ini}. Indeed, as soon as a transient contact occurs, the system becomes much more stiff and nonlinear, requiring a more accurate time step and tracking of events. For the present problem, the two following sets of such parameters are proposed:

- For the simulation phases where the "cam/follower" pair exhibits a constrained contact ("established"), we choose: $\Delta t_{min} = 10^{-5}$; $\Delta t_{max} = 10^{-3}$ and $\Delta t_{ini} = 10^{-1}$.

- For the simulation phases where the "cam/follower" pair has lost contact or is in a transient contact situation, we choose: $\Delta t_{min} = 10^{-9}$ $\Delta t_{max} = 10^{-3}$ and $\Delta t_{ini} = 10^{-9}$.

The required results are plotted hereunder.

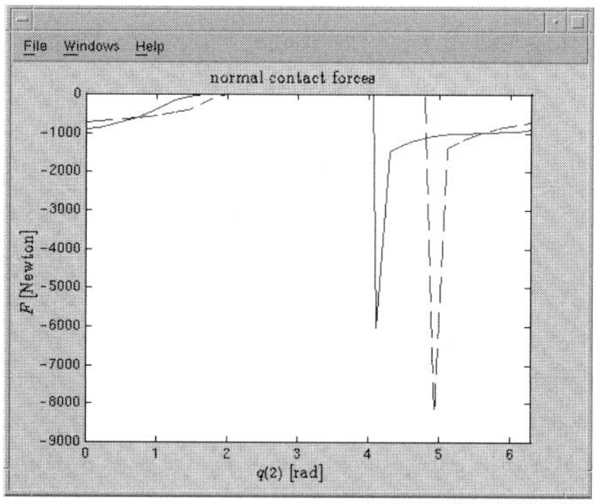

We observe loss of contact and therefore a model switch (contact $\Longrightarrow$ loss of contact $\Longrightarrow$ contact resumed) for the 2 profiles. The peaks obviously relate to the shock which occurs when contact is resumed.

Bibliography

[1] Bakker, E., L. Nyborg, and H.-B. Pacejka: 1987, 'Tyre Modelling for Use in Vehicle Dynamics Studies'. Technical report, SAE Technical Paper 870421, 1-15, Society of Automotive Engineers, Inc.

[2] Baumgarte, J.: 1972, 'Stabilization of Constraints and Integrals of Motion'. *Computer Methods in Applied Mechanics and Engineering* **1**, 1–16.

[3] Boltzmann, L.: 1902, 'Über die Form der Lagrangeschen Gleichungen für nichtholonome, generalisierte Koordinaten'. *Sitzungsberichte der Wiener Akademie der Wissenschaften* **111**, 1603–1614.

[4] Bourdeau, D.: 2000, *Modélisation et étude des performances routières d'une voiture équipée d'un nouveau système anti-roulis*. UCL, Louvain-la-Neuve, Belgium: Student Thesis.

[5] Cardona, A. and M. Géradin: 1993a, 'Kinematic and Dynamic Analysis of Mechanisms with Cams'. *Computer Methods in Applied Mechanics and Engineering* **103**, 115–134.

[6] Cardona, A. and M. Géradin: 1993b, 'Numerical Integration of Second-order Differential-Algebraic Systems in Flexible Mechanism Dynamics'. In: M.-F.-O. S. Pereira and J.-A.-C. Ambrósio (eds.): *Computer-Aided Analysis of Rigid and Flexible Mechanical Systems, NATO ASI Series, Series E: Applied Sciences, Vol. 268*. Dordrecht, the Netherlands, pp. 501–530.

[7] Chatelle, P., J. Duponcheel, and J.-C. Samin: 1984, 'Investigation on Non-Conventional Railway Systems through a Generalized Multibody Approach'. In: K. Hedrick (ed.): *The Dynamics of Vehicles on Roads and on Railway Tracks*. Lisse, pp. 43–57.

[8] Chenut, X., P. Fisette, and J.-C. Samin: 2002, 'Recurisve Formalism with a Minimal Dynamic Parameterization for the Identification and Simulation of Multibody Systems. Application to the Human Body'. *Multibody System Dynamics* **8**(2), 117–140.

[9] de Pater, A.-D.: 1988, 'The Geometrical Contact between Track and Wheelset'. *Vehicle System Dynamics* **17**, 127–140.

[10] Eichberger, A., C. Fuhrer, and R. Schwertassek: 1993, 'The Benefits of Parallel Multibody Simulation and its Application to Vehicle Dynamics'. In: W. Schiehlen (ed.): *Advanced Multibody System Dynamics : Simulation and Software Tools*. Dordrecht, the Netherlands, pp. 107–126.

[11] Ferrière, L., G. Campion, and B. Raucent: 2001, 'ROLLMOBS, a new drive system for omnimobile robots'. *Robotica* **19**, 1–9.

[12] Fisette, P.: 1994, 'Génération Symbolique des Equations du Mouvement de Systèmes Multicorps et Application dans le Domaine Ferroviaire'. Ph.D. thesis, Université catholique de Louvain, Louvain-la-Neuve, Belgium.

[13] Fisette, P., D.-A. Johnson, and J.-C. Samin: 1997, 'A Fully Symbolic Generation of the Equations of Motion of Multibody Systems containing Flexible Beams'. *Computer Methods in Applied Mechanics and Engineering* **142**, 123–152.

[14] Fisette, P., K. Lipinski, and J.-C. Samin: 1995, 'Dynamic Behaviour Comparison Between Bogies : Rigid or Articulated Frame, Wheelset or Independent Wheels'. *Supplement to Vehicle System Dynamics* **25**, 152–174.

[15] Fisette, P. and J.-M. Péterkenne: 1998, 'Contribution to Parallel and Vector Computation in Multibody Dynamics'. *Parallel Computing* **24**, 717–728.

[16] Fisette, P., J.-M. Péterkenne, B. Vaneghem, and J.-C. Samin: 2000, 'A Multibody Loop Constraints Approach for Modelling Cam/Follower Devices'. *Nonlinear Dynamics* **22**, 335–359.

[17] Fisette, P., T. Postiau, L. Sass, and J.-C. Samin: 2002, 'Fully Symbolic Generation of Complex Multibody Models'. *Mechanics of Structures and Machines* **30**(1), 31–82.

[18] Fisette, P., B. Raucent, and J.-C. Samin: 1996, 'Minimal Dynamic Characterization of Tree-like Multibody Systems'. *Nonlinear Dynamics* **9**, 165–184.

[19] Fisette, P. and J.-C. Samin: 1993, 'Robotran Symbolic Generation of Multibody System Dynamic Equations'. In: W. Schiehlen (ed.): *Advanced Multibody System Dynamics : Simulation and Software Tools*. Dordrecht, the Netherlands, pp. 373–378.

[20] Fisette, P. and J.-C. Samin: 1994, 'A New Wheel/Rail Contact Model for Independent Wheels'. *Archive of Applied Mechanics* **64**, 180–191.

[21] Fisette, P. and J.-C. Samin: 1996a, 'Symbolic Generation of a Multibody Formalism of Order N - Extension to Closed-Loop Systems Using the Coordinate Partitioning Method'. *Int. J. of Numerical Methods in Engineering* **39**, 4091–4112.

[22] Fisette, P. and J.-C. Samin: 1996b, 'Symbolic Generation of Large Multibody System Dynamic Equations using a New Semi-Explicit Newton/Euler Recursive Scheme'. *Archive of Applied Mechanics* **66**, 187–199.

[23] Fisette, P. and J.-C. Samin: 1997, 'The Modelling of Complex Mechanical Systems in the MATLAB/SIMULINK Environment'. In: *First Benelux MATLAB Usersconference*. Amsterdam, pp. 1–11 (chapter 4).

[24] Fisette, P., J.-C. Samin, and R. Valembois: 1994, 'Symbolic Generation of the Kinematic and Dynamic Equations of Vehicle Articulated Structures'. *European Journal of Mechanical Engineering* **40**, 9–17.

[25] Fisette, P., J.-C. Samin, and P.-Y. Willems: 1991, 'Contribution to Symbolic Analysis of Deformable Multibody Systems'. *Int. Journal of Numerical Methods in Engineering* **8**, 1621–1635.

[26] Fuhrer, C. and B. Leimkuhler: 1989, 'A New Class of Generalized Inverses for the Solution of Discretized Euler-Lagrange Equations'. In: E.-J. Haug and C.-D. Roderic (eds.): *NATO ASI Series, Series F: Computer and Systems Sciences, Vol. 69*. Berlin, pp. 143–154.

[27] Garg, V.-K. and R.-V. Dukkipati: 1984, *Dynamics of Railway Vehicle Systems*. Toronto: Academic Press.

[28] Gasch, R., D. Moelle, and K. Knothe: 1984, 'The Effect of non-Linearities on the Limit-Cycles of Railway Vehicles'. In: K. Hedrick (ed.): *The Dynamics of Vehicles on Roads and on Railway Tracks*. Lisse, pp. 207–224.

[29] Géradin, M. and A. Cardona: 1988, 'A beam Finite Element Non-Linear Theory with Finite Rotations'. *Int. Journal of Numerical Methods in Engineering* **26**, 2403–2438.

[30] Géradin, M. and D. Rixen: 1994, *Mechanical Vibrations - Theory and Application to Structural Dynamics*. Chichester: Wiley.

[31] Gilardi, G. and I. Sharf: 2002, 'Literature Survey of Contact Dynamics Modelling'. *Mechanism and Machine Theory* **37**, 1213–1239.

[32] Gillespie, T.-D.: 1992, *Fundamentals of Vehicle Dynamics*. Warrendale: Society of Automotive Engineers, Inc.

[33] Goldstein, H.: 1980, *Classical Mechanics*. Massachusetts: Addison-Wesley.

[34] Gonzalez-Palacios, M.-A. and J. Angeles: 1993, *Cam Synthesis*. Dordrecht: Kluwer Academic Publishers.

[35] Gorla, B. and M. Renaud: 1984, *Modèles des Robots Manipulateurs, application à leur commande*. Toulouse: Cepadues-Editions.

[36] Hairer, E. and G. Wanner: 1991, *Solving Ordinary Differential Equations II. Stiff and Differential-Algebraic Problems.* Berlin: Springer Series in Computational Mathematics 14, Springer-Verlag.

[37] Hamel, G.: 1904, 'Die Lagrange-Eulerschen Gleichungen der Mechanik'. *Zeitscrift für Mathematik und Physik* **50**, 1–57.

[38] Haug, E.-J.: 1989, *Computer Aided Kinematics and Dynamics of Mechanical Systems.* Boston: Allyn and Bacon.

[39] Haug, E.-J. and R.-C. Deyo: 1989, *Real-Time Integration Methods for Mechanical System Simulation, NATO ASI Series, Series F: Computer and Systems Sciences, Vol. 69.* Berlin: Springer-Verlag.

[40] Haug, J.-E. and J. Yen: 1990, 'Generalized Coordinate Partitioning Methods for Numerical Integration of Differential/Algebraic Equations of Dynamics'. In: E.-J. Haug and R.-C. Deyo (eds.): *NATO ASI Series, Vol. F69.* Berlin, pp. 97–114.

[41] Hiller, M., A. Kecskeméthy, and C. Woernle: 1992, 'Computer-aided kinematics and dynamics of multibody systems, Technical report, COMETT Course Computer Aided Analysis of Mechanical Systems'. Technical report, Liège.

[42] Huston, R.-L. and Y. Wang: 1993, 'Flexibilty Effects in Multibody Systems'. In: M.-F.-O. S. Pereira and J.-A.-C. Ambrósio (eds.): *Computer-Aided Analysis of Rigid and Flexible Mechanical Systems, NATO ASI Series, Series E: Applied Sciences, Vol. 268.* Dordrecht, the Netherlands, pp. 351–376.

[43] Jaschinski, A.: 1990, 'On the Application of Similarity Law to a Scaled Railway Bogie Model'. Ph.D. thesis, Delft University of Technology, Delft, The Netherlands.

[44] Jen, C.-W., D.-A. Johnson, and F. Dubois: 1995, 'Numerical Modal Analysis of Structures Based on a Revised Substructure Synthesis Approach'. *Journal of Sound and Vibration* **180**(2), 185–203.

[45] Kalker, J.-J.: 1990, *Three dimensional Elastic Bodies in Rolling Contact.* Dordrecht: Kluwer Academic Publishers.

[46] Kane, T.-R. and D.-A. Levinson: 1985, *Dynamics: Theory and Applications.* New-York: McGraw-Hill Inc.

[47] Kane, T.-R., P.-R. Ryan, and A. Banerjee: 1987, 'Dynamics of a Cantilever Beam Attached to a Moving Base'. *J. Guid. Control Dyn.* **10**, 139–151.

[48] Kirćanski, N., T. Petrović, and M. Vucobratović: 1993, 'Parallel Computation of Symbolic Robot Models and Control Laws: Theory and Application to Transputer Networks'. *Journal of Robotic Systems* **10**(3), 345–368.

[49] Kortüm, W. and R. Sharp: 1993, *Multibody Computer Codes in Vehicle System Dynamics*. Amsterdam: Swets & Zeitlinger.

[50] Kreuzer, E. and W. Schiehlen: 1990, 'Neweul - software for the generation of symbolical equations of motion'. In: W. Schiehlen (ed.): *Multibody System Handbook*. Berlin, pp. 181–202.

[51] Lemineur, V.: 1999, *Etude par simulation du comportement routier d'un véhicule à quatre roues directrices*. UCL, Louvain-la-Neuve, Belgium: Student Thesis.

[52] Luh, J.-Y.-S., N.-W. Walker, and R.-P.-C. Paul: 1980, 'On-line computational scheme for mechanical manipulators'. *Journal of Dynamics Systems, Measurements and Control* **102**, 69–76.

[53] Luré, L.: 1968, *Mécanique Analytique*. Louvain: Librairie Universitaire.

[54] M. Jahnke, K. P. and B. Dirr: 1993, 'Approximate Analysis of Flexible Parts in Multibody Systems Using the Finite Element Method'. In: W. Schiehlen (ed.): *Advanced Multibody System Dynamics : Simulation and Software Tools*. Dordrecht, the Netherlands, pp. 237–256.

[55] Maes, P., J.-C. Samin, and P.-Y. Willems: 1990, 'Robotran'. In: W. Schiehlen (ed.): *Multibody System Handbook*. Berlin, pp. 246–264.

[56] MATLAB, 'The Language of Technical computing'. Technical report, The MathWorks, Inc., Natick, MA.

[57] Meirovitch, L. and M.-K. Kwak: 1991, 'Rayleigh-Ritz based Substructure Synthesis for flexible Multibody Dynamics'. *AIAA Journal* **28**, 1709–1719.

[58] Melzer, F.: 1993, 'Symbolic Computation in Flexible Multibody Systems'. In: *Proceeding of the NATO-Advanced Study Institute (vol. 2)*. Troia, Portugal, pp. 365–381.

[59] Newmark, N.-M.: 1959, 'A Method of Computation for Structural Dynamics'. *Journal of the Engineering Mechanics Division, ASCE* pp. 67–94.

[60] Nikravesh, P.-E.: 1984, 'Some Method for Dynamic Analysis of Constrained Mechanical Systems : A Survey'. In: E.-J. Haug (ed.): *Computer Aided Analysis and Optimization of Mechanical System Dynamics NATO ASI Series, Series F: Computer and Systems Sciences, Vol. 9*. Berlin, pp. 351–367.

[61] Nikravesh, P.-E.: 1988, *Computer Aided Analysis of Mechanical Systems*. London: Prentice-Hall.

[62] Orlandea, N., M.-A. Chace, and D.-A. Calahan: 1977, 'A Sparsity-Oriented Approach to the dynamic analysis and Design of Mechanical systems, Parts 1 & 2'. *Journal of Engineering for Industry* **99**, 773–784.

[63] Pacejka, H.-B. and R.-S. Sharp: 1991, 'Shear Force Development by Pneumatic Tyres in Steady-State Conditions: A Review of Modelling Aspects'. *Vehicle System Dynamics* **20**, 121–175.

[64] Pars, L.-A.: 1965, *A Treatise on Analytical Dynamics*. London: Heinemann Ltd.

[65] Petzold, L.: 1982, 'Differential/algebraic equations are not ODE's'. *SIAM J. Sci. Stat.Comput.* **3**(3), 367–384.

[66] Petzold, L.: 1989, 'Methods and Softwares for Differential/Algebraic Systems'. In: E.-J. Haug and C.-D. Roderic (eds.): *NATO ASI Series, Series F: Computer and Systems Sciences, Vol. 69*. Berlin, pp. 127–140.

[67] Piedboeuf, J.-C.: 1995, 'Symbolic Manipulation of Flexible Manipulators'. In: *AAS/AIAA Astrodynamics Specialist Conference*. Halix, Canada, pp. AAS 95–357.

[68] Piedboeuf, J.-C. and B. Moore: 2002, 'On the Foreshortening Effects of a Rotating Flexible Beam Using Different Modelling Methods'. *Mechanics of Structures and Machines* **30**(1), 83–102.

[69] Postiau, T., P. Fisette, and J.-D. Legat: 1999, 'Fine Grain Parallelization of Multibody System Equations of Motion'. In: *Parallel Computing '99*. Delft, The Netherlands, pp. 193–200.

[70] Postiau, T., L. Sass, P. Fisette, and J.-C. Samin: 2001, 'High-Performance Multibody Models of Road Vehicles: Fully Symbolic Implementation and Parallel Computation'. *Supplement to Vehicle System Dynamics*.

[71] Raucent, B. and J.-C. Samin: 1993, 'Modeling and Identification of Dynamic Parameters of Robot Manipulators'. In: R. Bernhart and S. Albright (eds.): *Robot Calibration*. London, pp. 197–232.

[72] Renaud, M.: 1987, 'Quasi-Minimal Computation of the Dynamic Model of a Robot Manipulator Utilizing the Newton-Euler Formalism and the Notion of Augmented Body'. In: *IEEE International Conference on Robotics and Automation*. Raleigh, North Carolina, pp. 1677–1682.

[73] Roberson, R.-E. and R. Schwertassek: 1988, *Dynamics of Multibody Systems*. Berlin: Springer-Verlag.

[74] Samin, J.-C.: 1984, 'A Multibody approach for Dynamic Investigation of Rolling System'. *Ingenieur Archiv* **54**, 1–15.

[75] Samin, J.-C. and P.-Y. Willems: 1985, 'Multibody Formalism Applied to Non-Conventional Railway Systems'. In: G. Bianchi and W. Schiehlen (eds.): *Dynamics of Multibody Systems*. Berlin, pp. 237–248.

[76] Samin, J.-C. and P.-Y. Willems: 1993, 'Multibody Systems with Loops : A Pseudo-Constraint Approach'. *Journal of Dynamics Systems, Measurements and Control* **115**, 208–213.

[77] SAMTECH: 1991, 'SAMCEF User's Guide'. Technical report, SAMTECH S.A., Liège, Belgium.

[78] SAMTECH: 1994, 'Module d'Analyse de Mécanismes Flexibles MECANO'. Technical report, SAMTECH S.A., Liège, Belgium.

[79] Sayers, M.-W.: 1993, 'Autosim'. In: W. Kortüm and R.-S. Sharp (eds.): *Multibody Computer Codes in Vehicle System Dynamics*. Lisse, pp. 53–56.

[80] Schiehlen, W.: 1990, *Multibody Systems Handbook*. Berlin: Springer-Verlag.

[81] Schwertassek, R. and W. Rulka: 1989, 'Aspects of Efficient and Reliable Multibody Systems Simulation'. In: E.-J. Haug and R.-C. Deyo (eds.): *Real-Time Integration Methods for Mechanical System Simulation NATO ASI Series, Series F, Vol. 69*. Berlin, pp. 55–96.

[82] Shabana, A.-A., M. Berzeri, and J.-R. Sany: 2001, 'Numerical Procedure for the Simulation of Wheel/Rail Contact Dynamics'. *Journal of Dynamic Systems, Measurement and Control* **123**(2), 168–178.

[83] Shabana, A.-A. and K. Changizi: 1988, 'A Recursive Formulation for the Dynamic Analysis of Open Loop Deformable Multibody Systems'. *ASME Journal of Applied Mechanics* **55**, 687–693.

[84] Shi, P. and J. McPhee: 2000, 'Dynamics of Flexible Multibody Systems Using Virtual Work and Linear Graph'. *Multibody System Dynamics* **4**, 355–381.

[85] Shi, P., J. McPhee, and G.-R. Heppler: 2001, 'A Deformation field for Euler-Bernoulli Beams with Applications to flexible Multibody Dynamics'. *Multibody System Dynamics* **5**, 79–104.

[86] Simo, J.-C. and L. Vu-Quoc: 1986, 'On the Dynamics of Flexible Beams under Large Overall Motions - the Planar Case, Parts 1 and 2'. *ASME Journal of Applied Mechanics* **53**, 849–863.

[87] Strang, G.: 1986, *Linear Algebra and its Applications (third edition)*. Fort Worth: Hacourt Brace Jovanovich Int. Edition.

[88] Valembois, R.-E., P. Fisette, and J.-C. Samin: 1997, 'Comparison of Various Techniques for Modelling Flexible Beams in Multibody Dynamics'. *Nonlinear Dynamics* **12**, 367–397.

[89] Verlinden, O.: 1994, 'Simulation du Comportement Dynamique de Systèmes Multicorps Flexibles Comportant des Membrures de Forme Complexe'. Ph.D. thesis, Faculté Polytechnique de Mons, Mons, Belgium.

[90] Vigneron, P.: 1725, *Nouvelle Mécanique ou Statique.* Paris: Chez Claude Jombert.

[91] Wallrapp, O. and R. Schwertassek: 1991, 'Representation of Geometric Stiffening in Multibody System Simulation'. *Int. Journal of Numerical Methods in Engineering* **32**, 1833–1850.

[92] Wehage, R.-A. and E.-J. Haug: 1982, 'Generalized Coordinate Partitioning for Dimension Reduction in Analysis of Constrained Dynamic Systems'. *Journal of Mechanical Design* **134**, 247–255.

[93] Willems, P.-Y.: 1979, *Introduction à la Mécanique.* Paris: Masson.

[94] Wittenburg, J.: 1977, *Dynamics of Systems of Rigid Bodies.* Stuttgart: Teubner.

[95] Yang, G.: 1993, 'Dynamic Analysis of Railway Wheelsets and Complete Vehicle Systems'. Ph.D. thesis, Delft University of Technology, Delft, The Netherlands.

Index

Mechanics

SOLID MECHANICS AND ITS APPLICATIONS

Series Editor: G.M.L. Gladwell

Aims and Scope of the Series

The fundamental questions arising in mechanics are: *Why?*, *How?*, and *How much?* The aim of this series is to provide lucid accounts written by authoritative researchers giving vision and insight in answering these questions on the subject of mechanics as it relates to solids. The scope of the series covers the entire spectrum of solid mechanics. Thus it includes the foundation of mechanics; variational formulations; computational mechanics; statics, kinematics and dynamics of rigid and elastic bodies; vibrations of solids and structures; dynamical systems and chaos; the theories of elasticity, plasticity and viscoelasticity; composite materials; rods, beams, shells and membranes; structural control and stability; soils, rocks and geomechanics; fracture; tribology; experimental mechanics; biomechanics and machine design.

1. R.T. Haftka, Z. Gürdal and M.P. Kamat: *Elements of Structural Optimization.* 2nd rev.ed., 1990 ISBN 0-7923-0608-2
2. J.J. Kalker: *Three-Dimensional Elastic Bodies in Rolling Contact.* 1990 ISBN 0-7923-0712-7
3. P. Karasudhi: *Foundations of Solid Mechanics.* 1991 ISBN 0-7923-0772-0
4. *Not published*
5. *Not published.*
6. J.F. Doyle: *Static and Dynamic Analysis of Structures.* With an Emphasis on Mechanics and Computer Matrix Methods. 1991 ISBN 0-7923-1124-8; Pb 0-7923-1208-2
7. O.O. Ochoa and J.N. Reddy: *Finite Element Analysis of Composite Laminates.* ISBN 0-7923-1125-6
8. M.H. Aliabadi and D.P. Rooke: *Numerical Fracture Mechanics.* ISBN 0-7923-1175-2
9. J. Angeles and C.S. López-Cajún: *Optimization of Cam Mechanisms.* 1991 ISBN 0-7923-1355-0
10. D.E. Grierson, A. Franchi and P. Riva (eds.): *Progress in Structural Engineering.* 1991 ISBN 0-7923-1396-8
11. R.T. Haftka and Z. Gürdal: *Elements of Structural Optimization.* 3rd rev. and exp. ed. 1992 ISBN 0-7923-1504-9; Pb 0-7923-1505-7
12. J.R. Barber: *Elasticity.* 1992 ISBN 0-7923-1609-6; Pb 0-7923-1610-X
13. H.S. Tzou and G.L. Anderson (eds.): *Intelligent Structural Systems.* 1992 ISBN 0-7923-1920-6
14. E.E. Gdoutos: *Fracture Mechanics.* An Introduction. 1993 ISBN 0-7923-1932-X
15. J.P. Ward: *Solid Mechanics.* An Introduction. 1992 ISBN 0-7923-1949-4
16. M. Farshad: *Design and Analysis of Shell Structures.* 1992 ISBN 0-7923-1950-8
17. H.S. Tzou and T. Fukuda (eds.): *Precision Sensors, Actuators and Systems.* 1992 ISBN 0-7923-2015-8
18. J.R. Vinson: *The Behavior of Shells Composed of Isotropic and Composite Materials.* 1993 ISBN 0-7923-2113-8
19. H.S. Tzou: *Piezoelectric Shells.* Distributed Sensing and Control of Continua. 1993 ISBN 0-7923-2186-3
20. W. Schiehlen (ed.): *Advanced Multibody System Dynamics.* Simulation and Software Tools. 1993 ISBN 0-7923-2192-8
21. C.-W. Lee: *Vibration Analysis of Rotors.* 1993 ISBN 0-7923-2300-9
22. D.R. Smith: *An Introduction to Continuum Mechanics.* 1993 ISBN 0-7923-2454-4
23. G.M.L. Gladwell: *Inverse Problems in Scattering.* An Introduction. 1993 ISBN 0-7923-2478-1

SOLID MECHANICS AND ITS APPLICATIONS

Series Editor: G.M.L. Gladwell

24. G. Prathap: *The Finite Element Method in Structural Mechanics.* 1993 ISBN 0-7923-2492-7
25. J. Herskovits (ed.): *Advances in Structural Optimization.* 1995 ISBN 0-7923-2510-9
26. M.A. González-Palacios and J. Angeles: *Cam Synthesis.* 1993 ISBN 0-7923-2536-2
27. W.S. Hall: *The Boundary Element Method.* 1993 ISBN 0-7923-2580-X
28. J. Angeles, G. Hommel and P. Kovács (eds.): *Computational Kinematics.* 1993 ISBN 0-7923-2585-0
29. A. Curnier: *Computational Methods in Solid Mechanics.* 1994 ISBN 0-7923-2761-6
30. D.A. Hills and D. Nowell: *Mechanics of Fretting Fatigue.* 1994 ISBN 0-7923-2866-3
31. B. Tabarrok and F.P.J. Rimrott: *Variational Methods and Complementary Formulations in Dynamics.* 1994 ISBN 0-7923-2923-6
32. E.H. Dowell (ed.), E.F. Crawley, H.C. Curtiss Jr., D.A. Peters, R. H. Scanlan and F. Sisto: *A Modern Course in Aeroelasticity.* Third Revised and Enlarged Edition. 1995 ISBN 0-7923-2788-8; Pb: 0-7923-2789-6
33. A. Preumont: *Random Vibration and Spectral Analysis.* 1994 ISBN 0-7923-3036-6
34. J.N. Reddy (ed.): *Mechanics of Composite Materials.* Selected works of Nicholas J. Pagano. 1994 ISBN 0-7923-3041-2
35. A.P.S. Selvadurai (ed.): *Mechanics of Poroelastic Media.* 1996 ISBN 0-7923-3329-2
36. Z. Mróz, D. Weichert, S. Dorosz (eds.): *Inelastic Behaviour of Structures under Variable Loads.* 1995 ISBN 0-7923-3397-7
37. R. Pyrz (ed.): *IUTAM Symposium on Microstructure-Property Interactions in Composite Materials.* Proceedings of the IUTAM Symposium held in Aalborg, Denmark. 1995 ISBN 0-7923-3427-2
38. M.I. Friswell and J.E. Mottershead: *Finite Element Model Updating in Structural Dynamics.* 1995 ISBN 0-7923-3431-0
39. D.F. Parker and A.H. England (eds.): *IUTAM Symposium on Anisotropy, Inhomogeneity and Nonlinearity in Solid Mechanics.* Proceedings of the IUTAM Symposium held in Nottingham, U.K. 1995 ISBN 0-7923-3594-5
40. J.-P. Merlet and B. Ravani (eds.): *Computational Kinematics '95.* 1995 ISBN 0-7923-3673-9
41. L.P. Lebedev, I.I. Vorovich and G.M.L. Gladwell: *Functional Analysis.* Applications in Mechanics and Inverse Problems. 1996 ISBN 0-7923-3849-9
42. J. Menčik: *Mechanics of Components with Treated or Coated Surfaces.* 1996 ISBN 0-7923-3700-X
43. D. Bestle and W. Schiehlen (eds.): *IUTAM Symposium on Optimization of Mechanical Systems.* Proceedings of the IUTAM Symposium held in Stuttgart, Germany. 1996 ISBN 0-7923-3830-8
44. D.A. Hills, P.A. Kelly, D.N. Dai and A.M. Korsunsky: *Solution of Crack Problems.* The Distributed Dislocation Technique. 1996 ISBN 0-7923-3848-0
45. V.A. Squire, R.J. Hosking, A.D. Kerr and P.J. Langhorne: *Moving Loads on Ice Plates.* 1996 ISBN 0-7923-3953-3
46. A. Pineau and A. Zaoui (eds.): *IUTAM Symposium on Micromechanics of Plasticity and Damage of Multiphase Materials.* Proceedings of the IUTAM Symposium held in Sèvres, Paris, France. 1996 ISBN 0-7923-4188-0
47. A. Naess and S. Krenk (eds.): *IUTAM Symposium on Advances in Nonlinear Stochastic Mechanics.* Proceedings of the IUTAM Symposium held in Trondheim, Norway. 1996 ISBN 0-7923-4193-7
48. D. Ieşan and A. Scalia: *Thermoelastic Deformations.* 1996 ISBN 0-7923-4230-5

Mechanics

SOLID MECHANICS AND ITS APPLICATIONS

Series Editor: G.M.L. Gladwell

49. J.R. Willis (ed.): *IUTAM Symposium on Nonlinear Analysis of Fracture*. Proceedings of the IUTAM Symposium held in Cambridge, U.K. 1997 ISBN 0-7923-4378-6
50. A. Preumont: *Vibration Control of Active Structures*. An Introduction. 1997 ISBN 0-7923-4392-1
51. G.P. Cherepanov: *Methods of Fracture Mechanics: Solid Matter Physics*. 1997 ISBN 0-7923-4408-1
52. D.H. van Campen (ed.): *IUTAM Symposium on Interaction between Dynamics and Control in Advanced Mechanical Systems*. Proceedings of the IUTAM Symposium held in Eindhoven, The Netherlands. 1997 ISBN 0-7923-4429-4
53. N.A. Fleck and A.C.F. Cocks (eds.): *IUTAM Symposium on Mechanics of Granular and Porous Materials*. Proceedings of the IUTAM Symposium held in Cambridge, U.K. 1997 ISBN 0-7923-4553-3
54. J. Roorda and N.K. Srivastava (eds.): *Trends in Structural Mechanics*. Theory, Practice, Education. 1997 ISBN 0-7923-4603-3
55. Yu.A. Mitropolskii and N. Van Dao: *Applied Asymptotic Methods in Nonlinear Oscillations*. 1997 ISBN 0-7923-4605-X
56. C. Guedes Soares (ed.): *Probabilistic Methods for Structural Design*. 1997 ISBN 0-7923-4670-X
57. D. François, A. Pineau and A. Zaoui: *Mechanical Behaviour of Materials*. Volume I: Elasticity and Plasticity. 1998 ISBN 0-7923-4894-X
58. D. François, A. Pineau and A. Zaoui: *Mechanical Behaviour of Materials*. Volume II: Viscoplasticity, Damage, Fracture and Contact Mechanics. 1998 ISBN 0-7923-4895-8
59. L.T. Tenek and J. Argyris: *Finite Element Analysis for Composite Structures*. 1998 ISBN 0-7923-4899-0
60. Y.A. Bahei-El-Din and G.J. Dvorak (eds.): *IUTAM Symposium on Transformation Problems in Composite and Active Materials*. Proceedings of the IUTAM Symposium held in Cairo, Egypt. 1998 ISBN 0-7923-5122-3
61. I.G. Goryacheva: *Contact Mechanics in Tribology*. 1998 ISBN 0-7923-5257-2
62. O.T. Bruhns and E. Stein (eds.): *IUTAM Symposium on Micro- and Macrostructural Aspects of Thermoplasticity*. Proceedings of the IUTAM Symposium held in Bochum, Germany. 1999 ISBN 0-7923-5265-3
63. F.C. Moon: *IUTAM Symposium on New Applications of Nonlinear and Chaotic Dynamics in Mechanics*. Proceedings of the IUTAM Symposium held in Ithaca, NY, USA. 1998 ISBN 0-7923-5276-9
64. R. Wang: *IUTAM Symposium on Rheology of Bodies with Defects*. Proceedings of the IUTAM Symposium held in Beijing, China. 1999 ISBN 0-7923-5297-1
65. Yu.I. Dimitrienko: *Thermomechanics of Composites under High Temperatures*. 1999 ISBN 0-7923-4899-0
66. P. Argoul, M. Frémond and Q.S. Nguyen (eds.): *IUTAM Symposium on Variations of Domains and Free-Boundary Problems in Solid Mechanics*. Proceedings of the IUTAM Symposium held in Paris, France. 1999 ISBN 0-7923-5450-8
67. F.J. Fahy and W.G. Price (eds.): *IUTAM Symposium on Statistical Energy Analysis*. Proceedings of the IUTAM Symposium held in Southampton, U.K. 1999 ISBN 0-7923-5457-5
68. H.A. Mang and F.G. Rammerstorfer (eds.): *IUTAM Symposium on Discretization Methods in Structural Mechanics*. Proceedings of the IUTAM Symposium held in Vienna, Austria. 1999 ISBN 0-7923-5591-1

SOLID MECHANICS AND ITS APPLICATIONS

Series Editor: G.M.L. Gladwell

69. P. Pedersen and M.P. Bendsøe (eds.): *IUTAM Symposium on Synthesis in Bio Solid Mechanics*. Proceedings of the IUTAM Symposium held in Copenhagen, Denmark. 1999 ISBN 0-7923-5615-2
70. S.K. Agrawal and B.C. Fabien: *Optimization of Dynamic Systems*. 1999 ISBN 0-7923-5681-0
71. A. Carpinteri: *Nonlinear Crack Models for Nonmetallic Materials*. 1999 ISBN 0-7923-5750-7
72. F. Pfeifer (ed.): *IUTAM Symposium on Unilateral Multibody Contacts*. Proceedings of the IUTAM Symposium held in Munich, Germany. 1999 ISBN 0-7923-6030-3
73. E. Lavendelis and M. Zakrzhevsky (eds.): *IUTAM/IFToMM Symposium on Synthesis of Nonlinear Dynamical Systems*. Proceedings of the IUTAM/IFToMM Symposium held in Riga, Latvia. 2000 ISBN 0-7923-6106-7
74. J.-P. Merlet: *Parallel Robots*. 2000 ISBN 0-7923-6308-6
75. J.T. Pindera: *Techniques of Tomographic Isodyne Stress Analysis*. 2000 ISBN 0-7923-6388-4
76. G.A. Maugin, R. Drouot and F. Sidoroff (eds.): *Continuum Thermomechanics*. The Art and Science of Modelling Material Behaviour. 2000 ISBN 0-7923-6407-4
77. N. Van Dao and E.J. Kreuzer (eds.): *IUTAM Symposium on Recent Developments in Non-linear Oscillations of Mechanical Systems*. 2000 ISBN 0-7923-6470-8
78. S.D. Akbarov and A.N. Guz: *Mechanics of Curved Composites*. 2000 ISBN 0-7923-6477-5
79. M.B. Rubin: *Cosserat Theories: Shells, Rods and Points*. 2000 ISBN 0-7923-6489-9
80. S. Pellegrino and S.D. Guest (eds.): *IUTAM-IASS Symposium on Deployable Structures: Theory and Applications*. Proceedings of the IUTAM-IASS Symposium held in Cambridge, U.K., 6–9 September 1998. 2000 ISBN 0-7923-6516-X
81. A.D. Rosato and D.L. Blackmore (eds.): *IUTAM Symposium on Segregation in Granular Flows*. Proceedings of the IUTAM Symposium held in Cape May, NJ, U.S.A., June 5–10, 1999. 2000 ISBN 0-7923-6547-X
82. A. Lagarde (ed.): *IUTAM Symposium on Advanced Optical Methods and Applications in Solid Mechanics*. Proceedings of the IUTAM Symposium held in Futuroscope, Poitiers, France, August 31–September 4, 1998. 2000 ISBN 0-7923-6604-2
83. D. Weichert and G. Maier (eds.): *Inelastic Analysis of Structures under Variable Loads*. Theory and Engineering Applications. 2000 ISBN 0-7923-6645-X
84. T.-J. Chuang and J.W. Rudnicki (eds.): *Multiscale Deformation and Fracture in Materials and Structures*. The James R. Rice 60th Anniversary Volume. 2001 ISBN 0-7923-6718-9
85. S. Narayanan and R.N. Iyengar (eds.): *IUTAM Symposium on Nonlinearity and Stochastic Structural Dynamics*. Proceedings of the IUTAM Symposium held in Madras, Chennai, India, 4–8 January 1999 ISBN 0-7923-6733-2
86. S. Murakami and N. Ohno (eds.): *IUTAM Symposium on Creep in Structures*. Proceedings of the IUTAM Symposium held in Nagoya, Japan, 3-7 April 2000. 2001 ISBN 0-7923-6737-5
87. W. Ehlers (ed.): *IUTAM Symposium on Theoretical and Numerical Methods in Continuum Mechanics of Porous Materials*. Proceedings of the IUTAM Symposium held at the University of Stuttgart, Germany, September 5-10, 1999. 2001 ISBN 0-7923-6766-9
88. D. Durban, D. Givoli and J.G. Simmonds (eds.): *Advances in the Mechanis of Plates and Shells The Avinoam Libai Anniversary Volume*. 2001 ISBN 0-7923-6785-5
89. U. Gabbert and H.-S. Tzou (eds.): *IUTAM Symposium on Smart Structures and Structonic Systems*. Proceedings of the IUTAM Symposium held in Magdeburg, Germany, 26–29 September 2000. 2001 ISBN 0-7923-6968-8

Mechanics

SOLID MECHANICS AND ITS APPLICATIONS

Series Editor: G.M.L. Gladwell

90. Y. Ivanov, V. Cheshkov and M. Natova: *Polymer Composite Materials – Interface Phenomena & Processes*. 2001 ISBN 0-7923-7008-2
91. R.C. McPhedran, L.C. Botten and N.A. Nicorovici (eds.): *IUTAM Symposium on Mechanical and Electromagnetic Waves in Structured Media*. Proceedings of the IUTAM Symposium held in Sydney, NSW, Australia, 18-22 Januari 1999. 2001 ISBN 0-7923-7038-4
92. D.A. Sotiropoulos (ed.): *IUTAM Symposium on Mechanical Waves for Composite Structures Characterization*. Proceedings of the IUTAM Symposium held in Chania, Crete, Greece, June 14-17, 2000. 2001 ISBN 0-7923-7164-X
93. V.M. Alexandrov and D.A. Pozharskii: *Three-Dimensional Contact Problems*. 2001 ISBN 0-7923-7165-8
94. J.P. Dempsey and H.H. Shen (eds.): *IUTAM Symposium on Scaling Laws in Ice Mechanics and Ice Dynamics*. Proceedings of the IUTAM Symposium held in Fairbanks, Alaska, U.S.A., 13-16 June 2000. 2001 ISBN 1-4020-0171-1
95. U. Kirsch: *Design-Oriented Analysis of Structures*. A Unified Approach. 2002 ISBN 1-4020-0443-5
96. A. Preumont: *Vibration Control of Active Structures*. An Introduction (2nd Edition). 2002 ISBN 1-4020-0496-6
97. B.L. Karihaloo (ed.): *IUTAM Symposium on Analytical and Computational Fracture Mechanics of Non-Homogeneous Materials*. Proceedings of the IUTAM Symposium held in Cardiff, U.K., 18-22 June 2001. 2002 ISBN 1-4020-0510-5
98. S.M. Han and H. Benaroya: *Nonlinear and Stochastic Dynamics of Compliant Offshore Structures*. 2002 ISBN 1-4020-0573-3
99. A.M. Linkov: *Boundary Integral Equations in Elasticity Theory*. 2002 ISBN 1-4020-0574-1
100. L.P. Lebedev, I.I. Vorovich and G.M.L. Gladwell: *Functional Analysis*. Applications in Mechanics and Inverse Problems (2nd Edition). 2002 ISBN 1-4020-0667-5; Pb: 1-4020-0756-6
101. Q.P. Sun (ed.): *IUTAM Symposium on Mechanics of Martensitic Phase Transformation in Solids*. Proceedings of the IUTAM Symposium held in Hong Kong, China, 11-15 June 2001. 2002 ISBN 1-4020-0741-8
102. M.L. Munjal (ed.): *IUTAM Symposium on Designing for Quietness*. Proceedings of the IUTAM Symposium held in Bangkok, India, 12-14 December 2000. 2002 ISBN 1-4020-0765-5
103. J.A.C. Martins and M.D.P. Monteiro Marques (eds.): *Contact Mechanics*. Proceedings of the 3rd Contact Mechanics International Symposium, Praia da Consolação, Peniche, Portugal, 17-21 June 2001. 2002 ISBN 1-4020-0811-2
104. H.R. Drew and S. Pellegrino (eds.): *New Approaches to Structural Mechanics, Shells and Biological Structures*. 2002 ISBN 1-4020-0862-7
105. J.R. Vinson and R.L. Sierakowski: *The Behavior of Structures Composed of Composite Materials*. Second Edition. 2002 ISBN 1-4020-0904-6
106. Not yet published.
107. J.R. Barber: *Elasticity*. Second Edition. 2002 ISBN Hb 1-4020-0964-X; Pb 1-4020-0966-6
108. C. Miehe (ed.): *IUTAM Symposium on Computational Mechanics of Solid Materials at Large Strains*. Proceedings of the IUTAM Symposium held in Stuttgart, Germany, 20-24 August 2001. 2003 ISBN 1-4020-1170-9

Mechanics

SOLID MECHANICS AND ITS APPLICATIONS

Series Editor: G.M.L. Gladwell

109. P. Ståhle and K.G. Sundin (eds.): *IUTAM Symposium on Field Analyses for Determination of Material Parameters – Experimental and Numerical Aspects.* Proceedings of the IUTAM Symposium held in Abisko National Park, Kiruna, Sweden, July 31 – August 4, 2000. 2003 ISBN 1-4020-1283-7
110. N. Sri Namachchivaya and Y.K. Lin (eds.): *IUTAM Symposium on Nonlnear Stochastic Dynamics.* Proceedings of the IUTAM Symposium held in Monticello, IL, USA, 26 – 30 August, 2000. 2003 ISBN 1-4020-1471-6
111. H. Sobieckzky (ed.): *IUTAM Symposium Transsonicum IV*. Proceedings of the IUTAM Symposium held in Göttingen, Germany, 2–6 September 2002, 2003 ISBN 1-4020-1608-5
112. J.-C. Samin and P. Fisette: *Symbolic Modeling of Multibody Systems*. 2003 ISBN 1-4020-1629-8

Kluwer Academic Publishers – Dordrecht / Boston / London